About Allen Ginsberg

ALLEN GINSBERG was born in 1926 in Newark, New Jersey, the son of Naomi Ginsberg and lyric poet Louis Ginsberg. He died on April 5, 1997. In 1956 he published his signal poem "Howl," one of the most widely read and translated poems of the century. He was a member of the American Academy of Arts and Letters, was named a Chevalier de l'Ordre des Arts et des Lettres by the French Minister of Culture, and was a cofounder of the Jack Kerouac School of Disembodied Poetics at Naropa Institute, the first accredited Buddhist college in the Western world.

BY ALLEN GINSBERG

POETRY

Howl and Other Poems, 1956
Kaddish and Other Poems, 1961
Empty Mirror, Early Poems, 1961
Reality Sandwiches, 1963
Planet News, 1968
The Fall of America, Poems of These States, 1972
The Gates of Wrath: Rhymed Poems 1948–52, 1973
Iron Horse, 1973
First Blues, 1975
Mind Breaths, Poems 1971–76, 1978
Plutonian Ode, Poems 1977–1980, 1982
Collected Poems 1947–1980, 1984
White Shroud, Poems 1980–1985, 1986
Cosmopolitan Greetings, Poems 1986–1992, 1994
Selected Poems 1947–1995, 1996
Death and Fame: Poems 1993–1997, 1999

PROSE

The Yage Letters (with William Burroughs), 1963
Indian Journals, 1970, 1996
Gay Sunshine Interview (with Allen Young), 1974
Allen Verbatim: Lectures on Poetry, Politics, Consciousness, 1974
Chicago Trial Testimony, 1975
To Eberhart from Ginsberg, 1976
As Ever: Collected Correspondence Allen Ginsberg & Neal Cassady, 1977
Neal Cassady, 1977
Journals Early Fifties Early Sixties 1977, 1993

Composed on the Tongue: Literary Conversations 1967–1977, 1980
Straight Hearts Delight, Love Poems and Selected Letters 1947–1980 (with Peter Orlovsky), 1980
Howl, Original Draft Facsimile, Fully Annotated, 1986, 1995
The Visions of the Great Rememberer (with Visions of Cody, Jack Kerouac), 1993
Journals Mid-Fifties: 1954–1958, 1994
Luminous Dreams, 1997
Deliberate Prose: Selected Essays 1952–1995, 2000
Spontaneous Mind: Selected Interviews, 1958–1996, 2001

PHOTOGRAPHY

Photographs, 1991
Snapshot Poetics, 1993

VOCAL WORDS & MUSIC

First Blues, 1981
The Lion For Real, 1989, 1996
Howls, Raps & Roars, 1993
Hydrogen Jukebox (opera with Philip Glass), 1993
Holy Soul Jelly Roll: Poems & Songs 1949–1993, 1994
The Ballad of the Skeletons, with Paul McCartney, Philip Glass, 1996
Howl, U.S.A., Kronos Quartet, Lee Hyla score, 1996
Howl & Other Poems, 1998

ALLEN GINSBERG

Spontaneous Mind

Selected Interviews
1958–1996

With a Preface by Václav Havel
and an Introduction by Edmund White

EDITED BY DAVID CARTER

Perennial
An Imprint of HarperCollinsPublishers

A hardcover edition of this book was published in 2001 by HarperCollins Publishers.

HarperCollins books may be purchased for educational, business, or sales promotional use. For information please write: Special Markets Department, HarperCollins Publishers Inc., 10 East 53rd Street, New York, NY 10022.

First Perennial edition published 2002.

Designed by Lindgren/Fuller Design

The Library of Congress has catalogued the hardcover edition as follows:

Ginsberg, Allen, 1926–1997.
    Spontaneous mind : selected interviews, 1958–1996 / Allen Ginsberg ; with a preface by Václav Havel ; edited by David Carter.—1st ed.
        p. cm.
    ISBN 0-06-019293-3
    1. Ginsberg, Allen, 1926–1997—Interviews. 2. Ginsberg, Allen, 1926–1997—Political and social views. 3. Poets, American—20th century—Interviews. 4. Poetry—Authorship. 5. Beat generation. 6. United States—Intellectual life—20th century. I. Havel, Václav. II. Carter, David. III. Title.

PS3513.I74 Z476 2001
811'.54—dc21
[B]   00-040849

ISBN 0-06-093082-9 (pbk.)

02 03 04 05 06 WB/RRD 10 9 8 7 6 5 4 3 2 1

A long time ago I figured out that the interview and the media was a way of teaching. If you talk to people as if they were future Buddhas, or present Buddhas, that any bad karma coming out of it will be their problem rather than yours, so you can say anything you want, and you talk on about the highest level possible.

—ALLEN GINSBERG

CONTENTS

Preface / ix

Introduction / xi

Editor's Note / xix

INTERVIEWS

1950s

Interview by

Marc D. Schleifer, *Village Voice* / 3

1960s

Interviews by

Ernie Barry, *City Lights Journal* / 9

Tom Clark, *The Paris Review* / 17

Barry Farrell / 54

Bob Elliott, *Freelance* / 67

William F. Buckley, Jr., *Firing Line* / 76

Fernanda Pivano / 103

Michael Aldrich, Edward Kissam, and Nancy Blecker,
    "Improvised Poetics" / 124

Paul Carroll, *Playboy* / 159

Bill Prescott, *(untitled)* / 197

Chicago Seven Trial Testimony / 200

1970s

Interviews by

Mary Jane Fortunato, Lucille Medwick, and Susan Rowe,
    New York Quarterly / 245

Alison Colbert, *Partisan Review* / 259

Yves Le Pellec, "The New Consciousness" / 273

Allen Young, *Gay Sunshine Interview* / 303

John Durham, "The Death of Ezra Pound" / 343

Ekbert Faas, from *Towards a New American Poetics* / 355

Michael Goodwin, Richard Hyatt, and Ed Ward,
    "Squawks Mid-Afternoon" / 363

Peter Barry Chowka, *New Age Journal* / 377

Paul Portugés and Guy Amirthanayagam, "Buddhist Meditation
    and Poetic Spontaneity" / 398

1980s
Interviews by
Nancy Bunge, from *Finding the Words* / 421

Helen, *Flipside Fanzine* / 433

Michael Schumacher, *Oui* / 434

Steve Foehr / 444

Simon Albury / 452

John Lofton, *Chronicles* / 469

Josef Jařab / 499

1990s
Interviews by
Thomas Gladysz, *Photo Metro* / 523

Clint Frakes / 532

Steve Silberman, www.HotWired.com / 546

Afterword / 571

Acknowledgments / 577

Biographical List / 583

Permissions / 596

Index / 597

PREFACE

There has been a sensitive awareness of the Beat movement in our country since the nineteen fifties. The general revolt against the official establishment and the literary nonconformism of the Beat poetry and prose have most likely been perceived in our unfree conditions as even more rebellious than in the land of their origin. The literary works of the Beat authors were understood not only as a denouncement of the social establishment and as a quest for new attitudes and a new lifestyle, as a protest against the superficiality of our civilization openly manifesting all its tensions and strains as well as the tragic and reckless quality of life, but also as a potential instrument for resistance to the totalitarian system that had been imposed on our existence. And if those who knew the literature and, by fostering it, created through this common knowledge a brotherhood, a community of nonconformists, when they expressed their views, it was, understandably, more hazardous in our situation than it could have been in the United States of America.

I had the good luck to have enjoyed from my young days a close friendship with Jan Zábrana, the chief translator of Ginsberg and other Beat writers, so that I had access to the literature even if it was not and could not be published. I have to admit that in those years—through the fifties and sixties—I found the Beat authors' way of writing and their way of thinking very close to my heart as I was just a little younger than the original Beat Generation. I believe I understood their views and their protest as I could share much of it.

I first met Allen Ginsberg at the renowned student May Day festival at which he was elected king (*Král majáles*). After that I participated in one of the private gatherings with him in a Prague apartment. Then I had the luck to see him in Viola, a poet's café, and I believe it must have been at the moment when the notorious theft of his notebook took place. Not far from me Ginsberg was sharing a table with some young friends, and he seemed to be constantly occupied in looking for something all around the table space. I gather it must have been the notebook that he was missing; that was very likely as right next to his table there was seated a group of men with the undeniable appearance of plainclothesmen, who must have stolen it then or a while earlier.

Later, after 1989, when I had become president, I had the opportunity to see Ginsberg a few times. A couple of times we went to a pub together, and I also went to see his performance at the Chmelnice theater hall.

I have always held the poet in great esteem. I truly appreciated his "Howl" when I was a young man, and I was, of course, deeply moved by what I felt to be his untimely death. I have also greatly cherished his sophistication, his intellectual power, and his scope of vision.

—VÁCLAV HAVEL

Prague, The Castle, March 1999

INTRODUCTION

The Poetics of Breath

For Allen Ginsberg, granting an interview was a creative act. He came up with new thoughts on the wing and made new combinations of feelings and concepts to suit the occasion. As anyone knows who has ever been in a political campaign—or on a book tour—most interviewers ask the same three questions and most interviewees have memorized the same three sound bites. Indeed, mastering those sound bites and selling points is even called being "media-trained."

But Ginsberg chewed over his thoughts and more often than not was listening to his own mind at work. We overhear him actually thinking, which is rare enough, and he wanted to preserve the integrity of his complete thought. He relished rather than dreaded long interviews and once reprimanded an interviewer for cutting his answers. He didn't mind eliminating whole responses, but within a given answer he didn't want any editing.

He decided that giving an interview was a way of teaching and he was grateful to journalists for disseminating his ideas. Even hostility could serve his purposes. One of his best interviews in this collection is a dramatic confrontation with a born-again Christian named John Lofton. Ginsberg is patient and mannerly but also quite firm about not answering tendentious questions. He is equally supple and true to himself in his courtroom testimony at the Chicago Seven trial in December 1969 during which he was constantly harassed by the prosecutor Thomas Foran, whose objections were constantly sustained by judge Julius Hoffman.

His mannerliness even in adversity can be ascribed to one of his religious beliefs. As he wrote in 1978, "An early impulse to treat scholars, newsmen, agents, reporters, interviewers and inquirers as sentient beings equal in Buddha-nature to fellow poets turned me on to answer questions as frankly as possible," though he also admitted he came to understand that such openness might "lead to a hell of media self-hood . . ."

Quite obviously Ginsberg patterned his responses to the expectations and experience of his interlocutors. In this collection, for instance, he can be observed patiently starting at the beginning in order to explain the fundamentals of American literary life to a Czech interviewer, or citing the familiar names of Céline, Henri Barbusse, Rimbaud, and Artaud to a French interviewer and giving as reassuring examples Ungaretti and Marinetti to an Italian woman, Fernanda Pivano.

Not that Ginsberg ever betrayed his ideas in order to flatter his listener's preconceptions. He routinely pointed out in interviews that *Beat* was a facile label invented by journalists. With Ms. Pivano, for instance, he was building on her own literary references in order to bridge into a demanding technical discussion of American prosody in the work of Ezra Pound, Marianne Moore, and William Carlos Williams, among others.

I knew Ginsberg only slightly, but I can attest that his essential qualities come through in these transcribed interviews: His great personal sweetness and charm; his wide-ranging intellectual curiosity; his teacher-like clarity and patience, always devoid of pedantry; his utter lack of pretension, which permitted him to make his point with the simple words and humble examples at hand; his almost technological fascination with spiritual techniques; and his frank eroticism, free from boasting and the exploitation of others.

I had met him socially over the years in the 1970s and 1980s, but always in large crowds. I was initially surprised to see him in a coat and tie on these occasions, usually sporting the gold-and-purple rosette of the American Academy of Arts and Letters in his lapel—hardly the image of the Wild Man of American Poetry. In the 1970s I had a casual sexual friendship with a handsome Native American from Colorado, a wrestler and writer, who'd run away from home at age sixteen and been taken in by Allen Ginsberg. Although the guy was badly crippled by an unusually fertile and impervious form of paranoia, Ginsberg was always kind to him and always gave him a place to crash. Other drifters I met over the years had made pit stops at the Ginsberg Home for Wayward Boys in the East Village.

When I was researching my biography of Jean Genet I phoned

Ginsberg from Paris sometime in 1991. Ginsberg and Genet had spent time together at the Chicago Democratic Convention in 1968, and I wanted to know what had transpired between them. After all, as several references in the interviews that follow reveal, Genet had been an essential part of Ginsberg's prose pantheon, along with Dostoyevsky, Céline, Henry Miller, Artaud, Huxley, and Kerouac.

When I phoned, Ginsberg had only recently had a bout of congestive heart failure and had been instructed by his doctor to rest and refuse all demands on his time—but his life-loving and knowledge-fostering generosity got the better of him. He talked to me for an hour. He quoted verses by heart from Genet's "The Man Sentenced to Death" as well as prose passages from the opening pages of *Our Lady of the Flowers*. After giving me all the political details of the tumultuous events in Chicago, Ginsberg mentioned that he and Genet had gone to bed. As I wrote in my biography, "One night Genet invited Ginsberg to his room and got in bed with him. Ginsberg was offering warmth and affection that might possibly lead to sex, but Genet matter-of-factly felt for his crotch and when he found out that Ginsberg didn't even have an erection, he briskly got out of bed and went about his business." The contrast between Genet's unsentimental, Gallic realism and Ginsberg's Whitmanian adhesiveness could not be more telling.

Ginsberg's ceremonial sense of sex turns up several times in these interviews. For instance, as Ginsberg points out to Allen Young, he had slept with Neal Cassady (the model for Kerouac's "Dean Moriarty"), who'd slept with Gavin Arthur (President Chester Arthur's grandson), who'd slept with Edward Carpenter (the English Victorian champion of homosexual love), who'd slept with Walt Whitman, from whom he "received the Whispered Transmission, capital W, capital T, of that love."

Two or three years after the phone interview about Genet, Ginsberg called me in Paris and asked if he could drop in on me right away. I said sure, though I was surprised: why would he want to see me?

He showed up with an attractive young man. Soon I put it together—the young man was a fan of mine and had asked Ginsberg to arrange to meet me. Apparently this guy, a budding American writer living in Paris, had attended Allen's reading at Shakespeare and Company and slipped Allen a nude photo of himself with his phone number scrawled on the back. Allen called him, naturally, and they'd spent the week together.

When I said earlier that Allen didn't exploit these guys, I might have added that at this time in his life he was virtually impotent, due to diabetes, and that his sexual attention was almost a form of courtesy extended to all these runaways, layabouts, and unpublished poets, as if he were reassuring them that they did, after all, have something to offer him in exchange for the advice, help, introductions, pocket money, and shelter he gave them.

A handsome straight poet once told me that the only man he'd ever slept with was Allen Ginsberg. "Why him?" I asked. The poet looked astonished, as if the answer was self-evident: "Because he was fuckin' Allen Ginsberg, man."

Now, with me, Allen talked in a precise, melodious, well-modulated voice about his works, his travels, his health, Genet, William Burroughs (whom I knew slightly). . . . The young man fell asleep, perhaps because our attention wasn't focused on him. The tea turned cold in our cups and night fell, but I didn't turn on the lamps. Allen recited part of the Diamond Sutra from memory. It was the sort of unfettered talk that busy grown-ups seldom have a chance to enjoy. Later I heard someone complain of Ginsberg's "egotism." True, he spoke only of his own interests and activities but he shared those with such generosity and spontaneity that only a cold heart could label such artesian kindness "egotism."

When the young man was aroused to leave, he asked me if he could come later in the week to lunch, after Allen's departure. On the appointed day he rang the bell and was standing there in the hallway completely nude (he'd hidden his clothes in the corner). I felt my life was being touched just this once by the sort of lyrical good luck that Ginsberg must have enjoyed every day.

These interviews, which cover a forty-year period from the 1950s to the 1990s, show the evolution of Ginsberg's mind from a be-bop talking excitement to a calm connoisseurship of world culture of all sorts from all periods. In an early interview Ginsberg refers to "jazz prayer" and tells us that kids today "have a sexual awareness, openness and tolerance and compassionate tenderness that is absolutely ravishing." He was soon to abandon this sort of naïve counterculture utopianism in favor of a sober skepticism about drugs, the New Left, and the avant-garde. In 1978 he admitted:

Certain errors of judgment emerge by hindsight: advocacy of LSD legislation would now be accompanied by prescription for meditation practice to qualify its use. I would extend my selfhood less widely in sympathy with "movement" contemporaries whose subconscious belief in confrontation, conflict or violence encouraged public confusion and enabled police agents to infiltrate and provoke further violence and greater confusion. We were finding "new reasons for spitefulness," Kerouac explained.

But Ginsberg kept certain enthusiasms all his life—for Jack Kerouac as an artist, a poet, even a thinker, for instance. Again and again Ginsberg insists on Kerouac's position as the prime mover and chief inspiration to the entire Beat movement. Nor does he permit other people to condescend to Kerouac's late writing or political conservatism. He always finds a justification for remarks that seem indefensible, so great was Ginsberg's unswerving loyalty to his friends. (He proposed William Burroughs for the Nobel Prize, incidentally.)

Throughout these interviews Ginsberg returns to his high praise of William Blake and Walt Whitman. Ginsberg obviously loves Blake the visionary and Whitman the democratic sensualist, and indeed Ginsberg's own literary personality can be construed as a union of these two forces. Even the idea of being a legendary poet, of having "a large persona," is something he admitted he got from Whitman.

Ginsberg's intense relationship to Blake can be traced back to a seemingly mystical experience he had during the summer of 1948. The twenty-two-year-old Ginsberg was working a desultory job as a filing clerk and living in a stifling sublet in Harlem. He'd recently been rejected by Neal Cassady as a lover, who had just gotten married. Ginsberg had few opportunities to see Jack Kerouac, who was obsessed with his own writing and who'd just completed a 1000-page manuscript. Ginsberg was lonely and frustrated, artistically and sexually. He had yet to find his own distinctive style as a poet. His mother, Naomi Ginsberg, had recently gone mad and been confined to Pilgrim State Hospital in New Jersey. She wrote Allen constantly, begging him to help her get out of the asylum, but in fact it was Allen who eventually signed papers permitting the hospital to perform a prefrontal lobotomy on her.

In the midst of so much unhappiness, as Ginsberg recounts in his

Paris Review interview, he was reading Blake and masturbating one evening. After his orgasm he heard a deep voice, which he described as "a very deep earthen grave voice in the room, which I immediately assumed, I didn't think twice, was Blake's voice." This auditory hallucination (if that's what it was) changed Ginsberg entirely.

> Anyway, my first thought was this was what I was born for, and second thought, never forget—never forget, never renege, never deny. Never deny the voice—no, never forget it, don't get lost mentally wandering in other spirit worlds or American or job worlds or advertising worlds or war worlds or earth worlds. But the spirit of the universe was what I was born to realize. . . .

If a commitment to poetic mysticism remains central to Ginsberg's thought, other recurring themes in the interviews (and in his life) are ecology (Ginsberg warned of global warming two decades before a general alarm was sounded), mind-expansion through drugs and later yoga, a commitment to pacifism and interpersonal kindness, homosexuality, and the key role of spontaneity in making art. Each of these themes, however, received a special twist in Ginsberg's hands.

Take homosexuality. Ginsberg admitted that he was more attracted to young heterosexual men than to homosexuals, and homosexuals themselves he divided into those he liked ("heartfelt, populist, humanist, quasi heterosexual, Whitmanic, bohemian, free-love") and the sort he avoided ("the privileged, exaggeratedly effeminate, gossipy, moneyed, money-style-clothing-conscious, near hysterical"). Basically, what people in the sixties called Downtown guys and pissy East Side Queens.

True to his cult of Whitmanic democracy and frankness, Ginsberg solicited honesty and openness among homosexuals; he deplored the fearful, closed atmosphere of the usual uptight gay bar. Characteristically, when he met the bisexual Peter Orlovsky in 1954, the man with whom he would spend most of his life, they made a pledge of mutual "ownership" to each other, as if only such extreme terms of possession, such a violent commitment, might allay their deep insecurities.

> We made a vow to each other that he could own me, my mind and everything I knew, and my body, and I could own him and

all he knew and all his body; and that we would give each other ourselves, so that we possessed each other as property, to do everything we wanted to, sexually or intellectually, and in a sense explore each other until we reached the mystical "X" together, emerging two merged souls.

Probably no couple ever swore such desperate, literal marriage vows.

Ginsberg bore the traces of the general homosexual oppression of his epoch, but he did more than anyone else of his generation to overcome his gay self-hatred and to take a pro-gay militant stand. He was an apostle of tenderness among men. He never allowed his political and spiritual energies, however, to confine themselves to a gay ghetto. Like other big spirits of his time—Pasolini, Juan Goytisolo, Genet—Ginsberg was interested in the fate of the oppressed everywhere.

In 1965, four years before the beginning of gay liberation, Ginsberg visited Cuba and almost instantly became aware that Castro was consigning homosexuals to work camps, denouncing and outlawing homosexuality and censoring pro-gay statements in the press. Ginsberg responded fearlessly by criticizing the policy—and was hustled out of Havana on a plane bound to Prague. There he was crowned King of the May by Czech students, and again forced by worried authorities to leave the country. By staying true to his personal values of sexual and artistic freedom, Ginsberg confronted Communist authoritarianism a full decade before most other intellectuals in the west.

Ginsberg did not believe in revision. On the contrary, he'd learned from his guru the slogan, "first thought, best thought." No wonder he was attracted to the Chinese and Japanese arts of calligraphy, ink-and-brush painting, and the composition of haiku, all of which require years of preparation but only seconds to execute. He proclaimed the "bardic function" to be a form of meditation. He called for the "frank revelation of the heart." Writing, for him, was not a slow, agonizing process but rather a "natural expressive function" as automatic as breathing. He refused to censor his thoughts in order to isolate those suitable for poetry; no, he declared the whole spectrum of feeling to be the proper subject matter of art. He said that knowing how to walk across the street was the same thing as learning how to write. And he believed that if one wrote directly from personal experience one did not need to fear a loss of poetic power; as he put it, "any point on an autobiographical curve is interesting."

More than many American writers, Ginsberg had a sense of history—and the grace to see his cohorts as instant historical personages. His attitude reminds me of someone entirely different in a different country. Boris Kochno, Diaghilev's last secretary, once told me at the end of his life that he could still vividly recall a moment in the 1920s when Picasso had just left a café on the Rond-Point des Champs-Elysées where he'd spent a moment with Stravinsky, Diaghilev, and Kochno. Diaghilev had suddenly said to the others, "Look hard at Picasso—it's as if you were seeing da Vinci on the streets of Florence."

Ginsberg had the same precocious awareness of the significance of his fellow Beats and of himself. In one interview he said that for Columbia to have expelled Kerouac was as absurd as if Socrates had banned Alcibiades from the Symposium. Always the grandiose comparison. By the same token Ginsberg saw himself as a direct heir to Whitman and knew how important it was to pay a pilgrimage visit to Céline in France and Pound in Venice. Good career move? No, the forging of a lasting link to the artistic past.

Ginsberg possessed in abundance the gift of appreciation. He was a powerful admirer and in his interviews he summons up the names of those past artists he honored and those contemporaries he championed. In the following pages the reader will come across the names of Blake and Whitman but also of Pound, William Carlos Williams, Basil Bunting (who taught Pound—and Ginsberg—that poetry is the same thing as condensation), John Wieners (the great gay bard of the *Hotel Wentley* poems), Kenneth Rexroth (the elder statesman of San Francisco poetry who embraced and condemned the Beats in fits and starts), Gregory Corso, Gary Snyder, Peter Orlovsky, Lawrence Ferlinghetti, Herbert Huncke (the junky ur-figure of the Beat movement), Kerouac, William S. Burroughs, and many others. The reader will also find appreciations of Lenny Bruce, Timothy Leary, Carl Solomon, and Chögyam Trungpa. Thoughts about music and photography, war protest and rock music, drugs and meditation technique—it's all here, in a profusion as generous as the spirit of Ginsberg himself.

EDMUND WHITE

Editor's Note

The information provided at the beginning of each interview is as follows: the name of the interviewer; the date and location of the interview (if known); the title of the piece (if any, except that nondescriptive titles such as "Interview with Allen Ginsberg" are not given); and the name and date of the publication the interview appeared in (if any). Introductory notes by the current editor are in roman font and signed "DC"; introductions that were originally published with the interview, when used, whether in full or in part, are in italics.

DAVID CARTER

1950s

MARC D. SCHLEIFER
New York City

"Allen Ginsberg: Here To Save Us, But Not Sure From What"
Village Voice, October 15, 1958

October 7, 1955, Allen Ginsberg steps in front of a San Francisco audience at the Six Gallery and reads his new poem, "Howl," and Lawrence Ferlinghetti, founder of City Lights Press, immediately offers to publish it. Ginsberg takes a job as a yeoman storekeeper on a ship traveling to the Arctic Circle to raise funds to go to Morroco to help William Burroughs edit *Naked Lunch*. While on board ship he edits the proofs of *Howl and Other Poems*. Between the Six Gallery reading and giving the following interview to the *Village Voice* Allen commences a hurdy-gurdy of travel: Ginsberg and colleague Gary Snyder hitchhike from San Francisco to Seattle; with fellow poet Gregory Corso he visits Neal Cassady in Los Gatos, California; Allen and Corso and Allen's lover, Peter Orlovsky, and Lafcadio, Peter's brother, visit Jack Kerouac in Mexico City; from there Ginsberg, Peter, and Kerouac join Burroughs in Tangier via New York; from Tangier Allen travels throughout Spain, France, and Italy before settling with Corso in Paris, where he hopes to persuade Olympia Press to publish *Naked Lunch*; from Paris, Ginsberg visits Amsterdam and England before returning to New York in 1958.

During these travels the United States Customs Service seized 520 copies of *Howl* based on their estimate of it as "obscene and indecent," but released the books when the United States Attorney at San Francisco refused to institute condemnation proceedings against the book. The local police then took over, and Captain William Hanrahan of the juvenile department arrested Shigeyoshi Murao—the bookstore clerk who had sold plainclothes officers a copy of the book—and filed charges against Ferlinghetti as the book's publisher, reporting that the books were not fit for children to read. The American Civil Liberties Union defended Ferlinghetti and Murao, who were found innocent because of the book's literary value. Ginsberg was out of the country during much of this period of massive publicity and would continue to travel extensively throughout the early to mid-sixties. It is both because of Ginsberg's frequent travels and that a large alternative press did not fully emerge until the mid- to late sixties that there are so few interviews with Allen in his early career, and most of the early interviews that do exist are brief. We first encounter Allen in this volume upon his return to New York from Paris.

—DC

"Why have you come back, Allen," I said. "To save America," he answered. "I don't know what from."

Between the question-smile, answer-laugh, the first beer in the time and space between table and sawdust-covered floor, the order of an interview was lost: order that demands a stiffness one cannot long maintain when talking to Allen Ginsberg, digging Allen Ginsberg.

Data: Allen Ginsberg, 32, Paterson, N.J., Columbia College, Merchant Marine, Texas, Denver, Times Square, Mexico City, Harlem, Yucatan, Chiapas, San Francisco, "Howl," Rue Gît-le-Coeur, Lower East Side.

Ginsberg sat at the table in a Village bar wearing a colored T-shirt and faded wash pants. Also remember the breaks into time when I got the beer or he borrowed matches from three girls nearby. Sometimes I took notes and sometimes I didn't, and this is no *New Yorker* profile but a series of responses, thoughts, and phrases. If I were to write of Ginsberg instead of Ginsberg's sayings, this would not be an interview—it would be a litany.

Paris: "Eight months in Paris living with (William) Burroughs and Gregory Corso. Corso's poetry is really flowing now, he and Burroughs ('author of "Naked Lunch," an endless novel which will drive everybody mad'—*Howl*) are still living there, he's writing great perfect rich poems. Corso has extended the area poetry covers since 'Gasoline.' I'm too literary, you know, but Corso can write about moth balls or atom bombs . . . We went to visit (Louis-Ferdinand) Céline, you don't read anything about him anymore in Europe because of politics. He's an old gnarled man dressed in black, mad and beautiful and he thought we were newspapermen—'Ah, the press!'—until we told him we were poets."

Instinctive Style

Kerouac: "Jack is the greatest craftsman writing today. He writes continuously, can write a hundred words a minute, and gets better each time, reducing the grey-mush percentage that bugs every writer, with each effort . . . I dig your comparison of his spontaneous writing and Zen archery, but Jack's style was discovered—arrived upon instinctively, not copied theoretical-like from a theology."

Norman Podhoretz: (in the Spring 1958 issue of *Partisan Review*, Norman Podhoretz attacked Beat Generation writers, primarily Kerouac and Ginsberg, as "Know-Nothing Bohemians." Podhoretz charged that K. & G. were violent anti-intellectuals and that their cultivation of spontaneity destroyed "the distinction between life and literature.") "The novel is not an imaginary situation of imaginary truths—it is an expression of what one *feels*. Podhoretz doesn't write prose, he doesn't know how to write prose, and he isn't interested in the technical problems of prose or poetry. His criticism of Jack's spontaneous bop prosody shows that he can't tell the difference between words as rhythm and words as in diction . . . The bit about anti-intellectualism is a piece of vanity, we had the same education, went to the same school, you know there are 'Intellectuals' and there are intellectuals. Podhoretz is just out of touch with twentieth-century literature, he's writing for the eighteenth-century mind. We have a personal literature now—Proust, Wolfe, Faulkner, Joyce. The trouble is that Podhoretz has a great ridiculous fat-bellied mind which he pats too often."

Norman Mailer: "I read his 'White Negro' piece, it had a real grasp and kind of apocalyptic flip reality and is the only good definitive article I've run into. I'd love to talk to him. I hope he takes to pure poetry and becomes an angel poet; he has a great grasp of the Goof."

Flash, all news services! "I've been with an awful lot of beautiful juvenile delinquents. I've done my best to go to eternity with them."

A Renaissance

American Poets: "There's a renaissance in poetry going on. I'll give you a list of the 20 best poets in America; you know there's never been a list before of all the hip poets. These are poets who are mostly published underground because publishing in America is a trap, illusion, and fraud; Kerouac, yes. Jack's a poet; Corso, Ginsberg, Burroughs, we were together in San Francisco; Gary Snyder and Phil Whalen, also San Francisco, both now making the Zen scene; Robert Creeley, writes the small tight little poetry that you dig; Charles Olson, Denise Levertov, and Edward Marshall with Creeley, Black Mountain people,[1] Frank O'Hara and Kenneth Koch, New York painters' poets; John Ashbery, (Stan) Persky from Chicago; John Wieners, who publishes *Measure*; Paul Blackburn and Joel Oppenheimer, also Black Mountain; the recent Robert Lowell; Stuart Perkoff, Mike McClure; the old man is Robert Duncan. There are more like Raymond Bremser in Bordentown Reformatory and Ron Loewinsohn lost in Los Angeles, but I forget and must apologize for not giving the laurel halo to hundreds of unknown angels."

Flash. Reader's Digest: "So I told this guy on the radio that I liked marijuana and he put his hand on the controls and I said: 'Don't touch the button, if you cut me off, your audience will know why anyhow,' so he didn't do anything . . . Part of 'Howl' written under peyote. It's a vision of the St. Francis Drake Hotel in San Francisco."

The rest I cannot remember or can, but, leaving it at that, this is a piece on les pensées d'Allen Ginsberg, not about angels, Saints, (J.D.) Salinger, and "The Way of a Pilgrim." I cannot mock you, Allen, or deliver you described, pinpointed in a phrase or two. I can only listen, report some of it, and try to grasp the rest, for you are the only man I know who can discover *dharma* in a dentist's chair.[2]

---

[1] Andrew Rice founded Black Mountain College in 1933, near Asheville, North Carolina, as an experiment in community education. Among the contributing teachers, visitors, and students were Robert Creeley, Denise Levertov, Jonathan Williams, John Cage, and Robert Duncan. Josef Albers was rector, followed by Charles Olson, until it closed in 1956. Many of the associated poets became known as the Black Mountain School. *The Black Mountain Review* (1954–57), edited by Robert Creeley, featured many of the poets, as did the earlier magazine *Origin* (1951–56). See Martin Duberman, *Black Mountain: An Exploration in Community* (New York: Dutton, 1972).

[2] A reference to Ginsberg's experiments with laughing gas, administered to him by a dentist cousin to investigate alternative states of consciousness. See the reference to Ginsberg's use of laughing gas in the Clark interview, as well as the poem "Laughing Gas" in *Kaddish and Other Poems*.

1960s

ERNIE BARRY
October 28, 1963,   San Francisco

City Lights Journal, 1964

About one month after the *Village Voice* interview, Allen Ginsberg would sit down and in one marathon forty-hour session, sustained in part by Dexedrine pills, write "Kaddish." Two months later, Ginsberg participated in the making of the improvisational film *Pull My Daisy*, filmed by photographer Robert Frank, who would later influence Ginsberg's own work as a photographer (see Thomas Gladysz interview).

In the spring of 1959 Allen's interest in higher and altered states of consciousness would lead him to try LSD for the first time at Stanford University. After attending a writer's conference in Chile in January 1960 with Lawrence Ferlinghetti, Ginsberg traveled throughout South America both sightseeing and searching for the hallucinogenic drug *yage* or *ayahuasca*, made from a native plant and used by Amazon shamans (known as *cuaranderos* or *brujos*). After returning to New York, he completed "Kaddish" in the fall of 1960. On November 26, 1960, Allen took psilocybin under Harvard psychologist Timothy Leary's direction, and immediately after the dramatic experience the two of them began to plan a "psychedelic revolution."

In March of 1961 Ginsberg and Peter Orlovsky traveled to Paris to visit William Burroughs and Gregory Corso. When they arrived, Allen and Peter found that Burroughs had left for Tangier. A letter from Burroughs eventually arrived, inviting Ginsberg, Orlovsky, and Corso to join him there. When strains emerged among the circle gathered in Tangier, Peter Orlovsky set out on his own for Istanbul in July of 1961. On August 24 Ginsberg departed Morocco for Greece and then reconnected with Peter in Tel Aviv, where they met theologian Martin Buber and cabala scholar Gershom Sholem. From Israel Allen and Peter left for India, spending approximately one month in Africa on their way east. They finally sailed for India in early February 1962, arriving in Bombay on February 15. In India, they eventually joined up with fellow poet Gary

Snyder and his wife, Joanne Kyger, who had arrived from Japan, where they had been living for almost six years. After a long stay in India, recorded in the book *Indian Journals*, Allen flew by himself to Bangkok in May of 1963 and then on to Saigon on his way to see the ruins of Angkor Wat. While in Vietnam Ginsberg questioned journalists and others closely about the situation in Vietnam and the American role there.

On June 11 Allen flew to Tokyo on his way to a five-week visit with Snyder and Kyger in Kyoto before returning to Tokyo to fly back to North America to attend a poetry conference at the University of British Columbia in Vancouver in late July. (It was on the train ride back from visiting Snyder and Kyger that Ginsberg had the break-through experience that would lead him to write "The Change: Kyoto–Tokyo Express." The experience that inspired the poem is discussed in the Clark interview.) From Canada Allen returned to San Francisco. There, on October 28, Ginsberg attended his first political demonstration, a protest against an appearance by Madame Nhu, wife of Vietnam's chief of secret police and sister-in-law to President Diem, at the Sheraton Palace Hotel before a large audience of business and civic leaders. The crowd of five hundred protesters at a rally and picket line in San Francisco was the largest that greeted Madame Nhu while she was in the United States. Allen had walked the picket line and took time out for this interview in front of the Sheraton. The interview was published in the second issue of a periodical started by Lawrence Ferlinghetti, *City Lights Journal*.

—DC

Ernie Barry: What other political demonstrations have you been involved in, Allen?

(Shouting demonstrators swirling past us as we chat sitting on a raised stone strip on the side of the hotel.)

Allen Ginsberg: None. This is the first time I've taken a political stand.

EB: Is there any special reason for your new policy?

AG: Yes. Read my sign.

(Allen twisting his picket sign toward me so I could copy the poem on it.)

"Man is naked without secrets armed men lack this joy
How many million persons without names?
What do we know of their suffering?
'Oh how wounded, how wounded!' says the guru
Thine own heart says the swami
Within you says the Christ
Till his humanity awakes says Blake
I am here: saying seek mutual surrender tears
That there be no more hell in Vietnam
That I not be in hell here in the street

War is black magic
Belly flowers to North and South Vietnam
include everybody
End the human war
Name hypnosis and fear is the
Enemy—Satan go home!
I accept America and Red China
To the human race
Madame Nhu and Mao Tse-Tung
Are in the same boat of meat"

EB: You wrote it especially for the protest?

AG: Yes.

EB: Allen, I suppose you feel much of the tension in the world is due to a lack of compassion and love.

AG: Yes, Ernie, not enough people make love. (Allen putting his hands around me, caressing my shoulders, other parts of me.)

EB: Er. . . . Yes, Allen.

EB: Who do you make love to, Allen?

AG: Anybody I can sleep with, anybody who'll have me. (Allen starting to put his arms around me; me feeling the point is well-made without him having to physically demonstrate his policy. Trying to get his hands off me.)

AG: Young Rimbauds, young Monroes, as well as other human beings. And occasionally an android here and there.

EB: It appears that much of America's sex problems are due to a general sexual incompetence. You can't make love to everybody, at least I can't, because most people in the Western World just don't know how to fuck properly. And they don't wish to do anything to correct this incompetence or even admit it exists.

AG: I am sexually incompetent on account of all the different people that make my belly flutter would probably reject me if I looked at them crying. I'm sexually incompetent because I'm scared.

EB: Scared of what?

AG: Scared they won't have me.

EB: What about lack of sexual knowledge?

AG: Knowledge comes from doing what comes naturally.

EB: Surely you don't believe it's as simple as that. You're right only for the case of free individuals without emotional or psychological inhibitions and hang-ups. If they freely experimented with each other a sexual expertise and knowledge would result. But most of us do have inhibitions and hang-ups. What do you think is the cause of this?

AG: Blocked-up love.

EB: What do you mean by love, fucking?

AG: No, a feeling in the belly of trust, encouragement to Be that might lead to pushing bellies together and kissing ears and all sorts of delicious things including babies.

EB: Do you turn on to babies? Children?

AG: Yes, they don't hate me.

EB: Obviously a child's approach to life is more instinctive, and thus more natural, than anyone else's approach. As we get older I think we consciously avoid instinctive activities.

AG: Yes. Belly rubbing is instinctive. Tenderness between people is a normal instinctive relationship. Tenderness between men and men as well as men and women. And women and women. Read all about it in

Whitman. It took a lot of nature in him to expose that tenderness openly for the first time in America but that's the unconscious basis of our democracy and that's why I'm here today on the picket line trying to be tender to Madame Nhu and Mao Tse-Tung. Or better, asking them to be tender.

EB: How do you feel about Madame Nhu?

AG: She needs more love.

EB: She needs more sexual relationships or what?

AG: She needs more human contact in a friendly human universe. See, the basis of her paranoia is the assumption that all but a few of the SELVES out there are basically hostile to her existence and so she thinks she has to protect herself in the manner that we read about.

EB: I came out here to personally confront her with my anger over what she stands for.

AG: That would only make it worse. It's like a big booby trap of massed hatred and anxiety that cuts out all soft feelings in the body and ultimately results in a mass illusion of fear manifested in an H-bomb. However since H-bomb resolution of the problem is no longer acceptable to the body—even unconsciously—another alternative resolution of the conflict is apparent which is we all surrender to each other, all of us bankrupt, and find a friendly human universe where we can all completely exist at once. By completely exist I mean the liberation of all the blocked-up feelings of need for each other and reassurance and emotional ecstasy that is our birthright. That is to say it's built-in physiologically into the body but it's been or we've been so rejected, put down, refused, unloved and desperately hopeless we forget the old human tears.

EB: Do you pay rent to landlords?

AG: Yeah.

EB: Economic wealth is unevenly and what I would call unjustly distributed. Do you feel bad about helping to continue this situation?

AG: I can't think of any economic theory that satisfies modern history; everything is changing too fast. I suppose some form of community sharing or communism is appropriate to the future State of Man. However I don't see how that can work without first a sharing of feelings. Then material arrangements will fall into place. So I'll begin right now sharing my feelings. (Allen touching me for a second on the shoulders with his hands.)

EB: It is possible that the *american anarchist* might be banned either because of political unorthodoxy or frankness in discussing sex. How do you feel about this?

AG: Nobody should be banned, naturally.

EB: About 6 months ago an obscure (circulation below 500) Greenwich Village poetry magazine called *Fuck You* published an account of you and poet Peter Orlovsky making it together. Giving each other blow-jobs to be specific. Since it was co-written by you and published seemingly with your permission I assume you place some value to it.

AG: Orlovsky wrote (transcribed) that love scene in Tangier several years ago. At the time I was ashamed of it, we were supposed to be interviewing each other on world politics for City Lights' *Journal for Protection of All Beings*, and I thought Orlovsky was being irrelevant. However now I see he had the right heart: bringing everything mental right down back into the body and disclosure of the secrets of the body's feelings:

> Each man in his spectre's power
> til the arrival of that hour
> when his humanity awake
> and cast his spectre into the lake.

Orlovsky's humanity was waked while I was in the power of my spectre— thinking I could make great angry pronouncements about the universe. All I wanted was love. All anybody wants. He sent that sex-writing to *Fuck You* from India and I was happy to see it printed, since now I am *be-nakeded* in my own and other's eyes so have no more Secrecy to hide behind; no image to live up to but the sad and happy feels of my own being.

> Now all my charms are overthrown
> and what strength I have's my own

which is most faint, 'tis true
unless I pardoned am by you, etc.

EB: It seems to me that the anti-intellectual nature of American colleges would be an aid to, as you say, bringing everything mental right down back into the body.

AG: American colleges are not anti-intellectual, they're too intellectual: that is, the spectre of objective mental intellect and reason has them in its power. Until the human universe of direct feeling returns the colleges will feel like machines and people in them will be afflicted by impersonal, non-personal, coldwar subjectivity. However, everyone wants to feel, and wants to feel loved and to love, so there's inevitable Hope beneath every grim mask.

EB: Do you think any left wing philosophy or ideology can help us to see this problem clearer and to solve it?

AG: Anger and fury of left wing will only drive the humanoid bureaucrats and cops into deeper humanoidism. Only affection and tenderness will make the world safe for democracy. Be kind to cops; they're not cops, they're people in disguise who've been deceived by their own disguise.

EB: Last summer *Esquire* published an immensely distorted account of your stay in India. The title was, I think, "On the road with Allen and his Boys in India." Besides trying to create the impression that you were just a naive and unsophisticated dilettante the article inferred that most of your time was spent eating borscht and scrambled eggs and half-heartedly trying to be a Hindu. Since this article was based on an interview you gave them, why do you bother to even communicate with hack reporters from commercial magazines bent on distorting what you are for the sake of sensationalism?

AG: Because of what I said in answer to your last question. It is necessary to treat everybody with equal care, *the american anarchist* or *Esquire* photographers: No exclusions. Everybody will be reborn back to their bodies which they've been almost driven out of by atomic fear. But by hate, indignation, anger, resentment, carping and bugging hostility, nothing that feels good can be accomplished.

EB: In August you lectured at a Canadian University with a number of other beat poets. Teaching verse-writing skills in a formal manner or mainly communicating freely with the students?

AG: I taught with Charles Olson, Robert Creeley, Denise Levertov, Robert Duncan, and Philip Whalen at Vancouver. The University of British Columbia. We didn't teach verse-writing skills. We were all emotionally bankrupt and went around weeping and asking the students for love.

TOM CLARK
mid-May 1965,   Cambridge, England

"The Art of Poetry" (No. VIII in a series)
The Paris Review, Spring 1966

Allen Ginsberg left San Francisco at the end of November 1963 to return to New York, where he soon settled in the East Village, which would remain his primary residence for the rest of his life.

In January 1965 Allen attended a writer's conference in Havana. While in Cuba he was critical of the Castro regime, which resulted in his being thrown out of Cuba. Placed by the Cubans on a plane to Prague, Ginsberg used the opportunity of being so close to his mother's homeland to visit Russia, departing Prague on March 18 to go to Moscow by way of Warsaw. He returned to Czechoslovakia the day before May Day. His arrival was timely for, although the traditional celebration of May Day, *Majáles*, a festival which included the election of a May king by students and a beauty contest to select a May queen, had been outlawed, in 1965 the Communist government decided to allow the celebration of *Majáles* for the first time in twenty years. To the great consternation of the Communist authorities, the students elected Allen as *Král Majáles*, or May king. The celebration for this newly allowed *Majáles* turned out to be larger than even traditional May Day celebrations, and as Ginsberg was driven through the streets of Prague on the back of a flatbed truck, almost all of Prague turned out to watch. After government agents followed Allen for several days, assaulted him, and stole one of his notebooks, the Czech government threw him out of the country, placing him on a plane for London on May 7. During that plane ride Ginsberg wrote the poem "Kral Majales." (Allen discusses the Czech expulsion in detail in the Frakes interview; the Cuban expulsion is examined in both the Young and the Frakes interviews.)

It was while Ginsberg was in England that spring that he met up with Tom Clark, a young American poet studying in England and editing poetry for the *Paris Review*, resulting in what Allen always considered one of his best and most definitive interviews, largely because of the pivotal role in his life he attributed to the William Blake visions he had had while at Columbia University, which are so fully described here.

Ginsberg and Clark met in Bristol when they both took part in an informal, round-robin poetry reading at an art gallery. The two then hitchhiked together to Wells Cathedral and Glastonbury, where Allen picked a flower from King Arthur's grave. Leaving Glastonbury, they were caught in a rainstorm and took a bus to Bath, and then resumed hitchhiking to London. Two weeks later Ginsberg went to Cambridge, where Clark was studying, for a reading and to examine the Blake manuscripts in the Fitzwilliam Museum.

The interview took place in Clark's flat at 24 Newmarket Road in Cambridge in two two-hour sessions interrupted by a meal. It was recorded on a tape recorder and transcribed by Tom Clark. In 1999 Clark wrote that "I'd had a chance to get to know Allen, so I knew what to ask. There was a nice empathy. . . . He was conscious of the *Paris Review*'s large international literary audience, and I was conscious of him being conscious of that. There was the clear feeling that he was stating his credo, his poetics, his testament of belief: a sense almost of the Historic Moment, for him. . . . I think the shadow-presence of Blake on the landscape was also a contributing factor."

—DC

Tom Clark: I think Diana Trilling, speaking about your reading at Columbia, remarked that your poetry, like all poetry in English when dealing with a serious subject, naturally takes on the iambic pentameter rhythm. Do you agree?

Allen Ginsberg: Well, it really isn't an accurate thing, I don't think. I've never actually sat down and made a technical analysis of the rhythms that I write. They're probably more near choriambic—Greek meters, dithyrambic meters—and tending toward de DA de de DA de de . . . what is that? Tending toward dactylic, probably. Williams once remarked that American speech tends toward dactylic. But it's more complicated than dactyl because dactyl is a three, three units, a foot consisting of three parts, whereas the actual rhythm is probably a rhythm which consists of five, six, or seven, like DA de de DA de de DA de de DA DA. Which is more toward the line of Greek dance rhythms— that's why they call them choriambic. So actually, probably it's not really technically correct, what she said. But—and that applies to certain poems, like certain passages of "Howl" and certain passages of "Kaddish"—there are definite rhythms which could be analyzed as corresponding to classical rhythms, though not necessarily English classical rhythms; they might correspond to Greek classical rhythms, or Sanskrit prosody. But probably most of the other poetry, like "Aether" or "Laughing Gas" or a lot of those poems, they simply don't fit into that. I think she felt very comfy, to think that that would be so. I really felt quite hurt about that, because it seemed to me that she ignored the main prosodic technical achievements that I had proffered forth to the academy, and they didn't even recognize it. I mean not that I want to stick her with being the academy.

TC: And in "Howl" and "Kaddish" you were working with a kind of classical unit? Is that an accurate description?

AG: Yeah, but it doesn't do very much good, because I wasn't really working with a classical unit, I was working with my own neural impulses and writing impulses. See, the difference is between someone sitting down to write a poem in a definite preconceived metrical pattern and filling in that pattern, and someone working with his physiological movements and arriving at a pattern, and perhaps even arriving at a pattern which might even have a name, or might even have a classical usage, but arriving at it organically rather than synthetically. Nobody's got any objection to even iambic pentameter if it comes from a source deeper than the mind—that is to say, if it comes from the breathing and the belly and the lungs.

TC: American poets have been able to break away from a kind of English specified rhythm earlier than English poets have been able to do. Do you think this has anything to do with a peculiarity in English spoken tradition?

AG: No, I don't really think so, because the English don't speak in iambic pentameter either; they don't speak in the recognizable pattern that they write in. The dimness of their speech and the lack of emotional variation is parallel to the kind of dim diction and literary usage in the poetry now. But you can hear all sorts of Liverpudlian or Geordian—

that's Newcastle—you can hear all sorts of variants aside from an upper-tone accent, a high-class accent, that don't fit into the tone of poetry being written right now. It's not being used like in America—I think it's just that British poets are more cowardly.

TC: Do you find any exception to this?

AG: It's pretty general, even the supposedly avant-garde poets. They write, you know, in a very toned-down manner.

TC: How about a poet like Basil Bunting?

AG: Well, he was working with a whole bunch of wild men from an earlier era, who were all breaking through, I guess. And so he had that experience—also he knew Persian, he knew Persian prosody. He was better educated than most English poets.

TC: The kind of organization you use in "Howl," a recurrent kind of syntax—you don't think this is relevant any longer to what you want to do?

AG: No, but it was relevant to what I wanted to do then, it wasn't even a conscious decision.

TC: Was this related in any way to a kind of music or jazz that you were interested in at the time?

AG: Mmm . . . the myth of Lester Young as described by Kerouac, blowing eighty-nine choruses of "Lady Be Good," say, in one night, or my own hearing of Illinois Jacquet's *Jazz at the Philharmonic*, Volume 2; I think "Can't Get Started" was the title.

TC: And you've also mentioned poets like Christopher Smart, for instance, as providing an analogy—is this something you discovered later on?

AG: When I looked into it, yeah. Actually, I keep reading, or earlier I kept reading, that I was influenced by Kenneth Fearing and Carl Sandburg, whereas actually I was more conscious of Christopher Smart, and Blake's Prophetic Books, and Whitman and some aspects of Biblical rhetoric. And a lot of specific prose things, like Genet, Genet's *Our Lady of the Flowers* and the rhetoric in that, and Céline; Kerouac, most of all, was the biggest influence I think—Kerouac's prose.

TC: When did you come onto Burroughs's work?

AG: Let's see . . . Well, first thing of Burroughs's I ever read was 1946 . . . which was a skit later published and integrated in some other work of his, called *So Proudly We Hail*, describing the sinking of the Titanic and an orchestra playing, a spade orchestra playing "The Star Spangled Banner" while everybody rushed out to the lifeboats and the captain got up in woman's dress and rushed into the purser's office and shot the purser and stole all the money, and a spastic paretic jumped into a lifeboat with a machete and began chopping off people's fingers that were trying to climb into the boat, saying, "Out of the way, you foolth . . . dirty thunthufbithes." That was a thing he had written up at Harvard with a friend named Kells Elvins. Which is really the whole key of all his work, like the sinking of America, and everybody like frightened rats trying to get out, or that was his vision of the time.

Then he and Kerouac later in 1945—forty-five or forty-six—wrote a big detective book together, alternating chapters. I don't know where that book is now—Kerouac has his chapters and Burroughs's are somewhere in his papers. So I think in a sense it was Kerouac that encouraged Burroughs to write really, because Kerouac was so enthusiastic about prose, about writing, about lyricism, about the honor of writing . . . the Thomas Wolfe-ian delights of it. So anyway he turned Burroughs on in a sense, because Burroughs found a companion who could write really interestingly, and Burroughs admired Kerouac's perceptions. Kerouac could imitate Dashiell Hammett as well as Bill, which was Bill's natural style: dry, bony, factual. At that time Burroughs was reading John O'Hara, simply for facts, not for any sublime stylistic thing, just because he was a hard-nosed reporter.

Then in Mexico around 1951 he started writing *Junkie*. I've forgotten what relation I had to that—I think I wound up as the agent for it, taking it around New York trying to get it published. I think he sent me portions of it at the time—I've forgotten how it worked out now. This was around 1949 or 1950. He was going through a personal crisis, his wife had died. It was in Mexico or South America . . . but it was a very generous thing of him to do, to start writing all of a sudden. Burroughs was always a very *tender* sort of person, but very dignified and shy and withdrawn, and for him to commit himself to a big autobiographical thing like that was . . . at the time, struck me as like a piece of eternity is in love with the . . . what is it, "Eternity is in love with the productions of Time"? So he was making a production of Time then.

Then I started taking that around. I've forgotten who I took that to but I think maybe to Louis Simpson who was then working at Bobbs-Merrill. I'm not sure whether I took it to him—I remember taking it to Jason Epstein who was then working at Doubleday I think. Epstein at the time was not as experienced as he is now. And his reaction to it, I remember when I went back to his office to pick it up, was, well this is all very interesting, but it isn't really interesting, on account of if it were an autobiography of a junkie written by Winston *Churchill* then it'd be interesting, but written by somebody he'd never heard of, well then it's *not* interesting. And anyway I said what about the *prose*, the prose is interesting, and he says, oh, a difference of opinion on that. Finally I wound up taking it to Carl Solomon who was then a reader for A. A. Wynn Company, which was his uncle; and they finally got it through there. But it was finally published as a cheap paperback. With a whole bunch of frightened footnotes; like Burroughs said that marijuana was nonhabit-forming, which is now accepted as a fact, there'd be a footnote by the editor, "Reliable, er, responsible medical opinion does not confirm this." Then they also had a little introduction . . . literally they were afraid of the book being censored or seized at the time, is what they said. I've forgotten what the terms of censorship or seizure were that they were worried about. This was about 1952. They said that they were afraid to publish it straight for fear there would be a Congressional investigation or something, I don't know what. I think there was some noise about narcotics at the time. Newspaper noise . . . I've forgotten exactly what the arguments were. But anyway they had to write a preface which hedged on the book a lot.

TC: Has there been a time when fear of censorship or similar trouble has made your own expression difficult?

AG: This is so complicated a matter. The beginning of the fear with me was, you know what would my father say to something that I would write. At the time, writing "Howl"—for instance, like I assumed when writing it that it was something that could not be published because I wouldn't want my daddy to see what was in there. About my sex life, being fucked in the ass, imagine your father reading a thing like that, was what I thought. Though that disappeared as soon as the thing was real, or as soon as I manifested my . . . you know, it didn't make that much importance finally. That was sort of a help for writing, because I assumed that it wouldn't be published, therefore I could say anything that I wanted. So literally just for myself or anybody that I knew personally well, writers who would be willing to appreciate it with a breadth of tolerance—in a piece of work like "Howl"—who wouldn't be judging from a moralistic viewpoint but looking for evidences of humanity or secret thought or just actual truthfulness.

Then there's later the problem of publication—we had a lot. The English printer refused at first I think, we were afraid of customs; the first edition we had to print with asterisks on some of the dirty words, and then the *Evergreen Review*[3] in reprinting it used asterisks, and various people reprinting it later always wanted to use the *Evergreen* version rather than the corrected legal City Lights version—like I think there's an anthology of Jewish writers, I forgot who edited that, but a couple of the high-class intellectuals from Columbia. I had written asking them specifically to use the later City Lights version, but they went ahead and printed an asterisked version. I forget what was the name of that—something like *New Generation of Jewish Writing*, Philip Roth, et cetera.

TC: Do you take difficulties like these as social problems, problems of communication simply, or do you feel they also block your own ability to express yourself for yourself?

AG: The problem is, where it gets to literature, is this. We all talk among ourselves and we have common understandings, and we say anything we want to say, and we talk about our assholes, and we talk about our cocks, and we talk about who we fucked last night, or who we're gonna fuck tomorrow, or what kind love affair we have, or when we got drunk, or when we stuck a broom in our ass in the Hotel Ambassador in Prague—anybody tells one's friends about that. So then—what happens if you make a distinction between what you tell your friends and what you tell your Muse? The problem is to break down that distinction: when you approach the Muse to talk as frankly as you would

---

[3]The *Evergreen Review* featured many influential twentieth-century writers. Issue number 2, "The San Francisco Scene," helped put the Beat Generation on the map. Edited by Barney Rosset and Donald Allen from 1957 to 1973, it published seventeen volumes, numbers 1–97.

talk with yourself or with your friends. So I began finding, in conversations with Burroughs and Kerouac and Gregory Corso, in conversations with people whom I knew well, whose souls I respected, that the things we were telling each other for real were totally different from what was already in literature. And that was Kerouac's great discovery in *On the Road*. The kinds of things that he and Neal Cassady were talking about, he finally discovered were the subject matter for what he wanted to write down. That meant, at that minute, a complete revision of what literature was supposed to be, in his mind, and actually in the minds of the people that first read the book. Certainly in the minds of the critics, who had at first attacked it as not being . . . proper structure, or something. In other words, a gang of friends running around in an automobile. Which obviously is like a great picaresque literary device, and a classical one. And was not recognized, at the time, as suitable literary subject matter.

TC: So it's not just a matter of themes—sex, or any other one—

AG: It's the ability to commit to writing, to write, the same way that you . . . are! Anyway! You have many writers who have preconceived ideas about what literature is supposed to be and their ideas seem to exclude that which makes them most charming in private conversation. Their faggishness, or their campiness, or their neurasthenia, or their solitude, or their goofiness, or their—even—masculinity, at times. Because they think that they're gonna write something that sounds like something else that they've read before, instead of sounds like them. Or comes from their own life. In other words, there's no distinction, there should be no distinction between what we write down, and what we really know, to begin with. As we know it every day, with each other. And the hypocrisy of literature has been—you know like there's supposed to be formal literature, which is supposed to be different from . . . in subject, in diction, and even in organization, from our quotidian inspired lives.

It's also like in Whitman, "I find no fat sweeter than that which sticks to my own bones"—that is to say the self-confidence of someone who knows that he's really alive, and that his existence is just as good as any other subject matter.

TC: Is physiology a part of this too—like the difference between your long breath line and William Carlos Williams's shorter unit?

AG: Analytically, ex post facto, it all begins with fucking around and intuition and without any idea of what you're doing, I think. Later, I have a tendency to explain it, "Well, I got a longer breath than Williams, or I'm Jewish, or I study yoga, or I sing long lines. . . ." But anyway, what it boils down to is this, it's my *movement*, my feeling is for a big long cranky statement—partly that's something that I share, or maybe that I even got from Kerouac's long prose line; which is really, like he once remarked, an extended poem. Like one long sentence page of his in *Doctor Sax* or *Railroad Earth* or occasionally *On the Road*— if you examine them phrase by phrase they usually have the density of poetry, and the beauty of poetry, but most of all the single elastic rhythm running from beginning to end of the line and ending "mop!"[4]

TC: Have you ever wanted to extend this rhythmic feeling as far as, say, Artaud or now Michael McClure have taken it—to a line that is actually animal noise?

AG: The rhythm of the long line is also an animal cry.

TC: So you're following that feeling and not a thought or a visual image?

AG: It's simultaneous. The poetry generally is like a rhythmic articulation of feeling. The feeling is like an impulse that rises within—just like sexual impulses, say; it's almost as definite as that. It's a feeling that begins somewhere in the pit of the stomach and rises up forward in the breast and then comes out through the mouth and ears, and comes forth a croon or a groan or a sigh. Which, if you put words to it by looking around and seeing and trying to describe what's making you sigh—and sigh in words—you simply articulate what you're feeling. As simple as that. Or actually what happens is, at best what happens, is there's a definite body rhythm that has no definite words, or may have one or two words attached to it, one or two key words attached to it. And then, in writing it down, it's simply by a process of association that I find what the rest of the statement is—what can be collected around that word, what that word is connected to. Partly by simple association,

---

[4]Bebop, and mop-mop, which derived from it, was an onomatopoetic term for a style of jazz. Slim and Slam, who performed often in the mid-1940s, the time when Ginsberg was hearing quite a bit of jazz, are credited with coining the style and term "mop-mop." This might suggest Ginsberg's familiarity with the genre.

the first thing that comes to my mind like "Moloch is" or "Moloch who," and then whatever comes out. But that also goes along with a definite rhythmic impulse, like DA de de DA de de DA de de DA DA. "Moloch whose eyes are a thousand blind windows." And before I wrote "Moloch whose eyes are a thousand blind windows," I had the word, "Moloch, Moloch, Moloch," and I also had the feeling DA de de DA de de DA de de DA DA. So it was just a question of looking up and seeing a lot of windows, and saying, Oh, windows, of course, but what kind of windows? But not even that—"Moloch whose eyes." "Moloch whose eyes"—which is beautiful in itself—but what about it, Moloch whose eyes are what? So Moloch whose eyes—then probably the next thing I thought was "thousands." O. K., and then thousands what? "Thousands blind." And I had to finish it somehow. So I hadda say "windows." It looked good *afterward*.

Usually during the composition, step by step, word by word and adjective by adjective, if it's at all spontaneous, I don't know whether it even makes sense sometimes. Sometimes I do know it makes complete sense, and I start crying. Because I realize I'm hitting some area which is absolutely true. And in that sense applicable universally, or understandable universally. In that sense able to survive through time—in that sense to be read by somebody and wept to, maybe, centuries later. In that sense prophecy, because it touches a common key . . . what prophecy actually is is not that you actually know that the bomb will fall in 1942. It's that you know and feel something which somebody knows and feels in a hundred years. And maybe articulate it in a hint—concrete way that they can pick up on in a hundred years.

TC: You once mentioned something you had found in Cézanne—a remark about the reconstitution of the *petites sensations* of experience, in his own painting—and you compared this with the methods of your poetry.

AG: I got all hung up on Cézanne around 1949 in my last year at Columbia, studying with Meyer Schapiro. I don't know how it led into it—I think it was about the same time that I was having these Blake visions. So. The thing I understood from Blake was that it was possible to transmit a message through time which could reach the enlightened, that poetry had a definite effect, it wasn't just pretty, or just beautiful, as I had understood pretty beauty before—it was something basic to human existence, or it reached something, it reached the bottom of

human existence. But anyway the impression I got was that it was like a
kind of time machine through which he could transmit, Blake could
transmit, his basic consciousness and communicate it to somebody
else after he was dead—in other words, build a time machine.

Now just about that time I was looking at Cézanne and I suddenly
got a strange shuddering impression looking at his canvases, partly
the effect when someone pulls a Venetian blind, reverses the Venetian—
there's a sudden shift, a flashing that you see in Cézanne canvases.
Partly it's when the canvas opens up into three dimensions and looks
like wooden objects, like solid-space objects, in three dimensions
rather than flat. Partly it's the enormous spaces which open up in
Cézanne's landscapes. And it's partly that mysterious quality around
his figures, like of his wife or the cardplayers or the postman or who-
ever, the local Aix characters. They look like great huge 3-D wooden
dolls, sometimes. Very uncanny thing, like a very mysterious thing—
in other words, there's a strange sensation that one gets, looking at
his canvases, which I began to associate with the extraordinary sensa-
tion—cosmic sensation, in fact—that I had experienced catalyzed by
Blake's "Sun-flower" and "Sick Rose" and a few other poems. So I
began studiously investigating Cézanne's intentions and method, and
looking at all the canvases of his that I could find in New York, and all
the reproductions I could find, and I was writing at the time a paper
on him, for Schapiro at Columbia in the fine-arts course.

And the whole thing opened up, two ways: first, I read a book on
Cézanne's composition by Earl Loran, who showed photographs,
analyses and photographs of the original motifs, side by side with the
actual canvases—and years later I actually went to Aix, with all the
postcards, and stood in the spots, and tried to find the places where
he painted Mont-Sainte-Victoire from, and got in his studio and saw
some of the motifs he used, like his big black hat and his cloak. Well,
first of all, I began to see that Cézanne had all sorts of literary symbol-
ism in him, on and off. I was preoccupied with Plotinian terminology,
of time and eternity, and I saw it in Cézanne paintings, an early paint-
ing of a clock on a shelf which I associated with time and eternity, and
I began to think he was a big secret mystic. And I saw a photograph of
his studio in Loran's book and it was like an alchemist's studio,
because he had a skull, and he had a long black coat, and he had this
big black hat. So I began thinking of him as, you know, like a magic
character. Like the original version I had thought of him was like this

austere dullard from Aix. So I began getting really interested in him as
a hermetic type, and then I symbolically read into his canvases things
that probably weren't there, like there's a painting of a winding road
which turns off, and I saw that as the mystical path: it turns off into a
village and the end of the path is hidden. Something he painted I
guess when he went out painting with Bernard. Then there was an
account of a very fantastic conversation that he had had. It's quoted in
Loran's book: there's a long long long paragraph where he says, "By
means of squares, cubes, triangles, I try to reconstitute the impres-
sion that I have from nature: the means that I use to reconstitute the
impression of solidity that I think-feel-see when I am looking at a
motif like Victoire is to reduce it to some kind of pictorial language, so
I use these squares, cubes, and triangles, but I try to build them
together so interknit [and here in the conversation he held his hands
together with his fingers interknit] so that no light gets through." And
I was mystified by that, but it seemed to make sense in terms of the
grid of paint strokes that he had on his canvas, so that he produced a
solid two-dimensional surface which when you looked into it, maybe
from a slight distance with your eyes either unfocused or your eyelids
lowered slightly, you could see a great three-dimensional opening,
mysterious, stereoscopic, like going into a stereopticon. And I began
discovering in "The Cardplayers" all sorts of sinister symbols, like
there's one guy leaning against the wall with a stolid expression on his
face, that he doesn't want to get involved; and then there's two guys
who are peasants, who are looking as if they've just been dealt Death
cards; and then the dealer you look at and he turns out to be a city
slicker with a big blue cloak and almost rouge doll-like cheeks and a
fat-faced Kafkian-agent impression about him, like he's a cardsharp,
he's a cosmic cardsharp dealing out Fate to all these people. This
looks like a great big hermetic Rembrandtian portrait in Aix! That's
why it has that funny monumentality—aside from the quote plastic
values unquote.

Then, I smoked a lot of marijuana and went to the basement of the
Museum of Modern Art in New York and looked at his water colors
and that's where I began really turning on to space in Cézanne and the
way he built it up. Particularly there's one of rocks, I guess "Rocks at
Garonne," and you look at them for a while, and after a while they
seem like they're rocks, just the rock parts, you don't know where
they are, whether they're on the ground or in the air or on top of a

cliff, but then they seem to be floating in space like clouds, and then they seem to be also a bit like they're amorphous, like kneecaps or cockheads or faces without eyes. And it has a very mysterious impression. Well, that may have been the result of the pot. But it's a definite thing that I got from that. Then he did some very odd studies after classical statues, Renaissance statues, and they're great gigantesque herculean figures with little tiny pinheads . . . so that apparently was his comment on them!

And then . . . the things were endless to find in Cézanne. Finally I was reading his letters and I discovered this phrase again, *mes petites sensations*—"I'm an old man and my passions are not, my senses are not coarsened by passions like some other old men I know, and I have worked for years trying to," I guess it was the phrase, "*reconstitute the petites sensations* that I get from nature, and I could stand on a hill and merely by moving my head half an inch the composition of the landscape was totally changed." So apparently he'd refined his optical perception to such a point where it's a real contemplation of optical phenomena in an almost yogic way, where he's standing there, from a specific point studying the optical field, the depth in the optical field, looking, actually looking at his own eyeballs in a sense. The attempting to reconstitute the sensation in his own eyeballs. And what does he say finally—in a very weird statement which one would not expect of the austere old workman—he said, "And this *petite sensation* is nothing other than *pater omnipotens aeterna deus*."

So that was, I felt, the key to Cézanne's hermetic method . . . everybody knows his workman-like, artisan-like, pettified-like painting method which is so great, but the really romanticistic motif behind it is absolutely marvelous, so you realize that he's really a saint! Working on his form of yoga, all that time, in obvious saintly circumstances of retirement in a small village, leading a relatively nonsociable life, going through the motions of going to church or not, but really containing in his skull these supernatural phenomena, and observations . . . you know, and it's very humble actually, because he didn't know if he was crazy or not—that is a flash of the physical, miracle dimensions of existence, trying to reduce that to canvas in two dimensions, and then trying to do it in such a way as it would look—if the observer looked at it long enough—it would look like as much three dimension as the actual world of optical phenomena when one looks through one's eyes. Actually he's reconstituted the whole fucking universe in his canvases—it's like a fantastic thing!—or at least the appearance of the universe.

So. I used a lot of this material in the references in the last part of the first section of "Howl": "sensation of Pater Omnipotens Aeterna Deus." The last part of "Howl" was really an homage to art but also in specific terms an homage to Cézanne's method, in a sense I adapted what I could to writing; but that's a very complicated matter to explain. Except, putting it very simply, that just as Cézanne doesn't use perspective lines to create space but it's a juxtaposition of one color against another color (that's one element of his space), so, I had the idea, perhaps overrefined that by the unexplainable, unexplained nonperspective line, that is, juxtaposition of one word against another, a gap between the two words—like the space gap in the canvas—there'd be a gap between the two words which the mind would fill in with the sensation of existence. In other words when I say, oh . . . when Shakespeare says, "In the dread vast and middle of the night," something happens between "dread vast" and "middle." That creates like a whole space of, spaciness of black night. How it gets that is very odd, those words put together. Or in the haiku, you have two distinct images, set side by side without drawing a connection, without drawing a logical connection between them the mind fills in this . . . this space. Like

> O ant
>   crawl up Mount Fujiyama,
>   but slowly, slowly.

Now you have the small ant and you have Mount Fujiyama and you have the slowly, slowly, and what happens is that you feel almost like . . . a cock in your mouth! You feel this enormous space-universe, it's almost a tactile thing. Well, anyway, it's a phenomenon-sensation, phenomenon hyphen sensation, that's created by this little haiku of Issa, for instance.

So, I was trying to do similar things with juxtapositions like "hydrogen jukebox." Or . . . "winter midnight smalltown streetlight rain." Instead of cubes and squares and triangles. Cézanne is reconstituting by means of triangles, cubes, and colors—I have to reconstitute by means of words, rhythms of course, and all that—but say it's words, phrasings. So. The problem is then to reach the different parts of the mind, which are existing simultaneously, choosing elements from both, like: jazz, jukebox, and all that, and we have the jukebox from that; politics, hydrogen bomb, and we have the hydrogen of that, you see "hydrogen jukebox." And that actually compresses in one instant like a whole series of things. Or the end of "Sunflower" with "cunts of wheelbarrows," whatever that all meant, or "rubber dollar bills"—"skin of machinery"; see, and actually in the moment of composition I don't necessarily know what it means, but it comes to mean something later, after a year or two, I realize that it meant something clear, unconsciously. Which takes on meaning in time, like a photograph developing slowly. Because we're not really always conscious of the entire depth of our minds—in other words, we just know a lot more than we're able to be aware of, normally—though at moments we're completely aware, I guess.

There's some other element of Cézanne that was interesting . . . oh, his patience, of course. In recording the optical phenomena. Has something to do with Blake: with not through the eye— "You're led to believe a lie when you see with not through the eye." He's seeing through his eye. One can see through his canvas to God, really, is the way it boils down. Or to Pater Omnipotens Aeterna Deus. I could imagine someone not prepared, in a peculiar chemical-physiological state, peculiar mental state, psychic state, someone not prepared who had no experience of eternal ecstasy, passing in front of a Cézanne canvas, distracted and without noticing it, his eye traveling in, to, through the canvas into the space and suddenly stopping with his hair standing on end, dead in his tracks, seeing a whole universe. And I think that's what Cézanne really does, to a lot of people.

Where were we now? Yeah, the idea that I had was that gaps in space and time through images juxtaposed, just as in the haiku you get two images which the mind connects in a flash, and so that *flash* is the *petite sensation*; or the *satori*, perhaps, that the Zen haikuists would speak of—if they speak of it like that. So, the poetic experience that (A. E.) Housman talks about, the hair-standing-on-end or the hackles-rising, whatever it is, visceral thing. The interesting thing would be to know if certain combinations of words and rhythms actually had an electrochemical reaction on the body, which could catalyze specific states of consciousness. I think that's what probably happened to me with Blake. I'm sure it's what happens on a perhaps lower level with Poe's "Bells" or "Raven," or even Vachel Lindsay's

"Congo": that there is a hypnotic rhythm there, which when you introduce it into your nervous system, causes all sorts of electronic changes—permanently alters it. There's a statement by Artaud on that subject, that certain music when introduced into the nervous system changes the molecular composition of the nerve cells or something like that, it permanently alters the being that has experience of this. Well anyway, this is certainly true. In other words, any experience we have is recorded in the brain and goes through neural patterns and whatnot: so I suppose brain recordings are done by means of shifting around of little electrons—so there is actually an electrochemical effect caused by art.

So . . . the problem is what is the maximum electrochemical effect in the desired direction. That is what I was taking Blake as having done to me. And what I take as one of the optimal possibilities of art. But this is all putting it in a kind of bullshit abstract way. But it's an interesting—toy. To play with. That idea.

TC: In the last five or six months you've been in Cuba, Czechoslovakia, Russia, and Poland. Has this helped to clarify your sense of the current world situation?

AG: Yeah, I no longer feel—I didn't ever feel that there was any answer in dogmatic Leninism-Marxism—but I feel very definitely now that there's no answer to my desires there. Nor do most of the people in those countries—in Russia or Poland or Cuba—really feel that either. It's sort of like a religious theory imposed from above and usually used to beat people on the head with. Nobody takes it seriously because it doesn't mean anything, it means different things in different countries anyway. The general idea of revolution against American idiocy is good, it's still sympathetic, and I guess it's a good thing like in Cuba, and obviously Viet Nam. But what's gonna follow—the dogmatism that follows is a big drag. And everybody apologizes for the dogmatism by saying, well, it's an inevitable consequence of the struggle against American repression. And that may be true too.

But there's one thing I feel certain of, and that's that there's no human answer in communism or capitalism as it's practiced outside of the U.S. in any case. In other words, by hindsight, the interior of America is not bad, at least for me, though it might be bad for a spade, but not too bad, creepy, but it's not impossible. But traveling in countries like Cuba and Viet Nam I realize that the people that get the real evil side effects of America are there—in other words, it really is like imperialism, in that sense. People in the United States all got money, they got cars, and everybody else starves on account of American foreign policy. Or is being bombed out, torn apart, and bleeding on the street, they get all their teeth bashed in, tear gassed, or hot pokers up their ass, things that would be, you know, considered terrible in the United States. Except for Negroes.

So I don't know. I don't see any particular answer, and this month it seemed to me like actually an atomic war was inevitable on account of both sides were so dogmatic and frightened and had nowhere to go and didn't know what to do with themselves anymore except fight. Everybody too intransigent. Everybody too mean. I don't suppose it'll take place, but . . . Somebody has got to sit in the British Museum again like Marx and figure out a new system; a new blueprint. Another century has gone, technology has changed everything completely, so it's time for a new utopian system. Burroughs is almost working on it.

But one thing that's impressive is Blake's idea of Jerusalem, Jerusalemic Britain, which I think is now more and more valid. He, I guess, defined it. I'm still confused about Blake, I still haven't read him all through enough to understand what direction he was really pointing to. It seems to be the naked human form divine, seems to be Energy, it seems to be sexualization, or sexual liberation, which are the directions we all believe in. He also seems, however, to have some idea of imagination which I don't fully understand yet. That is, it's something outside of the body, with a rejection of the body, and I don't quite understand that. A life after death even. Which I still haven't comprehended. There's a letter in the Fitzwilliam Museum, written several months before he died. He says, "My body is in turmoil and stress and decaying, but my ideas, my power of ideas and my imagination, are stronger than ever." And I find it hard to conceive of that. I think if I were lying in bed dying, with my body pained, I would just give up. I mean, you know, because I don't think I could exist outside my body. But he apparently was able to. Williams didn't seem to be able to. In other words Williams's universe was tied up with his body. Blake's universe didn't seem to be tied up with his body. Real mysterious, like far other worlds and other seas, so to speak. Been puzzling over that today.

The Jerusalemic world of Blake seems to be Mercy-Pity-Peace. Which has human form. Mercy has a human face. So that's all clear.

TC: How about Blake's statement about the senses being the chief inlets of the soul in this age—I don't know what "this age" means; is there another one?

AG: What he says is interesting because there's the same thing in Hindu mythology, they speak of This Age as the Kali Yuga, the age of destruction, or an age so sunk in materialism. You'd find a similar formulation in Vico, like what is it, the Age of Gold running on to the Iron and then Stone, again. Well, the Hindus say that this is the Kali Age or Kali Yuga or Kali Cycle, and we are also so sunk in matter, the five senses are matter, sense, that they say there is absolutely no way out by intellect, by thought, by discipline, by practice, by sadhana, by jnanayoga, nor karma yoga—that is, doing good works—no way out through our own will or our own effort. The only way out that they generally now prescribe, generally in India at the moment, is through bhakti yoga, which is Faith-Hope-Adoration-Worship, or like probably the equivalent of the Christian Sacred Heart, which I find a very lovely doctrine—that is to say, pure delight, the only way you can be saved is to sing. In other words, the only way to drag up, from the depths of this depression, to drag up your soul to its proper bliss, and understanding, is to give yourself, completely, to your heart's desire. The image will be determined by the heart's compass, by the compass of what the heart moves toward and desires. And then you get on your knees or on your lap or on your head and you sing and chant prayers and mantras, till you reach a state of ecstasy and understanding, and the bliss overflows out of your body. They say intellect, like Saint Thomas Aquinas, will never do it, because it's just like me getting all hung up on whether I could remember what happened before I was born—I mean you could get lost there very easily, and it has no relevance anyway, to the existent flower. Blake says something similar, like Energy, and Excess . . . leads to the palace of wisdom. The Hindu bhakti is like excess of devotion; you just, you know, give yourself all out to devotion.

Very oddly a lady saint Shri Matakrishnaji in Brindaban, whom I consulted about my spiritual problems, told me to take Blake for my guru. There's all kinds of different gurus, there can be living and nonliving gurus—apparently whoever initiates you, and I apparently was initiated by Blake in terms of at least having an ecstatic experience from him. So that when I got here to Cambridge I had to rush over to the Fitzwilliam Museum to find his misspellings in *Songs of Innocence*.

TC: What was the Blake experience you speak of?

AG: About 1945 I got interested in Supreme Reality with a capital S and R, and I wrote big long poems about a last voyage looking for Supreme Reality. Which was like a Dostoevskian or Thomas Wolfeian idealization or like Rimbaud—what was Rimbaud's term, New Vision, was that it? Or Kerouac was talking about a New Vision, verbally, and intuitively out of longing, but also out of a funny kind of tolerance of this universe. In 1948 in East Harlem in the summer I was living—this is like the Ancient Mariner, I've said this so many times: "stoppeth one of three. / 'By thy long grey beard . . . ' " Hang an albatross around your neck. . . . The one thing I felt at the time was that it would be a terrible horror, that in one or two decades I would be trying to explain to people that one day something like this happened to me! I even wrote a long poem saying, "I will grow old, a grey and groaning man, / and with each hour the same thought, and with each thought the same denial. / Will I spend my life in praise of the idea of God? / Time leaves no hope. We creep and wait. We wait and go alone." Psalm II—which I never published. So anyway—there I was in my bed in Harlem . . . jacking off. With my pants open, lying around on a bed by the windowsill, looking out into the cornices of Harlem and the sky above. And I had just come. And had perhaps hardly even wiped the come off my thighs, my trousers, or whatever it was. As I often do, I had been jacking off while reading—I think it's probably a common phenomenon to be noticed among adolescents. Though I was a little older than an adolescent at the time. About twenty-two. There's a kind of interesting thing about, you know, distracting your attention while you jack off—that is, you know, reading a book or looking out of a window, or doing something else with the conscious mind which kind of makes it sexier.

So anyway, what I had been doing that week—I'd been in a very lonely solitary state, dark night of the soul sort of, reading Saint John of the Cross, maybe on account of that everybody'd gone away that I knew, Burroughs was in Mexico, Jack was out in Long Island and relatively isolated, we didn't see each other, and I had been very close with them for several years. Huncke I think was in jail, or something. Anyway, there was nobody I knew. Mainly the thing was that I'd been making it with N. C., and finally I think I got a letter from him saying it was all off, no more, we shouldn't consider ourselves lovers any more on account of it just wouldn't work out. But previously we'd had an understanding that we—Neal Cassady, I said "N. C." but I suppose you can use his name—we'd had a big tender lovers' understanding. But I guess it got too much for him, partly because he was three thousand miles away and he had six thousand girlfriends on the other side of the continent, who were keeping him busy, and then here was my lone cry of despair from New York. So. I got a letter from him saying, Now, Allen, we gotta move on to new territory. So I felt this is like a great mortal blow to all of my tenderest hopes. And I figured I'd never find any sort of psychospiritual sexo-cock jewel fulfillment in my existence! So, I went into . . . like I felt cut off from what I'd idealized romantically. And I was also graduating from school and had nowhere to go and the difficulty of getting a job. So finally there was nothing for me to do except to eat vegetables and live in Harlem. In an apartment I'd rented from someone. Sublet.

So, in that state therefore, of hopelessness, or dead end, change of phase, you know—growing up—and in an equilibrium in any case, a psychic, a mental equilibrium of a kind, like of having no New Vision and no Supreme Reality and nothing but the world in front of me, and of not knowing what to do with that . . . there was a funny balance of tension, in every direction. And just after I came, on this occasion, with a Blake book on my lap—I wasn't even reading, my eye was idling over the page of "Ah, Sun-flower," and it suddenly appeared—the poem I'd read a lot of times before, overfamiliar to the point where it didn't make any particular meaning except some sweet thing about flowers—and suddenly I realized that the poem was talking about me. "Ah, sun-flower! weary of time, / Who countest the steps of the sun; / Seeking after that sweet golden clime, / Where the traveller's journey is done." Now, I began understanding it, the poem while looking at it, and suddenly, simultaneously with understanding it, heard a very deep earthen grave voice in the room, which I immediately assumed, I didn't think twice, was Blake's voice; it wasn't any voice that I knew, though I had previously had a conception of a voice of rock, in a poem, some image like that—or maybe that came after this experience.

And my eye on the page, simultaneously the auditory hallucination, or whatever terminology here used, the apparitional voice, in the room, woke me further deep in my understanding of the poem, because the voice was so completely tender and beautifully . . . ancient. Like the voice of the Ancient of Days. But the peculiar quality of the voice was something unforgettable because it was like God had a human voice, with all the infinite tenderness and anciency and mortal gravity of a living Creator speaking to his son. "Where the Youth pined away with desire, / And the pale Virgin shrouded in snow, / Arise from their graves, and aspire / Where my Sun-flower wishes to go." Meaning that there *was* a *place*, there was a sweet golden clime, and the *sweet golden*, what was that . . . and simultaneous to the voice there was also an emotion, risen in my soul in response to the voice, and a sudden *visual* realization of the same awesome phenomena. That is to say, looking out at the window, through the window at the sky, suddenly it seemed that I saw into the depths of the universe, by looking simply into the ancient sky. The sky suddenly seemed very *ancient*. And this was the very ancient place that he was talking about, the sweet golden clime, I suddenly realized that *this* existence was *it!* And, that I was born in order to experience up to this very moment that I was having this experience, to realize what this was all about—in other words that this was the moment that I was born for. This initiation. Or this vision or this consciousness, of being alive unto myself, alive myself unto the Creator. As the son of the Creator—who loved me, I realized, or who responded to my desire, say. It was the same desire both ways.

Anyway, my first thought was this was what I was born for, and second thought, never forget—never forget, never renege, never deny. Never deny the voice—no, never *forget* it, don't get lost mentally wandering in other spirit worlds or American or job worlds or advertising worlds or war worlds or earth worlds. But the spirit of the universe was what I was born to realize. What I was speaking about visually was, immediately, that the cornices in the old tenement building in Harlem across the back-yard court had been carved very finely in 1890 or 1910. And were like the solidification of a great deal of intelligence and care and love also. So that I began noticing in every corner where I looked evidences of a living hand, even in the bricks, in the arrangement of each brick. Some hand placed them there—that some hand had placed the whole universe in front of me. That some hand had placed the sky. No, that's exaggerating—not that some hand had placed the sky but that the sky was the living blue hand itself. Or that

God was in front of my eyes—existence itself was God. Well, the for-
mulations are like that—I didn't formulate it in exactly those terms;
what I was seeing was a visionary thing, it was a lightness in my
body . . . my body suddenly felt light, and a sense of cosmic con-
sciousness, vibrations, understanding, awe, and wonder and surprise.
And it was a sudden awakening into a totally deeper real universe than
I'd been existing in. So, I'm trying to avoid generalizations about that
sudden deeper real universe and keep it strictly to observations of
phenomenal data, or a voice with a certain sound, the appearance of
cornices, the appearance of the sky, say, of the great blue hand, the
living hand—to keep to images.

But anyway—the same . . . *petite sensation* recurred several min-
utes later, with the same voice, while reading the poem "The Sick
Rose." This time it was a slightly different sense-depth-mystic
impression. Because "The Sick Rose"—you know I can't interpret the
poem now, but it had a meaning—I mean I can interpret it on a verbal
level, the sick rose is my self, or self, or the living body, sick because
the mind, which is the worm "That flies in the night, In the howling
storm," or Urizen,[5] reason; Blake's character might be the one that's
entered the body and is destroying it, or let us say death, the worm as
being death, the natural process of death, some kind of mystical being
of its own trying to come in and devour the body, the rose. Blake's
drawing for it is complicated, it's a big drooping rose, drooping
because it's dying, and there's a worm in it, and the worm is wrapped
around a little sprite that's trying to get out of the mouth of the rose.

But anyway, I experienced "The Sick Rose," with the voice of
Blake reading it, as something that applied to the whole universe, like
hearing the doom of the whole universe, and at the same time the
inevitable beauty of doom. I can't remember now, except it was very
beautiful and very awesome. But a little of it slightly scary, having to
do with the knowledge of death—my death and also the death of being
itself, and that was the great pain. So, like a prophecy, not only in
human terms but a prophecy as if Blake had penetrated the very secret

---

[5]In William Blake's system of the Four Zoas, *Urizen* symbolizes "reason," often
referred to as "your-reason." See S. Foster Damon, *A Blake Dictionary* (University
Press of New England, Hanover, NH, 1988); also "The [First] Book of Urizen," *The
Complete Poetry and Prose of William Blake*, ed. David Erdman, Anchor Books, NY,
1988.

core of the entire universe and had come forth with some little magic formula statement in rhyme and rhythm that, if properly heard in the inner inner ear, would deliver you beyond the universe.

So then, the other poem that brought this on in the same day was "The Little Girl Lost," where there was a repeated refrain,

> Do father, mother, weep?
> Where can Lyca sleep?
> . . .
> How can Lyca sleep
> If her mother weep?
> . . .
> If her heart does ache
> Then let Lyca wake;
> If my mother sleep,
> Lyca shall not weep.

It's that hypnotic thing—and I suddenly realized that Lyca was me, or Lyca was the self; father, mother seeking Lyca, was God seeking, Father, the Creator; and " 'If her heart does ache / Then let Lyca wake' "—wake to what? *Wake* meaning wake to the same awakeness I was just talking about—of existence in the entire universe. The total consciousness then, of the complete universe. Which is what Blake was talking about. In other words a breakthrough from ordinary habitual quotidian consciousness into consciousness that was really seeing all of heaven in a flower. Or what was it—eternity in a flower . . . heaven in a grain of sand? As I was seeing heaven in the cornice of the building. By heaven here I mean this imprint or concretization or living form, of an intelligent hand—the work of an intelligent hand, which still had the intelligence molded into it. The gargoyles on the Harlem cornices. What was interesting about the cornice was that there's cornices like that on every building, but I never noticed them before. And I never realized that they meant spiritual labor, to anyone—that somebody had labored to make a curve in a piece of tin—to make a cornucopia out of a piece of industrial tin. Not only that man, the workman, the artisan, but the architect had thought of it, the builder had paid for it, the smelter had smelt it, the miner had dug it up out of the earth, the earth had gone through aeons preparing it. So the little

molecules had slumbered for . . . for Kalpas.[6] So out of *all* of these
Kalpas it all got together in a great succession of impulses, to be
frozen finally in that one form of a cornucopia cornice on the building
front. And God knows how many people made the moon. Or what
spirits labored . . . to set fire to the sun. As Blake says, "When I look
in the sun I don't see the rising sun, I see a band of angels singing
holy, holy, holy." Well, his perception of the field of the sun is differ-
ent from that of a man who just sees the sun sun, without any emo-
tional relationship to it.

But then, there was a point later in the week when the intermit-
tent flashes of the same . . . bliss—because the experience was quite
blissful—came back. In a sense all this is described in "The Lion for
Real" by anecdotes of different experiences—actually it was a very dif-
ficult time, which I won't go into here. Because suddenly I thought,
also simultaneously, Ooh, I'm going *mad!* That's described in the line
in "Howl," "who thought they were *only* mad when Baltimore gleamed
in supernatural ecstasy"—"who thought they were *only* mad. . . ." If it
were only that easy! In other words it'd be a lot easier if you just were
crazy, instead of—then you could chalk it up, "Well, I'm nutty"—but on
the other hand what if it's all true and you're *born* into this great cos-
mic universe in which you're a spirit angel—terrible fucking situation
to be confronted with. It's like being woken up one morning by
Joseph K's[7] captors. Actually what I think I did was there was a couple
of girls living next door and I crawled out on the fire escape and
tapped on their window and said, "I've seen God!" and they *banged* the
window shut. Oh, what tales I could have told them if they'd let me in!
Because I was in a very exalted state of mind and the consciousness
was still with me—I remember I immediately rushed to Plato and read
some great image in the *Phaedrus* about horses flying through the sky,
and rushed over to Saint John and started reading fragments of *con un
no saber sabiendo . . . que me quede balbuciendo*, and rushed to the
other part of the bookshelf and picked up Plotinus about The Alone—
the Plotinus I found more difficult to interpret.

But I *immediately* doubled my thinking process, quadrupled, and I

[6]A *kalpa* is a cycle, an endlessly long period of time and the basis for the Buddhist
framing of time. Within one *kalpa* a universe arises, beings arise, the sun and moon
are created, and then all is annihilated and returns to Chaos.
[7]Joseph K is the protagonist in Franz Kafka's novel *The Trial*.

was able to read almost any text and see all sorts of divine significance in it. And I think that week or that month I had to take an examination in John Stuart Mill. And instead of writing about his ideas I got completely hung up on his experience of reading—was it Wordsworth? Apparently the thing that got him back was an experience of nature that he received keyed off by reading Wordsworth, on "sense sublime" or something. That's a very good description, that sense sublime of something far more deeply interfused, whose dwelling is the light of setting suns, and the round ocean, and the . . . the *living* air, did he say? The living air—see just that hand again—*and* in the heart of man. So I think this experience is characteristic of all high poetry. I mean that's the way I began seeing poetry as the communication of the particular experience—not just any experience but *this* experience.

TC: Have you had anything like this experience again?

AG: Yeah. I'm not finished with this period. Then, in my room, I didn't know what to do. But I wanted to bring it up so I began experimenting with it, without Blake. And I think it was one day in my kitchen—I had an old-fashioned kitchen with a sink with a tub in it with a board over the top—I started moving around and sort of shaking with my body and dancing up and down on the floor and saying, "Dance! dance! dance! dance! spirit! spirit! spirit! dance!" and suddenly I felt like Faust, calling up the devil. And then it started coming over me, this big . . . creepy feeling, cryptozoid or monozoidal, so I got all scared and quit.

Then I was walking around Columbia and I went in the Columbia bookstore and was reading Blake again, leafing over a book of Blake, I think it was "The Human Abstract": "Pity would be no more. . . ." And suddenly it came over me in the bookstore again, and I was in the eternal place *once more*, and I looked around at everybody's faces, and I saw all these wild animals! Because there was a bookstore clerk there who I hadn't paid much attention to, he was just a familiar fixture in the bookstore scene and everybody went in the bookstore every day like me, because downstairs there was a café and upstairs there were all these clerks that we were all familiar with—this guy had a very *long* face, you know some people look like giraffes. So he looked kind of giraffish. He had a kind of a long face with a long nose. I don't know what kind of sex life he had, but he must have had something. But anyway, I looked in his face and I suddenly saw like a great tormented soul—and he had just been somebody whom I'd regarded as perhaps a not particularly beautiful or sexy character, or lovely face, but you know someone familiar, and perhaps a pleading cousin in the universe. But all of a sudden I realized that *he* knew also, just like I knew. And that everybody in the bookstore knew, and that they were all hiding it! They all had the consciousness, it was like a great unconscious that was running between all of us that everybody *was* completely conscious, but that the fixed expressions that people have, the habitual expression, the manners, the mode of talk, are all masks hiding this consciousness. Because almost at that moment it seemed that it would be too terrible if we communicated to each other on a level of total consciousness and awareness each of the other—like it would be too terrible, it would be the end of the bookstore, it would be the end of civ— . . . not civilization, but in other words the position that everybody was in was *ridiculous*, everybody running around peddling books to each other. Here in the universe! Passing money over the counter, wrapping books in bags and guarding the door, you know, stealing books, and the people sitting up making accountings on the upper floor there, and people worrying about their exams walking through the bookstore, and all the millions of thoughts the people had—you know, that I'm worrying about—whether they're going to get laid or whether anybody loves them, about their mothers dying of cancer or, you know, the complete death awareness that everybody has continuously with them all the time—all of a sudden revealed to me at once in the faces of the people, and they all looked like horrible grotesque masks, grotesque because *hiding* the knowledge from each other. Having a habitual conduct and forms to prescribe, forms to fulfill. Roles to play. But the main insight I had at that time was that everybody knew. Everybody knew completely everything. Knew completely everything in the terms which I was talking about.

TC: Do you still think they know?

AG: I'm more sure of it now. Sure. All you have to do is try and make somebody. You realize that they knew all along you were trying to make them. But until that moment you never break through to communication on the subject.

TC: Why not?

AG: Well, fear of rejection. The twisted faces of all those people, the faces were twisted by rejection. And hatred of self, finally. The internalization of that rejection. And finally disbelief in that shining self. Disbelief in that infinite self. Partly because that particular . . . partly because the *awareness* that we all carry is too often painful, because the experience of rejection and lacklove and cold war—I mean the whole cold war is the imposition of a vast mental barrier on everybody, a vast antinatural psyche. A hardening, a shutting off of the perception of desire and tenderness which everybody *knows* and which is the very structure of . . . the atom! Structure of the human body and organism. That desire built in. Blocked. "Where the Youth pined away with desire, / And the pale Virgin shrouded in snow." Or as Blake says, "On every face I see, I meet / Marks of weakness, marks of woe." So what I was thinking in the bookstore was the marks of weakness, marks of woe. Which you can just look around and look at anybody's face right next to you now always—you can see it in the way the mouth is pursed, you can see it in the way the eyes blink, you can see it in the way the gaze is fixed down at the matches. It's the self-consciousness which is a substitute for communication with the outside. This consciousness pushed back into the self and thinking of how it will hold its face and eyes and hands in order to make a mask to hide the flow that is going on. Which it's aware of, which everybody is aware of really! So let's say, shyness. Fear. Fear of like total feeling, really, total being is what it is.

So the problem then was, having attained realization, how to safely manifest it and communicate it. Of course there was the old Zen thing, when the sixth patriarch handed down the little symbolic oddments and ornaments and books and bowls, stained bowls too . . . when the *fifth* patriarch handed them down to the sixth patriarch he told him to hide them and don't tell anybody you're patriarch because it's dangerous, they'll kill you. So there was that immediate danger. It's taken me all these years to manifest it and work it out in a way that's materially communicable to people. Without scaring them or me. Also movements of history and breaking down the civilization. To break down everybody's masks and roles sufficiently so that everybody has to face the universe *and* the possibility of the sick rose coming true and the atom bomb. So it was an immediate messianic thing. Which seems to be becoming more and more justified. And more and more reasonable in terms of the existence that we're living.

So. Next time it happened was about a week later walking along in the evening on a circular path around what's now I guess the garden or field in the middle of Columbia University, by the library. I started invoking the spirit, consciously trying to get another depth perception of cosmos. And suddenly it began occurring again, like a sort of breakthrough again, but this time—this was the last time in that period—it was the same depth of consciousness or the same cosmical awareness but suddenly it was not blissful at all but it was *frightening*. Some like real serpent-fear entering the sky. The sky was not a blue hand anymore but like a hand of death coming down on me—some really scary presence, it was almost as if I saw God again except God was the devil. The consciousness itself was so vast, much more vast than any idea of it I'd had or any experience I'd had, that it was not even human anymore—and was in a sense a threat, because I was going to die into that inhuman ultimately. I don't know *what* the score was there—I was too cowardly to pursue it. To attend and experience completely the Gates of Wrath—there's a poem of Blake's that deals with that, "To find a Western Path / Right through the Gates of Wrath." But I didn't urge my way there, I shut it all off. And got scared, and thought, I've gone too far.

TC: Was your use of drugs an extension of this experience?

AG: Well, since I took a vow that this was the area of, that this was my existence that I was placed into, drugs were obviously a technique for experimenting with consciousness, to get different areas and different levels and different similarities and different reverberations of the same vision. Marijuana has some of it in it, that awe, the cosmic awe that you get sometimes on pot. There are certain moments under laughing gas and ether that the consciousness does intersect with something similar—for me—to my Blake visions. The gas drugs were apparently interesting too to the Lake Poets, because there were a lot of experiments done with Sir Humphrey Davy in his Pneumatic Institute. I think Coleridge and Southey and other people used to go, and De Quincy. But serious people. I think there hasn't been very much written about that period. *What went on* in the Humphrey Davy household on Saturday midnight when Coleridge arrived by foot, through the forest, by the lakes? Then, there are certain states you get into with opium, and heroin, of almost disembodied awareness, looking down back at the earth from a place after you're dead. Well, it's not the same, but it's an interesting state, and a useful one. It's a normal state also, I mean it's a holy state of some sort. At times. Then, mainly, of course, with the hallucinogens, you get some states of consciousness which subjectively seem to be cosmic-ecstatic, or cosmic-demonic. Our version of expanded consciousness is as much as *un*conscious information—awareness comes up to the surface. Lysergic acid, peyote, mescaline, psilocybin,[8] Ayahuasca. But I can't stand them anymore, because something happened to me with them very similar to the Blake visions. After about thirty times, thirty-five times, I began getting monster vibrations again. So I couldn't go any further. I may later on again, if I feel more reassurance.*

However, I did get a lot out of them, mainly like emotional understanding, understanding the female principle in a way—women, more sense of the softness and more desire for women. Desire for children also.

TC: Anything interesting about the actual experience, say with hallucinogens?

AG: What I do get is, say if I was in an apartment high on mescaline, I felt as if the apartment and myself were not merely on East Fifth Street but were in the middle of all space time. If I close my eyes on hallu-

---

[8]In the original published version the word is spelled "sylocidin."

*Between occasion of interview with Thomas Clark June '65 and publication May '66 more reassurance came. I tried small doses of LSD twice in secluded tree and ocean cliff haven at Big Sur. No monster vibration, no snake universe hallucinations. Many tiny jeweled violet flowers along the path of a living brook that looked like Blake's illustration for a canal in grassy Eden: huge Pacific watery shore, Orlovsky dancing naked like Shiva long-haired before giant green waves, titanic cliffs that Wordsworth mentioned in his own Sublime, great yellow sun veiled with mist hanging over the plant's oceanic horizon. No harm. President Johnson that day went into the Valley of Shadow operating room because of his gall bladder & Berkeley's Vietnam Day Committee was preparing anxious manifestoes for our march toward Oakland police and Hell's Angels. Realizing that more vile words from me would send out physical vibrations into the atmosphere that might curse poor Johnson's flesh and further unbalance his soul, I knelt on the sand surrounded by masses of green bulb-headed Kelp vegetable-snake undersea beings washed up by last night's tempest, and prayed for the President's tranquil health. Since there has been so much legislative miscomprehension of the LSD boon I regret that my unedited ambivalence in Thomas Clark's tape transcript interview was published wanting this footnote.

—Your obedient servant
Allen Ginsberg, *aetat* 40
June 2, 1966

cinogens, I get a vision of great scaly dragons in outer space, they're winding slowly and eating their own tails. Sometimes my skin and all the room seem sparkling with scales, and it's all made out of serpent stuff. And as if the whole illusion of life were made of reptile dream.

Mandala also. I use the mandala in an LSD poem. The associations I've had during times that I was high are usually referred to or built in some image or other to one of the other poems written on drugs. Or after drugs—like in "Magic Psalm" on lysergic acid. Or mescaline. There's a long passage about a mandala in the LSD poem. There is a good situation since I was high and I was looking at a mandala—before I got high I asked the doctor that was giving it to me at Stanford to prepare me a set of mandalas to look at, to borrow some from Professor Spiegelberg who was an expert. So we had some Sikkimese elephant mandalas there. I simply describe those in the poem—what they look like while I was high.

So—summing up then—drugs were useful for exploring perception, sense perception, and exploring different possibilities and modes of consciousness, and exploring the different versions of *petites sensations*, and useful then for composing, sometimes, while under the influence. Part II of "Howl" was written under the influence of peyote, composed during peyote vision. In San Francisco—"Moloch." "Kaddish" was written with amphetamine injections. An injection of amphetamine plus a little bit of morphine, plus some Dexedrine later on to keep me going, because it was all in one long sitting. From a Saturday morn to a Sunday night. The amphetamine gives a peculiar metaphysical tinge to things, also. Space-outs. It doesn't interfere too much there because I wasn't habituated to it, I was just taking it that one weekend. It didn't interfere too much with the emotional charge that comes through.

TC: Was there any relation to this in your trip to Asia?

AG: Well, the Asian experience kind of got me out of the corner I painted myself in with drugs. That corner being an inhuman corner in the sense that I figured I was expanding my consciousness and I had to go through with it but at the same time I was confronting this serpent monster, so I was getting in a real terrible situation. It finally would get so if I'd take the drugs I'd start vomiting. But I felt that I was duly bound and obliged for the sake of consciousness expansion, and this insight, and breaking down my identity, and seeking more direct contact with primate sensation, nature, to continue. So when I went to India, all the way through India, I was babbling about that to all the holy men I could find. I wanted to find out if they had any suggestions. And they all did, and they were all good ones. First one I saw was Martin Buber, who was interested. In Jerusalem, Peter (Orlovsky) and I went in to see him—we called him up and made a date and had a long conversation. He had a beautiful white beard and was friendly; his nature was slightly austere but benevolent. Peter asked him what kind of visions he'd had and he described some he'd had in bed when he was younger. But he said he was not any longer interested in visions like that. The kind of visions he came up with were more like spiritualistic table rappings. Ghosts coming into the room through his window, rather than big beautiful seraphic Blake angels hitting him on the head. I was thinking like loss of identity and confrontation with non-human universe as the main problem, and in a sense whether or not man had to evolve and change, and perhaps become nonhuman too. Melt into the universe, let us say—to put it awkwardly and inaccurately. Buber said that he was interested in man-to-man relationships, human-to-human—that he thought it was a human universe that we were destined to inhabit. And so therefore human relationships rather than relations between the human and the nonhuman. Which was what I was thinking that I had to go into. And he said, "Mark my word, young man, in two years you will realize that I was right." He was right—in two years I marked his words. Two years is sixty-three—I saw him in sixty-one. I don't know if he said two years—but he said "in years to come." This was like a real terrific classical wise man's "Mark my words, young man, in several years you will realize that what I said was true!" Exclamation point.

Then there was Swami Shivananda, in Rishikish in India. He said, "Your own heart is your guru." Which I thought was very sweet, and very reassuring. That is the sweetness of it I felt—in my heart. And suddenly realized it was the heart that I was seeking. In other words it wasn't consciousness, it wasn't *petites sensations*, sensation defined as expansion of mental consciousness to include more data—as I was pursuing that line of thought, pursuing Burroughs's cutup thing—the area that I was seeking was heart rather than mind. In other words, in mind, through mind or imagination—this is where I get confused with Blake now—in mind one can construct all sorts of universes, one can construct model universes in dream and imagination, and with the speed of light; and with nitrous oxide you can experience several million universes in rapid succession. You can experience a whole gamut of possibilities of universes, including the final possibility that there is none. And then you go unconscious—which is exactly what happens with gas when you go unconscious. You see that the universe is going to disappear with your consciousness, that it was all dependent on your consciousness.

Anyway, a whole series of India holy men pointed back to the body—getting *in* the body rather than getting out of the human form. But living in and inhabiting the human form. Which then goes back to Blake again, the human form divine. Is this clear? In other words, the psychic problem that I had found myself in was that for various reasons it had seemed to me at one time or another that the best thing to do was to drop dead. Or not be afraid of death but go into death. Go into the nonhuman, go into the cosmic, so to speak; that God was death, and if I wanted to attain God I had to die. Which *may* still be true. So I thought that what I was put up to was to therefore break out of my body, if I wanted to attain complete consciousness.

So now the next step was that the gurus one after another said, Live in the body: this is the form that you're born for. That's too long a narration to go into. Too many holy men and too many different conversations and they all have a little key thing going. But it all winds up in the train in Japan, then a year later, the poem "The Change," where all of a sudden I renounce drugs. I don't renounce drugs but I suddenly didn't want to be *dominated* by that nonhuman anymore, or even be dominated by the moral obligation to enlarge my consciousness anymore. Or do anything anymore except *be* my heart—which just desired to be and be alive now. I had a very strange ecstatic experience then and there, once I had sort of gotten that burden off my back, because I was suddenly free to love myself again, and therefore love the people around me, in the form that they already were. And love myself in my own form as I am. And look around at the other people and so it was *again* the same thing like in the bookstore. Except this time I was completely in my body and had no more mysterious obligations. And nothing more to fulfill, except to be willing to die when I am dying, whenever that be. And be willing to live as a human in this form now. So I started weeping, it was such a happy moment. Fortunately I was able to write then, too, "So that I do live I will die"—rather than be cosmic consciousness, immortality, Ancient of Days, perpetual consciousness existing for-ever.

Then when I got to Vancouver, Olson was saying "I am one with my skin." It *seemed* to me at the time when I got back to Vancouver that everybody had been precipitated back into their bodies at the same time. It seemed that's what Creeley had *been* talking about all along. The *place*—the terminology he used, the *place* we are. Meaning this place, here. And trying to like be real in the real place . . . to be aware of the place where he is. Because I'd always thought that that meant that he was cutting off from divine imagination. But what that meant for him was that this place would be everything that one would refer to as divine, if one were really here. So that Vancouver seems a very odd moment, at least for me—because I came back in a sense completely bankrupt. My energies of the last . . . oh, 1948 to 1963, all completely washed up. On the train in Kyoto having renounced Blake, renounced visions—renounced *Blake!*—too. There was a cycle that began with the Blake vision which ended on the train in Kyoto when I realized that to attain the depth of consciousness that I was seeking when I was talk-ing about the Blake vision, that in order to attain it I had to cut myself off from the Blake vision and renounce it. Otherwise I'd be hung up on a memory of an experience. Which is not the actual awareness of now, now. In order to get back to now, in order to get back to the total awareness of now and contact, sense perception contact with what was going on around me, or direct vision of the moment, now I'd have to give up this continual churning thought process of yearning back to a visionary state. It's all very complicated. And idiotic.

TC: I think you said earlier that "Howl" being a lyric poem, and "Kad-dish" basically a narrative, that you now have a sense of wanting to do an epic. Do you have a plan like this?

AG: Yeah, but it's just . . . ideas, that I've been carrying around for a long time. One thing which I'd like to do sooner or later is write a long poem which is a narrative and description of all the visions I've ever had, sort of like the *Vita Nuova*. And travels, now. And another idea I had was to write a big long poem about everybody I ever fucked or slept with. Like sex . . . a love poem. A long love poem, involving all the innumerable lays of a lifetime. The epic is not that, though. The epic would be a poem including history, as it's defined. So that would be one about present-day politics, using the methods of the Blake *French*

Revolution. I got a lot written. Narrative was "Kaddish." Epic—there has to be totally different organization, it might be simple free association on political themes—in fact I think an epic poem including history, at this stage. I've got a lot of it written, but it would have to be Burroughs's sort of epic—in other words, it would have to be dissociated thought stream which includes politics and history. I don't think you could do it in narrative form, I mean what would you be narrating, the history of the Korean War or something?

TC: Something like Pound's epic?

AG: No, because Pound seems to me to be over a course of years fabricating out of his reading and out of the museum of literature; whereas the thing would be to take all of contemporary history, newspaper headlines and all the pop art of Stalinism and Hitler and Johnson and Kennedy and Viet Nam and Congo and Lumumba and the South and Sacco and Vanzetti—whatever floated into one's personal field of consciousness and contact. And then to compose like a basket-like weave a basket, basket-weaving out of those materials. Since obviously nobody has any idea where it's all going or how it's going to end unless you have some vision to deal with. It would have to be done by a process of association, I guess.

TC: What's happening in poetry now?

AG: I don't know yet. Despite all confusion to the contrary, now that time's passed, I think the best poet in the United States is Kerouac still. Given twenty years to settle through. The main reason is that he's the most free and the most spontaneous. Has the greatest range of association and imagery in his poetry. Also in "Mexico City Blues" the sublime as subject matter. And, in other words the greatest facility at what might be called projective verse, if you want to give it a name. I think that he's stupidly underrated by almost everybody except for a few people who are aware how beautiful his composition is—like (Gary) Snyder or (Robert) Creeley or people who have a taste for his tongue, for his line. But it takes one to know one.

TC: You don't mean Kerouac's prose?

AG: No, I'm talking about just a pure poet. The verse poetry, the "Mexico City Blues" and a lot of other manuscripts I've seen. In addition he has the one sign of being a great poet, which is he's the only one in the United States who knows how to write haikus. The only one who's written any good haikus. And everybody's been writing haikus. There are all these *dreary* haikus written by people who think for weeks trying to write a haiku, and finally come up with some dull little thing or something. Whereas Kerouac thinks in haikus, every time he writes anything—talks that way and thinks that way. So it's just natural for him. It's something Snyder noticed. Snyder has to labor for years in a Zen monastery to produce one haiku about shifting off a log! And actually does get one or two good ones. Snyder was always astounded by Kerouac's facility . . . at noticing winter flies dying of old age in his medicine chest. Medicine cabinet. "In my medicine cabinet/the winter flies/died of old age." He's never published them actually—he's published them on a record, with Zoot Sims and Al Cohn, it's a very beautiful collection of them. Those are, as far as I can see, the only real American haikus.

So the haiku is the most difficult test. He's the only *master* of the haiku. Aside from a longer style. Of course, the distinctions between prose and poetry are broken down anyway. So much that I was saying like a long page of oceanic Kerouac is sometimes as sublime as epic line. It's there that also I think he went further into the existential thing of writing conceived of as an irreversible action or statement, that's unrevisable and unchangeable once it's made. I remember I was thinking, yesterday in fact, there was a time that I was absolutely astounded because Kerouac told me that in the future literature would consist of what people actually wrote rather than what they tried to deceive other people into thinking they wrote, when they revised it later on. And I saw opening up this whole universe where people wouldn't be able to lie anymore! They wouldn't be able to correct themselves any longer. They wouldn't be able to hide what they said. And he was willing to go all the way into that, the first pilgrim into that new-found land.

TC: What about other poets?

AG: I think Corso has a great imaginative genius. And also amongst the greatest *shrewdness*—like Keats or something. I like (Philip) Lamantia's nervous wildness. Almost anything he writes I find interesting—for one thing he's always registering the forward march of the soul, in

exploration; spiritual exploration is always there. And also chronolog-
ically following his work is always exciting. (Philip) Whalen and Sny-
der are both very wise and very reliable. Whalen I don't *understand* so
well. I did, though, earlier—but I have to sit down and study his work,
again. Sometimes he seems sloppy—but then later on it always seems
right.

(Michael) McClure has tremendous energy, and seems like some
sort of a . . . seraph is not the word . . . not herald either but a . . .
not demon either. Seraph I guess it is. He's always moving—see when
I came around to, say, getting in my skin, there I found McClure sit-
ting around talking about being a mammal! So I suddenly realized he
was way ahead of me. And (John) Wieners . . . I always *weep* with him.
Luminous, luminous. They're all old poets, everybody knows about
those poets. Burroughs is a poet too, really. In the sense that a page of
his prose is as *dense* with imagery as anything in St. John Perse or
Rimbaud, now. And it has also great repeated rhythms. Recurrent,
recurrent rhythms, even rhyme occasionally! What else . . . Creeley's
very stable, solid. I get more and more to like certain poems of his
that I didn't understand at first. Like "The Door," which completely
baffled me because I didn't understand that he was talking about the
same heterosexual problem that I was worried about. Olson since he
said, "I feel one with my skin." First thing of Olson's that I liked was
"The Death of Europe" and then some of his later Maximus material is
nice. And (Ed) Dorn has a kind of long, *real* spare, manly, political
thing—but his great quality inside also is tenderness—"Oh the graves
not yet cut." I also like that whole line of what's happening with (John)
Ashbery and (Frank) O'Hara and (Kenneth) Koch, the area that
they're going for, too. Ashbery—I was listening to him read "The
Skaters," and it sounded as inventive and exquisite, in all its parts, as
"The Rape of the Lock."

TC: Do you feel you're in command when you're writing?

AG: Sometimes I feel in command when I'm writing. When I'm in the
heat of some truthful tears, yes. Then, complete command. Other
times—most of the time not. Just diddling away, woodcarving, getting a
pretty shape; like most of my poetry. There's only a few times when I
reach a state of complete command. Probably a piece of "Howl," a

piece of "Kaddish," and a piece of "The Change." And one or two moments of other poems.

TC: By command do you mean a sense of the whole poem as it's going, rather than parts?

AG: No—a sense of being self-prophetic master of the universe.

February 11, 1966,   On the Road Near Lawrence, Kansas

In September of 1965 Ginsberg began writing what he conceived of as a "Poem of These States," which his biographer Michael Schumacher described as "an ambitious project designed to update and rethink Whitman's celebration of America." To write this work, Allen traveled across America in a Volkswagen Microbus, noting whatever struck him, from newspaper headlines to bits of conversation to billboards to music and news he heard on the radio. His composition technique (discussed in the Aldrich interview) was also influenced by a gift, a state-of-the-art, portable, reel-to-reel Uher tape recorder he received from Bob Dylan that same year. Ultimately a series of poems came from these travels, and many were published in *The Fall of America*, but the most notable perhaps is "Wichita Vortex Sutra," originally published in *Planet News*.

The following interview itself took place in a car as Allen and Peter Orlovsky traveled through Lawrence and Kansas City on their way to a February 16 reading in Wichita. The Farrrell interview thus came precisely at the apogee of Allen's survey of the body of America, when Ginsberg was writing "Wichita Vortex Sutra." As the radio crackled with news reports of the escalation of America's involvement in Vietnam, Allen journeyed on to Wichita, speaking with Barry Farrell, telling him among other things of his first encounter with Carl Solomon, the man to whom Allen dedicated "Howl."

—DC

Peter Orlovsky (*shouting from Volkswagen camper*): Do you know where you're going?

Barry Farrell: No, I'm following you.

PO: Then don't use your signals. You had your signal on.

BF: Does the extent of deep personal confession in your poems bring you any anxiety?

Allen Ginsberg: No.

BF: Did it once?

AG: Occasionally, on some particular point. Like when I wrote "Howl," I didn't expect to publish it. I was concerned about my father seeing all that about cocksucking. That was the source of pressure—my father's disapproval. It took me a while to get over that before I realized that it didn't make any difference.

BF: Only a week to get past that?

AG: Yeah. Actually, I wrote it with the idea of turning aside from the kind of poetry I'd been writing and writing what I really felt. Tapping the sources of what I really felt outside literature and outside the social possibilities of communication. Funny wrinkles of my own awareness, little secret funny awareness that everybody has, not realizing that that is precisely the area where literature becomes literature, where writing becomes really art. Very few people ever get to that understanding—that art is something discovered from your own real nature. "Fool, said the Muse, look in thy heart—and write." Spenser, I guess.[9] (Herbert) Huncke's things are like really looking in the heart. "Howl" was really looking in the heart and writing. But I wasn't wise enough at the time to realize that that alone would make it right. Like automatically given a certain facility. That's what Kerouac discovered: that we're stepping forth in time to an irrevocable statement: speak now or forever hold thy tongue. The realization that the exact area which is classic art is what everybody's interested in, our secret personal doodlings.

BF: What actually happened with your father?

AG: Nothing.

BF: Nothing?

AG: No. Nothing at all. One cop tried to bust the book. My father wrote me and said he thought there were a few unnecessary words in it.

---

[9]The quote is actually: "Fool, said my Muse to me, look in thy heart, and write," by Sir Philip Sidney, not Spenser.

Probably "cocksucker"—unnecessary and artistically not good. But he welcomed the area of investigation. Trilling didn't like it: "the same old rhetoric handed back on a plate." A horrible lapse of taste.

BF: When did your father really recognize you as an artist?

AG: Before then, I think. I've been writing since I was 15. But the point I really got through was when I sent him "Kaddish" and he said he wept. I mean, to make statements and open his awareness to the point of tears! That was all that was necessary. It was years later, of course. And probably he'd responded much earlier. I'd wept for poems of his— so it works out all right. I still fight with him about other matters: Vietnam. It's residue hang-over aggression.

BF: Anyway, confession doesn't seem to you an act of courage.

AG: No. Not anymore.

BF: It once did?

AG: Not really. It seemed like humor. It seemed like the natural thing to do. If you have firm ground with your own body and feelings, then there's no threat from the outside.

BF: Have you always had this kind of firm ground?

AG: Literally, what happened is that until I was 18 I was a virgin. Like I felt a love that (chuckle) "dare not confess its name," and so was unable to reach out to anybody's body, to reach out to desire. So I just felt chained. There was a kid I was in love with at Columbia, who even tried overtly to bring me out of my shell, seduce me. It was somebody I was really in love with and still am, a real deep, passionate, desirous adoration, but I was too locked in on myself to admit to and respond to him. He took me to his room and took off all his clothes and said, "Get in *bed*," and I sat there like a frightened creep. I guess Jack was the first person I really opened my mouth to and said, "I'm a homosexual." From then on, I have been able to just reach out and speak to people.

  One of the common qualities of the young kids I meet now is this sense of tolerance and acceptance without some moral whip hanging over your head. If I were to say something shocking or shameful, or that would make them reject me, they wouldn't, and that's interesting, important, and helpful. Like the guru: "Oh, how wounded! How wounded!" So it was a quality of compassionate feeling that left it open for me to come out, feel, communicate. But that was just my retrograde hangover of early Paterson provincial fear. Very soon after that I seemed to realize that the whole thing was a fear-trap, illusory, that in a sense nobody was really shocked by anything, so long as you aren't out murdering people. Nobody was really shocked by an expression of feeling.

BF: This is something that has developed, though, isn't it?

AG: Yeah. I find it more and more reinforced. The more expression of feeling there is, the more response there is, the more tolerance there is, the more openness there is, and the more feeling there is all the way around. I feel much more strongly about it now than I did 10 years ago, and more strongly at, say, 26, than I did at 20. It deepened and grew with the exercise of feeling.

BF: Don't you think it's deepening in the public, too?

AG: Yeah, sure. Through this thing. If I run around doing what I feel, if I grab some beautiful young boy and reach out and kiss him—he'll kiss me back! Or even young kids coming up to seduce me now! Something that as a kid I would have been terrified to do. They have a sexual awareness, openness and tolerance and compassionate tenderness that is absolutely ravishing. It's going to save the world. Ask any psychiatrist. He'll say it's fine. The kids are coming out from under the anxiety barrier. I don't think they have it even, simply because there's been so much public sign that it's all right to have feeling and express it publicly. So the "private has become public" in one way or another. Burroughs, (Henry) Miller, (Jean) Genet—there's now a tradition, an authority to appeal to. The Beatles now becoming triumphant kings of woodle, with that high tender-voiced cry from the male. That's the really significant thing about the Beatles. They're able to express in that very high voice—not falsetto, just a high, sweet, tender sound that previously was barred from consciousness as being effeminate.

BF: Well, Johnny Ray got about a third of the way to it.

AG: Yeah, Ray was a breakthrough. The first time I got on peyote in '52 I heard a Johnny Ray record and noted down the breakthrough. It's like a great line in Blake:

> Children of a future age,
> reading this indignant page,
> know that in a former time,
> love, sweet love, was thought a crime

There's the prophet. Blake also says:

> Futurity, I, prophetic see,
> Grave the sentence deep,
> That the earth, from sleep,
> Will arise and seek,
> For her maker meek.

Isn't it lovely? Blake is really the great radiant source of awareness.

BF: So is Whitman to you . . .

AG: Yeah—but Blake more. Whitman also. Whitman read aloud to a college class opens up all sorts of thrilling sensations in the room. I've done it a few times. He says "I celebrate myself," just like Huncke.

> I celebrate myself and sing myself
> What I shall assume, you shall assume . . .
> For every atom of me belongs to you.

So Whitman, then, is defining or articulating democracy as based on adhesiveness, tender comradeship, and says it cannot succeed without it.

BF: In a way, though, that's just a nice, soft thing to believe. It's never been like that, after all.

AG: Well, of course we're in an especially difficult age of materialism in the sense that we are really surrounded by machines and electricity and wires and are sitting in cars in drive-ins and to some extent are isolated from direct contact we might have that comes from living in large families and farmhouses and country dances and adventures in the haybarn. But because of overpopulation and because of this highly centralized network of artificial communications it becomes increasingly necessary to have a breakthrough of more direct, satisfactory contact that is necessary to the organism. This all can be understood from the fact that if children are not handled by human flesh they can't develop, they just die, at least according to that film, *Grief*. Tenderness is a food to the human organism, without which he perishes: That is the sentence pronounced on mankind.

BF: This is Darwinian, huh? Who is tender survives.

AG: Yeah—well, it's a "hard-edged," "hard-nosed" way of looking at it. Otherwise the organism won't survive is what you say, and that puts everybody's back up against the wall: shit or get off the pot. If you want to live then you're going to live according to conditions of existence in this human form which require compassion, tolerance, tenderness, opening of feeling, warmth and—maturity, racial maturity. Or, Vishnu[10] coming back, saying, "Each time I rescue this way!" He has such a sweet voice to be able to get over like that—Vishnu. Then there's all those people starving in India! Like here we are, sitting on top of this fantastic 60-billion-dollar-a-year war outlay, when the rest of the world is waiting for the United States to turn its heart inside out and do something, and be the America it was supposed to be according to its founding fathers, its founding heart, its early traditions, like the good, grey poet, Whitman, or Emerson or Thoreau. That's America. That's the *Tradition*. And it's the tradition that because of the anxieties of the Cold War is now being denigrated by the more isolated or paranoid or anxiety-stricken of the politicians or thinkers. Like during the 40s or 50s when the whole academic community didn't consider Whitman the poet making any sense. Or the mass media, the reviews of books on Whitman's homosexuality. *Time!* Really fantastic how they rejected him when that came out. There was a book about Whitman as a fairy and they put him down as a neurotic and everything and like there was a new edition of Thoreau on civil disobedience and individuality and the necessity of integral individual sensibility and citizenship, identity, and like they were finally discounting Thoreau's conception of the citizen as being archaic or nineteenth-century or whimsical or not equal to the immense IBM-formulated hard realities which more "shrewd" and experienced "practical" bull fighters in the Pentagon were dealing with.

---

[10]One of the three principal Hindu deities, Shiva and Brahma being the other two. He is known through his ten incarnations or avatars, returning in a different form each time. Krishna is considered an avatar of Vishnu.

BF: How many collaborators do you count?

AG: I count on the collaboration of anybody that wants to have a good time or anybody who wants to continue the existence of the race.

BF: How about hard-edged psychiatry as a collaborator?

AG: The trouble with this generation of psychiatrists is that many of them were put into medical training by the army. They had their whole formulation founded on military necessities, their medical school paid for and their internship and first practice being involved with the army—and that made quite a crew-cut psychiatrist. And it wasn't real psychiatry. If you ask any of the old Freudians, what they'll say is that 90 per cent of American psychiatrists are oriented toward orienting the patient to suppress his feelings and simply conform to social convention. These psychiatrists—well, they aren't interesting people (*chuckle*), and they aren't doing any good and are certainly not following classical traditional-cultured Freudian depth analysis. I had experience with one such analyst.

BF: An old culture analyst or one of the crew-cut kind?

AG: Well, just before I left New York, I was going to him. I started with him in the Village while I was in the Psychiatric Institute as a patient. I was there eight months, about '49, just soon after this Blake vision thing.

BF: That was an annihilating illumination?

AG: What do you mean?

BF: Had the illumination had a shattering effect on you?

AG: Not really, no, because I was working as a copy boy at AP, and I was able to manage things. But Huncke came out of jail again. He'd been living with me just after I had this Blake vision, just after, and I was so distracted I didn't know what was going on, and like he was stealing all the books in the house and selling them to buy junk. And I was living in somebody's apartment that I'd rented for the summer and like these weren't my clothes or my books. And suddenly I woke up one day and found that like the bookshelves were bare of all these beautiful volumes—the complete works of Thomas Aquinas, Plotinus, St. John of the Cross—and so I began feeling very bad and said to Huncke, "You're never going to live with me again." It was the third time something like that had happened.

Well, about a year later I was living in an apartment on York Avenue and Huncke came by. I hadn't seen him for a long time but I heard he was in jail, Riker's Island. It was a snowy day, winter, snow all over the ground. He came by, tattered shoes, feet bleeding. He'd been all over, walking all over New York, didn't have a dime. He'd been down to Times Square to see if he could score or hustle or something and the cops there recognized him and chased him away from Times Square calling him a creep. They didn't want to bother arresting him, they didn't want him. They just said, "Get out of here, Huncke. We don't want you around here no more." Driven out of his chapel in Bickfords[11] where he'd been hanging around with young hustlers just in from Chicago and getting a little love there. Or help.

He arrived at my door in such a dreadful state that I couldn't reject him, and I let him in. Well, I washed his feet. He went to bed and he stayed in bed for three weeks, unable to get up out of sheer depression and defeat and dislocation. That's where the whole thing about my feeling about the Narcotics Bureau comes in. He should have gone to a doctor all along. He was capable of working and he did work very often. But the social dislocation of having to go out and hustle, to have to make 20 or 30 or 40 dollars a day for junk, which is impossible, or find a kind of job which was the only kind he could get, which was like sort of moving glassware in the stockrooms of whatever.

So finally he got up and went out one day. He had been talking in such a depressed way that I was really afraid he'd kill himself. But this one day he went out and pulled off a burglary of an auto. Signs of springtime! I was really grateful, he was coming back to life, something! Regrettable though it was, it meant he was going to survive.

But that scene built up unfortunately until it was like a *Beggar's Opera* scene in my house with a few other people that were friends of his, friends of mine, various people, and it all blew up one day with a big thing in the *News*. On page one of the *News:* my picture, Huncke's, Vicki's. Like "Gang Of Thieves Caught, Loot Found In Apartment." And all the rest.

BF: This really did happen—it's not a vision you're reporting?

---

[11]A chain restaurant. Huncke and other hustlers used to hang out at the Bickfords in Times Square.

AG: No, really. What happened was a very funny story. The place was being filled with silverware and very beautiful oaken furniture taken from apartment building lobbies. So I figured that things were getting too hot, but I didn't want to stop him in his course, because I didn't know what direction he'd take. But I figured I'd better get out of there and go down to New Orleans and see Burroughs. Get out from under. I decided there was too much anxiety for me and that I didn't quite approve of the thieving. So I piled all my manuscripts in a car with a friend of his who was one of his fellow *Beggar's Opera* robbers, and I was going to take all these papers out to my brother's, settle everything, seal things up, quit my job, then get a bus and go down to see Burroughs. It was about the time of the visit of Kerouac and Neal (Cassady) to Burroughs in *On the Road*. The guy that was driving us, little Jack Melody, his name was, had been in and out of bughouses, a little, short, thin Italian guy, very sensitive, had delusions of gangster grandeur at the time, but very intelligent or sensitive. He was making it with this girl Vicki (Russell) who had been an older friend of mine, and Huncke's, years ago. I got in Melody's car with all my papers. He wanted to take a side route out to see his mother, and I'd never met them, an Italian family, and I thought it would be interesting.

So we started out to Long Island. He had in the back like a whole bunch of stolen suits, suitcases, silverware, and he turned the wrong way up a one-way street, at the end of which was a police car, out somewhere near Utopia Boulevard on Long Island. Then he panicked and the cop waved him down and he sort of swerved trying to get out. The cop panicked too thinking he was trying to run him down, which he wasn't, and then he stepped on the gas and sped down Utopia Boulevard and the cop jumped in his car and sped out after us, so there all of a sudden we were riding down Utopia Boulevard at 90 miles per hour chased by a policeman. He swerved his car to take a side road out, then skidded, smashed into a telephone pole, turned the car over, the papers flying, my eyeglasses lost, all the clothes upside down. He jumped out and ran one way, and Vicki ran another, and I jumped out, looked around, then went back in the car to get my notebooks. I only found one. I couldn't find my glasses so I couldn't find my papers and the crowd was gathering around. Chaos, nobody bothering anybody.

Finally I just walked away from the car with my papers. One notebook, no glasses, seven cents. Sooooo. I managed to hustle up a telephone call to Vicki who had finally reached her house. I sort of wandered off from the scene. I had a funny little visionary thing. Just before we turned into that street I was singing Hebraic mantras to them: "Lord God of Israel, Isaac, and Abraham"—Handel, I think. When I walked out of the crash I saw that Jehovah had come in judgment with intolerance of thieving and fucking around. Anyway, a great chain of causes leading to my being in the ridiculous spot of a crash, with all my papers, all the letters from Burroughs, everything floating inside the wrecked hot car.

I phoned through to get someone to go out and tell Huncke to please clean out the house because the police were soon to be arriving. They had all my addresses, everything identifiable. I finally got home and found Huncke there cleaning up—dusting. He was in a hopeless condition, like it was inevitable he was going to get busted, there was too much stuff to clean out. Where was he going to throw all this stolen furniture and silverware? About five minutes after I arrived, a knock on the door, and about six big cops walk in and ask for Allen Ginsberg, and from then on it was jail. The last thing I reached for was a copy of the *Bhagavad Gita*, but they wouldn't let me take it to jail because it wasn't a religious book.

Well, I got out of it because I didn't do any stealing. They originally wanted to get me on receiving stolen goods. But that wasn't true. The easiest way for me to get out of the whole scene, as it turned out, was just to go to the Psychiatric Institute on 168th Street.

BF: So you weren't in a state of deep personal crisis?

AG: No, apart from the reality crisis, and that brought a little anxiety, and there was a metaphysical crisis because I was preoccupied with this Jehovah thing coming down and, like, what is reality? I really felt that in a funny way Huncke's victimage overweighed his actions as a robber and that the whole justice thing was getting screwed up. He had been criminalized by one part of the state and forced into a position of acting illegally.

BF: Of becoming a victim-maker himself.

AG: Exactly. So I didn't feel it was good tactics for him to steal, but I didn't think it was any outrage.

But it was a real bringdown for my teachers at Columbia, Trilling and Van Doren. Because the *Daily News* thing was fantastic. The front page, my picture, me coming out of the police station. And like I was billed as "Columbia College Criminal Genius"—being held captive by this gang of lowlife people because of craving for drugs: they were feeding me drugs, and I was making their plans. Wholecloth stereotypes straight out of Sherlock Holmes. Boy, was that a mess. Like was my father ever upset—it really brought him down. He was a nice, middle-class teacher, and the idea of anybody going to jail was horrifying to him. It was so shocking to him—and it cost him some money. It was certainly a traumatic thing.

BF: Did that end your studies at Columbia?

AG: No, no, I just went back and finished later on. I went to the bug-house and there I just had two papers to finish—one for (Jacques) Barzun on cycles of history. It was a project I'd invented for a seminar course in history, which was my minor, reading Spengler, Korzybski, Yeats, Petro, (Jacob) Burckhardt, trying to make like a unified field theory of what state of cycle our culture was in. From which I came to the conclusion that we were due for a second religiousness. I think Spengler called it a second religiousness. It was what Toynbee got so hung up on. I don't think it's a good terminology because it is not a religion, unless you consider God waking to his own nature as religion. Really, what the whole point is, coming back twelve pages earlier, the whole point is man is now God. He can change the universe, change the molecular structure, turn it inside out, blow it up, make it disappear, alter himself, alter his brain chemistry, alter his body. So that's like a big responsibility. It's a very interesting situation—to be in such control. All on account of somebody got busted for junk. (*chuckle*) I sound like Lenny Bruce. That kind of association is very beautiful when Lenny does it. Typical schizophrenic thinking pattern.

BF: Speaking of schizophrenia, what finally happened when you went into the Institute?

AG: Oh, yeah. I was going to a doctor there named Brooks who was somebody I could talk to. After I got out—I was in eight months and met Carl Solomon there; he was thinking about the void, also—such problems!—and we sort of recognized each other and started talking. He came up from shock and asked me, "Who are you?" and I said, "I'm Myshkin" and he said, "I'm Kirilov." So we sat around for eight months asking each other whether the authority of the doctors or their sense of reality was right for us or whether we were right, or what was happening. Carl was having problems because he was getting shocked. I didn't have any of that, no medicines, no shock, it was like a hotel for me, weekends and evenings off, it was like a very convenient monastery for me.

BF: Did you come out differently aimed?

AG: No, no. For a while I had a period of fear. I figured I really should not be seeing Huncke and Burroughs and everybody for a while until I got myself all straight. I began questioning my sense of my own reality versus the social sense that was being imposed on me, and for a while I toyed with the idea of trying it and seeing what it felt like. But when I started acting it out a little bit I found it necessary to reject everything I held dear, to reject Burroughs, reject Huncke, reject—you know, reject my own homosexual feelings, reject any compassion I might have felt toward my friends, and it was all interesting but not very rewarding. It gave me a tolerance toward doctors. But the position itself—it was a position many in that hospital came out with, like a total self-rejection, a rejection of their own universe, an acceptance of some formulation laid down on them as being the safest way to handle their conduct in the outer world. Lip service, really, to supposedly acceptable social patterns.

Years later I went to the Langley-Porter Clinic in San Francisco and had a year of psychotherapy with a Dr. Hicks. During that time I gave up that market research thing and wrote "Howl," partly to his good influence. He kept saying, "What do you want to do? Follow your heart's desire." So, I said I'd really like to quit all this, the apartment I had on Nob Hill, the tie and suit, and the job with several secretaries which was beginning to give me pleasure—the pleasure of knowing I could do it and was no longer intimidated by the social forms, to be able to make it in the very dry business world—analyzing the market, motivation research was what my training was—but I said I just wanted to quit and go off and do what I wanted to do, which was get a small room with Peter and devote myself to writing and contemplation, Blake, and fucking and smoking pot and doing whatever I wanted. So he said why don't you do it, then? So I said what'll happen to me if I grow up with "pee-stains on my underwear" in some furnished room and nobody will love me and I'll be white-haired and won't have any money and there'll be bread crumbs falling on the floor? He said, "Oooh, don't worry about that, you're very charming and lovable and people will always love you." Such a relief to hear that! His own liking of me, the transference—which is what it's all about, really, is feeling, the passage of feeling between the doctor and the patient until you cease to be doctor and patient and become friendly beings. As long as it doesn't include feeling, it doesn't include the transference and it isn't Freudian. As long as it's a role-playing thing of doctor and patient with the doctor not emanating any feeling, any organic feeling toward the patient, then it's not a human relationship, period. It's just an intellectual exercise, which is why psychoanalysis always takes so long: they're just shuffling around rearrangements of ideas. So I said, "What would the rest of the psychoanalytic fraternity have to say? What would the APA have to say about this advice you're giving me?" And he said, "We have no party line." He was of the interpersonal relations thing, Harry Stack Sullivan. I was worried about homosexual orientation, if it was all right to stay that way. And he said, "Well, as long as you're enjoying it, as long as you don't internalize the fear or the anxiety." And I realized, well, sure—and I went at it with a vengeance.

BOB ELLIOTT
March 3, 1967,   St. Louis, Missouri

Freelance, 1967

Allen Ginsberg began giving lectures and readings, often for free, on college campuses in the 1950s, and continued to do so until his death. This contact with colleges and the nation's youth often resulted in articles about Ginsberg in student and other university publications. The following interview, an early example of Allen's work on the nation's campuses, took place during a visit by Allen to Washington University.

—DC

Bob Elliott: In anticipating your coming here, the reaction of many people was not so much to you as the poet, but rather as a spiritual leader. How do you react to this?

Allen Ginsberg: I don't know, it's kind of pleasant. I begin to get the hallucination I'm a holy man after a while. The thing is that everybody in the United States is so confused and living in a state of hallucination, with television, radio and the Vietnam war and a false sense of identity. These are senses of our own identity that arise about ourselves, rather than actually being ourselves. So that anybody who comes along breathing in his belly who doesn't have any particular idea of who he is, just exists, probably is a holy man at this point. It's not very hard to be a holy man in America anymore. All you have to have is a sense of who you are and a certain amount of compassion for those trapped in their role playing. Sure, I might as well be a holy man. I'm holier than Cardinal Spellman probably.

BE: In your poetry, writing and conversation, your eclectic kind of total approach to religion, to ritual, is apparent. Why do you handle religion in this way?

AG: Ritual is useful. Anthropological knowledge of ritual, and ritual derived from various different traditions is probably useful now because the United States is just getting to a point of exploring its own identity. Young kids are beginning to break through the shell of consciousness that has grown up in America in the last half century, and find themselves born into a new space-age universe. So that can be a little scary when people discover they aren't the people they thought they were and their teachers and parents aren't the people they grew up with and that maybe the whole universe doesn't exist even. Some kind of stable rituals are useful, whether they're American Indian or Jewish or Christian or Tibetan or Zen Buddhist or Islamic. All of these ancient traditions have specialized, at one time or another, in exploration of subject, of consciousness. So now that young cats are beginning to explore the ground of their own consciousness, they can make use of the older disciplines too. They'll find them having some kind of meaning suddenly. All the old myths will begin "whispering" again. Also, the use of various rituals will prevent any single monolithic religion from being taken too seriously, as being the unique representative image of divinity. So that religious nationalism is somewhat avoided, by acquaintanceship with the similarity of all the different *sadhanas*[12] or disciplines or rituals. So that people begin to get a glimpse of what the Yorubas were up to, and what the Balubas[13] were up to, and what were the South American herb doctors up to, or what were the Hindus up to, or what were the Zen Buddhists doing sitting around on their behinds.

BE: You once said you thought American jazz was kind of a translation of African worship ritual into a twentieth-century language.

AG: Yes. The reason I got that idea was that I went to Cuba, a year and a half ago last January, and there I ran into survivals, intact, of the Yoruba tribe of Nigeria. They have a thing called Santeria, which is the worship of various gods: Chango, who is a phallic god, very similar to the Indian god Shiva, and Yemaya, the blue-bodied goddess of the ocean, female. They have a pantheon very similar to the Hindus. And

---

[12]In Vajrayana Buddhism, *sadhana* is a tantric initiate's meditation liturgy employing practices of visualization, mantra, and completion stage practice, transmitted from guru to disciple as methods to accomplish realization in this very lifetime.
[13]Bantu-speaking peoples of south-central Zaire. Also known as Luba, they have a strong belief in a Supreme Being, ancestor worship, and natural spirits.

they worship their gods through the drum. Chango is the drum. The priests make patterns of rhythm on the drum. The worshippers dance before the drum. The thing I noticed was something very similar to what I noticed in the caves of Liverpool where electric rock groups were dancing. First of all, both the drum and the electronic instruments set up vibrations which penetrate the lower abdomen, penetrate the body physically, as in the enclosed space of a hut where people were drumming and dancing. The drums have a physiological impact on the body just as the electric rock instruments do, unlike other Western instruments. Secondly, the dancing patterns were the same, communal rather than individual. That is, in the Yoruba-type ceremonies there would be a line of men and women mixed, as in a family-communal type group, dancing, and in Liverpool there'd be like a circle of 2 boys and 5 girls or 8 boys and 3 girls, so that it was done as a community thing. Then also in the Santeria Yoruba rituals, people flip out and go into trance states for Chango just like kids flip out and go into trance states for George, or John Lennon.

I noticed that the cave action was similar to the African style state of exaltation and openness and hysteria, so I began thinking. Well, it is obvious: the Africans brought jazz or brought music to America and originally it was part of their worship rituals. Their worship involved dancing and chanting and singing. It was brought over with the slaves to America. It evolved into modern syncretistic forms with other instruments, but the rhythms remained similar with historic gradation and change. The terminology always indicated it to be of a divine nature, like "soul" music or "spirituals." It simply went upriver from New Orleans to Chicago, and went across the states to New York and then to Liverpool and emerged as the children's sacramental assembly of worship because rock's the one thing the flower children do take seriously. And because in America and England, it's the one major social form that has qualities of invocation of the divine, as in Dylan or Donovan or the Beatles, or any of the white rock 'n' roll, a call to the community, toward adoration of some higher thing, like in "Strawberry Fields Forever" or any of the songs in "Rubber Soul." The earlier songs are humanistic things like "I Want to Hold Your Hand," or "We Can Work It Out." I mean they're all preaching messages about tolerance, transcendency, and ecstasy. What I think has happened in sum is that the basic religion of the slaves, because it was genuine, because it was real "soul," is finally being accepted in disguised form by the white children and is wakening them up spiritually. So it's like the revenge of Africa on the white middle class. And it fits in this way: since we have this puritanical, anti-human, anti-sexual, anti-Personal, anti-Subject, anti-introspective culture, the particular vibrations of jazz are just those vibrations which bring people to recognize themselves as soul, as body, twisting and shouting. In the Indian scheme of things, in yoga, you waken different chakras or centers of the body, beginning around the sphincter with the muladhara chakra, the sex chakra. Well, it's almost as if these worship rituals were specifically intended to make everybody shake his ass and waken the muladhara chakra. So it seems to me one can see the new development of electronic rock as a spiritual thing. It obviously is so in the new songs, like those by the Buffalo Springfield.[14] Jazz has always been a spiritual thing with the Negroes. I mean Monk is a monk. And Charlie Parker was a bird, a spiritual bird.

BE: Is it possible that many white people have dug jazz for all the wrong reasons?

AG: They may "mis-take" something, but I think the nice thing is that they take it at all and that it changes them. And that Negro religious art forms have had such an enormous impact on white culture. I think it's saved America. If America is to survive the next few decades, I think it will be the Negro who has saved America by introducing, by having preserved "soul," and having found the right forms for the penetration of "soul" into the consciousness of the younger kids.

BE: Apart from ritual being a unifying force, what about the purging elements of ritual?

AG: What I was saying is that people discover that they are person and subject, with feeling, affect, affections, tears, fears. So they hit a whole new level of consciousness when their emotions are unlocked by jazz or prayer, by jazz prayer. So in that sense it's purging. Whatever was hidden comes to the surface. Well, you take a stiff-backed date out in a taffeta dress to a wild rock 'n' roll concert and before the evening is over, she's sobbing and happy and having a ball. And all that character armor and conditioned response of negativity and coldness and fear is

---

[14]Buffalo Springfield was a mid-sixties California rock band that included Neil Young and Stephen Stills, among others who went on to successful solo careers.

broken down and she's swinging her ass, acting like a human being. Purged of the white middle-class horror of repression of self, repression of Person, fear of Person, fear of feeling.

BE: Do you think there are other forms of communication besides the verbal?

AG: Among the younger kids, they don't seem to rely on language as much or they begin to see language as a vehicle for feeling as well as ideational content. They begin to use language as mantra, prayer, also, in which the word can be pronounced gently, tenderly, softly, with awareness, or the word can be pronounced with robot intensities. But I don't think the kids rely so much on words. I don't anyway. (Andrei) Voznesensky asked me something very funny in Russia last year. He said, "What language do you think in?" I wasn't sure how smart he was, so I said, "Sometimes I think in Spanish and sometimes in English." And he said, "Do you always think in words?" And I was amazed and delighted, and said, "Well, what do you think in?" And he said, "I think in rhythms." And that's very hip, for anywhere, especially Russia, where everything is so conceptual and ideological.

BE: Do you think catch phrases like "white backlash" or "black power" have any basis in reality?

AG: Well, they do make people conscious of large waves of public psyche activity, or large waves of propaganda activity at any rate, in the mass media. Like (Marshall) McLuhan's phrase, "the medium is the message." That does make one suddenly conscious of the fact that television is selling itself, machinery is selling itself, radio is selling itself, newspapers are selling newspapers and they're not selling news. Those phrases provide a "rational hook" to make people aware of things. There has been a white backlash, I think. It's a good thing people are aware of it. See how stinky the whites really are.

BE: William Carlos Williams, in his preface to your earlier poems, *Empty Mirror*, was saying your poetry at that time read like a newspaper.

AG: Yes, I was purposely using newspaper language to see what the common language was like and see if that could be recombined to a higher power to make something more intense than a newspaper lead paragraph. I was working in newspapers then. Actually working on AP in New York, on the *Telegram* at the time. But that was Williams' idea more than mine actually. I wasn't thinking of it so consciously as "using newspaper language." I was thinking of using his kind of language, the common dialect of the tribe, the speech and rhythms common to Paterson and Rutherford, New Jersey. He had the idea that it was like spraying it back. William Burroughs does that in his later prose, using *Time* magazine rhetoric, which he simply chops up and recombines and sprays back into the machine that sprayed it at him. The reason for doing this is that the main instrument of brainwash and hallucinatory political control in America at the moment is language. That is you can get the American people, apparently, to fight gentle Indochinese by labeling them "communists." And you can get the American people to accept their desecration and deaths, torture and tear gas, as "legitimate weapons" as long as they're used against the "reds." So that if you have a newspaper lead story or television bulletin saying "75 reds" or "75 communists" were killed today in an attack on Bong-Son, then it doesn't seem like any human beings have been hurt, just "reds." It depersonalizes them. So that the use of labels makes atrocities committed more acceptable to the American consciousness which doesn't want to see them as atrocities. Like when the C.I.A. infiltrated the student organizations, *Time* in explanation said, "Well, it's just normal business after all. If we want to fight the communists, we have to take appropriate measures and these are appropriate measures." In other words, governments can use language to make people feel something's not abnormal when the situation is quite plainly and simply very abnormal. Or I can use language to say it is abnormal.

BE: Earlier, you were talking about breathing in your belly, and mimes and singers sometimes talk about a "column of song."

AG: Well, people who play saxophone, people who play clarinets, girls who are going through natural childbirth exercises, Zen Buddhists, all use that lower abdominal breathing. What it's useful for is to place the Self (capital S) in a reliable place in your body, in the center of your body, instead of letting the Self wander around in conceptual—in language—imagination rattling around in the head. In other words, if I'm sitting here with my mind operating, flowing and murmuring and leaving YOU, *absent*-minded—so that I'm fantasizing about Chicago or Be-ins in San Francisco or Testify-ins in Washington—then I'm not really here and the Person is trapped in ideas about himself. You know, like "I am Allen Ginsberg, a poet, who has been this and that."

So I'm not really the being sitting here, I'm the person of definitions and images. I'm an image. So in order to combat the hypnosis of imagery and the wanderings of the mind, which distract from present reality, present consciousness and present situation, it's useful to collect the mind's attentions so that what you are doing is paying attention to the breathing. Locating the breathing in a specific solid area like the lower abdomen, going through the exercise of pulling your fantasy back into the body from all the geographical directions of space thoughts wander to, and putting it down here (gesturing to belly) so that the Self gets to be in here where it's safe and where it can't be pushed out by hypnosis. This is just an image, but it works. If your mind is concentrating on breathing, it doesn't have time to be running around in the Arctic Circle.

BE: You were saying that the present reality is mystical and divine, but apparently we've lost touch with it?

AG: One, we've lost touch with our senses. On a very simple level, the cities are covered with smog. Our ecology has changed, so there no longer are animals, mammals with smell in the forests. All our smells now are artificial and chemical so that you have cities covered with black gas. And nobody can smell each other or smell the delicacies going on and when you look up you don't see the moon or the stars anymore, if you're in any of the large megalopolises where most people live now. Sense of touch has been forbidden so that people don't go around, as in the Garden of Eden, touching and embracing and kissing each other. Taste has been assaulted by plastics and the homogenization and mechanization of food. And cigarettes have destroyed the taste buds. At the same time, as McLuhan points out, the optical thing has been over-emphasized by print and television imagery, so that more and more of people's universes are involved with the visual and the other senses atrophy, through lack of use. In fact, McLuhan said he was getting a lot of money to try to do some experiments to measure, quantitatively, how much the other senses have shrunk through overuse of the visual, overextension of visual universe.

BE: Isn't it extremely difficult for a person oriented in Western thought and philosophy to have a sufficient breakdown of these methods to accept the kinds of consciousness Zen Buddhism or other Oriental disciplines present?

AG: Except, you know, Western thought has led that way up through Wittgenstein. That breakdown has already taken place. But Western thought through Einstein and Wittgenstein has already come to that same kind of relativity. And it's also in the tradition of Western thought with people like Blake and Whitman and Thoreau. So it's been there all along. It's just simply that historically there's a place for it now. There's a crying need for it now, because we've all gotten so locked up in our language that the language no longer is real. We no longer can see what is in front of our faces because we're seeing it through the eyes of "75 communists" killed today, instead of 75 living souls. What I think is that historically, as of the last few generations, there's been a natural precipitation of understanding, a natural discovery of the limits of language—as through the development of semantics—theories of semantics in the higher intellectual circles in the West. There's been a natural breakthrough of visionary experience, with *and without* drugs. So that at this point, Zen or Tibetan Buddhism or Hinduism comes along naturally as another technique, developed by another tribe, which you can use or adapt any way you want. I don't think America will ever "go Zen Buddhist," (I'm using the practice of classical Japanese Zen as an example) nor do I think it would be a good idea if it were adopted, because the Zen aesthetic is built around formal gardens and tiny wooden houses and little mountains and overcrowding, which wouldn't be appropriate for America. On the other hand, sitting and breathing in the lower abdomen and exploring the inside sensory universe that's something which is really basic in Zen which is applicable in America today. I just think all the different techniques are adaptable. Who would have thought Westerners could adapt African techniques? Well, now look at the Beatles, if what I've been saying is at all accurate. The syncretistic form is very effective. This has been spoken of many times before. A. B. Spellman in his book on jazz, *Four Lives in Be-Bop*,[15] talks about jazz as sacred African yoga and traces its ritual origins and is interested in it as an American syncretistic development of early African ritual origins. If a lot of people begin to dig jazz that way, their perception will become beautifully sharp. I mean they'll begin to take it for what it really is, instead of worrying if

---

[15]*Four Lives In the Bebop Business*, by A. B. Spellman (New York, Pantheon Books, 1966); biographical sketches and quotations from four black musicians: Herbie Nichols, Cecil Taylor, Ornette Coleman, and Jackie McLean; includes photographs.

it's good or bad, or worrying that they like it so much. If they suddenly wake up one day and realize that all along they've been worshipping God everybody's going to be really happy. Like all the old folks will be happy that all the young children are out there at the Fillmore Auditorium[16] in San Francisco, dancing before the "Ark of the Lord." Nobody will be able to complain anymore. And the seriousness and intensity of the soul of the younger people will be understood, as well as the saintliness of the older musicians. Then we'll have a really sacred society, because people will see things as sacred, as they really are. (Leopold) Stokowski is interesting: He was asked years ago, when the Beatles first came here, what he thought of them. He gave a very sympathetic answer. He said, "They are seeking ecstasy, which is old human search. Legitimate." He legitimatized them, like he was one of the few classical musicians who had the breadth of sympathy to lay an approving word down. To lay that particular word down. What else is everybody after? Especially in the rock scene. That's what they're getting too, in San Francisco anyhow.

BE: From what you've been saying, about how you keep bringing existence back down to Self, what do you think this will do to Western art forms?

AG: Well, it will do just what's happening. Things called "happenings"[17] are happening. Western art forms are beginning to get more existential. People are beginning to see, like household, as a tea ceremony. People begin to do kitchen yoga when they're washing the dishes. People begin sacramentalizing all relationships, because the purpose of art is to sacramentalize life, I think. That's a reasonable statement that I heard Swami Bhaktivedanta say recently. He said he thought the purpose of art was to bless and make sacred everything, so that people could see it that way. That is, to reveal the feeling in things, so they become more of a ball.

---

[16]Popular music venue in San Francisco, famous for early Grateful Dead, Jefferson Airplane, and other seminal psychedelic bands, created by music promoter Bill Graham. See Bill Graham and Robert Greenfield, *Bill Graham Presents* (New York: Doubleday, 1992).

[17]Multimedia performance art events emerging in the late fifties championed by the Fluxus movement. Allen Kaprow is credited for coining the term and staged happenings with Red Grooms as early as 1959.

"The Avant-Garde"
Firing Line, broadcast September 24, 1968

When conservative ideologue William Buckley invited Ginsberg to be on his PBS television program as part of a series on the avant-garde, it is not surprising that the exchange was vigorous, especially since the interview took place in 1968, a year now seen as pivotal for the decade. It is also not surprising that the strong disagreements between the two centered on three of the defining issues of the 1960s: drugs, censorship, and the Vietnam War. While the history of the Vietnam War is generally known and the issue of drugs remains current, it is probably worth reviewing Allen's significant role in the fight against censorship of the arts, especially as so many important breakthroughs in freedom of expression, now generally taken for granted, were won in the mid-1960s.

In 1962 the New York Coffee House Law was passed, which required any restaurant without a liquor license to obtain a coffeehouse license if it presented any kind of live entertainment other than background music. The law was used to harass cafés where poets and other artists read and performed. When the Café Le Metro's owner was issued a court summons in February 1964 for having a poetry reading, Ginsberg's work on behalf of Le Metro eventually resulted in the American Civil Liberties Union assisting with the defense, and the court ruled in the café's favor.

Allen spent much of his time and energy in the mid-1960s fighting such censorship. He wrote a petition in support of Lenny Bruce stating that his work was "social satire in the tradition of Swift, Rabelais, and Twain." Ginsberg succeeded in getting an impressive roster of names to sign it, including Bob Dylan, Woody Allen, James Baldwin, John Updike, George Plimpton, Gore Vidal, and theologian Reinhold Neibuhr. He testified in defense of Jonas Mekas and William Burroughs, helped organize committees, phoned civic leaders and attorneys, hounded important journalists to give the fight for artistic freedom news coverage, helped to coordinate the efforts against censorship, and shaved off his beard and mailed it to the assistant district attorney on behalf of Lenny Bruce in his own Dada form of protest.

When I discussed this interview with Allen, he suggested that I mention that Buckley had eventually come around to the positions Allen had argued for with regard to drugs. The reversal in position Allen was referring to was the cover story of the February 12, 1996, issue of the magazine Buckley edits, the *National Review*: "The War on Drugs Is Lost." The story was written in the form of a symposium, with contributions from Buckley and six other writers. Buckley's article, an address he had delivered to the New York Bar Association, included the statement that "it is outrageous to live in a society whose laws tolerate sending young people to life in prison because they grew, or distributed, a dozen ounces of marijuana. I would hope that the good offices of your vital profession would mobilize . . . to recommend the legalization of the sale of most drugs, except to minors."

The videotape of this interview was transcribed by the editor of this volume. Note that in the contentiousness of the dialogue the two men often spoke over each other, rendering some words and phrases inaudible.

—DC

William F. Buckley, Jr. *(After a brief introduction)*: I should like to begin by asking Mr. Ginsberg whether he considers that the hippies are an intimation of the new order.

Allen Ginsberg: New order? I'm hoping it will be orderly and gentle, yes. "Hippies" is kind of a stereotype, generally a newspaper stereotype, so that it's hard to generalize except to the extent that what is called the hippie movement involves an alteration of consciousness towards some greater awareness and greater individuality which you might even sympathize with?

WB: Yes.

AG: Then hopefully the future will see like a spread of that gentleness and consideration coming through, politically and artistically and *(looking pointedly at Buckley)* maybe even on television.

WB: Not quite yet. *(laughter)*

AG: Well, it's slowly approaching. The problem, I think, with television which I think is interesting—see, before I came on, one of your—Mr. Stiebel, who is your helper here, producer, asked me not to say any dirty words, what he considered dirty words, on the program, which presents a moral problem, you know, in that there is a political function to the language of everyday use—the language we actually speak to each other and off the air. There is a communication that's involved. And there's a classical use of all sorts of what are called off-color words in art, as well as images; so our problem here, or what I've been posed with, is having in a sense to censor my thought patterns.

WB: But you can overcome this through the process of love, can't you?

AG: No, I . . .

WB: Your love is, in part, consideration, isn't it?

AG: Awareness, awareness is what I would say.

WB: Yes, but there's also consideration. And therefore if it would be offensive to some people to hear those words spoken, then you would presumably assume the burden of expressing yourself without using them.

AG: I think I could probably pronounce them in such a way that it would be inoffensive. In fact, everybody would shout with joy on the television screen watching. But the real problem is . . .

WB: That might denature them.

AG: Well, good, it's time to denature them, it's time to denature rather than hold back the old Nobodaddy[18] Mystery.

WB: So I don't see . . .

AG: See, one of the problems is that, for instance, politically speaking, no one can understand the problem of police brutality in America, or the police-state scene we are going through as I see it, without understanding the language of the police. The language that the police use

---

[18]William Blake's term for the "false God." See S. Foster Damon, *A Blake Dictionary* (University Press of New England, Hanover, NH, 1988).

on, say, hippies or Negroes is such that I can't pronounce it to the middle-class audience, so the middle-class audience doesn't really have the actual data or some portion of the data to judge the emotional situation between the Negroes and the police.

WB: Oh, I think the middle-class audience can simultaneously maintain two things, one, that they don't want a series of words used on a television screen which may be being watched by their old aunt or their ten-year-old daughters, and, on the other hand, not be at all surprised to find truck drivers or intellectuals or policemen using these words simply as a matter of course.

AG: Yes, but what I'm saying is the reason the old aunt or ten-year-old daughter doesn't understand the problem of the police state in America is that the data is not given to them on the medium of mass communication, see?

WB: Well, no I, I . . .

AG: So that's why they remain old aunts.

WB: May I, may I . . . ?

AG: In other words, they're sort of like conditioned to be old aunts?

WB: They're born to be old aunts.

AG: No, they're conditioned to be old aunts by the media and in a sense by your insistence on protecting them from realizing the clubs of language and the blood of words that's running out of peoples' heads in America now.

WB: Well, I think I . . .

AG: Besides which it's a censorship on the spontaneity and delight of actual poetic language, which is a problem here on television. As I said, you know, because I was posed with it, I don't necessarily intend to violate it. But I . . .

WB: Well, I'll describe it.

AG: Well, why not?! But if I don't violate that, in other words, if I censor myself, and you know I'm censoring myself then, please don't give me—or please don't *insist* that it's *right* . . .

WB: OK, OK. In other words, I won't . . .

AG: for me to insist "no censorship" . . .

WB: Okay, okay, I won't say that I've made a moral point.

AG: It's a morally equivocal position that you're taking there.

WB: Yeah, yeah, I won't . . . But, however, I do want to make an artistic point . . .

AG: Yeah, fine.

WB: namely that unless certain of these words are used with great frequency, rather, when they are used with great frequency they have a completely different effect from the effect they have when they are used infrequently. So that even a relatively chaste word, if it is said leeringly and for effect . . .

AG: Yeah.

WB: in a context that makes it highly conspicuous, has a shock value that it oughtn't to have intrinsically. Whereas, when you hear certain words, let's say, uttered by a sergeant majors greeting the marines coming in for basic training, . . .

AG: Or the police dealing with

WB: it's entirely meaningless.

AG: the professors at Columbia.

WB: Which is why I'm rather surprised that you brought it up about the police because you've just finished admitting that it's meaningless when they use it.

AG: No, I'm not saying . . . you were the one that was saying it's meaningless, and I was saying it would take *some* of the shock value out of it in terms of the media. Those old aunts would be less shocked if they heard it a few more times . . .

WB: Well, I . . . yeah, I suppose . . .

AG: and the tone of voice with which it's pronounced . . .

WB: Well, well, maybe . . .

AG: In other words, if I use, if I use . . . fornication words in poetry or fellatio[19] words in poetry or—what is it?—depends on tone of voice with which it's pronounced, and if I say it in a nice way, it might be like just an ordinary human activity which everybody recognizes as being their own.

WB: Well, speak for yourself! *(laughter and clapping from audience)*

AG: Well, I'm also speaking properly. I presume you don't fornicate? I won't speak for you. Fellatio, I wouldn't know about you, but I think the Kinsey report says that like a large enough group of people . . .

WB: Now, you're changing the subject, you're changing the subject. Let's stick to the artistic viewpoint, not get off on Kinsey.

AG: Well, artistically the point is that . . .

WB: Your point really is that the maiden aunts and everybody else should hear these words frequently enough so that they wouldn't have any shock value when they're heard.

AG: No, that's not my point, that's not my real point. My real point is that it's part of our normal consciousness. If one is building a work of art out of one's actual consciousness instead of an artificial consciousness, that language will enter in as part of the building blocks, just as sensory perceptions, Pound's phanopoeic—that is, mind-eye images—will enter in because they're part of . . . part of our sensory field and part of our sensory field is the language running through our head, the language we speak in every day, those are the building blocks of art. So television can't be arty until television comes around to the old human usages.

WB: Well, [I have covered it.?] There are certain plays, seem to be pretty arty without using Shakespeare, for instance.

AG: Well, I've found, for instance, I can't read my poetry on television.

WB: You'd rather be [inaudible].

AG: Well, no! I have the same problem as Henry Miller or D.H. Lawrence had or Jean Genet or every poet in America, or almost every

---

[19]Ginsberg gives "fellatio" a Latin-style pronunciation: fell-ah-tee-o.

poet, all the avant-garde poets, and that's our subject here. The avant-garde in a sense has no medium to communicate with you or . . .

WB: No, but . . .

AG: directly.

WB: Surely the point, Mr. Ginsberg, is that we should [inaudible].

AG: I can get up here and blah, blah, blah, about fellatio, but I can't talk in a way that a truck driver will understand that I'm making sense.

WB: Well, I'll interpret for you, how's that?

AG: Okay, you interpret.

WB: No, no, the point I'm trying to make *is* that, you know, this continuum that you just assaulted us with, it began with Henry James and ended with you.

AG: No, no, no! I was saying Henry Miller. I'm sorry. Henry James? I meant Henry Miller.

WB: Oh, Henry Miller! Oh! Not James.

AG: I think I said Henry Miller. Did I?

WB: No, you did. Yeah, yeah. No, that's different.

AG: You hallucinated Henry James. Let's hallucinate Henry Miller. Have you read Miller, by the way?

WB: Yes, I have, yeah. I've read everything, roughly speaking. *(laughter)*

AG: How did you find Henry Miller?

WB: Well, I think, I think he's a very important writer. I think his philosophy is the greatest bore.

AG: I mean as prose writer, style?

WB: Oh, I think it's tremendous, yeah, tremendous.

AG: What I'm curious about is what odd literature have you read that would be considered like "subversive" that you like. In other words, what are your real literary tastes?

WB: Well, they're eclectic.

AG: Yeah. You've read Genet, for instance.

WB: No.

AG: You're not interested? Very delicate prose.

WB: I haven't read Genet.

AG: Céline?

WB: I suppose . . . Who?

AG: Louis-Ferdinand Céline.

WB: No. I even know who introduced you to Céline: Burroughs.

AG: Yeah, old teacher. Burroughs and Kerouac . . .

WB: Bill, Kerouac, yeah.

AG: Kerouac likes your style.

WB: Kerouac wants me to be president.

AG: Because he likes your style. Of speech. And so I think he's camping a little bit.

WB: I do too, yeah. *(laughter)* You believe—actually, you *have* in a curious sort of way a standard which says that there ought to be no standards.

AG: No censorship . . .

WB: It's interestingly . . .

AG: no, I'm not saying no standards, I'm saying no, that the police can't come in with guns or arms and take me to jail for speaking. I'm speaking of old constitutional free speech. Or that the state should not be intruding on discourse in public, public discourse. And the state has a tendency to do that. I mean I've experienced it in my own work, say my first work, and I've experienced it occasionally—well, I'm experiencing it *now* in these FCC regulations, okay? We're experiencing it *now*, the octopus of the state intruding on our language consciousness.

WB: Well, I'm not experiencing it . . .

AG: *You* are because you're being denied my actual speech.

WB: There's the poet in you because you really are exaggerating.

AG: No, I couldn't, I can't read what—you were speaking of "Howl" as a poem and I can't read that here. So that's my ax I can't bring forth. I have to hide my dear [delight?].

WB: But, however, you weren't invited to read it here.

AG: "Howl"?

WB: I say you weren't invited to read it here.

AG: No, no, I wasn't invited [to]. I was invited to . . .

WB: Let's simply talk about whether or not we have a police state which severely impinges on your freedom. It's true, I think, we did have the successful prosecution of the comic tragedienne, uh, the fellow that died.

AG: Lenny Bruce?

WB: Yeah. But it has since been reversed.

AG: He's dead!

WB: It has not been rather numerous.

AG: Oh, in the avant-garde, which is our deal, it has decimated the avant-garde in New York. This is information you probably don't know, but I would say that over the last five years, since the sixties [began], probably more money has been put into fighting the state and the intrusions of the state than has been put in the Lower East Side into creation of movies and plays. In other words, for instance, with the great rise of the avant-garde group Filmmaker Cinématheque, an enormous amount of money was put out defending "Flaming Creatures," and Genet's movie, "Un Chant D'Amour," and all sorts of . . .[20]

WB: But an enormous amount of money was spent in prosecuting it.

AG: Even more—that's what's so horrible—taxpayer's money is being used up to put down these artists!

---

[20]In the actual broadcast Ginsberg says "defending *Flaming Creatures*, Genet's movie, and all sorts of . . ." which could give the impression that Genet and not Jack Smith made *Flaming Creatures*, but I have a version of the transcript that lists both *Flaming Creatures* and *Un Chant D'Amour*, so Allen must have communicated this to Buckley or his staff, or they happened to know this.

WB: But, see, the whole point in the end is the taxpayers take the position that they ought to be free to protect certain kinds of privacy.

AG: That's the police take that position, not the taxpayers. The taxpayers really believe . . .

WB: The police enforce the laws that the taxpayers pass.

AG: No, they enforce them arbitrarily, like they thought they were enforcing "the laws" about Lenny Bruce, but they weren't, they were interpreting it to put down Lenny Bruce, and the law and Constitution actually said they shouldn't have persecuted him . . .

WB: Well, in the first place, it was . . .

AG: so they were wasting taxpayer money.

WB: Now, wait a minute, wait a minute: It was a split decision.

AG: Well, anyway . . .

WB: [inaudible] came to a different conclusion.

AG: all the artists in New York got together and put out a big petition . . .

WB: To which saying?

AG: saying Lenny Bruce should be protected, to which the police did not yield, to which the district attorney did not yield, to which the lower courts did not yield, to which nobody yielded for years and years until the poor man was *dead!*

WB: Well, you're suggesting that there was a relationship? He died prematurely, you'll agree?

(Station Break)

WB: You say he died prematurely. I think it's much more likely that he died from excessive use of the kinds of things that you're always urging on people, rather than because the DA got after him was what caused it.

AG: No, I think like part of the shutdown on his general activity was the police activity which prevented him from coming to New York even physically—[so] as to [not] be arrested—much less working and leading a more or less healthy, active life, and he got all hung up on drugs, researching his—like a carbolic scholar—researching his legal case. It was really a Kafkian thing with like a state-organized conspiracy against him, it was very sad.

WB: I don't really think you're going to make a very plausible case, Mr. Ginsberg, of suggesting that Lenny Bruce got hung up on account of being stopped halfway through a scatological tirade, which in itself suggests that he's slightly off.

AG: No, no! Scatological tirade it was not; it was a political speech. You know, like he—what you call "scatological tirade," was the language he was using, paraphrasing redneck southern sheriffs. In other words, he was doing old Swiftian satire, making use of the actual *language* of what he felt was the psychic enemy, he was presenting it in public in front, and he was arrested for presenting that political evidence so to speak.

WB: No, I don't think he was arrested for that . . .

AG: Yes, yes, yes, yes!

WB: because he was trying to make a political point. He's really . . .

AG: No, well, no: the texts of his nightclub routines were very similar to Burroughs's actually and were primarily social-political criticism.

WB: Uh-huh, uh-huh, uh-huh. Actually . . .

AG: Yes?

WB: Well, I know. And some people say that De Sade was engaged in political, primarily, in political criticisms that, as a matter of fact, there's no question that he was committed by people who had political motives in part in mind. My point about Lenny Bruce is that he undertook to challenge publicly certain conventions which in my . . .

AG: language and . . .

WB: judgment which perfectly civilized people are entitled to maintain.

AG: And which perfectly civilized people are entitled to challenge. We are a democratic discourse.

WB: The question is whether you're civilized. I think that'd be what you're talking about.

AG: How do you determine that somebody is uncivilized and say they shouldn't talk?

WB: I think it's worthwhile—now wait a minute: You can determine that somebody is uncivilized without necessarily determining that he shouldn't talk. Can't you?

AG: Well, the problem with him was that, like, I think that the district attorneys in the state thought he was so barbaric he should not be allowed to present his evidence in public. Or I'm so barbaric I can't read "Howl" in public. God knows what would happen if we tuned in to your secret consciousness or anybody's here. Oh, why don't I read a poem? I mean . . .

WB: Sure, go ahead.

AG: I mean, a legal poem.

WB: Do you have one? (laughter)

AG: Yes, I have one: an interesting project which is a poem which I wrote on LSD.

WB: Under the influence?

AG: Under the influence of LSD. It's longish.

WB: How long?

AG: Oh, long enough to be entertaining.

WB: I mean just roughly speaking, just so I can gauge it.

AG: One, two, three pages? It's a long text but it's a good text and it's a piece of interesting evidence . . .

WB: Can I, can I . . .

AG: You can interrupt it.

WB: Yeah, fine.

AG: We can all interrupt and discuss the text.

WB: Fair enough, fair enough.

AG: "Wales Visitation"

WB: W-h-?

AG: W-a-l-e-s, written in Wales, high on LSD. Fifth hour of LSD for those who are specific technologists in this. (*reads:*)

White fog lifting & falling on the mountain-brow
        Trees moving in rivers of wind
                               The clouds arise
  as on a wave, gigantic eddy lifting mist
        above teeming ferns exquisitely swayed
                            along a green crag
        glimpsed through mullioned glass in valley raine—

WB: Nice opening. (*laughter, clapping*)

AG: (*continues*)

Bardic, O Self, Visitation, tell naught
  but what seen by one man in a vale in Albion,
        of the folk, whose physical sciences end in Ecology,
              the wisdom of earthly relations,
        of mouths & eyes interknit ten centuries visible
          orchards of mind language manifest human,
  of the satanic thistle that raises its horned symmetry
        flowering above sister grass-daisies' small pink
                  bloomlets angelic as lightbulbs—

Remember 160 miles from London's symmetrical thorned tower
        & network of TV pictures flashing bearded your Self
  the lambs on the tree nook hillsides this day bleating
  heard in Blake's old ear, & the silent thought of Wordsworth in eld
                        Stillness
  clouds passing through skeleton arches of Tintern Abbey—
        Bard Nameless as the Vast, babble to vastness!

All the valley quivered, one extended motion, wind
              undulating on mossy hills
  a giant wash that sank white fog delicately down red runnels
              on the mountainside
  whose leaf-branch tendrils moved asway
             in granitic undertow down—

and lifted the floating Nebulous upward, and lifted the arms of the trees
    and lifted the grasses an instant in balance
       and lifted the lambs to hold still
   and lifted the green of the hill, in one solemn wave

A solid mass of Heaven, mist-infused, ebbs through the vale,
  a wavelet of Immensity, lapping gigantic through Llanthony Valley,
the length of all England, valley upon valley under Heaven's ocean
                 tonned with cloud-hang,
Roar of the mountain wind slow, sigh of the body,
    One Being on the mountainside stirring gently
        Exquisite scales trembling everywhere in balance,
one motion on the sky's cloudy floor shifting through a million
         footed daisies,
one Majesty the motion that stirred wet grass quivering
    to the farthest tendril of white fog poured down
       through shivering flowers on the mountain's head—

No imperfection in the budded mountain,
    Valleys breathe, heaven and earth moving together,
  daisies push inches of yellow air, vegetables tremble,
           green atoms shiver in grassy mandalas
    bright farmland,
        horses dance in the warm rain,
Out, out on the hillside, into the ocean sound, into Delicate gusts of
         wet air,
Fall on the ground, O great Wetness, O Mother, No harm on thy body!
Stare close, no imperfection in the grass,
     each flower Buddha-eye, repeating the story,
       the myriad-formed soul—
Kneel before the foxglove, raising green buds mauve bells drooped
    doubled down the stem trembling antennae,
  look in the eyes of the branded lambs that stare
    breathing stockstill under dripping hawthorn—
I lay down mixing my beard with the wet hair of the mountainside,
    smelling the brown vagina-moist ground, harmless,
      tasting the violet thistle-hair, sweetness
One being so balanced, so vast, that its softest breath

moves every floweret in the stillness on the valley floor,
lifts trees on their roots, grown through breast and neck,

A great O, to earth heart
        Calling our Presence together
  The great secret is no secret
    Senses fit the winds,
      Visible is visible,
  rain-mist curtains wave through the bearded vale,
Crosslegged on a rock in dusk rain, mind moveless,
 breath trembles in white daisies by the roadside,
     Heaven breath and my own symmetric
  Airs wavering through antlered green fern
drawn in my navel, same breath as breathes through Capel-Y-Ffn,
    Sounds of Aleph and Aum
      through forests of gristle,
  my skull and Lord Hereford's Knob equal,
      All Albion one.

WB: I kind of like that. *(clapping)*

AG: I'm amazed to be able to get through a poem of such length on tele-
vision! I wonder, were you able to follow it at all?

Audience: No.

AG: I wonder what the television audience would make of that?

WB: Well, a lot of modern poetry one doesn't really follow in any narra-
tive sense, does one?

AG: Well, the public could follow my hand gestures and my eye-face
gestures. You were very responsive, that was nice.

WB: Yeah. Well, I enjoyed it, I think it's very beautiful. Now, are you
suggesting that because you took that while under LSD you were able to
do it? Or, or . . .

AG: Well, I think it could be. It's a natural thing. I cited Blake and
Wordsworth as having that natural vision. I think the LSD thing is not
an unnatural vision as evidenced by the particular details like "sheep

speckle the moutainside, revolving their jaws with empty eyes:" Things we recognize . . .

WB: Yeah.

AG: but are human things that, you know, like everybody has seen in nature. I think the LSD clarified my mind and left it open to get that sense of giant, vast consciousness.

WB: Yeah.

AG: Specifically, the one thing I noticed which I was trying to describe was that the ocean of heaven—the atmosphere is like an ocean—and at the bottom of that ocean of air there are all sorts of rivulets—the winds, breezes—and that those rivulets, like at the bottom of a fish tank, move all the trees, move the people around, the hair, the beards, the grass, the lambs' wool, hung rain-beaded in the grass. So that it's like one system, like one giant being breathing—one giant beating being that we are a part of and which we always forget that we are together the God of the universe.

WB: You say "together" but your poems sometimes are saying that it's not really together—it's you as individual . . .

AG: Yeah.

WB: who is God, or anyone who "discovers his consciousness."

AG: I would say, going back to that first question you asked like, "are the hippies the New Order," I think the primary hippie, and beatnik originally, back '56—Kerouac, Burroughs—perception was a recognition of that unity of being and a recognition of that great consciousness which we all were identical with, see? And that's what I think is the meaning of that "flower power"—you know, like "make love not war" ultimately is, like, it's grounded in an understanding of the nature of the universe in that you can't fight *us*, we can't fight *us*. That force finally is, you know, like—at least at this level of civilization—is just like a mess and is not going to resolve the problem, and the problem is see-ing our unity, particularly black and white, particularly square and hippie, particularly police and student, particularly Birchite and faggot individualist. See?

WB: Well, you know what in my judgment is unsatisfactory about this analysis, I really don't think—incidentally it is not so much analytical as it is poetical . . .

AG: Oh, poesy is the oldest form, . . .

WB: Yeah, yeah.

AG: analysis is a later form.

WB: The trouble is that it has philosophical pretensions and that's when it becomes a pity. My own notion in reading most hippie literature is that it succeeds in making love utterly rancid because . . .

AG: [inaudible] the advertisements on television, which are about rancid . . .

WB: Now, now, now, look: I've had some butter, rancid butter too, but come on. As Norman Mailer says, allow people their metaphors.

AG: But Mailer's version of love . . .

WB: Now, now, we're not talking about Mailer. At this moment we're talking about what I'm trying to say, and . . .

AG: Okay, I've talked long enough.

WB: The point is that the hippies, who have, as you know, been compared to the Adamites[21] of the second century . . .

AG: Or the Diggers, the English Digger[22] group.

WB: Yeah, and those were . . . Precisely by their failure to understand the limitations of love really don't give you any idea of the potential of love or even the efficacy of love. If love is just something, a mood, whether it's just a mood that you crank on—preferably under some narcotic impulse—then as I understand it you lose any true understanding of human beings, of the ambiguity of the human temperament.

---

[21]Essentially nudists; ancient and modern sects based themselves on Adam's nakedness.

[22]In mid-seventeenth-century England, during the English Civil War, a section of Levelers, an early form of communists, adapted their communal principles to the land and began to dig and plant the public parks, hence diggers.

AG: Don't worry: *that* would certainly be a loss of understanding. But I don't think that, at best, was what was intended, or what comes through actually at best in Haight Ashbury[23] is that at all. I think . . . first of all, love is not what was ultimately proposed, it was a widening of awareness . . .

WB: Yeah.

AG: from the feeling that we were separate, alienated, isolated individuals to be regimented and made into advertising workers or soldiers going off with short Prussian haircuts to napalm in Vietnam.

WB: If that's all it was, after all, the fraternity of Christianity is *exactly* about *that*, to say that we are, we are a community. *(Bell announcing station break.)* Excuse me.

(Station Break)

WB: I mean, in what sense have the hippies or the Beats said something that isn't, after all, the essence of the whole Christian idea—that we're all related to each other?

AG: In no sense have they said anything new, except that they've simply brought forward and out front, both like politically and sociologically and consciousness and avant-garde artwise, the old gnostic tradition which had been somewhat suppressed by the Whore of Babylon, that is to say, the organized, rigidified, militarily crusading Church.

WB: As should have been, too.

AG: Well, after all, it was Cardinal Spellman here who was one of the key figures in starting the Vietnam War, it was his backing of Diem— you know, he was one of the people who got together, I believe it was with Nixon and . . .

WB: Do you really believe that myth?

AG: Oh, I think the pope had to rebuke him for being so bloodthirsty toward the end of his life one Christmas, didn't he?

---

[23]Intersection of Haight and Ashbury Streets in San Francisco and the name given to the immediate neighborhood; the center of hippie activity in San Francisco in the 1960s.

WB: Uh, no, no.

AG: The pope said, "Don't scream for victory, there is no victory in the Sacred Heart."

WB: Don't tell us that you got some visitation under drugs and suggest that it's true.

AG: Spellman . . . uh, President Diem . . . CIA man . . .

WB: Spellman said that he was praying for the victory of our forces in Vietnam.

AG: And also saying that we had to have victory, too, rather than like reconciliation with the Vietcong.

WB: No, no, he was always in favor of . . .

AG: But also like he was one of the people that was involved originally with Nixon and (John Foster) Dulles and probably Leo Cherne in proposing Diem as our man in Vietnam to substitute for the election. And you know what proposing the man . . .

WB: I don't know where you . . .

AG: for the secret police?

WB: I don't know where you get your demonology. The last person I talked to, and that was the man from . . .

AG: I get my demonology from the *New York Times*.

WB: Well, maybe you should . . .

AG: Also, [I get my information] from his speeches. Spellman was one of the people who was calling for a rollback from censorship, too. Like he didn't like *Tropic Of Cancer*—he made a big speech in Denver denouncing it.

WB: Well what do you want to do—talk about Spellman?

AG: We're talking about the Whore of Babylon—the Christian tradition—and I was saying we have brought forth the gnostic open tradition instead of the old bringdown.

WB: Do you want to say Spellman was short and ugly while you're at it?

AG: No, no, no, no: That his ideas were negative and unsympathetic in terms of the growth of spirit when he was supposed to be representing the Sacred Heart.

WB: I disagree with you very strongly as witness the fact that when Cardinal Spellman died hundreds of thousands of people who knew him best mourned the passing of a great benevolence. Now . . .

AG: Om raksha, raksha, Hum, Hum, Hum, Phat Svaha!

WB: I don't know that language.

AG: Well, that's Tibetan. It's like an exorcism.

WB: Alright. Now. Well, one of the reasons why I'm rather glad that you made this salient is because it precisely dramatizes my point . . .

AG: Yeah.

WB: that your kind of love, for instance, is no use at all against the Vietcong, whereas my kind of love is.

AG: You mean napalm love?

WB: That's right, that's right.

AG: Napalm.

WB: In other words, the use of force.

AG: Actually, why don't you try some LSD love on them? Why don't you try some LSD love on them, it'd be less chromosome damage than napalm.

WB: This, this is what—a concrete suggestion?

AG: Yes. Maybe, if you have any power.

WB: I'm all in favor of chemical warfare.

AG: With LSD though, I'm saying.

WB: The . . .

AG: That would be the only kind of chemical warfare [I'm in favor of.?] Well, you'd have to drop it on our side too.

WB: The point is you can't effectively love anybody unless you recognize . . .

AG: Have you turned on to anything?

WB: No.

AG: Do you drink?

WB: Sure. The point is, you can't effectively love anybody unless you are prepared to recognize the fact of sinfulness.

AG: (*Starts to make a humming, Om-like sound, leading into*) Oh, that's [okay?].

WB: Any attempt to implementize "flower-power" is just going to lead you right to Buchenwald.

AG: Well . . . no, I think it was an original sin shot that led to Buchenwald.

WB: I grant, I grant. I grant. But . . .

AG: Are you a Catholic?

WB: Yes, I am. Are you saying . . .

AG: I believe in the Sacred Heart but I'm not so sure about the original sin thing you want to push on everybody. It's so much of a guilt producer, or as it's been used: you know, "there's original sin, therefore we've got to use napalm," *come on!* I mean, it's a large jump you're making!

WB: Would you, for instance—let me ask you this: suppose you had the following two exclusive alternatives by, let's say, having run temporarily out of LSD—to drop napalm on a Gestapo unit that was on its way to Buchenwald to, let's say, kill 100,000 Jews or not use it?

AG: I would say . . .

WB: And that's . . .

AG: bring it right up to the present, buy off Ho Chi Minh and Mao Tse Tung.

WB: I said if you had those exclusive alternatives.

AG: Yeah, the goose is out of the bottle. You're putting a goose into a bottle with a narrow neck and saying how do you get the goose from the bottle. The goose is out of the bottle, sir.[24] To get the goose out of the bottle in Vietnam buy off Ho Chi Minh, buy off Mao Tse-tung. You've got lots of money, spend it. Use it for something sensible instead of like trying to escalate the fighting and conflict. It's interesting that by using conflict, you way escalate conflict.

WB: Would that be a good thing if we bought him?

AG: It would be better than what we are doing.

WB: OK. It would be a good thing. Therefore somebody has to produce money, right?

AG: Well, we've got 70 billion a year that's floating around if you want to buy them.

WB: But you're against the system that made that surplus available.

AG: No, no, no.

WB: You've counseled your disciples, "don't work for dollars, don't work for money."

AG: Don't work for money if the money . . .

WB: Yeah. How are we going to have money with which to buy off people that you want to buy off unless somebody gets the money?

AG: Oh, the same way we get the money to buy napalm, man. We've got that money there. We could buy off the production.

WB: But you're against production, aren't you?

AG: Pardon me?

WB: You're against production, aren't you?

AG: No, I'm in favor of production as long as the production is dedicated to something constructive instead of just destructive: more bombs, more atom bombs, more chemical warfare, more negative. It's obvious!

---

[24]See page 487, where in the John Lofton interview Allen explains the famous Zen koan of how to get a goose that is inside a bottle out without breaking the bottle or injuring the goose.

WB: The only obvious thing . . .

AG: We've got 800 million people to face. We do have a race war despite what Professor Kahn[25] thinks. In the sense of our motivation, it seems to me, our motivation is fear of the yellow life form virus taking over the planet. I don't think we're going to make it with the Chinese unless we display a certain amount of acceptance of their existence on the planet and helpfulness and cooperation. The cause of their paranoia is our paranoia. It's a nerve reflecting system, the two images are just reflecting each other, there's paranoia building up—like in a barroom brawl—the more we think in terms . . .

WB: If you keep this up I'm going to ask you to read more poetry. (laughs)

AG: Why don't I sing instead!

WB: It's just that in politics you are a little bit naive.

AG: I come from an old tradition, sir. Do you know the Hare Krishna chant?

WB: No, go ahead.

AG: For the preservation of the universe instead of its destruction, Krishna returns, in the Bhagavad Gita, every time there's fire, original sin leading to atom bombs.

(Sings, accompanying himself on the harmonium:)

Hare Krishna, Hare Krishna
Krishna Krishna, Hare Hare
Hare Rama, Hare Rama
Rama Rama, Hare Hare

(Ginsberg repeats the verses several times.)

WB: That was the most unhurried Krishna I ever heard.

---

[25]Herman Kahn (1922–1983) was an American strategist, futurologist, and physicist known for controversial studies of nuclear warfare. Author of Thinking about the Unthinkable and The Emerging Japanese Superstate.

AG: Well, now, if you armed 700 or 500 thousand Americans with lit-
tle, tiny, fifteen dollar harmoniums and sent them in after the Viet-
cong you'd probably get better results, well, as a matter of practical,
unnaive politics.

WB: You mean by contrast with the present result?

AG: Well, by contrast with the present bringdown fear, bad feeling,
negative commie paranoia.

WB: Whose paranoia?

AG: This is the world's paranoia, yours, mine, certainly I think (Lyn-
don) Johnson and (Dean) Rusk's. Certainly Dulles's paranoia.

WB: You say we're becoming a police state no different from East
Europe. Yet East Europe has kicked you out, but we haven't kicked
you out.

AG: Yeah. Well, I've had a taste of both, you know. I've been kicked out
of, say, Cuba . . .

WB: Yeah.

AG: and Prague and like there have been attempts to set me up for
arrests by the New York Narcotics Bureau for my political speech
against the marijuana illegalization laws.

WB: How do you know?

AG: Well, a friend of mine, Huncke, who's an old-time junkie from way
back and a very interesting literary figure—he appears in Kerouac's
books occasionally—was busted in New York a couple of years ago, and
the police threatened to throw the book at him unless he brought some
pot to my apartment so that they could, under the Rockefeller no-
knock law, bust into my apartment and bust me. Huncke called my
brother who is a lawyer and I went and complained to my representa-
tive, Frobstein, Congressman Frobstein, in Washington, Congressman
Joelson from Paterson, and we went to Senator Kennedy's office and
got a story in the *Times* about it and got a story in the *Washington Post*
finally. I had enough, you know, like energy to run around to protect
myself by making enough stink about it. The same—the detective

involved in that was a Detective Imp in New York who was later busted for pushing junk. How's that!

WB: So there's hope for everybody? (laughter)

AG: Well, no, it's hopeless unless the public realizes that, for instance, in this area of narcotics—drugs, marijuana—the police have been acting illegally all along and the police have been running a police state in that area which is causing immense social damage.

WB: You mean, acting illegally by . . . by what? By not running people in who in fact are violating the law?

AG: Well, the police themselves have been peddling a lot of dope in New York, particularly. There was a little series in the *New York Times* dealing with that from . . .

WB: At the instructions of Cardinal Spellman? (laughter)

AG: Well, now, I don't know what Cardinal Spellman was . . . No, I presume Spellman . . .

WB: [inaudible]

AG: I think probably Spellman was like very much morally against people smoking pot lest they do damage to themselves socially, and so he would prefer that they go to jail.

WB: I do think that . . .

AG: You know how many people are in jail for pot now?

WB: you ought to pay a little attention to what it is that you say, it's rather reckless. You have these sort of categorical definitions . . .

AG: You talking about the police? You talking about the Narcotics Bureau in New York?

WB: You can find police who are crooks, you can find popes who are crooks.

AG: The whole Narcotics Bureau . . . (After rummaging in a bag, searching for newspaper clippings, he reads headline and article:) "Shake-up in Police Shifts 3 Top Aides in Narcotics Unit." February 17, 1968: "The sudden removal of *Chief* Inspector Bluth and his two top assistants stunned many police officials. Officers at the headquarters said it was the most dramatic shakeup they could remember in the Bureau."

WB: So what?

AG: "Last December 13, three detectives of the Bureau of Special Investigating Unit were indicted by a federal grand jury on charges of selling narcotics to peddlers. A trial date has not been set."

WB: Now, wait a minute, wait a minute.

AG: "Two city detectives . . .

WB: No, no! Wait a minute, wait a minute!

AG: one of whom was indicted this week by a federal grand jury on charges of selling drugs, were charged in a civil law suit yesterday with stealing $2,783 from a suspect during a heroin raid. Two captains,

WB: Wait a minute!

AG: [Inaudible] and the two captains . . .

WB: Mr. Ginsberg. Hey!

AG: were transferred out of the police department's . . .

WB: Hey! Hey! Hey! Wait a minute. (*Buckley puts his clipboard over news-paper clippings to prevent Ginsberg from continuing to read.*)

AG: Narcotics Bureau for pushing." It's all *New York Times*, the last three months.

WB: Yes, well . . .

AG: Well, anyway, they tried to set me up. That's why I'm trying to get the word out before I get busted.

WB: I hope everybody recognizes what has just happened, you see. We start off by reading a gaudy headline about somebody very high up . . .

AG: [inaudible]

WB: a chief inspector, get *that*, "*chief* inspector," you see! He gets fired and then you get down to the end of the story and you find that three detectives have been caught selling drugs.

AG: No, no.

WB: So he invites you to assume that the chief inspector was very likely implicated in the selling of this stuff whereas a much more reasonable implication is that he was precisely fired because he didn't catch these people sooner, you know, so therefore, your attempt simply to reason that the whole fraternity is involved in this . . .

AG: *(reading:)* "As the department continues to investigate charges that the bureau is riddled with corruption . . ." *(Ginsberg continues to read, at which audience laughs, fading out over a station break.)*

FERNANDA PIVANO
November 22, 1968,   New York City

Published in Italian as the preface to an anthology of poems from
Allen Ginsberg's *Reality Sandwiches* and *Planet News*, titled *Mantra Del
Re Di Maggio* (*Mantras of the King of May*) (1973, Arnoldo Mondadori)

Fernanda Pivano was among the first of Allen Ginsberg's transla-
tors. Allen would spend many hours sitting with "Nanda" answering
translation questions and later writing out notes for her. As Allen
considered these notes to be the basis for all subsequent transla-
tions, the Pivano notes would be sent all over the world. Pivano and
her husband, Ettore Sottsass, were personal friends to many artists
of the avant-garde, and Pivano accompanied Ginsberg on one of his
visits to Ezra Pound. Pivano has played an incomparable role in
contemporary Italian literary culture, having introduced Heming-
way, Fitzgerald, Faulkner, and Gertrude Stein to the Italian public.
After the war, Pivano introduced Kerouac, Ginsberg, Burroughs,
and Corso. A popular novelist and essayist, Pivano wrote in 1999
that "interviewing Allen was an honor and a joy. It made me learn
endless things. He was very generous while speaking to me. What he
was saying was more than any school lesson: it was a communication
between faithful friends."

—DC

Fernanda Pivano: I always remember our first meeting in Paris, when
you told me a beautiful story, but I was so excited that I hardly under-
stood what you were saying. Are you willing to say that again so that I
can write it down?

Allen Ginsberg: Tell me exactly what it was about?

FP: You were making a portrait of what the poetry was when you
started, as far as prosody was concerned.

AG: It was related to composition?

FP: We were speaking of what poetry was, because I asked you: "Tell me what is it that you did to prosody" and that's what you answered.

AG: The material that should be covered is the fact that Ezra Pound in the early part of the century, with William Carlos Williams and Marianne Moore, followed the path given by Whitman in America in breaking up the old rhythmic form. The old rhythmic form was a count of accents, basically four or five to the line. By accents I mean stress, or "hów mŭch stréss yŏu pút ŭpón ă sýllăbĺe"—old form meant regular stress.

FP: What's the difference between accent and stress?

AG: Same thing here. The accent is stress. Accentual prosody, is what they call it, a count of stress. Stress, meaning the emphasis given to a syllable; the line is composed of a regular number of stressed syllables varied with unstressed syllables. Usually iambic. A common form was trochaic, which was stressed-unstress, stressed-unstress, stressed-unstress.[26] An example of that would be Longfellow's "Téll mĕ nót, ĭn móurnfŭl númbĕrs, lífe ĭs bút ăn émpty dréam! / Fór thĕ sóul ĭs déad thăt slúmbĕrs / Ánd thíngs ăre nót whát thĕy séem." That's the nineteenth-century classical poetry taught in high schools all through America. And because it was so metronomic and so regular people stopped paying attention to what was being said and just got hypnotized by the rhythm, and pretty soon everything that was being said was the same.

FP: I will find an example with an Italian poet . . .

AG: Well, the rhythm probably may be different in Italy.

FP: No, but we have the same. I will find a good example and I will put it there so that everybody can understand.

AG: Same time, same place, in history there is a universal revolution. In Russia (Vladimir) Mayakovsky broke up all verse forms, all the prosody forms. I think in Italy probably there were people like (Giuseppi) Ungaretti and (Filippo Tommaso) Marinetti: the Futurists[27] broke into

---

26 ˊˇˊˇˊˇˊˇ

27 F. T. Marinetti wrote the first Futurist manifesto in Italy in 1909. Up through 1930 many manifestos appeared in France, Germany, and Russia, often merging into

prose-poetry forms or shouted manifesto forms. In France there was Apollinaire and International Dada,[28] following Rimbaud and Laforgue, who returned poetry to rhythms which were much more varied. The key to early twentieth-century prosody change is that wherever there was this breakthrough beyond old form, the inspiration was actual speech, the rhythm of actual speech. People began realizing that the rhythm of actual speech didn't fit into the simplified automatic metronomic rhythmic series that had been used in the nineteenth century. But in the nineteenth century it was assumed that this was the eternal order of poetry; although, at least in English, it was an order that had been used for only a hundred years or so. So there was the same problem everywhere. The basic problem was that "traditional" rhythmic form, old prosodaic rhythmic form, no longer reflected the emotional variation (the emotions and the speed variations) of speech.

To resolve this problem Pound went back into history and researched more ancient forms of prosody. Like he tried Anglo-Saxon measure wherein the line is divided into two stresses on one side of the caesura and two or three stresses on the other side and no count of unstressed syllables. It was all stress and alliteration. If you look at Pound's early [translation of the Anglo-Saxon] "Seafarer," it's measured that way.

> May I, for my own self, song's truth reckon,
> Journey's jargon, how I in harsh days
> Hardship endured oft.
> . . . .
> He hath not heart for harping, nor in ring-having
> Nor winsomeness to wife, nor world's delight
> Nor any whit else save the wave's slash.

It's a musical thing.

---

other movements such as Cubism and Dada, intended as a rebellion against the whole tradition of the nineteenth century and an attempt to express the dynamic modern life of the twentieth century. To some extent Apollinaire experimented with the idea, and in Russia Khlebnikov and Mayakovsky were the best-known Futurists.
[28]A movement centered primarily in Zurich, Paris, New York, Berlin, Hanover, and Köln. Formed in part out of disgust for the war and bourgeois values, Dada championed the ridiculous and antiaesthetic nonsense. Officially named in Zurich in 1916, Dada lasted only into the mid-1920s. Artists included Man Ray, Morton Schamberg, Francis Picabia, and Richard Hülsenbeck. Tristan Tzara wrote the Dada manifestos.

FP: Just give me the accents without words.

AG: I can't because it's all positive alliteration. Also Pound went through Provencal musicians, the troubadours, to find out how they made such complex varied stanza forms. Minstrels hung their prosodaic structures on music they were putting words to, so that the music provided the structure around which the line is organized, instead of nineteenth century's automatic count of accents. Then he also finally concluded that in the future American prosody would tend, generally speaking, towards some sort of approximation of Latin quantitative, a count of the lengths of vowels: "Pay attention to the tone leading of vowels." What was being said was important, not the *automatic* count of the stress. If you wanted to find the structure of a line and analyze a line down to its elements, instead of analyzing it by stress, you can analyze it (hear it) according to length of the vowels to find out accurate emotional emphasis. And, as an example, you analyze by the length of vowels: ánalyze by the léngth of the vówels. In that way, you find you have:

ánalyze—by the léngth—of the vówels.

There are just three long vowels there in English. So you find a kind of otherwise unanalyzable secret music and regularity in Pound's lines—line after line, they don't look like "regular" poems, regular old-fashioned measured lines of the page. But they do have an equivalence, the same length line by line, if you count the vowels, long vowels and short vowels. Now in Latin and in Greek the vowels had definite assigned length, in English they don't, and in modern Italian they don't, so it became a question of judging them by ear according to the finesse of the poet's ear, i.e.: Williams told me Pound had a "mystical ear." Pound's ear was "mystically" sensitive to musical qualities and the length of vowels, musical qualities immanent in the length of vowels. So if you examine any of Pound's *Cantos* you'll find particular attention, or emphasis placed, to what he called "the tone leading of vowels." One vowel leading into another. As with *Canto XLV* the lines "with usura the line grows thick / with usura is no clear demarcation." They are roughly equivalent. "line grows thick" is three solid vowels, somewhat equivalent to its answering line, "no clear demarcation." The older accentual measure would have been "wíth ŭsúră thĕ líne

grows thick" (⌣⌣ˊ/⌣⌣ˊ/⌣⌣) and "with usúra is nŏ cléar dĕmárcátiŏn"
(⌣⌣ˊ⌣/⌣⌣ˊ⌣/⌣ˊ⌣).

FP: Yes, he makes a dactyl with four syllables instead of three.

AG: Yes, but Pound was not counting by dactyls, or that is to say by
accents; a dactyl, as iamb, is measured by accents; what he was count-
ing or hearing was the length of the vowels; his poems were composed
by a more thick musical sense than the thinner staccato count of
stress.

    Also that ear led to poets paying more attention to the content of
what was being said, because they had to be conscious of how they will
pronounce phrases, how the line was to be pronounced, and what
emphasis would be placed on what vowels. And therefore they began
thinking syllable by syllable—as speech; so the vowel-length preoccu-
pation led back to awareness of poetry as speech, with speech as
breath from the body, instead of something to be read and counted
automatically by the repetitive stress of vowels in iambic or dactylic
patterns.

    So, put in oversimplified form, speech-prosody was Pound's con-
cern. His main research was into all the different classical forms like
Greek and Roman prosody, Provençal, German, Anglo-Saxon, Chi-
nese. If you look up Pound's translations you see that he specifically
translated poets that were involved with this problem of transition
from classical to demotic speech; or specimens that add information
to English tradition on how you can arrange language in musical
units without depending on a symmetric count of accent only.
Because we had come to depend completely on an even kind of
accent, it became a sort of automatic, mechanic, robotic sing-song
formula. Important to remember that if the accent becomes auto-
matical mechanical the rhythm becomes automatical mechanical,
and the emotions therefore are automatical and mechanical and the
intellect is automatical mechanical, because you just fill in a form
which is made up in advance, like a bureaucratic form. With Pound's
new method, it meant the poet actually had to pay attention to what
was being said. Each rhythm had to rise out of a real emotion and be
a living articulation of feeling, because it wasn't repeating somebody
else's old emotion-rhythm-count.

    Now W. C. Williams solved the problem in another way. He lis-

tened, with his ear, to the language that was spoken around him in the
street, to hear exactly what little rhythms occurred in actual speech.
He didn't count vowels, he didn't count stress, he counted, if any-
thing, "breath stop," where people began and ended a little rhythm
while talking. So now:

"he counted, if anything, then,
                    breath stop,
        where people began and ended the run of speaking."

Breath stop is where you stop the phrase to breathe again. Stop to
think and breathe. And so he wrote what he called a "relative mea-
sure." By "relative" he meant that the line may be shorter or longer,
but relatively equal as measured by intensity of breath and the time
taken for emphasis in speaking the line, and pause between phrasings
and other vocal matters.

I remember when visiting him in 1948 he said he had heard, on
the street, a little fragment of speech: "I'll kick yuh eye." "I'll kick yuh
eye" had been a local, rougher way for saying "I will kick you in the
eye" or "I will kick your eye." So he had written that down on a pre-
scription pad, and he asked me: "How could that be fitted into the old
iambic stress count, 'I'll kick yuh eye?'" That's too subtle a gradation
of rhythm to be measured by the relatively crude standard count of
stress that was practiced. If you wanted to compose a poem out of the
elements of natural speech heard around you, perhaps heightened
speech and exaggerated street rhythm, but nonetheless out of spoken
elements, you need a much more subtle measuring instrument than
the "traditional" count of accented syllables. Because there was a
funny kind of syncopation in the speech just as I pronounce *that*. But
you *can* "analyze" it: *because*-there-is-a-funny-kind-of-syncopation-
in-the-speech.[29] To practice this art you would actually have to attune
your ear to listen to the actual speech around you: and hear the little
rhythm of spoken sound in provincial ways of talking that people
have, in order to make your rhythmic process on the page equally

[29]because there is a
        funny kind of syncopation
                    in the speech

above marked by stress.

subtle. Just the same conclusion many modern musicians are coming to when they say that old musical notation is too crude because *all sound* is music, and how can you measure, like, electronic sounds and machinery sounds and garbage can sounds and truck sounds on the street, which are musical to their ears—how can you measure them by the notes of Western scale? So you have to adopt a different measure, or a different notation, or a different way of thinking about sounds (if you are an electronic musician) than you used in the past. So we will use together a different way of thinking about rhythm: and the main art of rethinking rhythm is practicing and *listening* to rhythms and being conscious of rhythms in *our own mouths* and other people's mouths.

And that's how traditional poetic "progress" has gone on, it's always gone on, actual speech has always refreshed the ancient traditional practice of poetry. A few hundred years ago Wordsworth began to listen to actual speech around him and then changed the traditional method of writing poetry and introduced the new words that were coming into circulation. Or the new rhythms. Like in his preface to *Lyrical Ballads*, Wordsworth pointed out that he wanted to change the *diction* of his poetry and use words taken from common diction, everyday-life speech, instead of time-worn poetical, conventional expressions. The evolutionary poet finds a way of including unconventional rhythms. These rhythms were in practice part of the conventional oral articulation, oral speech, but had not yet been included in so-called conventional literary composition. Now, everybody's talking all the time, the rhythmic materials of speech change from year to year, decade to decade, century to century. However we'd been stuck with rigid measure, the accentual count, the stress count, which is centuries old. It may have been practical when first used in Shakespeare's time, but iambic pentameter went through many changes through Alexander Pope, through Wordsworth, and became looser and looser until by the end of the nineteenth century that count was a mechanical thing; and ever since, children in schools have been taught this mechanical count, as if that's all poetry was.

---

because there *iiii*s a
     *fuuuu*nny kind of syncop*aaaa*tion
          in the sp*eeee*ch.

---

above marked by vowel emphasis-length.

FP: How about Whitman?

AG: Oh, Whitman opened up the line entirely, like he started thinking in kitchen speech terms. He began going back to an older oral bardic tradition as in the Bible, to biblical prose; biblical prosodaic arrangements, including prose-poetry sentences, balanced phrases, "The Lord is my shepherd, I shall not want." Here a syntactical balance is used to give a measure to the line rather than an academic accent count: "I celebrate myself, and sing myself, and what I shall assume you shall assume." Whitman's line is very similar to biblical sentences in psalms. Whitman actually went to the psalms for a speech tradition, a tradition of speech *out loud* instead of speech to be read by the eye silent on the page.

FP: So in a way it is less refined and cultivated than Pound, less intellectual somehow.

AG: Yes, because he began it, Whitman broke the ground, opened the soil up for first cultivation. Whitman *also* made use of diction from everyday speech, and some rhythms of everyday speech. His own speech probably was a bit oratorical, he was conscious of being Whitman prophet, just as Pound was a little bit too conscious of being Pound aesthete. But on the other hand it was a very healthy voice that Whitman had, a healthy and very ancient voice that he had, to speak in. So it fitted what he had to say. The work of the later writers like Williams and Marianne Moore was to find systems to measure this new healthy speech and to refine and update it, and to do historical research to see why it happened in English-U.S. tradition that most poets finally got frozen into the ice of stress count.

An image Williams used is that when any form has been broken, such as the atom, when any old, ancient form is split, a lot of energy is released; and this is what happened when the old prosodaic form was broken. All these practical emotions that had been piling up for an industrial century suddenly got released into the poetry following Pound and Williams, all the perceptions of pollution and household emotions that never could fit in exactly before. They weren't considered "poetic" feelings, because they couldn't fit into old sing-song rhythms the nineteenth century used. Suddenly forms opened up and all their new energy flooded out.

Marianne Moore counted syllables, not vowels or accents: what she was going to say was her own way of saying it anyway, and she listened to a factual speech, very very factual.

FP: What's the difference between "factual" and "everyday" speech?

AG: Well, she is a librarian and she is listening to, like, business and industrial accounting talk about tax deduction statistics, baseball commentators talk about batting averages, zoologists talking about twitchings of elephants' ears. She's like a somewhat toneless lady librarian, very pragmatic factual Yankee, shrewd, or, in a sense, not too emotional, but very clear, very basic, very down to earth. Maybe she said to herself that it doesn't make any difference how I divide the line, I can divide each line by a different exact count of syllables and invent stanza forms according to arbitrary patterns suggested by the ear; with each stanza form in one poem identical, it would be like a lacework pattern, or snowflakes, each poem different. And the interesting thing would be the movement of the actual speech against the rigidity of the syllabic count of the line. So she got a very funny kind of music sound that way: making rigid patterns, counting syllables. To count syllables means that, I say a sentence like this: "I—say—a—sen—tence like this." That's seven syllables. "I will say a—no—ther one," it's another seven syllables. Those two seven-syllable lines, "I say a sentence like this / I will say another one," rhythmically are not the same, there are slightly different rhythms. But Moore put them together and would oppose them or balance them as being the same line because both count seven syllables. Then the music in this set would come from the difference, the opposition between the two rhythms in the same form of seven syllables. And so she attuned her ear to hearing the varying rhythms in equivalent numbers of syllables.

This would be a whole possible way for all poetry to go. Like, if everybody in high school were taught to listen and count the number of syllables in each line and then to listen to the opposition of the rhythms to those syllables, you could build the whole total universal poetry on that too. Because the rules of the game would be instead of counting accents or vowels, count syllables. In fact French prosody as built on the alexandrine, is basically built on a count of syllables. So what I am fantasizing actually has taken place, already happened. We rarely had those syllables counted in English, so that was a way of making the ear conscious of what it was hearing in American tongue. Because we had lost consciousness of what we were hearing when we were counting regular accents.

So that was what was going on at the turn of the century, that's almost sixty years ago, say. So actually experimental prosody has been the main tradition in American and English poetry for the better part of this last century. And so one may say that it is *the* "Tradition" that the younger poets in America are working on, it's the "real tradition." And the paradox is that these younger poets who were working in this tradition have been accused of being aesthetic anarchists, of *not working in any "tradition" at all*. Unfair! Ignorant accusation! And the problem was that most of the people in the academies, as Pound pointed out very early, were so backward technically, that they didn't know what was happening to prosody, and naturally it was the poets that were inventing new forms, the academy didn't catch up with them, the academy itself didn't study hard enough to find out what was happening. And most professor-critics were not prepared, the ears in the academy were not tuned to recognize what specific forms were being used. So old formalists were not refined enough to be able to recognize and judge new forms and to hear them, much less analyze them, because academic types didn't recognize anything as "formal" unless it sounded like a familiar nineteenth-century type of form—rhymed accentual quatrains.

FP: Yes, but this is what was happening in the first twenty years of the century.

AG: It was still happening in any of the high schools and grammar schools and colleges when I was going to college in 1945. In America, when I was at college, the English Department still considered that William Carlos Williams had no formal preoccupations and was some sort of embarrassing provincial, awkward, primitive . . . uncouth, naive, senseless, not "aesthetic," or high class, like T. S. Eliot.

FP: And what did they say of Ezra Pound?

AG: Nobody really understood exactly what Pound was doing, school teachers just assumed he was very great, but they didn't know why. They thought he was great, high class, because he read a lot, I guess. But his techniques and texts were not taught or known intimately.

FP: Because he was doing the thing with the old French and Chinese . . .

AG: Yes, what he did was so powerful on every level that nobody could resist awe. Well, the actual understanding of his prosody is to this day not clearly known or taught in the schools . . . simple understanding of what he was doing, in terms of prosody as vowel awareness. To this day if you ask a college professor of English here at the City College of New York he won't know, because I had a conversation on this a week ago with a professor of modern poetry at C.C.N.Y. We were talking just on the subject of Pound's quantitative vowel prosody, and the professor said that he never really tried to understand it. And when I recited the line "with usura the line grows thick" aloud he said he understood it a little better. The ear had not yet been trained from grammar school up to understand it. You see in grammar school they are still teaching everybody singsong verse.

FP: Non la quantita delle vocali.

AG: No, they are not teaching that in grammar school, they are not used to that. Because to do that you really have to pay attention, to listen physically, and pronounce lines physically, as real talk. It's like doing a waltz, like everybody can fall into a waltz, easily as if they had heard it all the time. It's almost as if what ancient musicians had written was nothing but waltz time, and nobody could recognize any other music except waltz time, and they had no way of notating anything but waltz time.

FP: And so what happened was that you started writing poetry?

AG: No, what happened was that simultaneously around the early forties Williams' particular physical yoga began being more commonly picked up all around the country. By physical yoga I mean his yoking his ear to the actual sounds of conversation heard around him. Yoga means yoke, actually, in Sanskrit: to yoke the mind. His yoga was listening to the actual physical sounds as usable poetic speech rhythms, then writing them, notating them in very simple ways like his little poem to the postman: "Why dontcha bring me a letter with some money in it: 'Atta boy! Atta boy!'" It's a whole new modern emotion. Two neighbors in a small town joking with each other in their actual language, which had never been an emotion admissible into U.S.

poetry before, because there was no rhythm set-up to put their words into. So Charles Olson and Williams were writing poems like that . . . like his wife sitting on a toilet, he said and wrote: "I kissed her while she pissed." You know that poem?

FP: I don't remember it.

AG: It has got to be an old man talking to himself, it's his sixty-eighth birthday or something like that, or fifty-eighth birthday, and his wife is taking a piss on the toilet, he's shaving, and she said to him that it's his fifty-eighth birthday, and he touches her breast and she didn't even push him away. "I kissed her while she pissed" is the refrain in the poem, "I kissed her while she pissed." Both rhythm and diction in a line like that is something so intimate, personal and homely, that it's archetypal, but at the same time never before represented in poetry. "It's your birthday!"

So around 1945 almost everybody among the younger poets in the avant-garde began picking up on their own intimate reality, and by the end of 1950 they had recognized that Williams had been doing that all along. And so I went, like Robert Creeley, to see Williams. I had been writing to Williams in the late forties and began going to see him.

FP: You and Creeley were the ones who started . . .

AG: No, it was *everybody*, 1948, independently: Gary Snyder and Philip Whalen met Williams at Reed College in Portland, and they talked about speech prosody without even knowing who we were and vice versa. There was just sort of a ripening personal new culture, a break-through at that time.

Simultaneously, Kerouac on his own, in late forties and early fifties, from his own ear and from his own preoccupation with the changes of rhythm that were going on in Bebop music, was beginning to write long prose sentences, similar to Negro breathing and Negro rhythm in bop music as exemplified by Charlie Parker's Bird-flight-noted runs of horn music. Have you heard the long complicated alto-saxophone cadenzas that roll through one chorus half way into the next on one breath? That's why he was called "Bird," because of the bird flight speed and altitude of these choruses. So Kerouac very con-sciously began imitating that, and thus Kerouac arrived at his "spon-taneous bop prosody," paying attention to the rhythm of what he was writing in his own athletic speech and to the breath-runs of it and to the lyrical quality of his own natural tongue. So when he met Neal Cassady, they had long exciting conversations. Kerouac was struck by the lyrical rhythmical quality of the Denver provincial western-twang explanations, that Neal was playing with. You know, long explanations of driving a car down the Rocky Mountains while making love going to Denver, like a long Proustian sentence for the many explanations in it and many parentheses in it. That was a pretty purely western characteristic of Neal Cassady's speech, which turned Kerouac on to listening to American speech and writing American speech, and that was precisely what Williams was interested in doing, writing in American prosody.

Robert Duncan had for many years been in correspondence with Pound and was a close student of Pound's circle—William Carlos Williams, Louis Zukofsky, H. D., and others—and Duncan himself was perhaps the most learned scholar of vowels of all—see his poem "An Owl Is an Only Bird of Poetry," wherein he notices that vowels make a continuous eternal tone shaped and interrupted by consonants.

(Robert) Creeley in the forties corresponded with Pound—on the basis of Creeley's interest in breath stop, and syllables as building blocks of lines. Creeley asked Pound for advice on how to edit a magazine, *Black Mountain Review*.

Frank O'Hara and Kenneth Koch and the "New York School" poets had also learned from Williams, mostly the practice of personal speech. O'Hara in the sixties in fact wrote a witty Manifesto of Poetics called *Personism*, saying (in sum) that poetry should be personal and anything we poets *said* was Poetry, because we were Poets in History. So our daily speech was Poetry, telephone conversations, etc.

Gregory Corso also had perceived Williams' downright homely doctor's talk as so real it was surrealistic—first time I met Gregory in 1951 he quoted W.C.W.'s line from a dead baby poem—"Sweep the baby under the bed . . ."

(Charles) Olson had been in touch with Pound from the very beginning, being an older man, and when Pound was brought back from Italy, Olson I think was one of the first persons to meet him here and he acted as his secretary for a while before Pound got into St. Elizabeth's Mental Hospital. And Olson had been in contact with Williams, and he had also been a secretary to Edward Dahlberg, who was in that prose tradition (originally of the "proletarian novel") which made prose use of Williams' natural speech idea.

Williams used to write prose too; of course he was concerned with American-style paragraphs. Hemingway through Stein—yes, the same Gertrude Stein who was preoccupied with how people actually talk and think. What are the rhythms of the little sentences we use, especially certain types of people like truck drivers, gangsters, spinsters, people of laconic means of language, who speak, you know, like, minimally? So this is a basic preoccupation in all American literature since 1905, meaning since Pound and Stein. In a rough way, this problem of natural-speech mind occupied everybody interesting in America in one way or another, once you begin thinking about it. Hemingway was also involved with the same problem. He resolved it early in a way which is very clear, easier than Gertrude Stein for people to understand. You know, those short stories. They are very simple, the presentation is straightforward.

FP: Because in a dialogue it's always clear to understand.

AG: It was more clear, though it was a different kind of dialogue than in nineteenth-century novels, because it was people saying things that seem to be so simple to each other, yet weighted with emotional significance by emphatic repetition of that simplicity.

FP: You know I didn't understand that until I heard him speaking, when he said "Do you, oh you do, yes you do." This way of working with the "to do" is incredible for us.

AG: This trick is in shifting attention to the *actual meaning* of a simple word being used in your mouth, (generally used in a robotlike, automatic way) some phrase so many people are using: "How do you do?" What's the word that Italians use for hello?

FP: Ciao, come va?

AG: Okay, so someone begins saying, *"Ciao, come va?"* intending each syllable *"Come va?"* Looking into your eye and saying *"Come va?"* until somebody realizes that *come va* was a question that God could ask. Stein's writing showed a shift to a new consciousness of language itself. And from that point of view Hemingway goes back to Gertrude Stein, a shift of consciousness of the language itself. If you take the alteration of consciousness or the widening of consciousness to include language, then you begin to get an idea of the importance of the literary change. For this you could even go back to Flaubert's *Dictionary* of Received Opinions—Dictionnaire des Idées Reçues. It's like he suddenly became conscious of the pop art quality of some of these mechanically repeated formulations of bourgeois opinion, and saw them as a form of pop art, and isolated them and presented them as "aphorisms" or apothegms in order to make people conscious of the language they were using; in this case, conscious of the complete disorientation, stupidity of the opinions, you know, the robotlike secondhand mechanicality of the words that carried the ideas. And you get that in Gertrude Stein, she would take some very banal phrases and repeat them over and over again, as a musical variation; and you have the same thing going in Eliot in certain poems, when he says: "Hurry up please, it's time" or when he uses the girls chattering in *The Wasteland* bar like:

> "When Lil's husband got demobbed, I said—
> I don't mince my words, I said to her myself,
> HURRY UP PLEASE IT'S TIME"

But Eliot never solved the verse form problem for us, because he went to England and wrote ultimately in the last plays in basically an old style of Shakespearean blank verse, written slightly adapted to (intelligent) modern speech. But he never solved the problem of how do you register American speech? So we postwar prosodists came in, then, basically on the coattails of the classicists, of Pound and Williams and Marianne Moore. Except that by our time, you see, Pound had stayed in Europe, and so he was researching in historical variations, ancient articulations of local speech. Williams stayed in our local area and Pound wrote to him: "You are interested in the raw material, and I am interested in the finished product." And Williams said he was interested in the raw material because an American poetry had yet to be invented. Marianne Moore was a very orderly mind as I said, so she stuck with this simple form, syllabic count, and she exploited it very beautifully. Others did other things. But those were the greater people, I guess, they were the most conscious of the traditional problem of vocalizable poetic forms. Other people, lesser poets like J. C. (John Crowe) Ransom, were trying to write "literature." But Pound and Williams were mostly working like scientists with basic materials.

FP: Then you worked in this direction.

AG: In this direction, I don't know if we added anything basic, because Pound's was the first great discovery of the change. The only thing I think is, *we learnt the lesson.* We were the first generation after them to learn the lesson and begin applying to *our own conditions,* our own provincial speeches, mouths of Denver and New Jersey, our own personal physiologies and personal breathing rhythms, and to our own police state science fiction postwar Buck Rogers Newspeak universal conditions of local ecstasy of god-realization.

FP: Buck Rogers, what's that?

AG: Hero of an old science-fiction radio show popular in the twenty-fifth century. I used to listen to it in Paterson in 1933.

I found for instance that my own breath, my own speech, was more like "Melvillian-Hebraic Bardic" style. Williams is more like a soft-spoken, wry-mouthed, middle-class doctor; I was more of an excitable visionary Jewish Buddhist so that my own language, my own line, was much longer than Williams'. So using Williams' basic perception, which I practiced in *Empty Mirror,* I did little spoken-speech imagist exercises, Williams-styled, short little phrases, the best of his speech. Once I got that down, straight, and could present a tiny little hard poem in personal speech form, then I began extending the line out, to fit my own breath a little better, so that's how "Howl" comes about. With a long line based on a peculiar kind of an excitement breathing that I have, or a speech alone on the Brooklyn Bridge, the kind of soulful Alone breathing that I do, which is the same in "Howl" as it is in "Kaddish" as it is in mantra chanting or singing.

(*After a detailed discussion of several of Allen's poems, Pivano and Ginsberg begin to discuss "Television Was a Baby Crawling Toward That Deathchamber":*)

AG: Re "TV Baby"'s line, "Six Billionaires that own all Time since the Gnostic Revolt in Aegypto," it's like a parody and paraphrase of fragments of Pound. Like another aspect of Pound, his surrealist use of economic facts. Time? *Time* magazine or "Time" itself. Actually the specific Pound line like this comes in *Canto XLVI* following "With usura"; remember his line "Ten empires fell on this grease spot?" Those billionaire sorcerers have been in there for long long time since

Gnostic revolt in Egypt. I mean the Capitalist Empire. I mean it's just a sort of unconscious echo of Pound, it's nothing serious.

FP: Yes, but this is exactly what I want very much to know.

AG: Well it's nothing you can prove anything with. It's more like fancy play, but it's just the same rhetorical ecstasy that Pound put in

"helandros   kai heleptolis   kai helarxe."

towards the end of that Canto, that kind of aesthetic messianic cry. So that kind of aesthetic rhapsodia can be built on political material in poetry: "Ten empires fell on this grease spot," or "Billionaires that own all Time since the Gnostic revolt in Aegypto." There are all those references in the *Cantos* to Fort Knox and the gold reserve. I guess it's just imitation of Pound's conspiracy theory poetic manner. So that the form has nothing to do with it, there's just an echo of his imagery. The other influence of Pound that accompanies this: after having visited him, I did a lot of revision of this book in terms of concision, condensation, with him and Basil Bunting in mind.

You don't know Basil Bunting. He was a friend of Pound and Yeats in Rapallo, also a great poet—still alive—who didn't write for a long time and he was not published until the last two or three years. Then suddenly Fulcrum Press London published him again, and he turned out to be one of the greatest poets alive in England. Well, this poem here I dedicated to him. Bunting was published by Stuart Montgomery in London, and his collected works is now published by Stuart's Fulcrum.

FP: The poem is "Studying the Signs."

AG: It's not a poem to translate, it's just . . . all the signs on Piccadilly Circus. A complete pop art, a sort of description of . . . It's an interesting sketch, it would be interesting as a background voice theme for a five-minute documentary movie with a camera turned around on Piccadilly.

FP: And it's not very long so we can put it here.

AG: Well, it wouldn't make any sense to anybody. It would be like saying, you know, like a description of everything in Via Veneto. Maybe someone might say, "I lived along Via Veneto for long enough, and I

would be interested in how this guy saw Via Veneto." This was particu-
larly what he noticed and thought it was funny to mention. It's really
more like descriptions in news books, a documentary film five min-
utes long, for an underground film. It's only got one poetic line, "Ah
where the cars glide slowly around Eros." You know the statue of Eros?
It's a pretty line. You know, *the Guinness, Café, Swiss Watch, London
Insurance, Underground, Boots, Revlon, Criterion Theatre, Players, Coca-
Cola, Cafe, Restaurant, Cartoons*. It's a little exercise in making a poem
out of nothing but facts.

FP: Oh, this was the idea.

AG: So, as concise as possible, and also based on the idea of complete
count of vowel: "White light's wet glaze on asphalt city floor, / the *Guin-
ness Time* house clock hangs sky misty." If you notice, except for the
trade name Cathay, every single word there is concrete: "Yellow,"
"food," "lamps," "blink," "rain," "falls." The line grows thick: "lamps
blink," "rain falls." Each syllable or vowel is . . .

FP: Is some *thing*, has a concrete value.

AG: Yeah, rather than just being fat. I cut out a lot of things. You know,
what I did here was try to make this *duro* as possible. In Bunting is pure
duro. All the Bunting is *duro*, more than Pound even. But this, with
condensation that finally put all the facts together, was all based on
"with usura the line grows thick," that one line, do you remember?

FP: Yes, yes. I think I remember.

AG: Oh, that's important. Remember in the Milan University record[30]
of Pound reading "Canto XLV," the little tiny record you had in Milan,
"usura" that I kept playing in your house? The one line in it that I kept
pointing to was "with usura the line grows thick."

FP: I remember when you were thinking of the usura thing.

AG: Yes, remember that line, I kept pointing to the solid vowel sound
of that line. I had years ago read it as "wĭth ŭsúră thĕ líne grŏws thíck,"
but Pound wanted it to be read another way. He was not saying: "with
ŭsúră thĕ líne gĕts thíck." He meant *grows* thick. "with usúra the líne

[30]Ezra Pound, *Con Usura* [Canto XLV] (Scrittori Su Nastro II, Universita Degli Studia
e Milano, EP 1V6001B, 1958).

grows thick," and he pronounced that *grows* long, like *thick*. Meaning that he intends *grows*. If it didn't mean that word to be active and *duro*, he would just say "gets" thick. If you are counting accents it would be "gĕts thíck." Small accent on "gĕts" and big accent on "thíck." But if you are counting vowels you say "with usu̱ra-the li̱ne-gro̱ws-thi̱ck." "Pay attention to the tone leading of vowels."

FP: And then you call it hard: "grows," you call it hard?

AG: Just to get you to hear it. "Hard" in the sense that "grows" has *intentional* content, intentional length of vowel, "grows" has intentionally a long vowel, to be pronounced aloud, as if it meant something. That's the key, that it has to be pronounced as if it meant something. And if an American says *gets thick*, the "gets" is not pronounced as if it means something particularly by "gets," which is only a helper word. Pound just says that "Over the centuries, with usura, the artist's line has become muddy, decade by decade it grows thick." In the museum you know, the paintings accumulating in museums with the lines growing thicker, more careless. You know what "the line grows thick" means?

FP: *Diventa spesso*.

AG: Yes, because people no longer are painting for their own pleasure, so they are not making fine delicate drawing, they are making a pasta, like the big murals—not Rembrandt, Rubens—and after ten thousand walls in Florence to fill up, five hundred feet of the Pitti Palace, with big thick lines. With usura as a motif for the painting commerce, rather than as in Angelico or primitive sacred drawing, the line grows thick. And in poetry "with usura" people no longer pay any attention to the words they are using: "with usura is no clear demarcation." He is talking about the decline of attention to the materials that the craftsman worked with, caused by money motives; these people are in a hurry to get work out as a *product*, so they are not spending enough time to make any pretty line. They're not doing the things for the thing itself, they do the thing to sell it. You know, if you can stamp a picture out on the machine, you can sell more copies and make more money.

FP: And he says *usura* for money?

AG: No, usura means the use of money to make money. Aside from the economic sociological implication of the line, the *ear* in that line sets a standard you can use: "with usura the line grows thick." It's a very good example of his use of quantitative prosody, it's the one that stuck in my mind. So you can use this as an example; if I ever tried to make a poem that didn't have *me* in it, that was all composed with hard images and concrete visual facts, in which there was no hot air, you know, I would take that line as model. Every word, not in excess, direct presentation, "no excess words." The main thing that I got from Pound is no excess words, or few excess words, because I have a tendency to be excessive. More than necessary.

FP: Who told you that?

AG: Bunting. In 1965 Morden Tower, Newcastle, after I read "Howl" and "Kaddish," three hours of poetry, Bunting said, "too many words." Very right of him. He reminded me he'd shown Pound a formula for poetry from an old German dictionary: "Dichten = Condensare": Poetry = To Condense. After I saw you I went through this long book to eliminate, I'll show you examples.

FP: I make you work too much like a dog.

AG: Oh, it's just a simple thing. "I prophesy: the pigs won't mind! I prophesy: Death will be old folks home" instead of "Death would be *the* old folks home." In English it is much better. It's more like actual speech, "Death will be old folks home." "Chango will prophesy on the National Broadcasting system," and I just made it "Chango will prophesy on National Broadcasting system." Just taking a lot of articles away. Exactly what it says over here on acknowledgments page of City Lights book, "there has been some revision for syntactical condensation."

FP: Am I going to try to do the same in Italian?

AG: As much as you can. Only where it makes sense. But I wanted to point out the revisions and changes. There is that one little sentence here that explains revisions in relation to Pound also. "Some revision for syntactical condensation towards directer presentation."[31] You know "direct presentation" is Pound's phrase? *Presentazione diretta.* Not by implications, by example. Direct means straightforward. "Presentation," as distinct from "reference."

This phrase is something that he wrote (like Kerouac wrote

---

[31]The quoted material is from the acknowledgments page of *Planet News*.

"Essentials of Modern Prose"). Pound around 1910 wrote "A Few Don'ts For Poets" in *ABC of Reading* or somewhere. In one of his early essays, he speaks of "direct presentation of the image." In my case, of the original spontaneous imagery. But there has been some revision, so this is the first time I am saying I am revising, in the acknowledgments: "These poems were printed first—in *TV Baby Poems,* among other places—(in forms slightly closer to original composition, i.e. there has been some revision for syntactical condensation toward directer presentation . . . a method similar to manicuring [cleaning] grass that is removal of seeds and twigs, ands, buts, ors, ofs, thes, alsos, that don't contribute to getting the mind high)." So that's the key.

FP: You say harder, tougher, thicker, but which is the English word for *duro?*

AG: Hard, harder material. The image "solid" is a phrase of Louis Zukofsky. Solid, matter, as distinct from gas; as in the image of ice, that when it gets too warm it goes watery, then into steam, it gets steamy, gaseous. So just make it a little bit more solid.

MICHAEL ALDRICH, EDWARD KISSAM,
AND NANCY BLECKER
November 26, 1968,   Cherry Valley, New York

"Improvised Poetics"
From the book *Composed on the Tongue* (1980, Grey Fox)

In 1966 Ginsberg formed the nonprofit Committee on Poetry (COP) to assist small and underground presses, impoverished poets, and similar projects, largely because Allen wanted to prevent the money he earned from being taxed and therefore helping to pay for the war in Vietnam. In 1968 COP bought a ninety-acre farm near Cherry Valley, located in upstate New York, both because Allen was drawn to the back-to-the-land movement popular in the late 1960s and also because Allen felt the farm could be a useful place for poets with substance problems—he had in mind both Jack Kerouac's addiction to alcohol and Peter Orlovsky's methedrine habit—to go to dry out.

Michael Aldrich had known Ginsberg since 1966 and wanted to do an interview with him on poetics, so Allen invited Aldrich along with several of his friends and Gary Snyder to the recently acquired farm over Thanksgiving. The interview took place in the living room of the farmhouse, where Allen used an old upright pump organ he had bought at a barn sale to play some of the poems from Blake's *Songs of Innocence and Experience* that he had recently begun to set to music. (See Colbert interview for an account of how Ginsberg was inspired to set Blake.)

Commentary in the original interview is in roman type; that added by the present editor is in italics.

—DC

Allen Ginsberg: The question is how to figure out where to break the line. Well, here's the problem of writings like the long airplane poem you were just looking at, "New York to San Fran."[32] Since it is written down on a page, silently, without everybody talking, the page determines the length of the line. It's an arbitrary thing—if I have a big enough notebook, it's a big long line, generally, because there's room for the hand to move freely across, and the mind to think freely in terms of, the long line. If it's a little pocket notebook that you stick in your back pocket then you tend to have smaller, choppier lines.

Michael Aldrich: So the literal page you're writing on . . .

AG: . . . very often determines the length of the line. Now poems like "New York to San Fran" generally fall into paragraphs of short lines which could be extended out on the page, when printed, to be like a long line, or a strophe, call it—for lack of a better word—the Whitmanic strophe.

But what also determines where the line breaks is: when the thought breaks. You know, if you are writing in a notebook and—give me a text—writing in a notebook: "Being filled with drum beats and total"[33]—and I didn't know what was going to come after "total," total, total what? Probably it was some sort of Hart Cranian thing like "total ascensions," but that didn't make airplane movie earphone sense—it was total "orchestra shaking Ascensions." So, "Being filled with drum beats and total" on a short page, on a little notebook page, and then indented after that, "orchestra shaking Ascensions."

Then the next line . . . ah . . . what am I referring to? Crane's "migrations that must needs void memory,/ inventions that cobblestone the heart" from "Atlantis" [section of The Bridge].[34] So we had "orchestra shaking Ascensions" and that was where the thought broke off, and I had to go back and footnote my thought, so to speak. So the next line was "Crane'd've come to Forever." Now I was interested in the music, so it was "orchestra shaking Ascensions / Crane'd've come to Forever" . . . and then I realized that wasn't really complete, so I put "if he could." (laughter) So:

---

[32]"New York to San Fran," *Airplane Dreams: Compositions from Journals* (Toronto: Anansi, 1968), p. 11.
[33]*Airplane Dreams*, p. 22.
[34]Hart Crane: *The Complete Poems & Selected Letters & Prose*, ed. by Brom Weber (New York: Doubleday-Anchor, 1966), p. 116.

> Being filled with drumbeats and total
> orchestra shaking Ascensions
> Crane'd've come to Forever
> If he could—

And that's four lines. And since the "if he could" is a subnote on "Crane'd've come to Forever," it's indented below the "Forever" as a modifier of that last line. So the little short phrases there that modify lines before, get hung out on the page a little to the right—which is just normal common sense. A little bit like diagramming a sentence, you know, the old syntactical method of making a little platform and you put the subject, verb and object on it and hang adjectives and adverbial clauses down . . .

MA: Does that affect your reading of the poem?

AG: Yeah, because actually the mind-breaks that you go through in composing are the natural speech pauses too, or are identical with natural speech pauses: after all, natural speech pauses indicate mind-breaks. This is a really important point. Though the natural speech pauses or breath stops, line stops, and end stops might not fit the way it would sound if someone were perorating, TV platform politics public style and running on, the pauses and stops *would* fit if someone were in intimate conversation . . . uh, saying *uh*, hesitating and then completing the little syntactical unit. Just like I just did.

MA: So it's a natural speech rhythm.

AG: Actually it is a kind of natural speech rhythm that comes when you are speaking slowly, interestedly, to a friend. With the kind of breaks that are hesitancies waiting for the next thought to articulate itself. So actually it was the composing—the breaks in composition with a short-line notebook that wind up functionally, as the equivalent of the way you not only think it out, but the way you might say it, if you were talking—at least that's the way I read finally. When I get up and read those units I just make believe I'm trying to think of the next phrase. And then I come out with the next phrase. And so actually that leads the listener, or reader, in an oral interpretation.

MA: Directly imitating the process of composition.

AG: Yeah, and so it leaves the mind of the reader or listener to be hovering with mine, with the next spurt. And then since the next spurt is always some kind of funny little change, there is always a constant surprise line to line. And the perhaps slight overemphasis of hesitancy on the end stop before the next spurt gives it a dramatic fillip.

MA: Now the way I phrase that is, if you are going to read a poem by its white spacing, if you're going to read a poem by the way it's actually physically arranged on the page, you pause every time there's a white space.

AG: Yeah.

MA: Is that generally the way you read poetry?

AG: Yeah. I pause at the end of a line; or else run-on purposely. So that if someone is reading the page, the text, and hears me read it aloud— and I don't pause, but I run-on, the music, so to speak, is in ignoring the possible pause and enjambing. And making a run and making like a fast rhetorical speedy swoop.

My basic measure is a unit of thought, so to speak. That's an interesting idea, say, as Williams' idea of a basic "relative measure" for the line. Corso's contribution to the whole thing was that the line was a unit of thought, so to speak, which is something I follow. And the reason it's a unit of thought is that's what you wrote down on that line. If you're writing a short line anyway. So it's not so much a unit of sound as a unit of thought. But it also turns out that if you vocalize the thought it's also a unit of sound and that somehow or other the squiggles for the units of sound are identical to the squiggles of thought. And they're just as interesting as units of sound as units of thought; if you pronounce them aloud they make a funny kind of rhythm.

MA: There's something about the difference between the "TV Baby"[35] poem on the page and, for instance, this "New York to San Fran" poem on the page.

AG: The "TV Baby" poem is like a long blast of rhetoric, with all sorts of language tricks and mind loops going on at one point or another in the

---

[35]"Television Was a Baby Crawling Toward That Deathchamber," *Planet News* (San Francisco: City Lights, 1968), pp. 15–32.

poem, in which everything gets mixed up as "a long Train of Associa-tions." But since it is like a train of associations, a train that goes chug-ging on until it hits high speed, it hits high speed toward the end.

MA: Like a raga.

AG: Sort of, yes. At the beginning, the strophes or long lines zap along heavily.

> it's a long Train of Associations stopped for gas in the desert &
>   looking for a drink of old-time H_2O—
> made up of molecules, it ends being innocent as Lafcadio afraid
>   to get up & cook his bacon—[36]

And sometimes they build up and get longer and longer. Now in reading them, they're to be read each strophe as one breath, if possi-ble. However there comes a point about five-sixths toward the end where the lines are slightly shorter and are broken up into halves, like in Biblical apposition or psalmic apposition to the point where:

> I am masturbating in my bed, I dreamed a new Stranger
>   touched my heart with his eye,
> he hides in a sidestreet loft in Hoboken, the heavens have
>   covered East Second Street with Snow,
> all day I walk in the wilderness over white carpets of City, we
>   are redeeming ourself, I am born, [37]

So it's like a series of staccato comes, spurts, within the line. So the breathing there, if read aloud, would be like a heavily labored breath-ing, with like a gasp for breath after each comma.

MA: With each comma.

AG: With each comma there. And I read it aloud about three times, so I know now what it sounds like that way. I read it in Buffalo. It works up toward the end where it's like a come, that begins the orgasm, the cli-max. The climax, literally, is:

---

[36]*Planet News*, p. 15.
[37]*Planet News*, p. 29.

Life is waving, the cosmos is sending a message to itself, its
    image is reproduced endlessly over TV
Over the radio the babble of Hitler's and Claudette Colbert's
    voices got mixed up in the bathroom radiator . . .
there is a mutation of the race . . . [38]

MA: "Life is waving"

AG: Yeah, the lines get slightly shorter and shorter till you get to . . .
the real climatic moment is I think, "the heavens have covered East
Second Street with Snow / all day I walk in the wilderness over white
carpets of City" . . . Which for me is the most Shelleyan line. This is
absolutely William Carlos Williamsish real, and at the same time
there's kind of a spectral thing.

    From then on "we are redeeming ourself, I am born"—it gets like
metaphysical screamings. To a point where—there are some new lines
in here that weren't in the original version.

the Messiah woke in the Universe, I announce the New Nation,
    in every mind, take power over the dead creation,
I am naked in New York, a star breaks thru the blue skull of the
    sky out the window, [breath]
I seize the tablets of the Law, [breath]
                   the spectral Buddha and the spectral
    Christ turn to a stick of shit in the void, a fearful Idea,
I take the crown of the Idea, and place it on my head, and sit a
King beside the reptile Devas of my Karma—

    And that's sort of like that—Ta da! Trumpets! And then a little like
a coda slightly going down.

Eye in every forehead sleeping waxy & the light gone inward—
to dream of fearful Jaweh or the Atom Bomb—[39]

And then a little spurt up again,

---

[38]Planet News, pp. 28–29.
[39]Planet News, pp. 29–30.

All these eternal spirits to be wakened, all these bodies touched
   and healed, all these lacklove
suffering the hate, dumbed under rainbows of Creation, O Man
   the means of Heaven are at hand, thy rocks & my rocks are
   nothing,
the identity of the Moon is the identity of the flower-thief, I
   and the Police are one in revolutionary Numbness![40]

Then it goes back to what the original text was. I inserted about eight
lines, to make the climax more Messianic. The lines I inserted were
lines written about the same time in 1961 in Leary's Newton-
Cambridge house.[41] They were referring to the same subject.

MA: Is there ever a time that comes to mind when you change the way
that a line is arranged on the page? Meaning, literally, in order to get
that syntactic condensation you are talking about?

AG: Yeah, if it's a short . . .

MA: Or jam a couple of lines that were looser, together?

AG: Yeah, Yeah. If it's a short-line poem. In some of this *Planet News* I
did that, when I found that there was a lot of bullshit in a line, and that
two lines could make one line if I put it all together. I did that in a little
poem about the Beatles in the Portland Colosseum.[42]

MA: Yes, I know the poem. Do you know the exact lines?

AG: Let's see what I did with it. I'd have to go back to the original man-
uscript to see exactly what I did, but I know I changed things. Like
"Hands waving like myriad snakes of thought" to "hands waving myr-
iad / snakes of thought." Ah . . . "The million children of the thou-
sand worlds," so I just changed "The million children / the thousand
worlds." In other words, I just kept the images and took out a lot of
syntactical fat and occasionally put two lines together, two short lines
together that had just images in them. I don't know if I can find
that . . . Now there's a reason for my putting the words in that order.

[40]*Planet News*, p. 30.
[41]See *Journals: Early Fifties Early Sixties*, ed. by Gordon Ball (New York: Grove Press,
1977), pp. 168–70.
[42]"Portland Colosseum," *Planet News*, pp. 102–103.

"bounce in their seats, bash" is one line, and "each other's sides, press" is another. But the way I would read it is: "bounce in their seats bash each other's sides!"—so there isn't much of a pause between "seats" and "bash" because you're excited. So you're saying "bounce in their seats, bash each other's sides, press legs together nervous"—all three phrases fast together. So I put "bash" on the same line with "seats" to indicate that the breath should continue. It's really a run-on. And if I said "bounce in their seats," and then in the next line "bash each other's sides" that would be too much of a pause, a halt in the rhythm run.

MA: So why not print that on the same line?

AG: It could be. As in "Howl" form it would be "bounce in their seats bash each other's sides press legs together nervous" in enjambment that would create a run-on effect. It could be done that way; alternatively, it could be sounded to get the humor of it, "bounce in their seats, bash/ each other's sides, press!/ legs together nervous" because each one of those is like a little haiku.

MA:—Is a thought, too.

AG: So to retain the haiku element for the mind's eye, "bounce in their seats, bash/ each other's sides, press/ legs together nervous," I keep them separate. It could be a run-on. But it also could be read run-on with a slight hesitancy after "bash"; so it's "bounce in their seats, bash!/ each other's sides, press/ legs together nervous." Which gives it like three different rhythms running at once.

MA: It's counterpoint.

AG: Yeah, so you have a counterpoint thing, which is the old use of enjambment, or run-on. In "Howl," there's that kind of run within the line; there are little breaks within the line, actually, depending on the humor in reading it, too. The short line form as presented just makes it a little more obvious. And since I started out with a little tiny notebook in my pocket going to a Beatles concert, this is the way I wrote it out anyway. But I could have . . . If you notice, in this poem, each "Howl"-type long line would begin at the margin. So it could be "A brown piano in diamond white spotlight" as one single line instead of the "white spotlight" subdependent, indented, from "diamond."

MA: Then what you said, right at the start, about the page size you're writing on controlling the poem's shape and form is really there.

AG: Yes, Creeley and I had a big long conversation about that, in "Contexts of Poetry,"[43] about how he writes. He was talking about how his writing was determined by the typewriter, neurasthenias of his habit; mine is determined by the physical circumstances of writing, i.e., literally that. And I got that actually from Kerouac, who was that simple and straight about it. If he had a short notebook he wrote little ditties and if he had a long . . . a big typewriter page, he wrote big long sentences like Proust.

MA: When you are looking at that later, do you have any selection principle for cutting things in or out?

AG: Principally cutting out. I do a lot of writing like that, so not all of it is as good, because not all of it is focused, not all of it is tied together by some emotional feeling-center. Unless there is an emotional feeling-center with a clear idea, like "I here declare the end of the War!"[44] which is unconsciously present all through that whole "Wichita" composition—from the beginning where I get into a bus with a microphone and say "all right, now we're going through the middle of America and we're going to dictate a giant poem on the middle of America and the war and everything that is happening"—so obviously it is Messianic, or could be. It was from the very beginning "Face the Nation," the first line [of Part II]: so "Face the Nation" confronted the Messianism.

MA: I noticed you added another section to the start of it here—why? And what is that from?

AG: Well, I did that for about a year, running back and forth across the country. The "Wichita Vortex" was the transcription of only one day.

MA: The "Face the Nation" part.

43Robert Creeley, *Contexts of Poetry: Interviews 1961–1971* (Bolinas, CA: Four Seasons Foundation, 1973), pp. 29–41.
44*Planet News*, p. 127.

AG: Yeah—I have similar types of composition, sections ranging over a year, covering the whole country, for a text which is like a whole book, about a hundred and fifty pages.

MA: "These States."

AG: "A Long Poem on These States."[45] The "Face the Nation" part was coming from Lincoln to Wichita. The proemic part—that's in *Planet News*—is from Wichita to Lincoln, on the way up to Lincoln. It's a shorter piece but as intense. And they make a complementary thing.

And there's other parts published in *Fall of America*. One is from Kansas City to East St. Louis and is the part about Hart Crane—which makes the third section. Then there's another Northwest piece, with Gary Snyder driving the Volkswagen from the Canadian border down through the east side of the Cascade Mountains and down through Pendleton and along the west shore of Pyramid Lake to San Francisco, which is called "Beginning of a Poem of These States."

MA: And hopefully they will all tie together the way that—

AG: You don't have to hope for it because they naturally tie together; they're all done the same way, during the same time period, by the same mind, with the same preoccupations and obsessions, during the same war. So I mean no matter which way they went they'd all go out from the same brain place. In other words the very nature of the composition ties them together. You don't really have to have a beginning, middle and end—all they have to do is to register the contents of one consciousness during the time period. Hopefully tie together in the sense that hopefully the consciousness has a bottom—or a top, or, you know, the consciousness comes to rest somewhere.

MA: Does the Pound language theme run through all of these poems?

AG: It comes in and out. It first appears in the going up to Lincoln, "language, language / over Big Blue River." Then it reappears again, "Language language / Communist / Language language soldiers"—later on it comes in the section to St. Louis, the thing about Senators talk— try to find a language, talk on their feet, saying:

---

[45]Published as *The Fall of America, Poems of These States, 1965–1971* (San Francisco: City Lights, 1972).

Language, language, uh, uh
from the mouths of Senators, uh
trying to think on their feet
Saying uhh, politely

That actually comes from Pound, not from Kerouac.

In an odd way Kerouac babbles beautiful American name language from *On the Road* in the long section about "Tarpaulin power" on the Wasatch snow peaks or something, just his use of names makes amazing music.

MA: Here's a question about the spacing, here:

That the rest of earth is unseen,
                            an outer universe invisible,

Then why indent . . .

Unknown except thru
                    language
                    airprint
                        magic images?

AG: The way this was determined was: I dictated it on this Uher tape recorder. Now this Uher microphone has a little on-off gadget here (click!) and then when you hear the click it starts it again, so the way I was doing it was this (click!); when I clicked it on again it meant I had something to say. So—if you listen to the original tape composition of this, it would be

That the rest of earth is unseen, (Click!)
                            an outer universe invisible, (Click!)
        Unknown (Click!) except thru
                    (Click!) language
                        (Click!) airprint
                            (Click!) magic images
    or prophecy of the secret (Click!)
        heart the same (Click!)
        in Waterville as Saigon one human form (Click!)[46]

---

[46]*Planet News*, pp. 123–24.

So when transcribing, I pay attention to the clicking on and off of the machine, which is literally the pauses, as words come out of my—as I wait for phrases to formulate themselves.

MA: Outta sight!

AG: And then, having paid attention to the clicks, arrange the phrasings on the page visually, as somewhat the equivalent of how they arrive in the mind and how they're vocalized on the tape recorder. And for that, I have some samples here I can play you—of composition in the car, using the clickings.

It's not the clicks that I use, it's simply a use of pauses—exactly the same as writing on a page: where you stop, you write, in the little notebook, you write that one line or one phrase on one line, and then you have to wait for another phrase to come, so you go on then to another line, represented by another click.

On a typewriter I can see the space that I can fill up.

Of course, this could be put on one line: "Unknown except thru language airprint magic images or prophecy of the secret heart" . . . Then you could have another line—"the same in Waterville as Saigon one human form." These could be rearranged. But these lines in "Wichita" are arranged according to their organic time-spacing as per the mind's coming up with the phrases and the mouth pronouncing them. With pauses maybe of a minute or two minutes between each line as I'm formulating it in my mind and the recording.

MA: Each pause comes as a click.

AG: Yeah. Like if you're talking aloud, if you're talking—composing aloud or talking aloud to yourself. Actually I was in the back of a bus, talking to myself, except with a tape recorder. So every time I said something interesting to myself I put it on tape.

Edward Kissam: So what we're taught to think of as complete thoughts aren't really complete thoughts as much as those kind of bits that come out.

AG: Yeah. Well if you try—let's see now. How do we think is the problem. How do we actually think? In other words this is like a form of Yoga: attempting to pronounce aloud the thoughts that are going through the head. But to do that you have to figure out how the thoughts go through your head. Do they go through as pictures? Or do they go through as a series of words, or do they go through your head as full sentences or as phrases?

Now Burroughs doesn't see words—words don't go through his head, pictures go through his head. So his method of composition is sitting before a typewriter sort of looking up in the middle distance at the wall, seeing pictures flash through his head like . . . mugwumps seated on a barstool slurping up honey with a long reptilian tongue . . .

MA: And then trying to put that into words.

AG: And then he simply transcribes it into pictures—words, picture words.

Edward Kissam: I was thinking of that formal phrase of yours where you said "a fearful idea," where I wouldn't very easily think of it as a fearful idea . . .

AG: No, well, the line was . . .

EK: "Fearful" even though that doesn't make any sense [it] might be a complete thought . . .

AG: Yeah. Well, let's see now. "I take the crown of the idea"[47] . . . See like I just said that Buddha is a stick of shit, and Christ is a stick of shit, which is actually not my idea, it's an old Zen koan—Buddha is a stick of shit.

EK: Spit on every image.

AG: No, literally, quote: "Buddha is a stick of shit. The Buddha is a stick of shit," quote unquote, is in Suzuki, as something that somebody said in the fifteenth century, some Zen master—which Gary quoted to me once. So, I say "the spectral Buddha and the spectral Christ turn to a stick of shit in the void"—and then, like, stop for that second, what a weird idea, what a fearful idea—it's a scary idea, "a fearful Idea."

MA: Where are we, what poem are we talking about?

AG: That's in "TV Baby"—"a fearful Idea." Well it could've arrived in the mind as War! fear—I mean it could have arrived as a shudder rather than as a word or a picture. Or, actually, it could've, it really arrived in the mind as uh . . . gee what would Lionel Trilling say? or what would

---

[47]*Planet News*, p. 30.

the pope say, that's kind of a presumptuous rather fearful thing—you can get crucified saying something like that—announcing something like that. It is a fearful idea. The shortest portmanteau phrase for that was, "a fearful Idea."

Then, "I take the crown of the Idea," okay, I'll take it on, "I take the crown of the Idea and place it on my head, and sit a King"—And where? "beside the reptile Devas of my Karma—"

MA: A very Medusan image.

EK: And there where you go on it shoves it all together. It all happens at the same time I guess.

[End of tape. Conversation resumes later.]

AG: We're just talking about whether it's written or pronounced on tape. Did you see the long poem called "Beginning of a Long Poem on These States" in *Fall of America?*

MA: I think so, yes.

AG: It's all about driving through the Canadian border, past Omak and Nespelem and Chief Joseph's grave.

MA: I do remember the poem, yes.

AG: Well, it's a long thing. That's done in paragraphical form. And that was done with a pencil and paper, sitting in the front seat, next to Gary, who was driving, and I wrote down maybe one or two phrases a day— just the key phrases for like, a little epiphany as we passed Omak with "red red apples bend their tree boughs props with sticks—" so it was just one or two little Rimbaud-like key phrases; then I simply added them, like a tapeworm, one to another—and when I'd get three or four that made an apposition I'd start a new paragraph. So it's maybe one paragraph a day for three weeks. Covering from the Canadian border down to San Francisco. That was *written* in pencil.

That written form is different from tape-poems when transcribed on the page. Though it's written down from the tape recorder, the tape process got me talking more—well, I was able to get more fugitive things going like "Face the Nation," "You're in the Pepsi Generation," pop signs signaling less, maybe less intense than inner poetizing, but . . .

let me show you . . . mental ephemera that I never published . . . over
the Rockies . . .

(Break. Discussion moves to sound articulation and Sanskrit.)

Nancy Blecker: Unaaah, it's not so much *yah*, but *nyahh*, *mmmahh*,
thinking of something like you use . . .

AG: Well, this is Bengali:

> *Jáya Jáya Dévi,*
> *Chára Cháro Sári,*
> *Kúcha-Juga Sóvita, Múkti Hári*
> *Vina Nándita, Pústaka Hástey*
> *Vághabati Bhárati, Dévi Namáste*

Which is like—sounds like a nursery rhyme. Well the thing about
Sanskrit is, they—
    We have for a beginning, say, to measure a line—we have the pos-
sibility of counting the accent, or else we have the possibility of
counting the length of the vowel, that is taking into account the length
of the vowel, which is the thing I was getting interested in, from hear-
ing the Pound tapes—which you heard, I guess.

MA: No.

AG: Oh you haven't heard those tapes of Pound?[48] I have them here—I
can play those. That's a revelation. Of the musical possibilities of the
vowels, or what Pound calls the "tone leading of vowels."

*(Ginsberg next explains, as in preceding interview with Fernanda Pivano,
how listening to a tape of Pound reading Con Usura [With Usury] made him
realize how every syllable in poetry must be intentional, and how following
Pound's principles plus Bunting's direction for the poet to condense "gives a
density to the line." He then goes on to say about Sanskrit:)*

Now, the Sanskrit thing is even deeper in a funny way, 'cause the gram-
mar is all built on yoga—on a physiological body yoga. So that the first
letter would be *a* of some sort or other, but there'd be four or five dif-
ferent kinds of *a*s apparently beginning at the back of the throat *awhhh*

---

[48]*Ezra Pound at Spoleto* (New York: Applause Prod., 1968).

and then there's the *ah* and then there's the *aw* or whatever would come up to the front of the mouth box . . . *a* . . . so it's *awhh* to *a*.

And then *bawhh* to *bee* and *sah* or *kah* to *cee*, and *da* or *duh*, *thuh*, to *dee* . . . I don't know exactly.

So the mantra formulas have what are called *bija* syllables, or seed syllables, because their deployment, physiologically in the body during their pronouncing, is crucial—because they have a whole schematic significance that we don't begin to have in our alphabet and in our combinations of the alphabet into word-sounds. In a funny way there's a superhuman onomatopoeia going on in Sanskrit. For instance, one of the mantra-yoga seed-syllables is *dhuh*. (Note: *Dha*, sound is a cross between *duh* and *thuhh*.) The reason is that, to pronounce it, first you have the *the* with the tongue between the teeth, the, but then it jumps to the back of the throat, *dhuhh*—and not only to the back of the throat, it jumps down to the middle of the chest to get the *uhh* the dh*uhh*!—so like there's a mantra-yoga singing teacher who was telling me that the next step I should learn, for instance, after doing A-OHM and things like that, is the famous DUH! sound, because that's the key to suddenly wakening up like a whole Reichian chain of muscular reactions, from the front of the lips to the heart center: DHUH!

And then if you use the *Duh* in combinations with others—with others like, I don't know,—*Dha-Phat!*—you're going through a whole physiological exercise, involving not merely vocal cords but the whole *prana* breathing apparatus.

MA: It's like getting hit in the chest, and you want to go uh!

AG: Yeah, and especially, it's touching special jujitsu pressure points on the body, by pronouncing them—so it's doing like a physical exercise or a yoga involved with the breathing and also the exhalation of the breathing. Breathing IN to a certain depth and exhaling in a certain way—through the ears or through the nose, or exhaling nasally.

Now the AUM is interesting because—as I've discovered in practice at this point, after Chicago—the *Ahh* is like an open sound, coming up from the center of the body, *Ahhhhh!*—like a sigh out of the mouth—but then the air gets held and imprisoned in the mouth to vibrate the palate and the bottom of the brainpan, skullbone: *Aawmmmm*, because the *mmmm* when you breathe it out nasally, actually literally makes a vibration in the bone, around the nasal cheekbone and the palate, which is what is upholding the brain. So what you're doing is setting up some vibration which is giving a massage . . .

MA: Right on the bottom of your brain!

AG: A thing on the bottom of your brain. And that's why, if you say, AUMMM, AUMMM, for a couple hours, after a while your brain begins vibrating. Or the vibration begins there and begins to affect the whole body, so it must be some physiological-electrical, alpha-rhythm tie-in that gets set up—because if you do that AUMMMM, and you do that MMMM for like five minutes, you begin to feel that buzzing throughout your whole physical skull . . .

Getting back to the Sanskrit prosody, if you have a prosody built on that, it's so complex—you can do anything with it—it's like having the basic patterns of physiological reactions built into the language, into the alphabet—and then making combinations of the alphabet you can play like an organ, to get different effects.

MA: The body is a literal violin, then.

AG: Yeah! An interesting statement in the *Bhagavadgita* is, Krishna is pointing out all his different forms and aspects, he's saying among gods I am Krishna, among directions I am, I don't know, the East and among colors I am blue, and "among poetic meters, I am the *Gayatri* meter."[49]

The universal meter is the Gayatri meter, apparently, and so— what is this Gayatri meter? There is the Gayatri mantra which is a very, very famous one, which I don't know all the way through, but it begins:

AUM, BHUR, BHUVAH SUAHA
TAT, SAVITUR, VARENYAM
BHARGO DEVASYA DHIMAHI,
DHIYO YO NAH PRACODAYAT.

It's some kind of thing that's asymmetrical. And once you examine it through, it's not symmetrical, not repeated, but it seems to cover like

---

[49]The section of the *Bhagavad-Gita* referred to is Chapter 10, sloka 35: see Radha-krishna edition, p. 226.

a whole long free line and be complete. It covers all body sounds possible.

The meaning of it is something about Hail to the first light, that begins all other lights, which is the Female Principle, also, Gayatri, Devasya.

MA: Is that meaning given in the *Gita?*

AG: No, it's not: It's from another Sanskrit text. Though that *Gita* line is written in Gayatri meter I would guess.

MA: I would think so, yes.

AG: But the Gayatri mantra is a very famous mantra which all Brahmins know, it's one of the—like when they do Aarti[50] or worship in the evening, or when they do any kind of a ritual thing usually at one point or other Gayatri meter is one of the things that's pronounced. So. That's all I know about the Sanskrit prosody—which is just a hint that there's this giant, extremely sophisticated and physiologically based system, that's as complicated as the nature of the human body, practically, or is fitted to the nature of the human body and touches all the key combinations. So probably a study of Sanskrit prosody would take us deeper into what Pound and Williams and everybody—and Rimbaud's alchemy of the word and color of the vowels—had all been hinting at, over the last century or so.

MA: Has your own use of mantra done anything that you can be very specific about, with your poetry?

AG: Yeah a lot, *now.* Mainly it's made me conscious of what I had been doing with long lines in "Howl." And . . . made me conscious of what I'd been doing with breathing as in the Moloch section of *"Howl,"* or parts of "Kaddish"—that the . . . rhythmic . . . units . . . that I'd written down . . . were basically . . . breathing exercise forms . . . which if anybody else repeated . . . would catalyze in them the same *pranic* breathing . . . physiological spasm . . . that I was going through . . . and so would presumably catalyze in them the same *affects* or emotions. That's putting it a little bit too . . . rigorously, but . . . that's the direction.

Doing mantra made me conscious of what I was doing in Poesy, and then made my practice a little more clear, because now I realize that certain rhythms you can get into, are . . . mean certain feelings. Well, everybody knew that anyway all along. But some rhythms mean something.

MA: If you're going Dumpty dumpty dumpty dumpty dumpty dum it's different than if you're going daahh, duhhh, dummm, duh-dummm.

AG: Or, if you're going Bum! ba-ta TUM, BUM: BUM, ba-da-DAA . . . / BOM, bata BOM BOM, BOM bata DAA . . . See, I'm writing a long poem that's got that rhythm now:

> DAT- dada- DA- dada
> DAT- dada- DA- dada
> DAT- dada- DA- didi-da, DON- dada- Da,
> DAT- didi, DAAA!

It'd be interesting again—those are like ancient Greek dance Dionysian rhythms. See, so I get more then . . .

And then, also, the vowel sounds—I've been digging the vowel sounds more, and the tone leading of vowels, out of Pound, as Pound talks about it—and that's led onto an examination of Blake. And the single syllable, or the individual syllables, in Blake, and then the possibility of putting them to music. So I'm getting all hung up now, from listening to that line, "with usura the line grows thick," as pronounced by Pound, I begin to get more sensitive to the fact that each syllable in a Blake poem is intentional and therefore has to be pronounced intentionally, as it was meant—like:

> Beneath them sit. The aged men. Wise
> guardians of the poor.

which means a great syncopation

> Beneath them SIT the AGEd MEN, WISE
> GUARDians of the POOR!

Or—

'Unseén, thĕy poúr bléssĭng,'

It's not—

'Unseén, thĕy poúr blĕssĭng,'

It's—

'Unseén, thĕy poúr bléssĭng,'

[Sings:]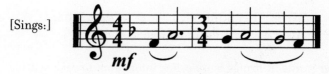

Un-SEEEN, they POUR BLESS-ing . . .

So by trying to fit them to music, I have to pay attention to the lines in Blake, I have to pay attention to the intonation of the syllables, whether they go up or down:

```
                       POUR
          SEEN,               BLESS
      Un-       they                -ing
```

Or whether they are going up or down emotionally, and musically, and whether they're to be skipped over, or whether they're to be pronounced emphatically, like:

ÁH! SÚN FLÓwer

instead of,

Ăh Súnflŏwér.

In other words it gets me out of the hangup of iambic stress into vowel-length consciousness, which is deliberate speaking voice awareness.

So I've been going through all of Blake, trying to understand a lot

of things I didn't understand before, for the first time—by simply pay-
ing attention to what's being said by mouth, on account of I'm having
to pay attention to figure out how each syllable would have its musical
note as an equivalent. See, otherwise, it would be:

> My mother bore me in the southern wild
> And I am black, but O! my soul is white;[51]

But—

> My MOTHer BORE ME in the SOUTHern WILD
> And I AM black but O! my soul IS white;

So in other words, it brings more color, intonation, to each syllable
pronounced.

MA: And what's your tune for that?

[Ginsberg sings.][52]

MA: In other words, the tunes grow right out of looking at what's on
the page.

AG: I'm looking at what's on the page, but to be a little more definite it
grows out of being conscious or aware of the meaning, intention, or the
significance of each syllable on the page—and recognizing that each
syllable has a place, AND a purpose, when it's really good poetry. When
it's sloppy dreamy poetry then there's a lot of syllables that don't have
any function—that don't have any *intention*—they're just there because
the guy was writing unconsciously and, you know, unconsciously hears
the . . . echoes of old iambic quatrains up in the alley or something.

MA: All right: what you were saying a minute ago about Sanskrit, and
about Gayatri meter, if that is in fact true, and I'm sure it is; that by
pronouncing certain syllables correctly, not meaning correctly in a
grammatical sense, but correctly to find the right place in the body

---

[51]William Blake, "The Little Black Boy," in "Songs of Innocence," *The Portable Blake*,
ed. by Alfred Kazin (New York: Viking, 1946), p. 86.
[52]William Blake, *Songs of Innocence and Experience*, tuned by Allen Ginsberg (N.Y.:
MGM-Verve FTS-3083 [Side 1, Band 4], 1970).

where they are to be pronounced from and where they most affect—then that means you can write a prosody, you can write poetry using a prosody even if you can't write about the prosody specifically.

AG: Oh, yeah! All you have to do is know what they're saying.

MA: Which is literally bringing all your physical universe into the thrust of that prosody.

AG: Yeah. And that's exactly what Olson has been talking about all along as *projective* verse, involving the complete physiology of the poet. That's what he meant.

MA: What does he mean by "The syllable is connected to the mind, but the line is connected to the heart"?—I think that is his schema.[53]

AG: Well, I don't think he was thinking in terms of AUM syllables; I think he was thinking in terms probably of Marianne Moore syllables being connected to the mind, because you can . . . divide them, like Marianne Moore did, just sort of automatically or arbitrarily—each syllable—five-syllable, seven-syllable lines, then repeat them. But the *line* itself is connected with the breath in that the whole body's intention is mobilized to pronounce the complete phrase (or complete line), in his projective conception. And if the whole body is mobilized, that means the whole single breath of the body is used, whether it's a shallow breath or a deep breath . . . and if it is a physical breath it means it's the whole metabolism and the feelings of the body and the *heart* spasm that's involved, so that the breath leads, so to speak, directly to the heart, the center of feeling. At least that's the way I interpret it.

MA: I'm interested in the gossip back of all the starting of that theory. In Kerouac's *Paris Review* interview[54] he says that he started that. So, what's the story?

AG: Well, I don't think that anybody can claim to have quote "started that" unquote, whatever "that" is—the idea of breath, because . . .

[53]Charles Olson, "Projective Verse," in *Selected Writings*, ed. by Robert Creeley (New York: New Directions, 1966), p. 19:
Let me put it baldly. The two halves are:
   the HEAD, by way of the EAR, to the SYLLABLE
   the HEART, by way of the BREATH, to the LINE
[54]Jack Kerouac, "The Art of Fiction XLI," *Paris Review*, no. 43, Summer 1968, pp. 60–105.

MA: The breath as control and measure of the line.

AG: Well, it's implicit in Apollinaire, it's articulate in Artaud, in Artaud's cries . . . It's developed, independently and theoretically, by Olson, I presume, from the early 40s, or earlier, out of Pound . . . but Kerouac's most clear use of it, and most available use of it—Kerouac used it in such a way as it became immediately apparent in a way, a popular way. That was arrived at independently by Kerouac in the early and late 40s. So I guess what he means is that he doesn't want to get all involved with complicated literary theories or terminology with which he had nothing to do . . . his is just simple common-sense practice. And since I learned mostly from Kerouac, and then put a patina of literary categorization over it later on, having dealt with Williams and Creeley and Olson—though actually I learned the simple spontaneous practice from Jack—without any of the complex labels involved. This is a very simple thing: *talk as you think*. And talk as you talk, instead of talking as a literary person would be taught to talk if he went to Columbia.

I think also Kerouac is very resentful of my trying to make a unified field of his practice, and Olson's practice, and trying to reconcile them all and say it's all one, community effort, when Kerouac *wasn't* exactly working in that large a community, or, you know, I guess he felt more like a private solitary Melvillean minnesinger or something.

MA: Just before "Howl" was written, the only people who were singing, that way, out of themselves completely, out of their bodies, were blacks. And then along came Elvis, and revolutionized white music— "Ah sing thuh way ah fee-ul."

AG: Yeah. Kerouac learned his line from—directly from Charlie Parker, and (Dizzy) Gillespie, and (Thelonious) Monk. He was listening in '43 to Symphony Sid and listening to "Night in Tunisia" and all the bird-flight-noted things which he then adapted to prose line.

MA: How?

AG: Kerouac comes autonomously from sitting in the middle of Manhattan listening to the radio and picking up vibrations of a new breath, from the spades, and he does give credit there.

MA: And so do you, in the liner notes in the back of *The New American Poetry* anthology.

AG: Yes. I really learned it from Jack. So—when it gets mixed up with more literary/literate discussions, he probably feels that's getting it too complicated, and it's slipping away from his actual sources. Or I guess he doesn't want to be literally categorized as a follower of Olson's projective verse, because he's not. I mean he's following Charlie Parker, and also following Thomas Wolfe, and (William) Saroyan—and Proust, and Céline, who also has that funny kind of speech extension.

I don't think that Olson would claim that Kerouac was writing projective verse. I think that Olson would say that "projective verse" is his terminology for this kind of writing.

MA: Sure.

AG: Which is a universal form—rising out of Gertrude Stein, rising in Céline independently in the 20s, rising in Kerouac and Wolfe. Olson really was attempting to formulate it in academic terms so professors could understand it—because that's all they—the only terms they can think in—so he found a categorical, terminological set that people who were hung up on categories could deal with—and he wrote a very literate essay to explain it in literary-essay terminology. Kerouac wrote his "Essentials of Spontaneous Prose," which was written, I dunno. When was the "Projective Verse" essay written?

MA: '50? '51? Fairly early.[55]

AG: I don't know when Kerouac wrote his "Essentials of Spontaneous Prose,"[56] but that's around the same time, because when I came to San Francisco in 1954, I had it pinned on the wall of my hotel room and Robert Duncan came to look at a little book[57] of poems I had and saw it on the wall, looked at it, looked twice and thrice, and said, "Who wrote that? Who wrote THAT?" because it's so, you know, right.

So that was a source of—the development was synchronous—the key is synchronicity, because it was darkly inevitable.

MA: It was coming out of the Head, hair, it was coming out of the universe.

55"Projective Verse" was first published in *Poetry New York*, no. 3, 1950.
56Jack Kerouac, "Essentials of Spontaneous Prose," in *New American Story*, ed. by Donald Allen and Robert Creeley (New York: Grove Press, 1965), pp. 270–71.
57Published as *Empty Mirror* (New York: Totem/Corinth, 1961).

AG: Just as the return of attention to actual images and observed fact, through Williams, was coming synchronously, at Reed College in '48 with Snyder and Whalen, who'd met Williams, and myself who went to see Williams that same year in Paterson—and in the Berkeley Renaissance.[58]

MA: Can we come back to Williams later? Basically it's his and Pound's sense of meter and music that's at the heart of my white space idea, but I wanted to see what you think about this remark about "Howl," that I made several years ago.

> —*Howl*, the most famous of his poems, is extremely rhythmical. The meter is sustained primarily by anaphora, the repetition of the same word or words at the beginning of two or more successive verses (lines), clauses, or sentences, a device also used by Shakespeare in Sonnet 66 and by Whitman throughout his poetry (see Sections 31, 33, and 43 of "Song of Myself," for example). To use musical terms (particularly appropriate because the lines of *Howl* use almost exactly the same methods of Charlie Parker's saxophone improvisations), a repeated cadence of anaphoric words like "who" and "Moloch" is taken off from the cadenzas, long swirling patterns of movement. The long line is sustained just because its movement is interrupted recurrently by one unit of that movement.[59]

A device also used by Shakespeare—throughout his poetry.

AG: More specifically by Christopher Smart, in *Rejoice in the Lamb*.

MA: To use musical terms—saxophone improvisations—

AG: Lester Young, actually, is what I was thinking about.

MA: Oh. Okay.

AG (*sings*):

---

[58]Gary Snyder and Philip Whalen met William Carlos Williams when he visited Reed College in November 1950.
[59]*The Nassau Literary Magazine*, February 1964, p. 18.

> Dadada DAT DAT DA, dada DA da.
> Dadada DAT DAT DA, dat da Da da,
> Dadada DAT DAT DA, dat da Da da,
> Dadada DAT DAT DA, dat da Da da,
> Dadada
> > dadada
> > > dada da dadah . . . [60]

"Lester Leaps In," "Howl" is all "Lester Leaps In." And I got that from Kerouac. Or paid attention to it on account of Kerouac, surely—he made me listen to it.

MA: A repeated cadence of anaphoric words, like *Who* and *Moloch*.

AG: *DAH*! da da DAT DAT DA, da da DA da, yes.

MA: Is taken off from, by cadenzas, long swirling patterns of that movement.

AG: Exactly. Yeah, *Dah* or *Who* was a base to return to and spurt out from again. I'm thinking of that little essay I wrote.[61] Cadenzas is a nice word. They're all cadenzas beginning with that one note . . . which is like the opera—you know the opera girl studying:

> ah ah ah AH ah ah ah, ah ah ah AH ah ah ah . . . (scales)

MA: What's interesting about that, is that *WHO* is never emphasized.

AG: What?

MA: *Who* is not accented. It's

> who poverty and tatters a la la la la la,
> who a la la la la la

but the *who* is not,

---

[60]Rhythmic paradigm of "Lester Leaps In"—awkwardly remembered.—A.G. [*Lester Leaps In* is both the name of a song and an album by Lester Young.—DC]
[61]Allen Ginsberg, "Notes on Howl," *Allen Ginsberg Reads Howl and Other Poems*, Fantasy Record #7006 (1959).

WHO! poverty and tatters a la la la la la—

it's never accented.

AG: Well. It's more like—no, it's not. Well the image I was thinking of is the bardic thing—where the bard has his strumming instrument, or his lyre, and gives . . .

> Plongggggggg . . .
> And Ulysses went forth on the ships and the ocean
> Plongggggggg . . .
> And the next thing they saw was the god Neptune rising
> up out of an island,
> Plongggggggg . . .
> And they went up and crawled on Neptune's beard,
> Plongggggggg . . .

So the *plong* was just something to get your mind going again, or the "Who."

MA [to Kissam, referring to an earlier conversation]: What were you saying about the nesting quality of "Who"?

EK: I was just saying that the "Who" brought everything, compacted it completely, almost as though it were all written vertically, to have the "Who" there, because syntactically it always brings you back to the start . . .

AG: Yeah. Like I was doing that one. It's such an easy thing to do, like once you get something going like that you can always come back to the same thing and add another one on.

MA: Well, I think as a rhetorical device that's your most common. You keep doing it.

AG: Yeah. Whomdoya call it does it also—Smart uses the word "and"— "And let me rejoice with the ass Oneocrotalus whose voice brays like" . . .

MA: That's Greek! That's the Greek *kai*.

AG: The Greek what?

MA: KAI. *Kai* is the Greek word for "and"—and they use it at the start of more sentences than not because it ties it right back into the . . .

AG: Pound uses it once, in the "And they heard the frogs singing against the fawns"—remember that?

EK *(laughing)*: Melina Mercouri uses it in the *Children of Piraeus*. *(laughter)*.

AG: I finally use it in "Kral Majales." "And I am the King of May, and I am the King of May, and I am the King of May"—I didn't know it came from the *kai* though—that's interesting. Well, what was necessary was a conjunction, or a junction, to link all those rhythmic cadenzas together, that's all.

MA: I noticed in the back of the Allen Anthology[62] you said that you're trying to do the same thing in "Sunflower Sutra",[63] but without using a specific word repeated as a base.

AG: Yeah, a lot of times what I try to do is—the important thing is to get that continuous locomotive rhythm going. So with "Sunflower" I simply eliminated the "who," figuring that—now when I was conscious that it was the locomotive rhythm I was after, then I could have one long cadenza that you know, didn't return, didn't interrupt itself by having to return to its base—

And this "TV Baby" poem is like one giant cadenza, literally, in the sense that it's all typed out as one continuous, practically all one, sentence that goes on and on, working all day and all night—that builds to a climax, without the need of returning to a rhetorical base. And the ideal would be, you know like, why get hung up on the conjunction. Why not just, you know, do it—the conjunction's sort of like a crutch to start it moving again. And then the interesting thing would be—to fly without crutches.

So like in "Sunflower" I was trying to fly without crutches. Or in "TV Baby," to fly all the way. Same thing also in a lot of *Kaddish*.[64] By

[62]"Notes on Howl," reprinted in *The New American Poetry, 1945–1960*, ed. by Donald Allen (New York: Grove Press, 1960), pp. 414–18.
[63]"Sunflower Sutra," *Howl & Other Poems* (San Francisco: City Lights, 1956), pp. 28–30.
[64]*Kaddish and Other Poems* (San Francisco: City Lights, 1961), pp. 7–36.

that time I was really quite conscious of the fact that the "Who" was a device, an interesting device but I wished that I could do without it and get the thing to—purely speed forward.

MA: When you say literally "make Mantra of American language now"[65] what do you mean? Do you mean we should literally start chanting the lines for mantra purposes?

AG: No, that's not what I mean. One function of a mantra is that the name of the god is identical with the god itself. You say Shiva or Krishna's name, Krishna is the sound of Krishna. It's Krishna in the dimension of sound—so if you pronounce his name, you, your body, is *being* Krishna; your breath is *being* Krishna, itself. That's one aspect of the theory of mantra.

So I wanted to—in the English language—make a series of syllables that would be identical with a historical event. I wanted the historical event to be the end of the war, and so I prepared the declaration of the end of the war by saying "I hereby make my language identical with the historical event, *I here declare the end of the war!*"[66]—and set up a force field of language which is so solid and absolute as a statement and a realization of an assertion by my will, conscious will power, that it will contradict—counteract and ultimately overwhelm the force field of language pronounced out of the State Department and out of Johnson's mouth. When they say "We declare war," their mantras are black mantras, so to speak. They pronounce these words, and then they sign a piece of paper, of other words, and a hundred thousand soldiers go across the ocean. So I pronounce *my* word, and so the point is, how strong is my word?

Well, since Shelley says that the poet's word is the strongest, the unacknowledged legislator's, the next thing is: let the president execute his desire [laughing], and the Congress do what they want to do, but I'm going to do what I want to do, and now it's—if one single person wakes up out of the mass hallucination and pronounces a contrary order, or declaration, contrary state, instruction to the State, to the Government, if one person wakes up out of the Vast Dream of America and says I here declare the end of the war, well, what'll happen? It was an interesting experiment, to see if that one assertion of language will

---

[65]"Wichita Vortex Sutra," *Planet News*, p. 127.
[66]Ibid.

precipitate other consciousnesses to make the same assertion, until it spreads and finally until there's a majority of the consciousnesses making the same assertion, until that assertion contradicts the other assertion, because the whole War is WILL-FULL-NESS, and the War is a Poetry, in the sense that the War is the *Happening,* the *Poem* invented and imagined by Johnson and Rusk and Dulles, Luce, and Spellman and all those people; so the *end* of the War is the *Happening*, the Poem invented by Spock, or myself, or Phil Ochs, or Dylan, or—

MA: (Ed) Sanders . . .

AG: Sanders, or A. J. Muste's ghost, or Dorothy Day, or David McReynolds, or Dave Dellinger, or anybody who wishes to make a contrary statement or pronouncement.

Now as I make a pronouncement, contrary to the Government's pronouncement, the question is—I give in to the desire—there's an explanation of that here, to (Paul) Carroll which he didn't understand, but what I wrote, is:

> Not only a question of legislator as Shelley's formula. Merely that the War has been created by language (as per Burroughs' analysis for his cut-ups) (or Olson's complaint about abuse of language in Maximus Songs) (or W.C.W.'s) & Poet can dismantle the language Consciousness conditioned to war reflexes by setting up (Mantra) absolute contrary field of will as expressed in language. By expressing, manifesting, his DESIRE (BHAKTI in Yoga terminology—'adoration').[67]

Now, my desire is for the end of the war, so I simply say, flatly, "I desire the end of the war"—or I even declare the end of the war. So in that sense, make a mantra of AMERICAN LANGUAGE. "Now I declare the end of the war." Make a magic phrase, which will stick in peoples' consciousnesses like a rock, just as the phrase "domino theory"— another phrase that stuck in peoples' consciousnesses like a rock—got them all confused.

MA: The problem seems to be translating that into anything more than your own single power.

---

[67]Paul Carroll, *The Poem in Its Skin* (Chicago: Follet/Big Table, 1968), p. 101.

AG: Well, I don't think that's such a problem. See—my own single power in saying "I declare the end of the war" isn't just my single power because it represents my desire—it's my unconscious as well as my conscious power. And that desire is archetypal, it saturates half the nation, according to the Gallup poll. Except "What oft was thought but ne'er so well expressed," so why doesn't somebody get up and say it? So once somebody gets up and says it, that precipitates the awareness, the same awareness of the same desire to end the war, in lots of other people—or, as Shakespeare says, "One touch of nature makes the whole world kin."

So it's a question of making that "touch of nature," or making the mantra, or expressing "what oft was thought," CLEARLY, publicly, consciously, to make that same unconscious awareness and desire appear up front in the public mind.

In other words it isn't necessary for my word to like quote *over-whelm* or *convince*, unquote, anybody else, it's just necessary for me to place my word out there, not to overwhelm but to clarify other people's sane thought, or to make it conscious or to bring it to the surface of their minds, so they say: *Oh yeah! That's what I think too! Why didn't I say that before? I didn't think you were supposed to say that, I thought you were supposed to think about it maybe, but not say it, pub-licly . . .*

[After a detailed discussion of technical aspects of modern poetry con-cerning the arrangement of text on the page:]

MA: All of these devices I've been dreaming up and seeing in poems, and that you've been talking about—are essentially a romantic, expres-sionistic way of organizing a poem. You don't organize a poem along lines of a story that you have to tell, like an epic; you don't even orga-nize a ballad type of poem along a narrative in which one stanza is one part of the story; you don't organize it logically like a sonnet; you don't even organize it logically like John Donne. It's not an intellectual logic or a narrative logic.

AG: I think it is an intellectual logic—because you're organizing it logi-cally, to follow the precise forms of the movement of the mind. So I don't see it as romantic-expressionistic at all—I see it as absolutely logical scientific notation of event.

EK: Isn't it logical the way that montage is? I mean—

AG: Yeah, well montage is logical—montage was at first considered to be illogical, and irrational, or surrealism was first considered to be irrational, until everybody realized that what really was irrational was a rearrangement of the actuality of mind consciousness into syntactical forms which didn't have anything to do with what was going on in the head! So that finally the practical, pragmatic, common-sense form of notation of thought, was the surrealistic one, because that's the way the mind works.

So it may turn out that we've been undergoing centuries of—this is what Blake was complaining about—of Newtonian thought. God save us from Newton's thought and somebody's—somebody else's—Bacon's funk. Or something. What Blake was saying is that they were unreal, in the sense that it was just the creation of an imaginary universe and what could be more illogical? And what could be more logical than the actual clear demarcation and definition of the way the mind works?

MA: Now we're moving; we've had this printed poem ever since—

AG: Since the invention of movable type, which is not very long. We've had the printed poem for a short period within 40,000 years of poetic history, now we've had the printed poem for approximately one one-hundredth of that time! The older tradition, the most ancient tradition, the conservative tradition, the actual tradition, is oral.

The oral tradition has all of the mnemonic devices and variations of rhythm and speech things . . . possible. The main structural guiding thing would be convenience in memorizing, I imagine, so that would be why alliterative repetition or rhythm or rhyme is used.

MA: That's why the *Rig-Veda*[68] comes down to us and that's why *Beowulf* comes down to us. Because those mnemonic devices were sufficient to allow people to remember them, generation to generation.

AG: It would be interesting to know what are the mnemonic devices of the Australian aborigines, who apparently have the largest and most

---

[68]The principal of the Vedas or sacred books of the Hindus. See *Rig-Veda sanhita: a collection of ancient Hindu hymns of the Rig-Veda: the oldest authority on the religious and social institutions of the Hindus*, translated from the original Sanskrit by H. H. Wilson (New Delhi: Cosmo Publications, 1977).

complex system of oral tradition of any cultural group. See, 'cause they don't have writing at all.

MA: Didn't know that.

AG: So the Australian aborigines are probably, in terms of the non-written culture, the most sophisticated. What they carry in their head; the most sophisticated memory'd linguistic group of any. In other words, apparently the Aborigines have fantastically long epics—and a dazzling aptitude for audiographic memory repetition of those epics. Like Homer—all of their culture, being oral, that means all of their creation myths, metaphysics, stories, narratives, histories—all of that is carried around in the head—and apparently they have an extremely extensive library in their head—because they think that way. We don't have to because we've got the crutch—written language is a mind crutch—we can put it down on a piece of paper so we don't have to remember anything. But they have to remember everything, so apparently they've specialized in not getting hung up on external things—the only possession that they have, for instance—their only tool—is a single-purpose tool which also is used as a headrest—like it's some sort of wooden stick-cane-spoon, which they can sleep on, eat with, kill with, wash themselves with, count on.[69]

MA: Brush their teeth with?

AG: Probably brush their teeth with. Now they reduced that all to a single multipurpose tool—no other possessions, no houses, practically—and so everything is in their consciousness. All other activity is in their consciousness. Like the whales, who are also probably in that situation, or the porpoises—like porpoises have a fantastic language, according to Gregory Bateson, mostly dealing with infinite gradations of interpersonal relations. Subtleties of interpersonal relations that we haven't conceived of—because that's all they've got, they don't have hands, machines, and they don't get hung up on building Empire State buildings—all of their attention is directed at each other.

---

[69]This piece of mythical anthropologic data is adapted after gossip heard from Harry Smith's lips, possibly a hippie bull-roarer (*churinga*) reinvented in Western eternal dream time. See A. P. Elkin, *The Australian Aborigines* (New York: Doubleday-Anchor, 1964), pp. 185–89.—AG

MA: No wonder John Lilly and people like that are really into whale communication!

AG: Yeah—Bateson was working with Lilly, and that was his conclusion.

MA: Directions away from the printed poem. Right there *(pointing at the Uher)*, you've got your direction away from the printed poem—in the tape recorder.

AG: Not really, it is the same thing—the tape recorder's not much different from writing in a notebook.

MA: Okay, I'm trying to think of ways in which—

AG: The way beyond the printed page is music! Bob Dylan. That's the inevitable . . . well. The first way out is simply platform chanting, like William Jennings Bryan, or Vachel Lindsay or Dylan Thomas or myself or whoever makes it on the platform—the vocalization. The bardic thing. Platform bardic—Aah! *(laughs to himself)* Then next—at least in America at this point—it seems historically to have led to a revival of poetry as *song*.

To some extent Dylan was influenced by the whole wave of poetry that went before and he got to thinking of himself as a poet, except a singing poet.

MA: Sure—Woody Guthrie. Singing poet.

AG: Yeah. So song, which fits in with Pound's famous scheme you know, where he says the trouble with poetry is that it departs from song, and the trouble with song is that it departed from dance. In the *ABC of Reading*, or somewhere—*Guide to Kulchur*? Pound says that what happened historically was a big goof. Poetry began—the ictus (beat in the foot) of poetry originally was the Greek footfall as the chorus *chanted* and *danced*. And so the measure originally was literally the physical *stance* or *foot measure*. And that's what the *ictus / hit* is and that's what the measure was.[70]

Then, when poetry left that and got to be just chanting it lost some of the physical base—it got a little bit more disembodied. When it left the physical chanting and went to song it got slightly more disembod-

---

[70]*Ictus* means struck, as with a lance (or foot?)—also remember "A Foot Is To Kick With"—Charles Olson's essay.

ied, when it left song and got to speech it got slightly more disembod-
ied, when it left speech and got to the printed page—and wasn't even
spoken aloud, it became completely disembodied and that's the nine-
teenth-century vagueness of "dim vales of peace" and nothing means
anything any more, words move no thing, it's totally abstract. So the
next step was to bring words back to actual speech; then the next step
after Pound modeling words from actual speech, and Williams' actual
speech, is to bring it to—chanting is the next step—Which is what we
did, chant—the next step was to bring it to song again, which is what
Dylan did, and then the next step will be what Jagger and the others
do, which is shamanistic dance-chant-body rhythm "I wanna go
hooome, no satisfaction!" So what's happening in rock and roll, with
all the body thing which is being laid on, and the dancing—or what the
spades were doing, with all those funny little dances while they're
singing (you know, a quartet singing words and then dancing) is actu-
ally a return cycle,[71] following Pound's original analysis, in a way.

So ultimately what you can expect is a naked, prophetic kid getting
up, on a stage, chanting, in a trance state, language, and dancing his
prophecies, all simultaneously in a state of ecstasy, which is, pre-
cisely, the return to the *original religious shamanistic prophetic priestly
Bardic magic!*

---

[71] "As the old Egyptian devotee sings, dances and perhaps plays some musical instru-
ment before his God . . . ," Sydney Lanier, *Science of English Verse* (New York: Scrib-
ner's, 1920), p. 265.

PAUL CARROLL
1968, Chicago

Playboy, April 1969

Allen considered the following interview to be one of his very best.

In February 1968, at the height of the Tet offensive in Vietnam, antiwar activist Abbie Hoffman called on Ginsberg at his Tenth Street apartment. He told Allen that he and Jerry Rubin wanted to emulate the 1967 Human Be-in in San Francisco (see Albury interview) by holding a Festival of Life in Chicago at the same time as the Democratic National Convention in August. As at the Be-in, Hoffman explained that he and Rubin had in mind a group of teachers, yogis, poets, musicians, and antiwar speakers to address the festival to extend "the feeling of humanity and compassion" present at the Be-in. To show their seriousness, they would even start their own party, the Youth International Party (whose members came to be known as Yippies).

Although the festival idea appealed to Allen, he was concerned about the potential for violence. As he had known Abbie Hoffman for only a few months, he called up Jerry Rubin with whom he had worked since 1965. Rubin reassured Allen that their intent was nonviolent and even asked Allen to refer him to a yogi or swami who could teach breathing exercises to calm crowds of people. (When Ginsberg found out much later that some of the Festival of Life's organizers, including Rubin, had been planning violence in Chicago from the very beginning, he was furious.) Rubin also told Allen that he would be traveling to Chicago to apply for all necessary permits. This made the Festival's success seem more plausible, as Rubin had shown himself to be a skilled organizer of a large peaceful gathering at the Be-in. Ginsberg finally decided to lend his name to the cause, and participated in a March 17 press conference to formally announce the Festival of Life.

By the time Allen flew to Chicago on August 24 to participate in the planned events, however, the atmosphere had turned ominous. The indications from the city of Chicago in particular were not good: When rioting had broken out in Chicago after the murder of Martin Luther King, Jr., Mayor Richard Daley had issued orders to the police to "Shoot to kill." Moreover, Daley's Democratic party machine had refused to cooperate at all with the Festival's organizers. Instead, Daley put all 11,500 Chicago police on twelve-hour shifts, mobilized 5,500 armed National Guardsmen, and airlifted in 7,500 troops under the command of President Johnson from Fort Hood, Texas, all supplemented by 1,000 intelligence agents from the FBI, CIA, army, and navy. Seeing how things were going in Chicago, Ginsberg discussed whether the Festival should be called off. Upon considering the matter, however, it seemed that many young people were planning to come to Chicago to protest the war no matter what happened, so Allen felt that having lent his name to the cause, he was obligated to be present, if for no other reason than to do all he could to lessen the probability of violence. (Details of his efforts are given both in this interview and in the Chicago Seven trial testimony.)

There were many confrontations during the convention, culminating in the week's bloodiest event when the police formed in opposition to a peace march against the Vietnam War inside Grant Park. Not only did the police and National Guardsmen prevent the marchers from going to the site of the convention, but after the march was over, they tried to prevent them from leaving the park at all. A portion of the frantic demonstrators finally found a way out of the park, luckily blending into the Reverend Ralph Abernathy's Poor People's Campaign march, which Daley had permitted in order to prevent a full-scale riot on Chicago's black South Side. For a while the police let the antiwar protesters march along with the Poor People's Campaign, but as they reached the Hilton Hotel, where most of the convention's delegates were staying, the police surrounded the demonstrators on three sides and then waded into them, resulting in over seven hundred persons injured.

Though Paul Carroll had first met Ginsberg in 1958, their relationship deepened in 1959, when Carroll and Irving Rosenthal started the magazine *Big Table* to publish work suppressed by the University of Chicago when it prohibited the publication of an issue of the *Chicago Review* that was to include Kerouac's prose-poem "Old Angel Midnight" and some chapters of *Naked Lunch*. Carroll died in August 1996. The introduction to the *Playboy* interview, after summarizing Ginsberg's role at the Democratic National Convention, explained that "it was in Chicago some time later that the poet was interviewed by fellow poet and critic Paul Carroll. His beard and hair in luxuriant disarray, Ginsberg stretched out on a sofa in Carroll's apartment on the city's North Side and talked for seven and a half hours."[72]

—DC

Playboy: In the past few years, it's become commonplace in newspapers, magazines and on TV programs to describe you as the onetime angry beat poet who's become the joyous leader, guru and elder statesman of the flower people. What do you think about that characterization?

Allen Ginsberg: It's stereotyping—objectionable because not quite humane—to take something living and changeable and fix one robotlike Orwellian image on it, reduplicated to cover all situations, modes and selves. The guru image may fit once or twice, but I don't want to be responsible for being a "nice man" all of the time. It doesn't fit when I'm irritable, bugged busy, want to run the gamut of any sexual desire, or when I just want to go to the movies and be left alone eating an ice-cream cone, or get angry because I'm afraid I'll be discovered secretly carrying a lot of money so I don't have to suffer street-starvation-vagrancy-jail like everybody else. Any stereotype image, like paterfamilias, imposes a role, freezing the life out of personal situations—just so there'll be a comfortable old-shoe guru for the readers of the *New Yorker* or *Time*. It's an image reassuring for them and presumably comfortable for me, but such a stereotype assumes that one's real-life situation has to be labeled all the time. It's the sort of thing that comes through electric mass media, homogenizing and reducing everything to dated lead-paragraph terms. "Beat" and "hippie" are all yesterday's headline bullshit.

Playboy: How *do* you see yourself?

AG: Not in a word or image; I see myself as a being who is being and being more and more. Sometimes that looks heroic, sometimes fucked-up heroic because of having to *be* at all. Such a bad karma!

---

[72]According to Allen Young's introduction to the book version of his interview with Ginsberg (*Allen Ginsberg: Gay Sunshine Interview*), the Carroll interview took place in 1968. This is confirmed by a handwritten note next to a list of interviews that was in Ginsberg's Union Square office that says merely "taped Chicago 1968" next to the entry for the Carroll interview.

Playboy: Some years ago, you wrote, "The message is: Widen the area of consciousness." Critics such as Leslie Fiedler feel that this is the key to both your life and your poetry. Would you describe what you mean by widening the area of consciousness?

AG: We are all blocked off from our own perceptions. The doors of perception have been closed, the gates of feeling shut, the paths of sensation overgrown, the roads of consciousness covered with smog. Blake said our five senses have been closed in, so that we're "moving about in worlds unrealized," as Wordsworth said. Everybody has momentary breakthroughs of consciousness—the vividness of comradely eye-glances, or the sisterliness of plant life, or the iron science-fiction enormity of a police van, or a cow in a glass cage standing silently being tended by an aluminum milking machine, or any crisis between child-hood and deathbed, like war, marriage, mountaintop, saved from drowning, got fired, walked the streets and shuddered at the Wrigley Building, or broke your hip and four ribs in a car crash.

The world that opens up seems strange, familiar but forgotten: more real than the usual place because of deeper feeling—doom significance—but at the same time, frightening. We forgot it's been there all along; it means we have been mad all along. So people, out of shame and fear of exploring the future, and fear of death—fear of life itself—close the doors and go back to their old Safety Habit, and call their own breakthrough a hallucination or freak-out abnormality.

Playboy: Why?

AG: Average young guys have been so heavily conditioned to living in the closed circle of night-club-money-machine-airplane-taxi-office-bank-roll-television-family that they distrust other modes of consciousness and pathways of existence—like knapsack-long-hair-farm-commune-picket-line-street-high-sign—that are viable and real. Here's the danger that the money man, thinking his security is dependent on money, afraid his supply will be cut off like junk from a junkie, may entirely reject his own unconscious, cutting himself off from his own nature and organic perceptions and becoming, as Williams Burroughs says, a "walking tape machine." That's a precise definition of square, because a limited and therefore defensive social consciousness is set up which shuts out other life forms. It causes fear of strange experiences and other people—suspicion of menace in black power and the yellow peril and flowery hippies and the purple virus from Venus—conspiracies to invade our consciousness. Thus rises the whole social paranoia, personal in nature and individual in each man, which is known as the cold war. Now, if it were really safe to stay inside the shell of the white American image—successful, protected, "viable," going up and down office buildings in carpeted elevators—it would be hard to get people out of that state of mind. But it's not even safe in that shell anymore. Limitations of perception imposed by such egotism are biologically and evolutionarily self-defeating. I mean that our refusal to coexist with other life forms is causing a planetary ecological crisis.

Playboy: You've frequently said that LSD is one of the means whereby we can widen the area of consciousness and perception and break out of whatever shell we're in. How, specifically, can LSD accomplish this?

AG: Dr. Jiri Rubicheck, Czechoslovakian psychiatrist, in his book *Artificial Psychosis*, wrote: "LSD inhibits conditioned reflexes." That's what's really significant—and so political—about LSD. That's why there's what's called an acid revolution, as well as why police are against its use. The same people who denounce acid are always calling other people "Communists" or "dope fiends" or "sex fiends" or "unwashed-hairy-nonhuman-creepo-un-Americans" or "intellectuals" or "beat-nik-nigger-Jew-capitalist-conspirators"! Anybody who's different from the Communist-capitalist police-state image is the "enemy"—and particularly anyone who sees through the robot hallucination. If acid helps people see through conditioned hallucinations, then acid's a threat to such police states as now exist in America and in Russia.

Playboy: How many times have you taken LSD?

AG: Not often. Ten or fifteen times.

Playboy: Would you describe what an acid trip is like for you—or is it largely indescribable, as some claim?

AG: LSD perceptions aren't indescribable. I've written some poems during trips I've taken—*LSD 25*, in 1959, and *Wales Visitation*, last year. Since the acid experience can be achieved by other means as well—such as meditation or the social breakthrough that almost everybody felt during the Democratic Convention, free and liberated on Grant Park grass, staring at the Hilton Hotel politics prison—LSD is really like some natural experiences. What does a trip feel like? A creeping sensation comes over your body, a change in the planetary nature of your mammal eyeballs and hearing orifices. Then comes sudden realization that you're a spirit inhabiting a vast animal body containing giant apertures, holes, circulatory systems, interior canals and mysterious back alleys of the mind. Any one of these back alleys can be explored for a long, long way, like going back into recollections of childhood or going forward into the future, imagining all sorts of changes in the body, in the mind or in the world outside, inventing imaginary universes or recalling ones that existed, like Egypt.

Then you realize that all these exist in your mind simultaneously. Slowly you approach the mysterious feeling that if all these histories and universes exist in your mind at the same time, then what about this one you're "really" in—or *think* you are? Does that also exist only in your mind? Then comes a realization that it *does* exist only in your mind; the mind created it. Then you begin to wonder, Who is this mind? At the height of the acid experience, your mind's the same mind that's always existed in all people at all times in all places: This is the Great Mind—the very mind men call God. Then comes a fascinating suspicion: Is this mind what they call God or what they used to call the Devil? Here's where a bum trip may begin—if you decide it's a demonic Creator. You get hung up wondering whether he *should* exist or not.

To get off that train of thought: You might open your eyes and see you're sitting on a sofa in a living room with green plants flowering on the mantelpiece. Outside the window, wind is moving through big trees; there's a huge being moving through the street in all of its forms—people walking under windy trees—all in one rhythm. And the more you observe the synchronous, animal, sentient details around you, the more you realize that *everything* is alive. You become aware that there's a plant with giant cellular leaves hanging over the fireplace, like a huge unnoticed creature, and you might feel a sudden, sympathetic and intimate relationship with that poor big leaf, wondering: What kind of an experience of bending and falling down over the fireplace has that stalk-blossom been having for several weeks now? And you realize that everything alive is experiencing on its own level a suchness existence as enormous to it as your existence is to you. Suddenly you get sympathetic, and feel a dear brotherly-sisterly relationship to all these selves. And humorous, for your own life experiences are no more or less absurd or weird than the life experience of that plant; you realize that you and plant are both here together in this strange existence where trees in the sunroom are blossoming and pawing toward the sky. Finally you find out that if you play them music, they grow better.

So, the widened area of consciousness on acid consists in your becoming aware of what's going on inside your own head cosmos—all those corridors leading into dreams, memories, fantasies—and also what's happening outside you. But if you go deep enough inside, you may find yourself confronted with the final problem: Is this all a dream-nature? Great ancient question: What *is* this existence we're in? Who are we? Then can come what Timothy Leary terms the "clear light" experience or, as they call it in South America, "looking into the eyes of the Veiled Lady"—looking to see who it is, doing or being all this. What's the self-nature of it all? This is the part of the acid experience that's supposed to be indescribable, and I'm not sure I've had the proper experience to describe it.

Everything turns out to be all one great conscious Self whose organs are every different living being, so that this Self conceives and perceives in every different possible way at once, vaster than words. But there's also a sensation that the entire universe is a Happening. Occasionally, the Happening seems a bit stagy: I mean that it could exist in the form of fireplaces with the plants hanging over them and police states and the glare of blue lights and Chicago tear gas actually drifting through floodlit Lincoln Park in great waves over Christ's cross, like an old World War One movie scene.

Playboy: Though LSD may have widened consciousness for you and others such as Dr. Leary, what do you think about attacks on the drug by a number of doctors and psychiatrists who argue that chromosome breakage in blood cells may occur after only two or three usages of the drug? Such breakage, they claim, could cause subsequent children of the users to be born abnormal, retarded—or both.

AG: It's a pile of unscientific crap. Or in the favorite words of bureau-cratic double talk, as one of the hydraheads of the Food and Drug Administration might put it: "No causal relationship between LSD use and chromosome breakdown has been experimentally established with scientific method other than the normal breakage equivalent to slight excess use of aspirins, coffee, Coca-Cola. Besides which, Portland, Oregon, antihedonist Professor Irwin, the original chromosome-

Frankenstein theorist, forgot that half his eight subjects were Meth freaks anyhow." Refreshing views from Dr. Goddard, ex-head of the FDA, or somebody using him intelligently for a ventriloquist dummy. Basically, the chromosome breakdown will turn out to be a spook story. Innumerable rigorous evaluations of the bibliography of learned scientific journals on the subject boil down to the conclusion that nothing special can or need be concluded except that everybody ought to do a lot more research on how to make acid 100 percent foolproof. The Government is not doing that; quite the opposite, in fact. It's spending appropriations in the Pentagon to produce bum-trip acid for military uses.

Playboy: How do you know this?

AG: One, I've read it in the underground newspapers; and two, it's been reported in the *New York Times*, under the terminology: "R and D appropriations allocated for investigation of military uses of psychotoxic and psychedelic substances with special subapplication to domestic riot control." Look it up in the *Times* index—or ask your Congressman before the credibility gap falls into the abyss.

Playboy: OK. In a recent interview, Dr. Leary recalled that during an LSD trip you took in 1960 at his home in Cambridge, you said how "this mushroom episode had opened the door to women and heterosexuality" and how you could see "womanly body visions and family life ahead." Is that true?

AG: Well, I get those feelings every time I take acid. On a trip, you enter corridors inside, and into the heart. Naturally, you'll come upon old feelings you didn't know were there and were ashamed of, like loving your mother and realizing that you and she were one and that you'd separated from her because you couldn't stand the fear of being one with her. And realizing that all women and your mother are one—for myself, at least—I cut myself off from all women because I was afraid I'd discover my mother in them, or that I'd have the same problems with them that I had with her.

In much the same way, the heterosexual man may discover during a trip the *natural* homosexual identity in himself—an identity suppressed by our culture but not by many others. As Whitman observed, if the natural love of man for man is suppressed, men won't be good citizens and democracy will be enfeebled. What Whitman prophesied was an adhesive element between comrades—the "sane, healthy love of man for man." But because of suppression of feelings in America, the overemphasis on competition and rivalry—a tough guy, *macho*, hard, sadistic police-state mentality—American men are afraid of relationships with each other. It's almost as if there's been a plot to separate man from his heart by making him afraid of being a fairy or a queer or a faggot or a queen. Real feeling can be recovered, though, because it's natural. But the official police form is that masculine tenderness is homosexuality, to be treated as a womanly weakness or poodle-dog-like perversion. So finally men are ashamed of themselves and as a result tend to torture each other. What you see recently, however, is the reappearance, in the form of long hair and joyful dress, of the affectionate feminine in the natural Adamic man, the whole man, the man of many parts.

Playboy: Would you explain what you mean when you say there's a natural element of homosexuality in every man?

AG: There's homosexuality in every *Playboy* reader. To say that in a *Playboy* interview is interesting because obviously every *Playboy* reader expects me to say that; so I'll say it and liberate him from his fear that somebody will say it sooner or later. So I hereby announce: Everybody is acknowledged not as a homosexual or heterosexual but as a complete person with all the aspects of that completeness—all the dreams, hard-ons, wet nightmares, anxieties, buddies, all secret masturbations and all refusals to masturbate. Any more rigid masculine ideal would be a perversion of human nature—heartbreaking because unsatisfiable.

Playboy: Have you been able to fully accept your own homosexuality?

AG: Homosexuality has been like a koan—a Zen riddle—for me. Whole areas with my mother were screwed up and conditioned me in this way sexually. The riddle was: How do I deal with my homosexuality? Do I accept it or reject it or freak out, or do I go into it and find out what it is? Another problem: Is it something public? Anything that common is public; anything that happens to us is as good or bad as anything else as a subject for poetry. It's actual. So I can write naturally about my own homosexuality. The poems get misinterpreted as promotion of homosexuality. Actually, it's more like promotion of *frankness*, about any subject. If you're a foot fetishist, you write about feet; or if you're a stock-market freak, you can write about the rising sales-curve erections in the Standard Oil chart. When a few people get frank about homosexuality in public, it breaks the ice; then anybody can be frank about anything. That's socially useful.

Playboy: Is that what you meant when you told *Life* that by announcing in public that you're a "homosexual, take drugs and hear Blake's voice, then people who are heterosexual, don't take drugs and hear Shakespeare's voice may feel freer to do what they want and be what they are"?

AG: Yes, then anybody who wants to can get up and say, like, "I fuck girls!" or "I'm not scared to wear a Brooks Brothers suit" or "I wear my hat indoors or out as I please," which Whitman said. But I don't stand up in public and suddenly announce, "I'm a bearded-beatnik-bohemian-faggot-dope-fiend" to boast about it. When somebody asks me: "Why don't you shave?" or "Why do you have so much homosexual imagery in your poems?" or "Are you willing to admit you smoke marijuana?" and "You look as if you have Communistic tendencies" or "You need a good bath!"—well, then, I say: "My beard just grows, I didn't plant it, I don't get up every morning and try to murder my hair and obliterate my human image. It's just Adam's hair. Yes, I like to make it with boys; I'm not sure whether it's good or bad; it feels all right so I describe it. And I admit I smoke dope. But I think police-state bureaucrats mounted their secret conspiracy to suppress marijuana in order to create police-state conditions. And I am a Communist of the heart, except that I've been bricked off the set by police in Communist Prague and Communist Havana—and 'Communist' Chicago. I was kicked out of Havana and Prague for *talking* about homosexuality."

Playboy: It doesn't sound as if you buy the psychoanalytic theory that homosexuality is a neurosis that cripples or limits a man's emotional growth.

AG: Homosexuality is a condition, and like all average things, it has advantages and disadvantages. Obvious disadvantages are that it keeps you from reproducing your own image, if that's biologically important anymore; and it shuts me off from full relations with women. Though unless a chick is really trying to make it with me, I'm affectionate and physical and sexy enough toward women to give out some normal social, happy cheer when I'm with them. The advantages are that homosexuality provides me with sufficient affection and gasoline to communicate on a tender level with my fellow citizens, especially the Prussian butch-crewcut freaky military types—the old Socratic situation. Also, because it alienated or set me apart from the beginning, homosexuality served as a catalyst for self-examination, for a detailed realization of my environment and the reasons why everybody else is different and why I am different. In a tank-military hyper-sadistic overmasculinized society fearful of sensitivity and the unconscious and the full man, my homosexual specialization made me aware of the rigid armoring, defensiveness, overcompensation and high camp put on by police-state police.

It's like the old shamans who are often androgynous or homosexual: Since they're outside normal routine, they're specialized social critics and have sensitivities that others don't have; they're men who see aspects of male history from a woman's point of view. That spectrum of experience is a useful information bank of supplementary intelligence that can be of real value in community self-understanding and awareness. Anyone in that position has enough troubles fulfilling such heavy duties to the society without being hit on the head for being a fairy; he should be kissed, instead. In fact, innumerable young men ought to offer their bodies to him in order to recompense him for the suffering solitariness of his freaky prophecyhood. And they should come up offering their bodies before I get too old to enjoy it.

Playboy: You mentioned that not having children is one of the disadvantages of being homosexual, and that you envision "family life" ahead during LSD trips. Do you still want to be a father?

AG: I did a while back, but I ran into a funny, long-haired Indian Vishnuite to whom I talked a lot about my problems. He said, oddly, "Give up desire for children." Which made me mad. Who was he to tell me to cut myself from that desire? Later, I realized what he meant: Give up attachment, compulsion to have children on account of you're a Jewish boy from New Jersey; if you want children or if they come, fine, but don't have children because you're *supposed* to. Anyway, there are already too many people and lost unattached children in the world today. So I'm an old cranky bachelor wanting to stay with my poetry and run around doing whatever thing I'm doing, and I think I might be satisfied to leave it at that. Still, it might be good to have this self-importance broken up by "a Zen master in the house all the time," which is how Gary Snyder, the poet, describes his first child.

Playboy: Do you think you'll ever marry?

AG: I don't yet feel enough of that erotic romance around the belly for a chick—not enough to want to contract to stick with one woman the rest of my life. I don't even have that kind of erotic heat anymore to want to sleep with just one man. But I certainly have more heat for men, so it would be a shame to hang up some chick just to have a child or a companionable marriage. Maybe if there were some chick I dug who had the same detachment as myself and who wouldn't suffer continually from being unsatisfied by my lack of erotic interest, a marriage would be all right. Certainly I wouldn't get married just to have the appearance of being married.

Playboy: Have you ever made love to women?

AG: Lots of times.

Playboy: Do you enjoy it as much as you do with a man?

AG: Well, sometimes it's just as good. I get into a deeper emotional intimacy if the chick is lissome and springy, skinny and pretty. I like little blonde furry fucky dolls.

Playboy: Is there any kind of sexual act that you'd consider a perversion—with a man or woman?

AG: I don't know what we mean by perversion. Some sex acts are "perverted" when they get self-destructive or obsessive; incest might be one. I've always had an anti-incest block, a hypersensitivity about that. I've had wet dreams about everybody in my family; I suppose everybody has, whether they remember them or not. But grooving incestuously at a very early age with brother or sister or father or mother would tend to close in the circle of contact, limit the expansion of social mobility and become a habit—like junk. Incest would really complicate the normal problems of independence that most people have, which they solve by leaving home or going on the road.

Playboy: Would you agree with Norman Mailer's claim in his *Playboy* interview that such acts as cunnilingus and fellatio are perversions because they substitute for the normal heterosexual act of penile orgasm within the vagina?

AG: Ideally, orgasm inside a woman is a complete act natural to the construction of the genitalia and pelvis and the whole interlocking muscular system from top of scalp to tip of toes. In a complete Reichean orgasm, one would presumably experience a total orgasmic conscious glow throughout the body—tingling everywhere. So when you get or give head, there are probably dysfunctions of cosmic glow.

But, like, "Let the crooked flower bespeak its purpose in crooked-ness, to seek the light / Let the straight flower bespeak its purpose in straightness, to seek the light." The anal-sphincter-prostate orgasm some men are capable of having is a great opening of feeling and delight and an extraordinarily beautiful experience, and rare. Possibly everybody should experience it for his own humanity, good judgment, tolerance and empathy in understanding feminine sensations as well as masculine nature in the human mammal and the universe itself. I don't know whether it's anything that needs to be recommended universally, but I do feel that whenever it happens it should be honored rather than despised, just as mountain climbing and courage in the boxing ring are honored—or any extension of natural faculties to experience high, luminous extremes of awareness of nature.

Playboy: In an interview in the *Paris Review* (see Tom Clark), you told how you felt despair in 1948 about the possibility of ever finding any "psycho-spiritual sexo-cock jewel fulfillment" in your life. Have you found it?

AG: Yes, I've found the lightness and liberty of experiencing the satis-faction of most of my sexual fantasies. But I've also found the resulting bad karma—like now, ten years after Peter Orlovsky and I became lovers, we've had to detach ourselves sexually from each other.

Playboy: Why?

AG: Our relationship was a big long fantasy that finally got played out. Time changes, the body turns to ashes.

Playboy: Will you and Peter stay together?

AG: Yeah, we like each other; we're old friends. As a matter of fact, once we'd reached a dead end homosexually, our relationship became lighter and happier. Now it's between two equals who've had a revolution within themselves that freed them from each other; we look at each other now as if we're newborn angels who shared an old history in another life.

Playboy: Has the fact that you're now an internationally famous poet changed your sex life in any way?

AG: When I was elected the King of May in Prague in 1965, I made it with all sorts of beautiful Middle European 17-year-old blond cats. Having a fame identity makes it easy to make it with young kids who are, like, friendly. I went through a big run of making it with young cats in San Francisco last July. I went to bed with almost everybody who'd stand still for it. About a month before, I'd written a long poem exploring the anal slave-master sexual-drama fantasy. In this poem, I wanted to be the slave. I'd already written another in which I was master; I wanted to try the whole thing.

At first, I wasn't sure I could read the new poem in public—it was so far-out and intimate and real. But I finally decided that this kind of fantasy is sufficiently universal to be of general interest, that it isn't a peculiar or private aberration, and that reading it wouldn't be an act of excessive exhibitionism. So I read it at the "Rolling Renaissance" poetry reading in San Francisco before a giant funny audience of squares, hippies, high school kids and old bohemian poetry lovers; they seemed to dig it. Later, I was in a gay bar on Grant Avenue—gay bars there are groovy now, all the kids have long hair and motorcycle jackets, they're friendly and first-rate and don't look like fairies but like strong young men—and I met this kid who said my poem had turned him on. So we made it. It was like living out the fantasy described in the poem.

Playboy: Why do you say that anal eroticism is a universal fantasy?

AG: When I was an adolescent, I'd have assumed it was just my particular Dr. Jekyll-Mr. Hyde scene, though everybody's got something going, whether it's foot fetishism, licking eyebrows or God knows what. But in 42 years, I've read books and made it with a lot of cats and talked with married friends, and realized that anal pleasures are so common that they're recognizable as part of almost everybody's secret mythology. Almost everybody enjoys what could be called the erotic pleasure of a good shit; being screwed in the ass is only an extension of that sensation. Since on our separate islands we all have that same coconut tree, there's no harm; and it might be a blessing to take the hex and bane and guilt off the subject with a "public" poem. Once it becomes no longer a secret, romantic thing, it turns into another common, charming quiddity—a "humour," in Ben Jonson's sense of the term.

Playboy: As a poet who's become famous for his erotic verse—and for his brutal candor—why were you so hesitant to read this poem before an audience?

AG: I don't know; it was the first time in years that I've really been scared to read something I'd written. When I get to a barrier of shame like the one I felt when writing this poem, I know it's the sign of a good poem, because I'm entering new public territory. I write for private amusement and for the golden ears of friends who'll understand and forgive everything from the point of view of *humani nihil a me alienum puto*—"Nothing human is foreign to me"—but it's fearsome to make private reality public.

Playboy: Are you saying that your poetry is an exorcism of shame?

AG: An exorcism of fear. Shame is just one aspect of fear. I felt much the same when I wrote "Howl" in 1955. That poem also refers to getting fucked in the ass, but only by allusive mention. In this poem, it's a deep-end description from the lips of the anus to the bottom of the bowels—what it feels like and what the fantasies are—all done in an ecstatic, rhetorical manner.

Playboy: In your poem *Death to Van Gogh's Ear!*, you wrote that you'd "die only for poetry that will save the world." What, exactly, did you mean?

AG: I meant that the only thing that can save the world is the reclaiming of the awareness of the world. That's what poetry does. By poetry I mean the imagining of what has been lost and what can be found—the imagining of who we are and the slow realization of it. First come prophetic images from the unconscious—like the scary image of Moloch, "eater of children," in "Howl"—and then the gradual realization that such an image isn't merely an "image" but an articulation of what one actually sees and experiences. See, back in 1959, Peter and Gregory Corso and I read at a benefit for *Big Table* magazine in Chicago. Not quite unconsciously, we said there was a god abroad in the land that ate children. At that time, it wasn't clear whether Moloch existed only in our imaginations. But today, that same Moloch, "whose eyes are a thousand blind windows," looks like Mayor Daley's main civic concern; he's building larger and larger, more demonic robotlike Moloch buildings in mid-Chicago till finally one sees this 100-story black John Hancock Tower of Babel, whose site and shape are by-products of usurious land speculation. What's sacrificed to such a Moloch are the care and cultivation of Chicago's tear-gassed children—and greenery, and the souls of men. In a more general sense, Moloch is the military-ward-heeling-IBM police state we've been living in for years without knowing it. What I didn't realize twenty and even ten years ago was that images from the unconscious that went into my poems, which I thought were visionary and transcendental, were really literal realism, simple common sense.

Playboy: If you say so, Allen. Do many other poets share your conception of poetry as an embodiment of prophetic perceptions rather than visionary imaginings?

AG: Yeah, I think most do. American poets have always been one of the real sources of news—news you couldn't get from *Time/Life*. A lot of poetry is coming true today, in the same way that a photograph reveals itself as it's being developed. Like Dylan's *Blowin' in the Wind*; that song could have been any little boy's lyric fancy, but when it was played one afternoon during the convention in Grant Park across from the Hilton, it revealed itself as prophecy all along, because it described what was going on right there on the grass. Crowds of strange children with long hair, who weren't afraid to have their bodies hit by police phantoms armed with billy clubs, were demanding reality and truth from business-delegates who were walking around in upstairs Hilton rooms scared of the stink of their own karma. Tear-gassed! That scene was, literally, blowing in the wind. Was it going to be a police state or a liberation from what had been a police state all along?

When I wrote "Howl," I thought it was like something in the Gnostic tradition, in that only a few companions of the Grail would recognize the humor of a lot of the rhythms and images. What I didn't anticipate was that there were so many companions of the Holy Spirit in America—or that *everybody* is really inhabited by the Holy Spirit. By Holy Spirit I mean the recognition of a common self in all of us and our acceptance of the fact that we're all the same one.

Playboy: Hundreds of thousands have read your poems and many have been disarmed by your frankness. But many critics and some of your fellow poets complain about what you conceded a few minutes ago was a degree of exhibitionism. What do you think of such charges?

AG: Exhibitionism as they use the word is a classification of so-called psychosis invented in a society so repressive that any frank revelation of what's going on has to be characterized as a form of madness. It would be inappropriate revelation if you'd seize on the revelation of your own genitals in the park, scaring people, as a symbol of that common self and desire to get through. But that's all exhibitionism is: somebody trying to communicate, get out of his shell, break out of this prison we're all in. As such, it should be acknowledged: A guy should be allowed to parade himself; in fact, the cure for exhibitionism would be to have a special day for walking along Michigan Avenue exhibiting one's genitals. That should satisfy an exhibitionist: he'd really be out of his prison then; we wouldn't have to worry about the problem anymore. What I was digging in poems like "Howl" was the humor of exhibitionism. You're free to say any damn thing you want; but people are so scared of hearing you say what's unconsciously universal that it's comical. So I wrote with an element of comedy—partly intended to soften the blow. At first, people think this is overexertive exhibitionism, but on second thought, they realize that it's entirely serious and perfectly normal, natural and real.

Playboy: Is that what you intended when you took off your clothes in 1957 during a reading in Los Angeles? The story goes that a heckler in the audience asked, "What are you trying to prove in this poem?" and you answered, "Nakedness," and when he demanded, "What do you mean by that?" you took off your clothes without a word.

AG: Yes, but the act was in context. After he'd asked what I meant by nakedness, I wondered, What *did* I mean? And I thought: Nakedness. That's what I meant. So I took off my clothes. That story gets retold with the implication that I take off my clothes at innumerable poetry readings or that nobody had asked me any question in the first place which takes out of focus the precision, clarity and normalcy of the gesture. Another anecdote that keeps recurring is about the question a lady asked me at that *Big Table* reading in Chicago: "Why is there so much homosexual imagery in your poems?" My answer was: "Because

I'm queer." I was writing about actual feelings; those images arose nat-
urally. I wasn't *boasting* that I was queer; I was simply answering her
question. But the story gets retold as if I intended some kind of Oscar
Wilde sensationalist answer. The point was: The lady didn't understand
that I *had* homosexual feelings; she seemed to feel that poetry meant
writing about flowers one never saw or places one never visited—like
the seacoast of Bohemia. Apparently it never occurred to her that I was
writing about something simple and real. She was probably also chal-
lenging me to be *ashamed* of being homosexual. If that lady read Shake-
speare's *Sonnets*, she'd probably feel that all his allusions to his young
boyfriend were some kind of literary conceit or flowery imagery. What
rose out of Shakespeare's soul was universal; he was angry because his
buddy went off with a Dark Lady on another motorcycle.

Playboy: Your longtime interest in Eastern religions—particularly
Zen—as well as your revelations about homosexuality, adds to your
image as one who stands outside the mainstream of American life.
How did your interest in Zen Buddhism begin?

AG: In 1948, as I mentioned earlier, I'd had some visionary experi-
ences while reading Blake, but I hadn't been able to find words that
seemed to articulate them. Then, in 1953, I saw a scroll painting by
Liang Kai called *Sakyamuni Coming Out from the Mountain*, which
showed Sakyamuni Buddha with long, tearful eyebrows and big ears,
looking as if he'd been on the mountain a long, ascetic year and had
experienced a comedown enlightenment of some kind. Around the
same time, I got a big, sorrowful, enthusiastic letter from Jack Kerouac
in San Jose about the *Diamond Sutra* and satori, or illumination. The
word "satori" seemed to fit my earlier spontaneous illuminations.
That led into Oriental poetry, yoga and travels.

Playboy: Have you had other visionary or mystical experiences?

AG: Well, I had a trance experience with mantra chanting in Chicago
during the Democratic Convention. You may know, a mantra is a short
magic formula or prayer with syllables consisting of the names of
deities—Hindu, Buddhist, Tibetan, Japanese—who can be interpreted
as aspects of one's unconscious hopes, desires, fears. For example:
Krishna is Hope for the Preservation of the World. Siva: Realization of
Enormous Changes. Tara: Mother's Tears and Compassion. Dharma:
Brothers' Justice. Buddha: Self's Throne-Power, OM. According to the

Hindus, there are three major aspects of experience to dig: One is the inconceivable, the unborn, the void—what we know is out there after death, and perhaps also before birth; the ground out of which the universe imagined itself. That's called Brahma.

Then comes the second aspect—the world of names and forms, preservation, stability and responsibility, and the hope that returns after the Flood. Noah, and Christ resurrected, return to save human beings every time human evil gets so heavy, as it has today, that the planet seems threatened with destruction. That's Vishnu. Vishnu has many forms and returns over and over again. Krishna is one of his forms—a blue-bodied cowboy with a flute. Creation and destruction, birth and death, change: That's the third aspect, Siva.

The Hare Krishna mantra is a round of the names of Vishnu repeated again and again: "Hare Krishna Hare Krishna, Krishna Krishna Hare Hare, Hare Rama Hare Rama, Rama Rama Hare Hare." If you sing it continually in a sweet tone, you can use it as a vehicle for any emotion you're feeling, and also as a method to regulate breathing and point your consciousness to one place where your body is. Repeated over and over, the mantra can lead to a regularization of all the body's rhythms in one even, tranquil, harmonious, continual, unending, pleasurable, recurrent, reassuring tune. You can take these mantras, sing them over and over again in any tune you want, and you'll find it's a way to involve your whole mind and body, through breathing, in one single-minded activity that's both contemplative and expressive. Although there are all sorts of intellectual-mystical-theological potentials involved in chanting a mantra, on its simplest, most Americanesque level, it's just like singing in the shower or an interesting phys. ed. that can get you high.

Playboy: High on what—your own consciousness?

AG: That would be too inaccurate a way to put it. High by means of focusing your consciousness in one place to deepen your awareness of that place, which is your body. By means of rhythmical regularization of breathing, there's a slow alteration of chemical metabolism in the body, which, in turn, awakens existent but unrealized physiologic electric sensations and densities and neural patterns, and clarifies the consciousness of the inner observer who's putting the body through such motions and training it and the mind to sit still and think.

Playboy: Who is the inner observer? One's self?

AG: Us. It. One. Chicago. The planet. Anyway, I had this extraordinary experience chanting OM here in Chicago. On Sunday afternoon—the day before the convention began—a lot of us were wandering around Lincoln Park when unexpectedly the police showed up with guns and clubs. Nobody knew why or if the police were going to attack. Panic—a few people freaked out. Some of the Maoists were acting insulting and revolutionary in their ideological prophetic style. Police fear everywhere. So I sat down and began chanting OM. I thought I'd chant for about 20 minutes and calm myself down, but the chanting stretched into hours, and a big circle surrounded me. A lot of people joined in the chanting. Then somebody passed me a note on which an Indian had written: "Will you please stop playing with the mantra and do it seriously by pronouncing the 'M' in OM properly for at least five minutes? See how it develops." I realized I'd been using the mantra as song instead of concentration, so I started doing it his way. After about 15 minutes, my breathing became more regular, even, steady—as if I were breathing the air of heaven into myself and then circulating it back out into heaven. After a while, the air inside and outside became the same—what the Indians call prana, the vital, silvery, evanescent air.

Then I began to feel a funny tingling in my feet that spread until my whole body was one rigid electrical tingling—a solid mass of lights. It was around eight P.M. now and I'd been facing the John Hancock Building, which was beginning to light up. I felt like the building, except I realized it wasn't alive and I was. Then I felt a rigidity inside my body, almost like a muscle armor plating. With all this electric going up and down and this rigid muscle thing, I had to straighten my back to make a clear passage for whatever flow there was; my hands began vibrating. Five or six people were touching them. Suddenly, I realized I was going through some kind of weird trance thing like I'd read about in books. But it wasn't mystical. It was the product of six continuous hours of chanting OM, regularizing breathing and altering rhythmic body chemistry.

Playboy: Did it feel good, Allen?

AG: Oh, yeah! Powerful, good, solid. I felt my body was *mine* in a funny way; I put my legs in a full lotus position, which I can rarely do. I realized that it was possible, through chanting, to make advances on the body and literally to alter states of consciousness. I'd got to euphorias, ecstasies of pleasure, years before; I'd gotten very far with feelings—but this was the first time I'd gotten into neurological body sensations, cellular extensions of some kind of cosmic consciousness within my body. I was able to look at the Hancock Building and see it as a tiny little tower of electrical light—a very superficial toy compared with the power, grandeur and immensity of one human body.

Another familiar thing I recognized during the trance was the animal, brown, snaky, sentient living presence of some big trees standing outside the circle of chanters. I realized that those trees had more going for them than the Hancock Building; they were *alive*, at least, and so to be respected, observed and communed with—in the sense of being noticed in one's consciousness as they hiply signified their own trunkhood and leafage. They looked like great big doggy-trees.

Playboy: How long did this go on?

AG: I kept chanting till ten P.M. Boy, what a thing! I'd never chanted for eight hours before and thank God I could do it on that occasion. It felt like grace. It felt harmoniously right that some psychophysical rarity should be happening on that political occasion as Sunday dusk fell on Lincoln Park and the Hancock Building lit up on the horizon. If there'd been panic and police clubs at that moment, I don't think I would have minded the damage. Clubbing would have seemed a curiously impertinent intrusion from skeleton phantoms—unreal compared with the natural omnipresent electric universe I was in; the cops would have been hitting only one form of electric. The fear of death was gone, in the sense that I recognized I was already dead; I was a revolving mass of electricity. I was in a dimension of feeling other than the normal one of save-your-own-skin. I was so amazed and gratified that I don't think I would have minded any experiment, including death. That was the most interesting thing that happened, for me, in Chicago—more interesting than marches and conventions and glittering Hiltons and giant Galbraiths moving like phantoms through the city. But I think everybody who watched television during the convention experienced a widening of consciousness.

Playboy: In the sense you spoke of earlier?

AG: Yes. Because of the social imagery they saw on the screen. Outright police brutality was shown so clearly that even TV and radio commentators were saying: "This is a police state!" Before Chicago, that would have been considered an impropriety, even though many already felt it was true, secretly. To make it official like that turns things over in people's minds; suddenly they wake up in a different country from where they thought they were. But it was there all along! People realized that they knew it was there but were afraid to recognize it, because that would mean being caught in a nightmare they didn't want to confront. It's like a smoking cough; you'd prefer to ignore it rather than face the fact that you're getting cancer. Or like not wanting to turn around to see if anybody's following you, for fear somebody really is. But when you do turn around to see, you widen your area of consciousness.

Playboy: What made you decide to go to Chicago in the first place? Surely you're not that interested in party politics.

AG: Well, the original fantasy was to hold a Festival of Life; I thought of it as a continuation of the Human Be-in that happened in January 1967 in San Francisco.[73] What would happen if all the psychic heroes of America—breakthrough artists and manifesters of consciousness like Tim Leary, William Burroughs, Ed Sanders, Buckminster Fuller, Paul Goodman and some of the great Digger-anarchist cats—were to assemble a universal academy in Chicago and give everything away free, mingle with each other and all the younger people whose consciousness is in tune on the new transcendental rock-'n'-roll revolutionary sexual aesthetic planet level? Jerry Rubin and Abbie Hoffman began the idea, and the Beatles and the Rolling Stones and Dylan were supposed to be invited to celebrate, as well as all the swamis and Hare Krishna singers in America and anybody else with a nongrasping nature and a constructive demeanor. Like, have a wild grass-roots planet academy festival where teeny-bopper poet-revolutionaries and technological prophets could be let loose to sing naked and exercise the great humane arts free—and blueprint a new nation.

Imagine all those mad folk together in a classical *Kumbh Mela*[74]—a gathering of the tribes. Every young kid in America who had any spiritual or artistic interests or any penetration into his consciousness

---

[73]The interview quotes Allen as saying January 1966, but the Be-in actually took place on January 14, 1967, on which see the Albury interview in particular as well as the Chicago Seven trial testimony.
[74]The *Kumbh Mela* is a periodic coming together of spiritual seekers and leaders in India.

would want to come, learn, contribute, get laid, amuse himself, turn on—and *think*. And it would be a marvelous way to expose young kids to traditional intelligence as it's been adapted to psychedelic consciousness by their elders, making them aware of the larger world of planet history, new attainments in articulation of community and the links with older traditions.

Playboy: Couldn't you have held your festival more peacefully in a less troubled place and time?

AG: The whole idea of a street-theater festival was to gas out the political-drag-Neanderthal scene in the Amphitheater. What kind of show could the Democratic National Convention stage that would have rivaled that? Old, tired theater—that's what the convention was, with its showbiz bunting, bands, flags, people making speeches written for them and acting roles ordered by city bosses. No joy or spontaneity at all.

Playboy: How do you feel your Festival of Life turned out?

AG: Well, it didn't come anywhere near its goal, obviously. For one thing, few people realized what a locked-up police state Chicago was, just like Prague—and how much stealing, guns and Mafia and outright illegality there was. Police and City Hall break laws and lie. Chicago has no government; it's just anarchy maintained by pistol. Inside the convention hall it was rigged like an old Mussolini strong-arm scene— police and party hacks everywhere illegally, delegates shoved around and kidnapped, telephone lines cut. And Daley himself mouthing curses at Ribicoff on TV.

Playboy: What do you think caused the violence in Lincoln Park and in front of the Hilton Hotel?

AG: None of those assemblies in the park or marches were violently provocative. Provocation was ultimately on the part of the city, whose threats, shows of force, tear gas and physical attacks triggered every specific instance of violence. The city also helped to provoke the violence by insisting on technical interpretations of law that only foment riots—such as refusing to make humane adaptations of rules to let people sleep in the park. And the city's restrictions prohibited free movement of people in what is, after all, their own territory. "The streets belong to the people," as Abbie Hoffman kept saying. "The streets belong to whoever we say can have them," the cops replied. Obviously, that's a confrontation.

Playboy: Did you see any acts of violence by the demonstrators?

AG: Sure. There was lots of violent language after people were beat up, but there weren't many kids who wanted outright revolutionary violence or bloody confrontation. Of course, there was a small group of "terrorists" who believed in stink-bomb sabotage as therapy, and a few youthful drunks who were violently resentful of all symbolic authority. But there were no more of these than in the normal population of any country club.

Actually, the tragicomedy of the Daley-police position was that the city was paranoiacally obsessed with any sign of inflammatory language or behavior by rare revolutionary birds. So they beat up an entire cross section of typical healthy American youths, avant-garde flowery longhairs, the entire mass media and all those in the "Clean for Gene" suite[75] at the Hilton. They also tear-gassed half of their own force on occasion. As Burroughs remarked, it wasn't that the police intended to beat up on citizens; it was that they literally did not see whom they were attacking in their hysteria.

It became a question of how to handle such confrontations. What kind of street theater would best make one's point? One possibility was a theater of resistance—by using your body to try to hold the ground and being ready to have your head clubbed.

Playboy: Isn't that what the police would call "asking for it"?

AG: Are Americans reduced to having to regain liberty by violence—like in the American Revolution? Obviously this is the question on everybody's mind. I'm convinced there's another way. Blake has some lines appropriate to armed resistance: "Thy brother has armed himself in steel / To avenge the wrongs thy children feel; / But vain the sword and vain the bow / They never can work war's overthrow." I'm willing to die for freedom, but I'm not willing to kill for it.

Playboy: Would you literally die for freedom?

AG: After this Chicago experience, yes. This police state's unreal.

---

[75]In an effort to help elect Eugene McCarthy, many antiwar youths cut their long hair off and shaved their facial hair.

Playboy: You mentioned resistance as one form of "street theater" with which the police could be confronted. That was just what happened in Chicago. How would you have preferred to confront them?

AG: Organized chanting and organized massive rhythmic behavior on the streets, shamanistic white magic, ghost-dance rituals, massive nakedness and distribution of flowers might have broken through the police-state hallucination-politics theater wall. Now, nobody got naked in Chicago, but the few times there was communal chanting of mantras, that proved helpful. I've described one of them. A few other times, when everybody felt trapped in confrontation-fear—lights glaring and tear-gas anxiety and Mace and police ready to attack at midnight and a few people up front on useless barricades shouting and banging on trash barrels, like some scene from a Warsaw-ghetto movie—a group came along chanting OM and transformed the scene into another kind of glorious pageant theater. It was like a religious service, chorales under the trees and midnight sky.

Playboy: But the chanting, of course, didn't stop the violence.

AG: As a matter of fact, it did stop a lot of violence; it really calmed several scenes where police didn't have remote-control orders to attack. But it didn't stop all the violence, because the chanting wasn't participated in universally by all the people in the park. To succeed, it would have required an unbroken circle or at least a majority of participants in order to set up mammal-vibrations strong enough to be irresistible.

Playboy: If that's true, why did you chant OM all alone one night from the gallery of the convention hall?

AG: Better one prayer than none at all—and by then lots of people knew what it was for. When the Wisconsin delegate asked that the convention be adjourned for two weeks because of all the beatings in the streets, I jumped to my feet in the balcony and began shouting OM. It was sort of a Phantom of the Opera feeling—like I was swinging down from the chandelier with my weapon *mudras*.

Playboy: Meaning?

AG: *Mudras* are Buddhist hand gestures, magic swords that cut down phantoms of illusion.

Playboy: Like what?

AG: Fake flag imagery, hallucinatory patriotic band music drowning out antiwar delegate language, ghostly cops listening everywhere on the convention floor, black-magic curses coming from Daley's throat, complete blackout of popularly expressed vote against the Humphrey-Johnson-Nixon war psychosis, takeover of convention-hall procedure by an elite group of ward heelers in the name of old-fashioned democratic procedure, rigged electronics and seating arrangements, dead telephones, galleries illegally packed with a party-hack audience, and physical intimidation of delegates and observers by secret police. Zap! The convention illusion was staged by self-interested power conspirators against the expressed wish of the majority of citizens of the nation; and their shoddy sleight-of-hand shell game was meant to make voters think they were getting a choice, while actually phantom manipulators were nominating a candidate who continued to apologize for a war the voters wanted to reject. The whole convention, in fact, was an exercise in black magic and mass hallucination—just as the Vietnam war has been.

So when the priest began to pronounce benediction on all this massacre and hypocrisy, it seemed to me that a complete formal exorcism of the convention was necessary. So I stood up and went into some kind of fit of total ecstasy and started chanting OM over and over again—louder and louder, until almost everybody in the convention heard it, and officials on the platform turned to look at the balcony where I was standing. I went at it for five solid minutes; nobody stopped me, they were all so guilty. It was a ritual to exorcise all demons in sight, including myself.

Playboy: It obviously didn't succeed.

AG: Well, it had exactly the same effect, on whoever heard it, as the priest praying and hypnotizing people into a state of stupefaction so that they believed they were all doing something traditionally moral together. Since my chant was untraditional but ritual, formal, everybody noticed an interruption opposite from what they expected. It did shake their consciousness a bit because it was out of context. It was white-magic theater; it helped to break up the mass hallucination of political respectability that the priest's prayer created for the convention as a whole—if only for a few minutes.

Playboy: How would you reply to the explanation that most of the people who heard your exorcism did nothing to stop you not because they felt guilty but because they thought you were some kind of nut?

AG: Perhaps, but I'm not.

Playboy: Were there any other "positive experiences" that came out of the confrontation between the establishment and the New Left?

AG: There was the sudden recognition of eyes between blacks and the whites who'd been beaten—the whites discovering with relief that blacks are allies, not enemies, and have been there all along. An enormous relief for everybody! Blacks got what they had hungered for: companionship instead of lip service and the usual brush-off. Several times during that week, the social fog lifted; there was a psychedelic clarity in the air. All week I felt touches of what were almost acid illuminations when I talked with people—everybody conscious of a crisis of public reality—or looked in their eyes on the street. Everybody was looking at one another, wondering: "Is he aware what's happening, or is he one of the brainwashed police-state robotlike forms? Is he a good German, a good Russian bureaucrat, or a free man?" A whole society of respectable people became beat.

Playboy: Do you mean beat as in beat-up, Beat Generation or beatific?

AG: I mean beat-illuminated the way Kerouac defined it over and over. After the Chicago convention, that word should be clear to the youth of America; they'd seen the shambles of authority as it authorized itself to issue its own image. Everyone who had been so hopeful—like the McCarthyites—wound up beat. Even Humphrey and Daley were beat, all dust and ashes. The day after Humphrey'd been nominated, beat McCarthyites came down from the Hilton and discovered that free territory in Grant Park. They mixed with the Yippies, all eyes glancing at each other, everybody bankrupt and suddenly discovering a holy community on the grass—so holy that Senators and delegates, and finally McCarthy himself, came down and mingled with the youngsters who had been considered dirty, despised, bathless, bearded terrorists. A marvelous afternoon. So, beat also in the sense of crazy new feelings of joy vibrations right in the middle of America—as Kerouac always linked beat with "beatitude." Now that enough people have had this

experience, it's time everybody stopped making believe they don't
understand the term.

Playboy: What do you feel will be the long-term results of the con-
frontation in Chicago?

AG: The populace is a bit cowed; the military takeover has already hap-
pened, in the sense that the tactics of takeover—arms, Mace and so
on—were prepared and tested in Chicago. Whether such tactics will be
used on a large scale in every city may not be up to mayors anymore,
because the country has already been taken over by the police, the
Birch Society, the CIA, the military industry, the party hacks. Exactly
the same type of hard-line people whose tanks rolled into Czechoslo-
vakia last August. That kind of communism really means a strong-arm,
authoritarian police state—involving occasional illegal violence,
mostly on the part of police agencies and their collaborators. The
Chicago Tribune, in U.S. context, is a collaborator, in that it's been
manipulating readers' minds for years to think that America needs
stronger, more powerful police and military agencies in order to com-
bat the "Red menace." But the police state itself is the Red menace.
Actually, it gets to be a question of two *versions* of communism: the
Russian and American police states. Both dig the same hot cold war.

Playboy: To what extent do you think the new "hip planetary con-
sciousness" you've mentioned can resist a police state?

AG: Part of that hip consciousness is the realization that authoritarian-
ism of any nature is a usurpation of human consciousness—open
manipulation, brutalization and arbitrary manhandling of bodies and
consciousness. Student rebellions in Prague and New York are the
same. More and more people will become alienated from the authori-
tarian state and its images, such as angry mayors and police. One way
to oppose such authoritarian images will be increased community
action. The problem here will be how to transform the "greasers"—the
blue-collar class which is always in favor of a strong police force and
the persecution of minorities. Naked street theater ain't going to
transform them. But giant rock 'n' roll might get greasers into rock
concerts and turn them on. The spread of psychedelics—which is a
person-to-person matter, like all grassroots procedures—might be
another weapon. So might new electronic art forms. And long, inten-
sive conversations and explanations among blacks and greasers and

squares and kids and Yippies and party hacks might be another. What would you do if you were in Prague? Fight? Propagandize? Infiltrate?

Playboy: How successful do you believe this re-education of "greasers and squares" will be?

AG: One of the big questions here, as in Prague, is how free the news media will remain. For the first time, there was a breakthrough in Chicago: Actual social reality, with unmistakable images of social truth, was shown on TV and reported in the papers. Minority groups have been beaten and clubbed and jailed for years—hippies, teaheads, Negroes, avant-garde literary folk, underground-newspaper editors. But now white middle-class McCarthyites were suddenly shown being beaten on the head—and the message got through. There'd been a breakthrough in graphic coverage of American armed violence in Vietnam; now there was one on American armed violence at home. A lot of people have begun putting the two images together and realizing they're the same: natives getting napalmed and citizens getting beaten in their own streets. Newsmen finally realized these things really happened, because it started happening to them, too. But then a big campaign began to convince newsmen and other people who saw or read about the beatings that it was all a subjective hallucination. It's like what happened in Prague. During the eight-month freedom there, everybody blew their tops and told the truth: then the *Putsch* set in. Authoritarian Communists like the *Tribune* and *Pravda*, Mayor Daley and the commissars, foster the hallucination by calling the demonstrators who'd been clubbed "agitators" and "terrorists."

Playboy: Though opinion is polarized on the subject, a great many people seem to agree with that description. What do you call them?

AG: Good, old-fashioned Americans trying to be free men on free territory, manifesting the American dream. And it's amazing how many of them there are. In Chicago, I met lots of kids who'd spent a year or two in Alabama working as civil rights workers, and then another year in SNCC and VISTA[76] and maybe two in the Students for a Democratic Society. Others are "professional revolutionaries" totally dedicated to

[76]SNCC, Student Nonviolent Coordinating Committee; VISTA, Volunteers in Service to America.

social activity and community work—like the post-Digger "Mother-fuckers." It's marvelous. And many of them are high a lot psychedeli-cally, so there seems to be a joining of forces between social-activist and hippie youths.

Playboy: What do you think about the fact that some of these groups are pledged to overthrow the U.S. Government?

AG: I think that there's no *legitimate* Government to overthrow in America. The military industry has *already* overthrown it by armed violence. We're living in an anarchic state. Already, the power struc-ture's upheld by guns, not equal law. To mistake Daley's *Putsch* in Chicago for democratic procedure would be a phantom illusion. Humphrey's nomination was taken by force. I wasn't in Miami for Nixon's theater.

Playboy: Do you agree with those who feel that these young people will "sell out" and turn into middle-class conformists as they grow older?

AG: No; impossible. *Time* or *Life* may think them misguided youths who'll straighten out in time and return to the ways of their collaborat-ing fathers. But how can they go back? The way's been barred by beat-ings and arrests. What bridges they haven't burned behind them have been burned for them with pot busts. What's happened to young peo-ple is a sudden breakthrough of cosmic consciousness catalyzed in part by psychedelic drugs. Another factor was the deconditioning caused by alienation from social authority as it proved itself completely incom-prehensible and mad and burned its own bridges from Hiroshima to Vietnam. What happened in Chicago is only one local example. Sud-denly a lot of people have awakened and asked: What in hell am I doing on this poisoned planet, where everybody else is running around wav-ing flags and shooting guns? In answer, many of the longhaired kids are turning into Adam and realizing—as they realized when they walked on the Grant Park grass across from the Hilton in Chicago—that they're walking on the green of antiquity, a "green and pleasant land," the ancient New Jerusalem Blake envisioned as possible in England. It's much like in a science-fiction story where suddenly this sponta-neous generation appears that realizes it's living in eternity, in the sense of having a whole new planetary consciousness. Now these kids can check back through ancient symbols and learn about the traditions from which they've sprung or to which they correspond. One source would be early Gnostic texts about the nature of man and the universe—in particular, the nature of the guardians of cosmic order who try to keep man locked in the body stump: the establishment.

Playboy: What kind of life do you think these young people want to lead?

AG: Here I can talk only about the life I'd like: more contact with nature; more and more occupation with exploration of subjective consciousness and enlargement of areas of inner and outer sensibility; more participation in rhythmic theater; and liberation of sexual energy from population reproduction. Since the Biblical injunction to "be fruitful and multiply" isn't sacramentally appropriate anymore, it seems to me that the time has come for the orgy to become a communal form of "adhesive" democratic festival. Ideally, what should have happened in Chicago, for example, was that the Festival of Life should have eclipsed the convention with the glow of thousands of naked bodies intertwined, making love, spurting semen all over the newsmen and TV cameras and one another on the grass, all to the accompaniment of *yab-yum*[77] mantras and naked rock-'n'-roll artists swinging through trees.

Playboy: You've got to be kidding.

AG: On the contrary. Life should be ecstasy. We need life styles of ecstasy and social forms appropriate to whatever ecstasy is available for whoever wants it.

Beyond man's natural ecstasy is total serenity and tranquillity: cessation of desire, which the Buddhists talk of, which is liberation from grasping and craving.

I think many of the younger children have already liberated themselves from grasping after the things of this world and have begun to grasp for wisdom; eventually they may even liberate themselves from grasping after wisdom. Everybody wants, needs, deserves and will have this free kingdom, which the police state, whether it triumphs or not, can never touch.

Playboy: Is that one of the reasons why you recently moved to a farm in upper New York State—to try to find such a life?

---

[77]*Yab-yum*, Tibetan, literally Mother and Father, most often refers to the masculine and feminine deities in sexual union, a frequent depiction in Tibetan art, which symbolizes the union of wisdom and emptiness.

AG: It's just a place of privacy. A retreat. A hermitage.

Playboy: Do you think there'll be an increase in communal retreats among the hip-psychedelic generation?

AG: The kids say so. Communal living's a good experiment; it provides another family. Most older people, on the other hand, are so sold on television sets and superfluous commercial gimmick amusements that they've forgotten about electric being and inside space and outside nature relationships in the human universe. Some garden, but too many of them are occupied in purely abstract mental worlds, like money power. They externalize their abstractions in creations like that John Hancock Building. Instead of using our space to create a humane construction overflowing with imagination and delight, a real-estate computer computed what the profitable angles, stresses, strains and maximum height should be, based on a big-money downtown building site. That black tower is the product of nonhuman automatism, total squareness built literally on greed—usury extended in robotic directions and with robotic tenacity.

Such desensitization has been closing in on human consciousness for centuries now, leading to complete disregard of the sensitive skin of the earth, the pollution of water and poisoning of the atmosphere. Who's grooving and profiting from disrupting our ecology? What alternative economic structures are inevitable if the planet isn't going to explode? The whole greed-money thing must be kicked like a junk habit. Obviously some type of equal use of moneys—as well as of sunlight and water—is appropriate, especially since in a mechanized society nobody really works. The guy who buys and sells the land on which the Hancock Building stands doesn't do any productive work. Instead of making money by producing some product, he only plays with paper, with telephones on his desk, and makes money out of money; that's usury, and usury creates these visible hells. The alternative is for the resources to belong to everybody.

Playboy: Are you talking about collectivism, socialism—or anarchy?

AG: Those are just words. To say that nature's resources—sun, water and air—should be available to all makes more sense. How that can be done is not the great problem. Lots of blueprints are available: Buckminster Fuller, Robert Theobald, Paul Goodman, (Charles) Fourier, Leary, Burroughs and (Mikhail) Bakunin offer practical directions.

The main problem will be how to dry out the money-property junkies. Try to break the money-property-power men of their habits and they'll act just like junkies—lie, steal, scream, rob, live in your house and swipe your last Tibetan tanka in order to sell it to the pawnbroker and buy themselves some more real estate, sell the ground from under unborn feet, cut down forests, raze hills, make uninhabitable box houses. Such acts literally destroy the billionfold delicate balance of relationships between man and nature on the planet in so many ways that today the main technological problem is in attempting to calculate what have been the effects of our intrusion on the planet. That's a problem so vast, scientific and manly that it staggers the imagination of most real estate operators. Wheelers and dealers of property are stuck in tiny little corners making piles, mortally ignorant of how the terrible feedback from their power plants affects all life around them.

I'm talking about the fact that this planet is in the midst of a probably fatal sickness; the by-products of that sickness include not only the political violence of real-estate developers but all the giant fantasies of the Cold War—the witch-hunts, race paranoia, projections of threat and doom. The sickness will end in our destroying our own planet. All sorts of presidential and scientific advisory commission reports, books on biology and ecology, prove that we are literally destroying the earth because of vulgarly applied technology, irresponsibly planned military-consumer industrialization and waste of resources. We've got about 30 years left to get straight—or else. The sickness may even be irreversible already; Burroughs thinks it is.

It's an ecological cancer. We're polluting more and more of the world's freshwater sources, for one example. Right now, Lake Erie's just a great dead pool of green goo, a toilet. If it were left alone today, it would take 500 years for it to return to any kind of freshness and normal balance of life. But we need that water now—not only to keep up with the population explosion but also to help freshen the atmosphere, not to mention for our own pleasure. Then there's the thermopollution of the oceans. Oceans are getting hotter and dirtier on account of all the atomic and DDT waste we're pouring into them; it's begun poisoning fish and some rare bird species. If all the atomic power plants now on the drawing boards are put into operation in the next 30 years, the atomic waste from them will alter the entire heat balance of ocean and land, change all marine forms and generate enough heat to melt the polar icecaps, causing a world-wide flood just

like in the Bible! "The icecaps are melting to wash away our sins," as
Tiny Tim sings.

Playboy: The possibility of a world flood seems remote to most people.
Can you honestly expect them to relate now to this almost-inconceivable
future catastrophe?

AG: They don't have to. Catastrophe is more immediate than that. Mis-
calculations about the rising strontium-90 count have already dam-
aged entire young pine forests in northern Canada, poisoned arctic
dolphins and probably changed the weather a little. Continuous use of
insecticides not only pollutes oceans and kills fish and birds, but it
accumulates in people's livers and poisons them and perhaps causes
freak-outs in subliminal ways. In the same way, the use of synthetic
nitrates in fertilizers depletes the soil. It's like people taking speed; it
may give more energy and productivity for a few years, but it ends by
depleting the system. In addition, such fertilizer eventually turns into
certain nitrogen forms which are poisonous in baby foods. And this is
continuing when the nitrate level in baby food has already passed
what's clinically considered the health level. We're poisoning our own
babies.

Playboy: Don't you feel that the technological strides now being taken
by medical science hold out some hope of coping with these health
hazards?

AG: The cancer is more than a purely medical problem. Car waste
products and industry gases not only pollute the atmosphere but they
cut us off from the clarity of Mediterranean azure by turning the sky
into shit-colored smog through which you can't see the moon and sun
and stars. People no longer know the procession of the seasons. And
they don't know they're on a planet anymore—much less in a vast
galaxy. They think they're in "Chicago" or "New York." Seventy per-
cent of them are. Their conceptual awareness of the world consists of
little piled blocks, streets, radios, politicians, TV stations, city halls.
We're living in somebody's comic science-fiction nightmare rather
than on a fresh planet.

Not only does smog cut off eyeball consciousness and mental
awareness but it also cuts in on your breathing, poisons the body,
causes suffocation, heart attacks and nervous anxiety. Add to that the
noise in the cities. Most city noise is subliminal; you get used to it, but

it's bugging you all the time. Noise attrition is considerable on the nervous system, but it also damages the heart—cardiovascular-cell freak-outs. If you've never lived for any length of time in the country, you think the city is the normal condition of existence. Pretty soon people are going to begin mutating and adapting to smog, noise, pollution of lakes and rivers, living in high-rise apartments and experiencing politics on TV and human relations through police-state stereotypes.

Most luxury products we use are useless and destructive—like the no-deposit, no-return bottles littering the planet; it will take hundreds of years to get rid of them. Aluminum throwaway cans! Everything's being turned into plastics and synthetics. Living forms are being turned into inorganic ones; cancer has exactly the same effect. Instead of continuous spontaneous creation, there's continuous spontaneous consumption of matter reduced to dead form. That's a pretty degraded human environment, I think. All I've been saying is symbolized by the fate of the very emblem of America—the bald eagle. The species is almost extinct.

Playboy: Some scientists believe that man's ability to control genetics and natural environment will lead to higher evolution of our species rather than to extinction. Do you think this is unrealistic?

AG: If technological mismanagement continues to take us in the direction we've been going—and there's reason to presume that it will, at a geometrically increasing speed—ecological damage, waste of resources and overpopulation will eventually create more and more fighting and anxiety and aimless Chicago beings and police states and grabbings of power and presumptions of authority and terrorism. It will all lead to giant genetic, intercontinental starvation—wars and rebellions in every direction which may ultimately end with people shooting arrows from windows of the Hancock Building and defending the last canned goods for some fat old banker's nephew.

Playboy: Do you seriously believe it's that bad?

AG: Absolutely—and young kids know it; that's the basis for their conduct. They're being cheated out of the Garden of Eden. While this science-fiction crisis is building, the old power and money freaks remain so self-involved that they don't even see the planet beginning to smoke under their feet. They're busy destroying on a more and more massive scale, polluting the atmosphere, defoliating Mekong Deltas and wreaking such vast havoc that newspapers get thicker and thicker, and larger and larger forests have to be cut down to print the news—all of it bring-down, or irrelevant.

Playboy: What do you think can be done?

AG: The first step toward a cure for the sickness is to realize it's there. The robot standardization of American consciousness is one side-effect feedback from a greedy, defective technology, just as ecological disorder is another feedback, and these systemic disorders reinforce each other fatally unless there is complete metabolic change. One could argue with the patient for years—forever. But the fact is that he is his own disease. That's what he must be taught to recognize. What we must first realize is the fact of our own diminished consciousness. "A new world is only a new mind," as William Carlos Williams said. Being willing to solve problems depends mostly on being aware that they exist.

  The specific cure probably lies in the direction of a technology that will restore living forms to the surface of the planet—revitalize the planet's skin—without setting up a feedback that further pollutes or disturbs that living, breathing skin. Then—assuming that we maintain our giant population—we'll face the terrific technological problem of finding a place for everybody to live. We'll have to tear down the cities—they stink—then decentralize and miniaturize our machinery. But these are just technological problems; there's no reason why they can't be solved. After all, we can split atoms, create giant computers, articulate the shapes of DNA molecules, go to the moon and wage vast wars. Why shouldn't we be able to solve our specific conservation and population problems?

Playboy: Even if we solve these problems and make continued human life at least feasible, what do you think can be done to enhance the *quality* of life?

AG: All sorts of propositions are available. Dr. Barry Commoner's books cover ecological reconstruction, and Gregory Bateson has the technical ideas. How the human community can reorganize and decentralize itself has been written about by Paul Goodman. Pound and others describe how to decentralize the monopolistic control of money by banks and individuals, and invent new forms of social

credit. Robert Theobald has suggested structures adaptable for eco-
nomic liberation from the domination of money. The Diggers have
told us how to organize Adamic communes. The Chinese Reds and
other Communists can teach us about some structures for group orga-
nization. Capitalists know a lot about certain aspects of production. We
should also listen to Dick Gregory and the macrobiotic people about
healthy foods; to the American Indians about seasonal and communal
rituals and how to respect the sacred body of the land we inhabit; to
shamans like Aldous Huxley about the appropriate use of natural and
synthetic psychedelics; to Leary and Gary Snyder about enlarged vari-
ant family structures; to Burroughs about the reorganization of educa-
tional systems in order to provide proper training in nonconditioned,
spontaneous consciousness; to scientologists on mind training to rub
out hang-ups like fear of death, sexual obsession and either love or
hate for the American flag; to poet Charles Olson about suggestions for
temporary reorganization of American police so they'll all be black or
female; and to a million Oriental Zen masters and yogis about physical
education and consciousness therapy. So plenty of prescriptions are
available. Practical problems can be solved in practical ways.

Playboy: Then why—in your view—aren't they being solved?

AG: Because our governments are addressing themselves in precisely
the opposite direction: how further to fuck up the surface of the earth.
Most of our national consciousness is invested in an 80-billion-dollar
"defense" budget still preoccupied with the completely hallucinatory
and paranoid problem of attack from the outside—by "the Communist
conspiracy"—and with competition for supremacy over colored races
and other life forms. Intermarriage would produce a golden race,
which obviously would be one solution to such hallucinatory problems,
as well as to the problem of race itself. Otherwise, we'll have a war
which will destroy the planet. Progress requires abolition of race ego,
national ego, boundaries; it requires planet-citizen consciousness.

    Although a minority is aware what that next step is, what about the
majority who are plunged in darkness, flood, apocalypse and destruc-
tion? How to redeem these "ignorant armies" who clash by night from
their own bad karma? Violent confrontation? Violence begets vio-
lence. Revolutionary violence begets fascist tyranny. So, though the
noble impetuosity of confrontation by some New Leftists may seem
appropriate to a situation in which long-haired angels are surrounded

by pigs, the problem remains: how to cast the Devil from the hearts of the swine?

Playboy: We'll bite. How do you?

AG: Since we're in an apocalyptic situation, old historical dialectics no longer apply. I prophesy that the only way to reverse the apocalypse is white magic, because the apocalypse itself is incarnate black magic. What would be the effect of total sacramental harmonious shamanistic ritual prayer magic massively performed in the American or Russian political theater?

Playboy: We're beginning to feel like a straight man. What would be the effect?

AG: Exorcism. We need a million children saints adept at high unhexings, technological vaudeville, rhythmic behaviors, hypnotic acrobatics, street trapeze artistries, naked circus vibrations—magic politics to exorcise the police state. Is there a kind of poetry and theater sublime enough to change the national will and to open up consciousness in the populace? If the direction of the will can be changed and consciousness widened, then we may be able to solve the practical problems outlined: ecological reconstruction and the achievement of clear ecstasy as a social condition. And once that is achieved, people could relax and start looking for the highest, perfect wisdom.

Playboy: And what is that?

AG: The realization that even visionary ecstasy is unnecessary because the universe was neither born nor will it be annihilated. The universe is an illusion. Once you realize this—with LSD or without it—you get into the ultimate science of ZimZum, as the cabalists call it. ZimZum is the science of the expansion and contraction of the universe—the appearance and disappearance of the universe itself. This is ultimate physics and also ultimate consciousness. But for this ultimate theater we need the advice of the Dalai Lama, Daitokuji Roshi, Lubovitcher Rebbe, Krishna and a hip Pope.

BILL PRESCOTT
December 3, 1969,   Annandale-on-Hudson, New York

[Untitled][78] 1970

In 1999, William Prescott related how he had been surprised one morning to see his "boyhood hero Allen Ginsberg walking through the snow toward my house. Pierre Joris, Jonathan Kaplan, and Stephen Kushner had convinced him to be interviewed for my program on the local Bard College radio station. However, I had been up all night and had no idea I was interviewing anybody, much less Allen Ginsberg. Even though I was unprepared, Allen was both gracious and generous with his time and thoughts. He had recently come from Jack Kerouac's funeral, and he spoke in a touching manner about his old friend." The interview took place in a cold and broken-down pre-Revolutionary War rental house, with Prescott and Ginsberg sitting on the floor near an inadequate heater, and Joris and Kushner also asking questions. The interview was taped and aired several times on the local Bard College station. Jonathan Kaplan then edited it for publication.

—DC

Bill Prescott: I have a question that I know you must've been asked a million times before: In a world as totally screwed up and spinning around as this one is, why do you write?

Allen Ginsberg: It's a form of meditation. I have to find out what I'm thinking. Literally, I find out what is on my mind. Or I articulate it out front, where I can see it, rather than a flash of mortal dream knowledge disappearing when I wake up in the morning 5 A.M.

Also, a certain kind of loneliness, wanting to communicate it,

---

[78]The untitled and undated publication the interview appeared in states: "This is the first issue of a magazine of prose published by and for members of the community of Bard College."

wanting to lay it out where others can touch it, touch me, where I can touch other people. And then just sort of superstition about how beautiful Rimbaud was—the old literary-glory-romanticism to be a poet. If it's a world that had people like Shelley, and Whitman, then that's the world I want to be in, or the universe I'd like to be in. It's a world that had Hakuin, and Han Shan, and Basho, and Issa . . . Buddha . . . Sangharakshita,[79] Milarepa, Shankara, Iamblechus, Proclus, and Plotinus . . . and William Blake. It's certainly where I feel a Call. And then it's the only thing I know how to do, it's the easiest way of making a living I know.

Also, in social terms, the bardic function, set side by side, or inside, the mass media, is a corrective to mass hallucination. Because the bardic thing is personal communication of actual data as far as you really *see* it or feel it from unconscious sources, uncensored. Whereas the mass media hallucination is a censored reality. So Poesy's a corrective to mechanical robot reduplication of castrated and manipulated news. So it has a heavy function now which the Russian poets show. I mean if you want news, all kinds of poetry is "news that stays news," *that's* news. It's kind of human consciousness newspapers. . . . I'm an old newspaper man—I used to work on the Associated Press, *New Jersey Labor Herald*, the *Columbia Review, Columbia Jester*, and the *New York World Telegram* . . . so I'm just practicing my fun and trade.

Also it's got to do with body rhythm, ecstatic body rhythm. It's an ecstasy path, yoga path to ecstasy. 'Cause I've received, experienced ecstasy through poesy, experienced visionary consciousness through poetry, briefly, occasionally. So that's my sadhana, as they call it, my path, or yoga, or practice. Also because I guess I loved Kerouac, and a few other people like Burroughs, wanted to please their consciousness. Now that Kerouac's dead, I sometimes find myself writing things forgetting he's dead and laughing to myself at what he would think at a funny phrase or *thing* I do—and sitting there "Ah, there's nobody there to see it and appreciate it," and "Ah, what's the use—who am I writing for?" And then when others go I don't know whether I'll

---

[79]Although the typescript of the interview, corrected by Allen, has "Sangharakshita" plus a notation on the typescript's first page that says "corrected May 1970 Dictated to J. Kaplan by Telephone June 6, 1970 AG," the printed version of the interview lists "Sakyapa-lama" before Milarepa and leaves out Iamblechus and Proclus. A last-minute change of heart by Ginsberg or, as seems more probable, Kaplan's own editing?

be motivated to write, if people that I thought I loved aren't there anymore, if people I was trying to please aren't there anymore—people I was trying to get close to through poesy. Be funny to . . . So I don't know what that will be like.

I was thinking of Pound last night, see all his close friends are dead. I'm sure Pound used to write lines thinking what T. S. Eliot would think of it, or Gaudier-Brzeska, or Brancusi, or Bunting, all the Activist-Vorticist[80] writers he was talking to. Pound wrote with Bunting in mind very often. *Very* often. In fact he always respected Bunting's perception of his poetry. So, I'm sure when Pound writes a line, and gets something that really "*presents*" the image very accurately, it may flash through his mind, a gleam of pleasure that Bunting will recognize and accept and approve that. 'Cause Bunting was the one who told him "Dichten = condensare." Pound said in conversation that he was depressed about the *Cantos* and about his poetry, because—and he quoted Bunting—"Bunting told me years ago that my poetry referred too much, there was too much "reference" and too little "presentation" in my poetry." So I'm sure even people like Pound did thinking through other people's eyes or ears or minds. So I would think a lot through Kerouac's mind. I always thought a lot through Burroughs's mind, and a few other people—Gregory (Corso), Peter (Orlovsky), Gary (Snyder) often, Phil Whalen often, (Robert) Duncan occasionally, (Robert) Creeley occasionally, thought through (Robert) Kelly's[81] mind once or twice, my father's mind, my own sometimes, also a particular cat I was trying to make it with. So that's one reason I write. To say what I could say when I was alive.

---

[80]An abstract art movement in England from about 1912–15 that embraced Cubist and Futurist concepts and was centered around novelist and polemicist Wyndham Lewis (1882–1957).

[81]Poet Robert Kelly was an English professor at Bard College. I am grateful to Simon Pettet for pointing this out to me. Pettet emphasized to me, as the sensitive reader will notice, how often Allen tailored his remarks to the person and the locale of the interview.—DC

The Chicago Seven Trial
December 11–12, 1969,   Chicago
From the book *Chicago Trial Testimony* (City Lights Books, 1975)

When the United States government determined to make an example of several individuals for the events that had taken place at the Democratic National Convention in Chicago, they found a cudgel in a law passed less than nine months before the convention that made it illegal to cross state lines to incite a riot. Although Lyndon Johnson's Justice Department drew up indictments for violation of the Anti-Riot Act, Johnson's attorney general, Ramsey Clark, refused to prosecute the case. When Richard Nixon came into office, however, his administration decided to pursue prosecution.

Indicted by the United States government were David Dellinger, Rennie Davis, and Tom Hayden of the National Committee To End the War in Vietnam (MOBE); Jerry Rubin and Abbie Hoffman of the Yippies; John Froines, a college professor; Lee Weiner, a graduate student; and Bobby Seale, national chairman of the Black Panther party.

The trial was assigned to federal judge Julius Hoffman, a seventy-three-year-old known both for his impatience as well as for his rulings sympathetic to the government. Representing the defense were William Kunstler and Leonard Weinglass, but at the trial's beginning Bobby Seale demanded to be represented by Charles Garry, his usual attorney, who had undergone surgery shortly before the trial began on September 24, 1969. Seale asked for a postponement of the trial until Garry could represent him, which motion Judge Hoffman denied. After Seale continually disrupted the proceedings—including by hurling epithets such as "pig" and "fascist" at Hoffman—the judge ordered Seale gagged and bound to a chair. When Seale struggled to free himself he was roughly manhandled by court attendants, prompting Kunstler to leap to his feet in loud protest: "Your Honor, are we going to stop this medieval torture that is going on in this courtroom? I think this is a disgrace." Hoffman then declared a mistrial with regards to Seale, thus severing his case from the other defendants and the group that started out as the "Chicago Eight" became the "Chicago Seven." (Hoffman simultaneously found Seale guilty of sixteen counts of contempt and sentenced him to four years' imprisonment.)

The following is an edited version of the transcript of Ginsberg's testimony for the defense at the Chicago Seven trial, published by Lawrence Ferlinghetti as *Chicago Trial Testimony*. Because the format here is an official court transcript and therefore meant to be both complete and verbatim, where any part of an answer or exchange has been deleted, it is indicated by an ellipsis contained in brackets: [. . .]; where an entire section of testimony has been left out, the deletion is indicated by asterisks: ***. The present editor's remarks are in brackets and italicized. The only other modifications to the transcript as published by City Lights Books are stylistic, such as changing some words from upper to lower case or eliminating the word "BY" before the name of an attorney when he began to question Allen.

—DC

UNITED STATES OF AMERICA
VS.
DAVID T. DELLINGER, et al.

DEFENDANTS: David T. Dellinger, Rennard C. Davis, Abbott H. Hoffman, Thomas E. Hayden, Jerry C. Rubin, Lee Weiner, John R. Froines, (Bobby G. Seale)

JUDGE: Julius J. Hoffman

PROSECUTION: Thomas Foran, Richard G. Schultz, Roger Cubbage

DEFENSE: William Kunstler, Leonard Weinglass

WITNESS FOR THE DEFENSE: Allen Ginsberg

ALLEN GINSBERG called as a witness on behalf of the defendants, having been first duly sworn, was examined and testified as follows:

Direct examination by Mr. Weinglass: Will you please state your full name?

A Allen Ginsberg.

Q What is your occupation?

A Poet.

Q Have you authored any books in the field of poetry?

A Yes.

Q Will you indicate to the jury the titles of the books you have authored?

A In 1956, "Howl and Other Poems," in 1960, "Kaddish and Other Poems," in 1963, "Empty Mirror," in 1963, "Reality Sandwiches," and in 1968, "Planet News."

Q Now, in addition to your writing, Mr. Ginsberg, are you presently engaged in any other activity?

A I teach, lecture, and recite poetry at universities.

Q Can you indicate to the jury without going extensively into your travels what your last trip in connection with teaching and lecturing consisted of?

A I was at Princeton University for three days at the invitation of Reverend John Snow, the Episcopal Chaplain, and Samuel, the Rabbi. [. . .]

Mr. Weinglass: And when were you at Princeton lecturing?

A Last week, the middle of last week.

Q Now, did you ever study abroad?

A Yes.

Q Where have you studied?

A In India and Japan.

Q Could you indicate for the Court and jury what the area of your studies consisted of?

A Mantra Yoga, meditation exercises, chanting, and sitting quietly, stilling the mind and breathing exercises to calm the body and to calm the mind, but mainly a branch called Mantra Yoga, which is a yoga which involves prayer and chanting.

Q How long did you study?

A I was in India for a year and a third, and then in Japan studying with Gary Snyder, a Zen poet, at DaiTokuji Monastery, D-a-i-T-o-k-u-j-i.

I sat there for the zazen sitting exercises for centering the body and quieting the mind.

Q Are you still studying under any of your former teachers?

A Yes, Swami Bhaktivedanta, faith, philosophy, Bhaktivedanta, B-h-a-k-t-i-v-e-d-a-n-t-a. I have seen him and chanted with him the last few years in different cities, and he has asked me to continue chanting especially on public occasions.

Q Have you received any special permission with respect to the chanting from the persons under whom you have studied?

A Yes, from Zen Master Roshi Suzuki, San Francisco Zen Buddhist Temple, who gave approval to my chanting of the Highest Perfect Wisdom Sutra, Prajna Paramita, P-r-a-j-n-a P-a-r-a-m-i-t-a, Prajna Paramita Sutra. And also from Swami Satchitananda of New York, also from the school of Dr. Rammurti Mishra, D-r R-a-m-m-u-r-t-i M-i-s-h-r-a, a yogi who was the adviser of the New York Yoga Society, by whose disciples I have been initiated as a Shivit, S-h-i-v-i-t. That is a branch of Hinduism. All of these involve chanting and praying, praying out loud and in community.

Q In the course of a Mantra chant, is there any particular position that the person doing that assumes?

A Any position which will let the stomach relax and be easy, fall out, so that inspiration can be deep into the body, to relax the body completely and calm the mind, based as cross-legged.

Q And is it, the chanting, to be done privately or is it in public?

Mr. Foran: Oh, your Honor, I object. I think we have gone far enough now to have established—

The Court: I think I have a vague idea of the witness' profession. It is vague.

Mr. Foran: I might indicate also that he is an excellent speller.

The Court: I sustain the objection, but I notice that he has said first he was a poet, and I will give him credit for all of the other things, too, whatever they are.

The Witness: Sir-

The Court: Yes, sir.

The Witness: In India, the profession of poetry and the profession of chanting are linked together as one practice.

The Court: That's right. I give you credit for that.

Mr. Weinglass: Mr. Ginsberg, do you know the defendant Jerry Rubin?

A Yes, I do. [. . .]

Q Do you recall where it was that you first met him?

A In Berkeley and San Francisco in 1965 during the time of the anti-Vietnam war marches in Berkeley.

Q Were you associated with Mr. Rubin in that anti-war march?

A Yes, we worked together.

Q Did you have any further occasion in the year of 1967 to be associated with Mr. Rubin?

A Yes. I saw him again at the Human Be-in in San Francisco. We shared the stage with many other people.

Q Would you describe for the court and jury what the Be-in in San Francisco was?

A A large assembly of younger people who came together to—

Mr. Foran: Objection, your Honor.

The Court: Just a minute. I am not sure how you spell the Be-in.

Mr. Weinglass: B-E—I-N, I believe. Be-in.

The Witness: Human Be-in.

The Court: I really can't pass on the validity of the objection because I don't understand the question.

Mr. Weinglass: I asked him to explain what a Be-in was. I thought the question was directed to that possible confusion. He was interrupted in the course of the examination.

Mr. Foran: I would love to know also but I don't think it has anything to do with this lawsuit.

Mr. Weinglass: Well, let's wait and find out.

Mr. Foran: This is San Francisco in 1967.

The Court: I will let him, over the objection of the government, tell what a Be-in is.

The Witness: A gathering together of younger people aware of the planetary fate that we are all sitting in the middle of, imbued with a new consciousness and desiring of a new kind of society involving prayer, music and spiritual life together rather than competition, acquisition and war.

Mr. Weinglass: Did you have occasion—and was that the activity that was engaged in in San Francisco at this Be-in?

A There was what was called a gathering of the tribes of all of the different affinity groups, political groups, spiritual groups, Yoga groups, music groups and poetry groups that all felt the same crisis of identity and crisis of the planet and political crisis in America, who all came together in the largest assemblage of such younger people that had taken place since the war in the presence of the Zen Master Suzuki that I mentioned before, in the presence of a number of Tibetan Buddhists and Japanese Zen Buddhists and in the presence of the rock bands and the presence of Timothy Leary and Mr. Rubin.

The Court: Now having had it explained to me, I will hear from you.

Mr. Foran: I object, your Honor.

The Court: I sustain the objection.

Mr. Foran: Your Honor, I will refrain from moving to have the jury directed to disregard it.

Mr. Weinglass: If your Honor please—

Mr. Kunstler: It isn't so funny that it has to be laughed at by the U.S. Attorney.

Mr. Foran: Your Honor—

Mr. Kunstler: I think the objection can be stated—

The Witness: Sir—

The Court: You will get another question.

Mr. Kunstler: I will let Mr. Weinglass—

Mr. Weinglass: In answering that objection, I think within the last hour the court has heard the prosecutor examine extensively a prior witness on another demonstration which occurred in October of 1967. I think we are talking about public meetings at which these defendants were present during the year 1967.

The Court: I will let my ruling stand. Ask another question.

Mr. Weinglass: Now during the—later on in the year of 1967 did you have occasion to meet again with the defendant Jerry Rubin?

A Yes.

Q And what was that meeting concerning?

A We met in a cafe in Berkeley and discussed his mayoral race for the City of Berkeley. He had run for mayor.

Q Did you have any participation in that campaign?

A I encouraged it, blessed it.

Q Now do you know the defendant Abbie Hoffman?

A Yes. [. . .]

Q For how long have you known Abbie Hoffman?

A Since late in 1967, I believe.

Q Now calling your attention to the month of February, 1968, did you have occasion in that month to meet with Abbie Hoffman?

A Yeah. [. . .]

Q Could you relate to the jury what was discussed between you and Mr. Hoffman at that meeting?

A We talked about the possibility of extending the feeling of humanity and compassion of the Human Be-in in San Francisco to the City of

Chicago during the time of the political convention, the possibility of inviting the same kind of younger people and the same kind of teachers who had been at the San Francisco Human Be-in to Chicago at the time of the convention in order to show some different new planetary life style than was going to be shown to the younger people by the politicians who were assembling.

Q Now when you say, "We discussed," did Mr. Hoffman indicate to you what his intention was with respect to that discussion?

Mr. Foran: Objection, leading and suggestive.

The Court: I sustain the objection.

Mr. Weinglass: Do you recall what Mr. Hoffman said in the course of that conversation?

A Yippie—among other things. He said that politics had become theatre and magic; that it was the manipulation of imagery through mass media that was confusing and hypnotizing the people in the United States and making them accept a war which they did not really believe in; that people were involved in a life style which was intolerable to the younger folk, which involved brutality and police violence as well as a larger violence in Viet Nam, and that ourselves might be able to get together in Chicago and invite teachers to present different ideas of what is wrong with the planet, what we can do to solve the pollution crisis, what we can do to solve the Viet Nam war, to present different ideas for making the society more sacred and less commercial, less materialistic, what we could do to uplevel or improve the whole tone of the trap that we all felt ourselves in as the population grew and as politics became more and more violent and chaotic.

Q Did he mention to you specifically any teachers that he had in mind for coming to Chicago?

Mr. Foran: Objection.

The Court: I sustain the objection.

Mr. Weinglass: Did you hear him mention the names of Mr. Burroughs, Mr. Olson and other teachers?

A Burroughs, Olson and Mr. Fuller.

The Court: Mr. Witness—

Mr. Foran: Objection to the leading and suggestive question.

The Court: I sustain the objection and I strike the answer and direct the jury to disregard it.

Mr. Weinglass: Mr. Ginsberg, does your prior answer exhaust your recollection as to what Mr. Hoffman said at that meeting?

A Yes.

Q Do you recall him mentioning anything about any rock and roll bands?

A Yes.

Q What did he say about rock and roll bands?

A Well, he said that he was in contact with John Sinclair who was the leader of the MC 5 rock and roll band and John Sinclair and Ed Sanders of the Fugs would collaborate together and invite a lot of rock and roll people, popular music such as Arlo Guthrie—Phil Ochs was also mentioned by Mr. Hoffman. Mr. Hoffman asked me if I could contact the Beatles or Bob Dylan and tell them what was afoot and ask them if they could join us so that we could actually put on a really beautiful thing that would turn everybody on in the sense of like uplift everybody's spirit and show actually what we were actually feeling as far as delight instead of the horror that was surrounding us.

Q Now did he ascribe any particular name to that project?

A Festival of Life.

Q Did he ask you to take any active role in the Festival of Life?

Mr. Foran: Objection, your Honor. Can't Mr. Weinglass—

The Court: I sustain the objection.

Mr. Foran: I mean, those are all leading and suggestive questions, your Honor, and I object to them. They are improper.

The Court: I have already sustained the objection.

Mr. Weinglass: After he spoke to you, what, if anything was your response to his suggestion?

A I was worried as to whether or not the whole scene would get violent. I was worried whether we would be allowed to put on such a situation. I was worried whether, you know, the Government would let us do something that was funnier or prettier or more charming than what was going to be going on in the convention hall.

Mr. Foran: I object and ask that it be stricken. It was not responsive.

The Court: Yes. I sustain the objection.

The Witness: Sir, this is—that was our conversation.

The Court: I direct the jury to disregard the last answer of the witness.

Mr. Weinglass: Your Honor, I would like to—

The Witness: How can I phrase that then because that was our—

Mr. Weinglass: Your Honor, I would like to be informed by the Court how that answer was not responsive to that question. It seemed to me to be directly responsive.

Mr. Foran: Your Honor, he asked him what he said and he answered by saying what he was wondering.

The Court: Worry.

The Witness: Oh, I am sorry, then. I said to Jerry that I was worried about violence—

The Court: I have ruled on the objection. Ask another question if you like.

Mr. Weinglass: Now during that same month, February of 1968, did you have occasion to meet with Jerry Rubin?

A I spoke to Jerry Rubin on the phone, I believe. [. . .]

Q Will you relate to the Court and jury what Jerry Rubin said to you.

A Jerry told me that he and others were going to Chicago to apply for permission from the city government for a permit to hold a Festival of Life and that he was talking with John Sinclair about getting rock and roll bands together and other musicians and that he would report back to me and try to find a good place near where we could either meet delegates and influence delegates or where we could have like some kind of central location in the city where people could sleep overnight so we could actually invite younger people to come or come ourselves with knapsacks and sleeping bags, somewhat as turned out at the Woodstock Festival this year.

Mr. Foran: I object to this, your Honor. He would have had a hard time saying that.

The Court: The reference to the Woodstock Festival may go out and the jury is directed to disregard it.

The Witness: Sir, the imagination that we had of it—

The Court: Will you excuse me, sir? I am not permitted to engage in a colloquy with the witness.

The Witness: I am just trying to clarify the data that I know.

The Court: A lawyer is asking you questions. You just answer them as best you can.

Mr. Weinglass: Did Mr. Rubin in the course of that conversation indicate what activities were planned for Chicago during the Democratic Convention?

A Yes. He said that he thought it would be interesting if we could set up tents and areas within the park where kids could come and sleep, and set up little schools like ecology schools, music schools, political schools, schools about the Viet Nam war, to go back into history, schools with yogis.

He suggested that I contact whatever professional breathing-exercise Yogi Swami teachers I could find and invite them to Chicago and asked if I could contact Burroughs and ask Burroughs to come also to teach non-verbal, non-conceptual feeling states.

Q Now you indicated a school of ecology. Could you explain to the Court and jury what that is?

A Ecology is the interrelation of all the living forms on the surface of the planet involving the food chain—that is to say, whales eat plankton; littler organisms in the ocean eat tiny microscopic organisms called plankton; larger fishes eat smaller fish; octopus or squid eat shellfish which eat plankton; human beings eat the shellfish or squid or smaller fish which eat the smaller tiny micro-organisms.

Mr. Foran: That is enough, your Honor.

The Court: You say that is enough?

Mr. Foran: I think that the question is now responsive. I think that—

The Court: Yes. We all have a clear idea now of what ecology is.

The Witness: Well, the destruction of ecology is what would have been taught. That is, how it is being destroyed by human intervention and messing it up with pollution.

Mr. Weinglass: Now you also indicated that Mr. Rubin mentioned non-verbal education. Will you explain what that is to the Court and jury.

A Most of our consciousness, since we are continually looking at images on television and listening to words, reading newspapers, talking in courts as this, most of our consciousness is filled with language, with a kind of matter babble behind the ear, a continuous yackety-yack that actually prevents us from breathing deeply in our bodies and sensing more subtly and sweetly the feelings that we actually do have as persons to each other rather than as to talking machines.

Q Now, Mr. Ginsberg, on March 17, 1968, where were you?

A I took part in a press conference at the Hotel Americana.

Q In what city is that?

A New York City.

Q Who else was present at this press conference?

A Abbie Hoffman and Jerry Rubin were there as well as Phil Ochs, the folk singer, Arlo Guthrie, some members of the USA Band, Bob Fass, who was a sort of Hip psychedelic radio announcer on the FM stations and a leader of the intellectual culture in New York was there; some members of the Diggers groups. [. . .]

Q Now at that press conference did you hear Jerry Rubin speak?

A Yes.

Q Could you indicate to the Court and jury what Jerry Rubin said?

A He announced the Yippie Festival of Life to the nation and said that he was going to Chicago during the convention time and hoped that a lot of younger people in America would come to the convention, come to Chicago during the convention and hold a Festival of Life in the parks, and he announced that they were negotiating with the City Hall to get a permit to have a life festival in the parks.

Q Did you hear Abbie Hoffman speak at that press conference?

A Yes.

Q Do you recall what Abbie Hoffman said?

A He talked a lot about—he said that they were negotiating with—were going to go to Chicago in a group to negotiate with representatives of Mayor Daley to get a permit for a large-scale gathering of the tribes and he mentioned the Human Be-in in San Francisco.

Q Did you yourself participate in that press conference?

A Yes. I stepped to the microphone also. [. . .]

Q Would you explain what your statement was.

A My statement was that the planet Earth at the present moment was endangered by violence, over-population, pollution, ecological destruction brought about by our own greed; that the younger children in America and other countries of the world might not survive the next 30 years, that it was a planetary crisis that had not been recognized by any government, nor the politicians who were preparing for the elections; that the younger people of America were aware of that and that precisely was what was called psychedelic consciousness; that we were going to gather together as we had before in the San Francisco Human Be-in to manifest our presence over and above the presence of the more selfish elder politicians who were not thinking in terms of what their children would need in future generations or even in the generation immediately coming or even for themselves in their own lifetime and were continuing to threaten the planet with violence, with war, with mass murder, with germ warfare, and since the younger people knew that in the United States, we were going to invite them there and that the central motive would be a presentation of a desire for the preservation of the planet. The desire for preservation of the planet and the planet's form, that we do continue to be, to exist on this planet instead of destroy the planet, was manifested to my mind by the great Mantra, from India, to the preserver God Vishnu—whose Mantra is Hare Krishna; and then I chanted the Hare Krishna Mantra for ten minutes to the television cameras and it goes:

"Hare Krishna, Hare Krishna, Krishna Krishna, Hare Hare,
Hare Rama, Hare Rama, Rama Rama, Hare Hare."

Q Now in chanting that did you have an accompaniment of any particular instrument?

Mr. Foran: Objection as immaterial. He wants to know if there was accompaniment of an instrument.

The Court: By an instrument do you mean—

Mr. Kunstler: Your Honor, I object to the laughter of the Court on this. I think this is a serious presentation of a religious concept.

The Court: I don't understand it. I don't understand it because it was—the language of the United States District Court is English.

Mr. Kunstler: I know, but you don't laugh at all languages.

The Witness: I would be glad to explain it, sir.

The Court: I didn't laugh. I didn't laugh.

The Witness: I would be happy to explain it.

The Court: I didn't laugh at all. I wish I could tell you how I feel. Laugh, I didn't even smile.

Mr. Kunstler: Well, I thought—

The Court: All I could tell you is that I didn't understand it because whatever language the witness used—

The Witness: Sanskrit, sir.

The Court: What is it?

The Witness: Sanskrit, sir.

The Court: Sanskrit?

The Witness: Yes.

The Court: Well, that is one I don't know. That is the reason I didn't understand it.

The Witness: There is a popular song put out by the Beatles with those words.

The Court: I am not interested in—

Mr. Foran: Your Honor, of course the laughter came from everybody that Mr. Kunstler is usually defending for laughing.

Mr. Kunstler: Your Honor, I would say—You mean from the press?

The Witness: Might we go on to an explanation.

The Court: Will you keep quiet, Mr. Witness, while I am talking to the lawyers?

The Witness: I will be glad to give an explanation.

The Court: I never laugh at a witness, sir. I protect witnesses who come to this court. They are entitled to the protection of the Court. But I do tell you that as I am sure you know, the language of American courts is English. The English language, unless we have an interpreter. You may use an interpreter for the remainder of this witness' testimony.

Mr. Kunstler: No. I have heard, your Honor, priests explain the mass in Latin in American courts and I think Mr. Ginsberg is doing exactly the same thing in Sanskrit for another type of religious experience.

The Court: No, no. You are mistaken.

Mr. Kunstler: Your Honor, I can't—

The Court: I don't understand Sanskrit. I venture to say the members of the jury don't. Perhaps we have some people on the jury who do understand Sanskrit, I don't know, but I wouldn't even have known it was Sanskrit until he told me.

Mr. Kunstler: Your Honor, I don't think it is being offered for people to understand the literal meaning of the words. It is being offered as an example of what he did before national television.

The Court: I can't see that that is material to the issues here, that is all.

Mr. Weinglass: Let me ask this: Mr. Ginsberg, I show you an object marked 150 for identification, and I ask you to examine that object.

A Yes. (Plays C chord)

Mr. Foran: All right. Your Honor, that is enough. I object to it, your Honor. I think that it is outrageous for counsel to—

The Court: You asked him to examine it and instead of that he played a tune on it.

Mr. Foran: I mean, counsel is so clearly—

The Court: I sustain the objection.

Mr. Foran:—talking about things that have no conceivable materiality to this case, and it is improper, your Honor.

The Witness: It adds spirituality to the case, sir.

The Court: Will you remain quiet, sir.

The Witness: I am sorry.

The Court: My obligation is to protect you, but my obligation is to see that you act in accordance with the law.

The Witness: I agree, sir.

Mr. Weinglass: Having examined that, could you identify it for the Court and the jury?

Mr. Foran: I object to it. It is immaterial.

Mr. Weinglass: I am not offering it. It is an exhibit marked for identification. We are entitled to have it identified.

The Court: You are entitled to have it identified. What is it?

The Witness: It is an instrument known as the Harmonium, which I used at the press conference at the Americana Hotel.

The Court: All you were asked was what is it, sir?

The Witness: It is a musical instrument which is used to accompany Mantra chanting, to accompany the chanting of the Hare Krishna Mantra, and other Mantra. It is commonly used in India—

The Court: You have answered that, sir.

Mr. Foran: I object to that.

The Court: I sustain the objection.

Mr. Weinglass: Now in the Mantra chanted at the press conference, were you accompanied with that instrument?

A I was accompanying myself on that instrument. I was chanting, rather than pronouncing it as I did before. I pronounced it before. At the press conference, I chanted.

Q Will you explain to the Court and to the jury what chant you were chanting at the press conference?

A I was chanting a Mantra called the Maha Mantra, the Great Mantra of Preservation of that aspect of the Indian religion called Vishnu, the Preserver, whom every time human evil, human evil rises so high that the planet itself is threatened, and all of its inhabitants and their children are threatened, Vishnu will return and preserve.

Mr. Foran: I object to that.

The Court: Oh, yes, I sustain the objection, and I strike the answer of the witness. I direct the jury to disregard it. When you offer anything in a foreign language, sir, and you think it is material, you must have an interpreter here so that the witness can be—

The Witness: Sir, it is a legal record here.

The Court: Did you hear what I said earlier to you?

The Witness: Yes.

Mr. Weinglass: If the Court please, I do have an interpreter. The interpreter happens to be the witness.

The Court: Oh, no, that would hardly be fair. An interpreter must be responsible to the Court, and he must take a special oath. I don't know whether you know that or not, but we have a special oath here for interpreters.

Mr. Weinglass: It is my understanding that an interpreter is only used when the witness is not proficient in the English language and requires the aid of an interpreter.

The Court: He used another language here.

Mr. Weinglass: And he has the capacity to explain it to the jury. Therefore, an interpreter is not necessary.

The Court: It is impossible to cross-examine a man when he is using Sanskrit which is a language—

The Witness: I am speaking English, sir.

The Court:—which is not used.
Now I have tried to be as kind as I could to you.

The Witness: I am trying to be kind to you.

The Court: I don't want you to interrupt me when I am speaking.
Ladies and gentlemen of the jury, we are going to recess now until ten o'clock tomorrow morning.

* * *

[After some testimony in which Allen identified stationery used by the Youth International Party Festival of Life that listed him under the category "religion," and recounted several conversations he had had with Abbie Hoffman and Jerry Rubin regarding their intentions about the Festival of Life and their unsuccessful negotiations with officials of the city of Chicago for permits of various kinds, he testified about his arrival in Lincoln Park, the location the Yippies wanted to use for the Festival of Life, on August 14, 1968:]

Mr. Weinglass: When you came to the park with Ed Sanders, what did you see in the park at that time as you entered the park?

A There were several thousand young people gathered, waiting, late at night. It was dark. There were some bonfires burning in trash cans. Everybody was standing around not knowing what to do.

Mr. Foran: Objection, your Honor.

The Court: "Standing around, not knowing what to do," those words may go out, and the jury is directed to disregard them.

Mr. Weinglass: Now, do you recall what, if anything, occurred at 10:30?

A There was a sudden burst of lights in the center of the park, and a group of policemen moved in fast to where the bonfires were and kicked over the bonfires.

Q Then what—

A There was a great deal of consternation and movement and shouting among the crowd in the park, and I turned, surprised, because it was early. The police were or had given 11:00 o'clock as the date or as the time—

Mr. Foran: Objection, your Honor.

The Court: Yes, "I was surprised," and the words which follow may go out. The jury is directed to disregard them.

Mr. Weinglass: When you observed the police doing this what, if anything, did you do?

A I turned to Sanders and said, "They are not supposed to be here until 11:00."

Mr. Foran: Objection.

The Court: I sustain the objection.

Mr. Weinglass: Without relating what you said to another person, Mr. Ginsberg, what did you do at the time you saw the police do this?

A I started the chant, O-o-m-m-m-m-m-m, O-o-m-m-m-m-m-m.—

Mr. Foran: All right, we have had a demonstration.

The Court: All right.

Mr. Foran: From here on, I object.

The Court: You haven't said that you objected.

Mr. Foran: I do after the second one.

The Court: After two of them? I sustain the objection.

Mr. Weinglass: If the Court please, there has been much testimony by the Government's witnesses as to this Om technique which was used in the park. Are we only going to hear whether there were stones or people throwing things, or shouting things, or using obscenities? Why do we draw the line here? Why can't we also hear what is being said in the area of calming the crowd?

Mr. Foran: I have no objection to the two Om's that we have had. However, I just didn't want it to go on all morning.

The Court: The two, however you characterize what the witness did, may remain of record, and he may not continue in the same vein.

Mr. Weinglass: Did you finish your answer?

A I am afraid I will be in contempt if I continue to Om. We walked out of the park.

Q How long did you—

A We walked out of the park. We continued chanting the Om for at least twenty minutes, slowly, gathering other people, chanting, Ed Sanders and I in the center, until there were a group maybe of 15 or 20 making a very solid heavy vibrational change of aim[82] that penetrated the immediate area around us, and attracted other people, and so we walked out slowly toward the street, toward the Lincoln Park Hotel.

Q Now, will you explain to the Court and jury what the chant Om is?

Mr. Foran: I object to that.

The Court: I sustain the objection.

Mr. Weinglass: What is the purpose of the chant Om?

Mr. Foran: Objection.

The Court: I sustain the objection.

Mr. Weinglass: Your Honor, why is the jury not being permitted to gain an understanding of this technique which is a technique used to control and quiet a crowd? We haven't objected to their testifying about stoning police cars and using obscenities. Now why are we being precluded from putting our defense in that there was an attempt on the part of this witness to calm and quiet the crowd? I don't know why we are not being permitted by the Court—

---

[82]It is tempting to wonder if what Allen really said here was "chant of Om."

The Court: The statement by Mr. Weinglass that the defense is being precluded from putting its defense in may go out and the jury is directed to disregard it. There have been several witnesses for the defense who have testified and this witness is a defense witness. He is testifying. The Court has ruled on one question, was obligated to rule on one question. It seems that every time there is an adverse ruling you get the impression or try to give the impression, Mr. Weinglass, that the Court is ruling unfairly. I am doing my best to rule according to the rules of evidence.

Mr. Weinglass: Well, your Honor, I am not—

The Court: That is my ruling, sir.

Mr. Weinglass: I am not trying to give an impression—

The Court: That is my ruling and I ask you to continue your direct examination.

Mr. Weinglass: The Government witnesses were permitted to testify about the purpose of Wash Oi, snake dancing, karate; now why can't our witness testify about the purpose of Om and other techniques which were intended to quiet the crowd.

Mr. Foran: Of course, your Honor, the Government never did testify to purposes. We testified to what was done in the park. We have no objection to this witness testifying to what he saw done or what he did himself, but I object to him talking about purposes and I ask your Honor that you direct the jury to disregard the statements of Mr. Weinglass.

The Court: I have already done so.

Mr. Weinglass: If Mr. Foran would represent to this court that his witnesses never said the purpose of snake dancing was to break through police lines, I will withdraw my objection.

Mr. Foran: Your Honor, the defendants told the purpose in the park.

The Court: I will direct you, sir, to continue with the direct examination of this witness if you have any additional questions. [. . .]

Mr. Weinglass: Now what could you observe was the condition of the people in the park at the time you began your Om chant?

A A great deal of swift and agitated motion in many different directions without any center and without any calm. When we began chanting, as it included more and more people, there was one central sound and one central rhythmic behavior vocalized by all the people who participated and a slow quietening of the physical behavior of the people that were slowly moving out of the park. They all moved in one direction, those who were involved in the chanting, out of the park and away from the police calmly without running and without physically agitated behavior.

Q At the end of your Om chant, where were you and the crowd?

Mr. Foran: I object to that, your Honor. We didn't have any testimony about a crowd.

The Court: I didn't hear that.

Mr. Foran: We had 15 or 20 people participating in this thing.

Mr. Weinglass: I will withdraw that question.

Mr. Foran: Now we have a crowd.

Mr. Weinglass: At the end of the Om chant, how many people were in your immediate vicinity?

A I remember at least a hundred.

Q Where was that group located at the end of your Om chant?

A That group as well as most of the people in the park arrived outside at the end of the park on the street that borders the park across from the Lincoln Park Hotel.

* * *

[Testimony continues, concerning a meeting in the Hilton Hotel on August 25, 1968, between Allen Ginsberg, David Dellinger, and a Mr. Baugher from Chicago's City Hall:]

Mr. Weinglass: Now what conversation in the presence of Mr. Dellinger ensued at that meeting?

A I believe that Mr. Baugher congratulated me on chanting Om and keeping the situation in the park peaceful.

Q And did you say anything in response to that?

A I said that if the city would grant a permit, then it would insure that there would be peace in the march because then there would be no conflict between the police and the kids who had come to Chicago for the Festival of Life and to observe the convention, and that it would be wise of the city even if it didn't grant a permit for the meetings in the park to at least grant a permit for a loud speaker system so that we could get on the loud speaker system and keep control of the crowd, large crowds of people, by chanting and by direction, so it wouldn't turn into an anarchic situation where people were running around like chickens with their heads off facing the police, not knowing what the police were going to do. If he would intervene with City Hall to see if we had a loud speaker system, it might insure more calm and more peace within the park.

Q What did Mr. Baugher say about that?

A He thought it was a good idea and he would talk to the people at City Hall and talk to Mr. Stahl and see where he could get a loud speaker for that afternoon when a rock festival was being planned.

Q Do you know, Mr. Ginsberg, whether Mr. Baugher actually did get a loud speaker?

Mr. Foran: Objection.

The Court: I sustain the objection.

Mr. Weinglass: Now at approximately three o'clock that afternoon which was Sunday, where were you?

A By the loud speaker in the center of the park where there was the MC5, a Detroit rock group led by John Sinclair who was at the microphone, so I came up to John Sinclair who had been arranging the music part of the day and asked him if I could do a bit of chanting on the microphone.

Q And what occurred at that time?

A I was introduced on the microphone and for about fifteen minutes chanted the Hare Krishna Mantra with the harmonium and then chanted a poem of William Blake in order to calm the crowd and to advise those who were of a violent nature—

Mr. Foran: Objection, your Honor.

The Court: I sustain the objection. The reference of the witness to his having spoken or chanted a poem of William Blake may go out and the jury is directed to disregard it.

Mr. Weinglass: Could you just state without chanting the poem[83] of William Blake to the jury?

A   " 'I die, I die!', the Mother said,
             'My Children die for lack of Bread.
             What more has the merciless Tyrant said?'
             The Monk sat down on the Stony Bed.

             "The blood red ran from the Grey Monk's side,
             His hands & feet were wounded wide,
             His Body bent, his arms & knees
             Like to the roots of ancient trees.

             "His eye was dry; no tear could flow;
             A hollow groan first spoke his woe.
             He trembled & shudder'd upon the Bed;
             At length with a feeble cry he said:

             "When God commanded this hand to write
             In the studious hours of deep midnight,
             He told me the writing I wrote should prove
             The Bane of all that on Earth I lov'd.

             "My Brother starv'd between two Walls,
             His children's Cry my Soul appalls;
             I mock'd at the wrack & griding chain,
             My bent body mocks their torturing pain.

             "Thy Father drew his sword in the North,
             With his thousands strong he marched forth;
             Thy Brother has arm'd himself in Steel
             To avenge the wrongs thy Children feel.

---

[83] The poem's title is "The Grey Monk."

"But vain the Sword & vain the Bow,
They never can work War's overthrow.
The Hermit's Prayer & the Widow's tear
Alone can free the World from fear.

"For a Tear is an Intellectual Thing,
And a Sigh is the Sword of an Angel King,
And the bitter groan of the Martyr's woe
Is an Arrow from the Almightie's Bow.

"The hand of Vengeance found the Bed
To which the Purple Tyrant fled;
The iron hand crush'd the Tyrant's head
And became a Tyrant in his stead."

Q Did you remain in the park after that?

A Yes.

Q How long did you remain in the park?

A I was in the park until about 11:30 that night. [. . .]

Q What, if anything, did you do for the remainder of the time that you were in the park?

A First I walked around away from the loud speaker system and the rock and roll music that was going on to the center of the park where suddenly a group of policemen appeared in the middle of the younger people. There was an appearance of a great mass of policemen going through the center of the park. I was afraid then, thinking they were going to make trouble—

Mr. Foran: Objection to his state of mind.

The Court: "I was afraid then—"

The Witness: Adrenaline ran in my body.

The Court: "I was afraid then" and all the words that followed may go out and the jury is directed to disregard them.

Mr. Weinglass: What did you do when you saw the policemen in the center of the crowd?

A Adrenaline ran through my body. I sat down on a green hillside with a group of younger people that were walking with me at about 3:30 in the afternoon, 4:00 o'clock, sat, crossed my legs and began chanting O-o-m—O-o-m-m-m, O-o-m-m-m, O-o-m-m-m.

Mr. Foran: I gave him four that time.

The Witness: I continued chanting for seven hours.

Mr. Weinglass: I am sorry, I did not hear the answer.

The Court: He said he continued chanting for seven hours. Seven hours, was it, sir?

The Witness: Until 10:30.

The Court: I wanted to know what your answer was. Did you say you continued chanting for seven hours?

The Witness: Seven hours, yes.

The Court: I wanted Mr. Weinglass to have your answer. He didn't hear you.

The Witness: About six hours I chanted Om and for the seventh hour concluded with the chant Hare Krishna, Hare Krishna, Krishna, Krishna, Hare, Hare, Hare Rama, Hare Rama, Rama Rama, Hare, Hare.

Mr. Weinglass: When you chanted during this period of time, were you joined in the chant?

A Yes, many people joined me. We formed a single—there were groups—it was a long period of time, the group shrank and increased as the day went on. Toward dusk there must have been—there were about 100, 200 people, shrinking, people coming and joining and going around, but there was a permanent group that stayed with me of about 50 people who continued chanting in unison.

Q Now directing your attention to Monday night, that is August 26, at approximately 11:30 in the evening, where were you?

A I had finished chanting and went down to talk to a police official who was in the center of a gathering of police at the side of the park.

Q In what park were you?

A Lincoln Park.

Q This is approximately what time of night?

A Around 11:00.

Q Now was there any kind of physical obstruction in the park on Monday night at that time?

A I went to talk to the police a little earlier than 11:00. At 11:00 I was by a barricade that was set up, a pile of trash cans and police barricades, wooden horses, I believe.

Q As you got there, what was occurring at that barricade?

A There were a lot of young kids, some black, some white, shouting and beating on the tin barrels, making a fearsome noise.

Q What did you do after you got there?

A Started chanting Om there.

Q And were you joined in that chant?

A For a while I was joined in the chant by a lot of other people who were there until the chant encompassed most of the people by the barricade and we raised a huge loud sustained series of Oms into the air loud enough to like include everybody. [. . .]

Q How long was the group that was there chanting as you just indicated?

A About fifteen minutes.

Q Then what happened?

A Just as it reached like a great unison crescendo when all of a sudden a police car came rolling down into the group, right into the center of the group where I was standing, and with a lot of crashing and tinkling sound of glass and broke up the chanting, broke up the unison and the physical—everybody was holding onto each other physically—broke up that physical community that had been built and broke up the sound chant that had been built.

Q Did this police car approach from the opposite side of that barrier or did they approach from the crowd side of the barrier?

A They came from behind the crowd.

Q And did you notice as you were standing there what, if anything, the crowd did with respect to that police car?

A People screamed, people stopped chanting, some people fled. I moved back. There was a crash of glass.

Q When you moved back, what did you do?

A I started chanting Om again.

Q Were you joined in your chant?

A Yes.

Q What occurred at that time?

A I started moving away from the scene. I started moving away from the scene because there was violence there.

Q When you say there was violence there, will you describe what you mean?

A The police car had come, had crashed right into the crowd.

Q And how long did you stay in that area after the police car came into the crowd?

A I started moving out immediately.

* * *

[After testifying about two mantra-chanting ceremonies on August 27, 1968—one at Lincoln Park's lakefront in the morning and one in the center of Lincoln Park in the afternoon—Allen is asked about events following the Un-birthday Party held for President Johnson at the Coliseum in Chicago that night, which he attended with writers Jean Genet, William Burroughs, Terry Southern, Esquire editor John Berendt, and Grove Press editor Richard Seaver:]

Mr. Weinglass: [. . .] Now, when you left the Coliseum, where, if anywhere, did you go?

A The group I was with, Mr. Genet, Mr. Burroughs and Mr. Seaver and Terry Southern, all went back to Lincoln Park.

Q What time did you arrive in the park?

A Eleven, 11:30.

Q What was occurring at the park as you got there?

A There was a great crowd lining the outskirts of the park and a little way into the park on the inner roads, and there was a larger crowd moving in toward the center. We all moved in toward the center and at the center of the park, there was a group of ministers and rabbis who had elevated a great cross about ten-foot high in the middle of a circle of people who were sitting around, quietly, listening to the ministers conduct a ceremony.

Q How many people were there?

A It must have been about a thousand.

Q Were you there while the ceremony was being conducted?

A Yes.

Q And would you relate to the court and the jury what was being said and done at that time?

A The ministers were telling whoever wanted to participate in the ceremony to sit down and be quiet, and when singing was done, to sing in unison. There were a few people who were making more disturbing noises. The ministers were trying to calm them down and have them sit down. Everybody was seated around the cross which was at the center of hundreds of people, people right around the very center adjoining the cross. Everybody was singing, "We shall overcome," and "Onward Christian Soldiers," I believe. They were old hymn tunes.

Q After the service was over, what if anything occurred at that place?

A The cross was lifted up. At the other side behind the circle that we were observing as we were seated—I was seated with my friends on a little hillock looking down on the crowd which had the cross in the center—and on the other side, there were a lot of glarey lights hundreds of feet away down the field. The ministers lifted up the cross and took it to the edge of the crowd and set it down facing the lights where the police were. In other words, they confronted the police lines with the Cross of Christ.

Q Where were you at the time that that occurred?

A I stayed sitting on the hill, watching the scene below me.

Q Could you see this?

A Yes.

Q And after the ministers moved the cross to another location which you have indicated, what happened?

A After, I don't know, a short period of time, there was a burst of smoke and tear gas around the cross, and the cross was enveloped with tear gas, and the people who were carrying the cross were enveloped with tear gas which began slowly drifting over the crowd.

Q Now prior to that happening, did you hear any announcement on any type of speaker equipment?

A No, none at all.

Q Were you told to get out of the park?

A I heard no announcement saying to get out of the park, no.

Q And when you saw the persons with the cross and the cross being gassed, what if anything did you do?

A I turned to Burroughs and said, "They have gassed the cross of Christ."

Mr. Foran: Objection, if the Court please. I ask that the answer be stricken.

Mr. Weinglass: Without relating what you said, Mr. Ginsberg, what did you do at that time?

A I took Bill Burroughs' hand and took Terry Southern's hand, and we turned from the cross which was covered with gas in the glarey lights that were coming from the police-lights that were shining through the tear gas on the cross—and walked slowly out of the park.

Q On Wednesday, the next day, at approximately 3:45 in the afternoon, do you recall where you were?

A Yes.

Q Where were you?

A Entering the Grant Park bandshell area, where there was a Mobiliza-
tion meeting or rally going on.

Q Did you enter the area alone or were you with other people?

A No, I was still with the same group of literary fellows, poets and
writers.

Q Where did you go in the band shell area?

A We had started out but couldn't get there early because all the
entrances to that area, overpasses, were blocked off by policemen say-
ing that we couldn't enter. So we had to go all the way down south to the
bottom of the park and enter over a wooden trestle. So when we
entered, we all went and sat down in the center of the crowd, waiting,
watching.

Q And did you at any time later get up to the bandshell stage?

A Yes. I walked up to the apron or front of the stage and saw David
Dellinger and told him that I was there and that Burroughs was there
and Jean Genet was there and that they were all willing to be present
and testify to the righteousness of the occasion and that we would like
to be on the stage.

Q Were you then invited onto the stage?

A He said, "By all means invite them on."

Q Were you then introduced?

A Yes. [. . .]

Q Did you speak?

A I croaked, yes.

The Court: What was that last? You say you what?

The Witness: I croaked. My voice was gone. I chanted or tried to chant.

*[After a series of questions from Leonard Weinglass that elicited an account
from Allen of his attempt to distract bohemian trickster candidate for presi-
dent Louis Abolafia, who had been telling the crowd that the police were
armed and dangerous, by tickling his leg:]*

Mr. Weinglass: Now when the rally was over, did you have occasion to talk with Mr. Dellinger?

A Yes. I went down off the platform into the crowd that was forming for the peace march then and saw Mr. Dellinger.

Q And what did Mr. Dellinger say to you at that time?

A He looked me in the eye, took my arm and said, "Allen, will you please march up front in the front line with me?"

Q And what did you say to him?

A I said, "Well, I am here with Burroughs and Genet and Terry Southern," and he said, "Well, all of you together, can you form a front line and be sure to stay behind me in the front line, be the first of the group of marchers?"

Q And did you form such a line?

A Yes. I went and told Burroughs and Genet and they all agreed and we went right up front.

Q Who was in front of you, immediately in front of you?

A Immediately in front was one fellow carrying on his head a portable loud speaker equipment, Mr. Dellinger, and immediately behind him a group of marshals and then ourselves as the head of the line of peace-manifesting marchers.

Q How were you walking?

A Slowly.

Q Were your arms—

A Our arms were all linked together and we were carrying flowers. Someone had brought flowers up to the back of the stage and so we distributed them around to the first rows of marchers so all of the marchers had flowers.

Q Now how far did you walk?

A We were still in the park. It was like a couple of blocks within the park along the side of a viaduct.

Q After you walked the several blocks, what occurred?

A We came to a halt in front of a large guard of armed human beings in uniform who were blocking our way, people with machine guns, jeeps, I believe, police, and what looked to me like soldiers on our side and in front of us.

Q And what happened at that point?

A Mr. Dellinger—the march stopped and we waited not quite knowing what to do. I heard—all along I had heard Dave Dellinger saying, "This is a peaceful march. All those who want to participate in a peaceful march please join our line. All those who are not peaceful, please go away and don't join our line." He continued that up to the point when we were stopped.

Q When you were stopped, did you see what Dellinger did at that time?

A He went forward to talk to the police heads that were there.

Q Did you go over with him or did you remain behind?

A Yes, I went over with him, took his arm for one moment and also brought a little armful of flowers that had been given us as we left the bandshell.

Q And what did you do with the flowers, if anything?

A Mr. Dellinger and the city agents, city officials and police heads were talking together, negotiating, and whenever they seemed to me to be agitated, I took the flowers and put them in between their faces and shook them around a little which made Mr. Elrod smile, I remember.

Q Do you know who Mr. Elrod is?

A He was the chief Corporation Counsel or the chief lawyer of the city, I believe.

Q Now did you return to the line that you had left to go up with Mr. Dellinger at any point in time?

A Yes. There was a time when there was a long wait for messages to be sent back and forth. Mr. Dellinger told me would I please go to the loud speaker system and the portable microphone and the bull horn that was being carried on the head of the kid, would I get up to the microphone and calm the crowd and calm the police by chanting because there was this long line behind us, there was a great mass of armed people in front of us, it was a tricky, scary moment. So Dave Dellinger asked me to get up on the microphone and begin chanting and chant as long as I pleased and chant as deeply as I pleased.

Q Did you do that?

A Yes.

Q Did anyone join you in chanting?

A Yes, many people.

Q At that point where were the people, standing or seated?

A They were mostly seated. Everybody had sat down for the long wait as we were blocked, so everybody was seated, some cross-legged on the ground. [. . .]

Q Now did there come a time when you left this area?

A Much later, yes. I stayed sitting there. There were other moves and other movements between the police and Mr. Dellinger and the line of march.

Q How long did you remain there?

A About three-quarters of an hour.

Q At the end of that three-quarters of an hour, what, if anything, did you do?

A I got up to walk away with Mr. Dellinger from the march. I think Mr. Dellinger announced that the march was over and had been victorious inasmuch as the Government had simply forced us to abandon our citizen's right to have a peaceful assembly for redress of grievances and there was nothing we could do about it at that point. We had offered ourselves to be arrested and were not being arrested and so we were going to disperse and move on. Mr. Dellinger declared the march over and so we began dispersing.

* * *

Mr. Weinglass: [. . .] I now show you what is marked D-153 for iden-
tification, which is a publication of yours that appeared in "Rat,"[84] I
believe you testified, three days before the convention began?

A Yes.

Q Could you read that to the jury?

A "Magic Password Bulletin. Psychic Jujitsu.

"In case of hysteria, solitary or communal, the magic password is Om,
same as Aum, which cuts through all emergency illusions. Pronounce
Om from the middle of the body, diaphragm or solar plexus. Ten peo-
ple humming Om can calm down one hundred. One hundred people
humming Om can regulate the metabolism of a thousand. A thousand
bodies vibrating Om can immobilize an entire downtown Chicago
street full of scared humans, uniformed or naked.

Signed, Allen Ginsberg, Ed Sanders.

"Om will be practiced on the beach at sunrise ceremonies with Allen
and Ed."

Q Now, Mr. Ginsberg, where you say a thousand people chanting Om
could immobilize an area in downtown Chicago, what did you mean by
that? [. . .]

A Immobilize an entire downtown Chicago street full of scared human
beings, uniformed or naked—by immobilize I meant shut down the
mental machinery which repeats over and over again the images of
fear which are scaring people in uniform, that is to say, the police
officers, or the demonstrators whom I refer to as naked, meaning
naked emotionally and perhaps, hopefully, naked physically.

Q And what did you intend to create by having that mechanism shut
down?

A A completely peaceful realization of the fact that we were all stuck in
the same street, place, terrified of each other, and reacting in panic

---

[84]New York underground newspaper. This piece appeared in *Rat: Subterranean News*,
vol. 1, no. 15 (Aug. 23–Sept. 5, 1968), p. 3.

and hysteria rather than reacting with awareness of each other as human beings, as people with bodies that actually feel, can chant and pray and have a certain sense of vibration to each other or tenderness to each other which is what basically everybody wants rather than fear.

Mr. Weinglass: I have nothing further.

The Court: Cross-examination.

Cross-examination by Mr. Foran:

[After cross-examining Allen about some of the marches and incidents involving the police and those attending the Festival of Life, Thomas Foran asked several questions about Allen's study of chanting, leading up to questioning him about the appearance of Allen's name on the Yippie stationery testified to above:]

Q You were named as kind of the Yippie religious leader on that letterhead. Do you think in whatever sense you would like to put it that is a fair designation of your connection with the Yippie organization?

A No, because the word leader was one we really tried to get away from, to get away from that authoritarian thing. It was more like—

Q Religious teacher?

A—religious experimenter, or someone who was interested in experimenting with that, and moving things in that direction.

Q In the context of the Yippie organization?

A Yes, and also in the context of our whole political life, too. [. . .]

Q Now, in the course of your work, itself, Mr. Ginsberg, you go around to various places and recite your poetry?

A Yes.

Q And chant?

A Yes.

Q You chant many times in conjunction with it?

A Yes, as I have done here.

Q Doing both of them at the same time?

A Generally.

Q Generally, for the same purpose, are they in combination?

A Yes, I try to begin by invoking some deeper spirit than intellectual language. I begin really to try to calm my own body, calm myself by chanting, and to calm the audience a little so that they're aware that I am here, that they are sitting there in their bodies, and that we are together in the same room and sharing our feelings.

Q So when you, like in Lincoln Park that Sunday afternoon, when you chanted, and when you recited some of William Blake's poetry, that combination of chanting and poetry, was within the context, as you see it, Mr. Ginsberg, of generating a spiritual and physical uplift to the audience?

A Not so much an uplift, but a calm, a feeling of ease and relaxation to eliminate tension, to eliminate anxiety, to eliminate hysteria, to eliminate the hallucination of scary images of police with—

Q Within that concept of yours as you designate yourself as the religious experimenter of the Yippie Organization, and I don't mean to be tricky at all about this, I mean just to use words that are perhaps—

A Yes.

Q—familiar, this concept of physical calmness and acceptance is a part of the religious experience that you were attempting to experiment with and teach?

A Yes.

Q Is that a fair statement?

A Yes.

Q And both your poetry and chanting are a part of that same religious experimentation concept?

A Using the term we have on the letterhead there, religion, yes.

Q Religion.

A Yes.

Q Now when you went out to the Coliseum and you met Abbie Hoffman, you said when you met him you kissed him?

A Yes.

Q Is he an intimate friend of yours?

A I felt very intimate with him. I saw he was struggling to manifest a beautiful thing, and I felt very good towards him.

Q And you do consider him an intimate friend of yours?

A I don't see him that often, but I see him often enough and have worked with him often enough to feel intimate with him, yes. [. . .]

Q [. . .] Now, you testified concerning a number of books of poetry that you have written?

A Yes.

Q One of them was the "Empty Mirror."

A Yes, a book of early poems written in 1946 to 1951.

Q You also talked about "In Society."

A Yes.

Q And that is another poem you wrote?

A Yes, it is the transcription of a dream that I had, a night dream. [. . .]

Q In the "Empty Mirror," there is a poem called, "The Night-Apple"?

A Yes.

Q Would you recite that for the jury? [. . .]

[The book Empty Mirror is introduced as Government's Exhibit No. 59.]

A Yes. "The Night-Apple."

> "Last night I dreamed
> of one I loved
> for seven long years,
> but I saw no face,

only the familiar
presence of the body:
sweat skin eyes:
feces urine sperm
saliva all one
odor and mortal taste."

Q Could you explain to the jury what the religious significance of that poem is?

A If you would take a wet dream as a religious experience, I could. It is a description of a wet dream, sir.

Q Now, I call your attention in that same Government's Exhibit No. 59, to page 14.

A Yes.

Q That has on it the poem, "In Society"?

A Right.

Q And is that one of the poems you have written, Mr. Ginsberg?

A Yes, it says on the bottom 1947, it was a dream.

Q Does that refresh your recollection of that poem?

A Yes.

Q After refreshing your recollection, can you recite that poem to the jury?

A Yes, I will read it.

"In Society"

"I walked into the cocktail party
room and found three or four queers
talking together in queertalk.
I tried to be friendly but heard
myself talking to one in hiptalk.
"I'm glad to see you," he said, and

looked away. "Hmn," I mused. The room
was small and had a double-decker
bed in it, and cooking apparatus:
icebox, cabinet, toasters, stove;
the hosts seemed to live with room
enough only for cooking and sleeping.
My remark on this score was under-
stood but not appreciated. I was
offered refreshments, which I accepted.
I ate a sandwich of pure meat; an
enormous sandwich of human flesh,
I noticed, while I was chewing on it,
it also included a dirty asshole.

More company came, including a
fluffy female who looked like
a princess. She glared at me and
said immediately: "I don't like you."
turned her head away, and refused
to be introduced. I said, "What!"
in outrage. "Why you shit-faced fool!"
This got everybody's attention.
"Why you narcissistic bitch! How
can you decide when you don't even
know me," I continued in a violent
and messianic voice, inspired at
last, dominating the whole room."

Dream 1947

It is a record, a literal record of a dream as the other was a literal
record of a dream.

Q Can you explain the religious significance of that poetry?

A Actually, yes.

Q Would you explain it to the jury?

A Yes, one of the major yogas, or yoking—yoga means yoke—is bringing
together the conscious mind with the unconscious mind and is an

examination of dream states in an attempt to recollect dream states no matter how difficult they are, no matter how repulsive they are, even if they include hysteria, sandwiches of human flesh, which include dirty assholes, because those are universal images that come in everybody's dreams.

The attempt in yoga is to enlarge consciousness, to be conscious that one's own consciousness will include everything which occurs within the body and the mind.

As part of the practice of poetry, I have always kept records of dreams whenever I have remembered them and have tried not to censor them so that I would have all the evidence to examine in light of day so that I would find out who I was unconsciously.

Part of Zen meditation and part of yoga meditation consists in the objective impersonal examination of the rise and the fall and disappearance of thoughts in the mind. All thoughts, whether they be thoughts of sleeping with one's mother, which is universal, or sleeping with one's father, which is also universal thought, or becoming an angel, or flying, or attending a cocktail party and being afraid of being put down, and then getting hysterical. In other words, the attempt is to reclaim the unconscious, to write down in the light of day what is going on in the deepest meditation of night and dream state so it is part of a yoga which involves bridging the difference between public, as in this courtroom, and private subjective: Public, which is conscious, which we can say to others in family situations, and private which is what we know and tell only our deepest friends.

Q Thank you. [. . .] You also wrote a book of poems called "Reality Sandwiches," didn't you?

A Yes.

Q In there, there is a poem called, "Love Poem on Theme by Whitman"?

A Yes.

Q I show you Government's Exhibit No. 60 for identification, and I call your attention to page 41, and I ask you whether or not that refreshes your recollection of that poem?

A Yes.

Q After having refreshed your recollection, would you recite that to the jury?

A "Love Poem on Theme by Whitman," Walt Whitman being our celebrated bard, national prophet. The poem begins with a quotation of a line by Walt Whitman—it begins with Walt Whitman's line:

"I'll go into the bedroom silently and lie down between the
        bridegroom and the bride,
those bodies fallen from heaven stretched out waiting naked and
        restless,
arms resting over their eyes in the darkness,
bury my face in their shoulders and breasts, breathing their skin,
and stroke and kiss neck and mouth and make back be open
        and known,
legs raised up crook'd to receive, cock in the darkness driven
        tormented and attacking roused up from hole to itching head,
bodies locked shuddering naked, hot lips and buttocks screwed
        into each other
and eyes, eyes glinting and charming, widening into looks and
        abandon,
and moans of movement, voices, hands in air, hands between
        thighs,
hands in moisture on softened hips, throbbing contraction of
        bellies
till the white come flow in the swirling sheets,
and the bride cry for forgiveness, and the groom be covered with
        tears of passion and compassion,
and I rise up from the bed replenished with last intimate gestures
        and kisses of farewell—
all before the mind wakes, behind shades and closed doors in
        darkened house
where the inhabitants roam unsatisfied in the night,
nude ghosts seeking each other out in the silence."

Q Would you explain the religious significance of that poem?

A As part of our nature, as part of our human nature we have many loves, many of which are suppressed, many of which are denied, many of which we deny to ourselves. He said that the reclaiming of those loves and the becoming aware of those loves was the only way that this nation could save itself and become a democratic and spiritual republic.

He said that unless there were an infusion of feeling, of tenderness, of fearlessness, of spirituality, of natural sexuality, of natural delight in each other's bodies into the hardened materialistic, cynical, life denying, clearly competitive, afraid, scared, armored bodies, there would be no chance for spiritual democracy to take root in America and he defined that tenderness between the citizens as, in his words, an Adhesiveness, a natural tenderness flowing between all citizens, not only men and women but also a tenderness between men and men, as part of our democratic heritage, part of the adhesiveness which would make the democracy function; that men could work together not as competitive beasts but as tender lovers and fellows.

So he projected, from his own desire and from his own unconscious, a sexual urge which he felt was normal to the unconscious of most people, though forbidden for the most part to take part.

"I will go into the bedroom silently and lie down between the bridegroom and the bride."

He projected as he did in another poem, Orgy, "City of Orgies", as he called New York, he projected physical affection even to the sexual— or his phrase is "physical affection and all that is latently implied" between citizen and citizen as part of the adhesiveness which would make us function together as a community rather than as a nation "among the fabled damned of nations", which was his phrase in the essay "Democratic Vistas."

Walt Whitman is one of my spiritual teachers and I am following him in this poem, taking off from a line of his own and projecting my own actual unconscious feelings, of which I don't have shame, sir, which I feel are basically charming actually.

The Court: I didn't hear that last word.

The Witness: Charming.

*   *   *

1970s

MARY JANE FORTUNATO, LUCILLE MEDWICK, AND SUSAN ROWE
December 17, 1970,   New York City

"Craft Interview with Allen Ginsberg"
New York Quarterly, Spring 1971

In 1999 Mary Jane Fortunato, one of the *New York Quarterly*'s founders, remembered that, "it was the intention of those of us who founded the *Quarterly* to obtain the ideas and approach to the writing of poetry from those poets whom we considered to be excellent, contemporary, meaningful—the best and the brightest, so to speak. The poet was either telephoned or written to, and, when the poet responded, the caller/writer would explain our approach, which was the craft approach. If the poet agreed to the interview, the choice of place was always up to the poet." The interview took place in the *Quarterly*'s office, located on the sixth floor of the Overseas Press Club on West 43rd Street, overlooking Bryant Park. "There was a large table in the middle of the room, and we gathered around the table. Allen was at the end of the table . . . facing all of us. . . . I was moved and touched by his gentleness. I was really impressed at being in his presence and felt strongly that the gentleness was a truly important part of him, one that I had not understood until then."

Fortunato transcribed the interview.

—DC

NYQ: You have talked about this before, but would you begin this interview by describing the early influences on your work, or the influences on your early work?

Allen Ginsberg: Emily Dickinson. Poe's "Bells"—"Hear the sledges with the bells—Silver bells! . . ." Milton's long line breath in *Paradise Lost*—

Him the almighty power
Hurled headlong flaming from the ethereal sky

> With hideous ruin and combustion down
> To bottomless perdition, there to dwell
> In adamantine chains and penal fire,
> Who durst defy the omnipotent to arms.

Shelley's "Epipsychidion"—"one life, one death, / one Heaven, one Hell, one immortality, / And one annihilation. Woe is me!" The end of Shelley's "Adonais"; and "Ode to the West Wind" exhibits continuous breath leading to ecstatic climax.

Wordsworth's "Intimations of Immortality"—

> Our birth is but a sleep and a forgetting:
> The soul that rises with us, our life's Star,
> Hath had elsewhere its setting,

Also Wordsworth's "Tintern Abbey" exhortation, or whatever you call it:

>   a sense sublime
> of something far more deeply interfused,
> Whose dwelling is the light of setting suns,
> And the round ocean and the living air,
> And the blue sky, and in the mind of man;

That kind of poetry influenced me: a long breath poetry that has a sort of ecstatic climax.

NYQ: What about Whitman?

AG: No, I replied very specifically. You asked me about my *first* poetry. Whitman and Blake, yes, but in terms of the *early* poems I replied specifically. When I began writing I was writing rhymed verse, stanzaic forms that I derived from my father's practice. As I progressed into that I got more involved with Andrew Marvell.

NYQ: Did you used to go to the Poetry Society of America meetings?

AG: Yes, I used to go with my father. It was a horrifying experience, mostly old ladies and second-rate poets.

NYQ: Would you elaborate?

AG: That's the PSA I'm talking about. At the time it was mainly people who were enemies of, and denounced, William Carlos Williams and Ezra Pound and T. S. Eliot.

NYQ: How long did it take you to realize they were enemies?

AG: Oh, I knew right away. I meant enemies of poetry, very specifically. Or enemies of that poetry which now by hindsight is considered sincere poetry of the time. *Their* highwater mark was, I guess, Edwin Arlington Robinson, "Eros Turannos" was considered the great highwater mark of twentieth-century poetry.

NYQ: Where did you first hear long lines in momentum?

AG: The texts I was citing were things my father taught me when I was prepubescent.

NYQ: Did he teach them to you as beautiful words or as the craft of poetry?

AG: I don't think people used that word "craft" in those days. It's sort of like a word that has only come into use in the last few decades. There were texts of great poetry around the house, and he would recite from memory. He never sat down and said now I am going to teach you: Capital C-R-A-F-T. Actually I don't like the use of the word craft applied to poetry, because generally along with it comes a defense of stressed iambic prosody, which I find uncraftsmanly and pedantical in its use. There are very few people in whose mouths that word makes any sense. I think Marianne Moore may have used it a few times. Pound has used it a couple of times in very specific circumstances—more often as a verb than as a general noun. "This or that poet has crafted a sestina."

NYQ: Would you talk about later influences on your work? William Blake? Walt Whitman?

AG: Later on for open verse I was interested in Kerouac's poetry. I think that turned me on more than anyone else, I think he is a very great poet and much underrated. He hadn't been read yet by poets.

NYQ: Most people associate Kerouac with prose, with *On the Road*, and not so much with *Mexico City Blues*. Or maybe they differentiate too strictly between prose and poetry.

AG: I think it's because people are so preoccupied with the use of the word craft and its meaning that they can't see poetry in front of them on the page. Kerouac's poetry looks like the most "uncrafted stuff" in the world. He's got a different idea of craft from most people who use the word craft. I would say Kerouac's poetry is the craftiest of all. And as far as having the most craft of anyone, though those who talk about craft have not yet discovered it, his craft is spontaneity; his craft is having the instantaneous recall of the unconscious; his craft is the perfect executive conjunction of archetypal memorial images articulating present observation of detail and childhood epiphany fact.

NYQ: In "Howl," at the end of Section One, you came close to a definition of poetry, when you wrote:

> Who dreamt and made incarnate gaps in Time & Space through
>   images juxtaposed, and trapped the archangel of the soul
>   between 2 visual images and joined the elemental verbs
>   and set the noun and dash of consciousness together
>   jumping with sensation of Omnipotens Aeterna Deus.

AG: I reparaphrased that when I was talking about Kerouac. If you heard the structure of the sentence I was composing, it was about putting present observed detail into epiphany, or catching the archangel of the soul between two visual images. I was thinking then about what Kerouac and I thought about haiku—two visual images, opposite poles, which are connected by a lightning in the mind. In other words "Today's been a good day; let another fly come on the rice." Two disparate images, unconnected, which the mind connects.

NYQ: Chinese poets do that. Is this what you are talking about?

AG: This is characteristic of Chinese poetry as Ezra Pound pointed out in his essay "The Chinese Written Character as a Medium for Poetry" nearly fifty years ago. Do you know that work? Well, way back when, Ezra Pound proposed Chinese hieroglyphic language as more fit for poetry, considering that it was primarily visual, than generalized language-abstraction English, with visionless words like Truth, Beauty, Craft, etc. Pound then translated some Chinese poetry and translated (from Professor [Ernest] Fenollosa's[85]

---

[85]See *The Chinese Written Character as a Medium for Poetry*, by Ernest Fenollosa. An ars poetica. With a foreword and notes by Ezra Pound, New York, Arrow Editions [1936].

papers) this philosophic essay pointing to Chinese language as pictorial. There is no concrete picture in English, and poets could learn from Chinese to present image detail: and out of that Pound hieroglyph rose the whole practice of imagism, the school which is referred to as "Imagism." So what you are referring to is an *old* history in twentieth-century poetry. My own thing about two visual images is just from that tradition, actually drawing from Pound's discovery and interpretation of Chinese as later practiced by Williams and everybody who studied with Pound or who understood Pound. What I'm trying to point out is that this tradition in American poetry in the twentieth century is not something just discovered. It was done by Pound and Williams, precisely the people that are anathema to the PSA mediocrities who were attacking Pound and Williams for not having "craft."

NYQ: In that same section of "Howl," in the next line, you wrote:

> to recreate the syntax and measure of poor human prose and
>   stand before you speechless and intelligent and
>   shaking with shame, rejected yet confessing out the
>   soul to conform to the rhythm of thought in his
>   naked and endless head.

AG: Description of aesthetic method. Key phrases that I picked up around that time and was using when I wrote the book. I meant again that if you place two visual images side by side and let the mind connect them, the gap between the two images, the lightning in the mind illuminates. It is the *Sunyata*[86] (Buddhist term for blissful empty void) which can only be known by living creatures. So, the emptiness to which the Zen finger classically points—the ellipse—is the unspoken hair-raising awareness in between two mental visual images. I should try to make the answers a little more succinct.

NYQ: Despite your feeling about craft, poets have developed an attitude towards your work, they have discovered certain principles of breath division in your lines—

AG: Primary fact of my writing is that I don't have any craft and don't know what I'm doing. There is absolutely no art involved, in the con-

---

[86]For *Sunyata*, see Ginsberg on emptiness, pg. 415, Portugés-Amirthanayagam interview.

text of the general use of the words art and craft. Such craft or art as there is, is in illuminating mental formations, and trying to observe the naked activity of my own mind. Then transcribing that activity down on paper. So the craft is being shrewd at flashlighting mental activity. Trapping the archangel of the soul, by accident, so to speak. The subject matter is the action of my mind. To put it on the most vulgar level (like on the psychoanalyst's couch is supposed to be). Now if you are thinking of "form" or even the "well made poem" or a sonnet when you're lying on the couch, you'll never say what you have on your mind. You'd be babbling about corset styles or something *else* all the time instead of saying, "I want to fuck my mother," or whatever it is you want. So my problem is to get down the fact that I want to fuck my mother or whatever. I'm taking the most hideous image possible, so there will be no misunderstanding about what area of mind you are dealing with: What is socially unspoken, what is prophetic from the unconscious, what is universal to all men, what's the main subject of poetry, what's underneath, *inside* the mind. So, how do you get that out on the page? You observe your own mind during the time of composition and write down whatever goes through the ticker tape of mentality, or whatever you hear in the echo of your inner ear, or what flashes in picture on the eyeball while you're writing. So the subject is constantly interrupting because the mind is constantly going on vagaries— so whenever it changes I have a dash. The dashes are a function of this method of transcription of unconscious data. Now you can't write down everything that you've got going on—half-conscious data. You can't write down *everything*, you can only write down what the hand can carry. Your hand can't carry more than a twentieth of what the mind flashes, and the very fact of writing interrupts the mind's flashes and redirects attention to writing. So that the observation (for writing) impedes the function of the mind. You might say "Observation impedes Function." I get down as much as I can of genuine material, interrupting the flow of material as I get it down and when I look I turn to the center of my brain to see the next thought, but it's probably about thirty thoughts later. So I make a dash to indicate a break, sometimes a dash plus dots. Am I making sense?—

Saying "I want to fuck my mother"—that's too heavy. It waves a red flag in front of understanding, so we don't have to use that as the archetypal thought. Like "I want to go heaven" may be the archetypal thought, instead of "I want to fuck my mother." I just wanted to get it down to some place that everybody knows where it is. If I say "I want to go to heaven" you might think it's a philosophic conception.

NYQ: How much do you revise your work?

AG: As little revision as possible. The craft, the art consists in paying attention on the actual movie of the mind. Writing it down is like a by-product of that. If you can actually keep track of your own head movie, then the writing it down is just like a secretarial job, and who gets crafty about that? Use dashes instead of semicolons. Knowing the difference between a dash—and a hyphen-. Long lines are useful at certain times, and short lines at other times. But a big notebook with lines is a helpful thing, and three pens[87]—you have to be shrewd about that. The actual materials are important. A book at the nightstand is important—a light you can get at—or a flashlight, as Kerouac had a brakeman's lantern. That's the craft. Having the brakeman's lantern and knowing where to use the ampersand "&" for swiftness in writing. If your attention is focused all the time—as my attention was in writing "Sunflower Sutra," "TV Baby" poem later, ("Wichita Vortex Sutra" later, in a book called *Planet News*)—when attention is focused, there is no likelihood there will be much need for blue penciling revision because there'll be a sensuous continuum in the composition. So when I look over something that I've written down, I find that if my attention has lapsed from the subject, I begin to talk about myself writing about the subject or talking about my irrelevant left foot itch instead of about the giant smog factory I'm observing in Linden, New Jersey. Then I'll have to do some blue penciling, excising whatever is irrelevant: whatever I inserted self-consciously, instead of conscious of the Subject. Where self-consciousness intervenes on attention, blue pencil excision means getting rid of the dross of self-consciousness. Since the subject matter is really the operation of the mind, as in Gertrude Stein, anything that the mind passes through is proper and shouldn't be revised out, almost anything that passes through mind, anything with

[87]The question was put to Bob Rosenthal of why "three pens"? He suspected it was because Allen always wanted a reserve since he wrote with fountain pens, which can get clogged or run out of ink. He checked his interpretation with Allen's brother, Eugene, who agreed.

the exception of self-consciousness. Anything that occurs to the mind is the proper subject. So if you are making a graph of the movements of the mind, there is no point in revising it. Because then you would obliterate the actual markings of the graph. So if you're interested in writing as a form of meditation or introspective yoga, which I am, then there's no revision possible.

NYQ: Your poem about the sunflower shows remarkable powers of concentration.

AG: "Sunflower Sutra," the original manuscript in pencil's somewhere at Columbia University Library.[88] In examining it you will see published poem deviates maybe five or ten words from the original penciled text, written in twenty minutes, Kerouac at the door, waiting for me to go off to a party, and I said, "Wait a minute, I got to write myself a note." I had the Idea Vision and I wanted to write it down before I went off to the party, so I wouldn't forget.

NYQ: Did it dictate the sense, or did you just do it for yourself?

AG: Observing the flashings on the mind. As somebody said, the craft is observing the mind. Formerly the "craft" used to be an idea of rearranging your package, rearranging. Using the sonnet is like a crystal ball to pull out more and more things from the subconscious (to pack into the sonnet like you pack an ice cream box). Fresher method of getting at that material is to watch mind flow instantaneously, to realize that all that is, is there in the storehouse of the mind within the instant any moment: that's the Proust of eternal recall, remember, the entire *Remembrance of Things Past* came to him just as he was dunking that little bit of madeleine cake into his tea. You know, the whole content of that one instant: that epiphanous instant, working with that instant—the mind then and there. That method I learned from Kerouac and I am interested in. That method is related to other "classical" methods of art composition and meditation like Zen Buddhist calligraphic painting, haiku composition also a spontaneous art, *supposed* to be spontaneous. People don't sit around revising haikus. They are supposed to be sitting around drinking saki, near a little hibachi (charcoal stove) with fireflies and fans and half moons through the

---

[88]Allen Ginsberg's papers were sold to Stanford University in 1994 and are no longer at Columbia.

window. And in the summertime you are supposed to say, "Ah, the firefly has just disappeared into the moon . . ." Make it up then and there. It's got to come from the perception of the moment. You can't go home the next day and send your friend the haiku and say "I thought of a funny one: the firefly just . . ." That wouldn't be real.

NYQ: Do you see time used as a unit of structure, as well as a point of view?

AG: Time of composition is the structure of the poem. That is the subject. What is going on in the mind during that moment is the subject. "Time is of the essence," said Kerouac in a very great little essay on writing poetry, one-page set of advice, *Essentials of Spontaneous Prose*,[89] in back of Don Allen's *The New American Poetry: 1945–1960*, the section is devoted to composition theory. I learned my theory from Kerouac. The preoccupations I have are Hindu, Buddhist, Hassidic—I spend every morning one hour sitting cross-legged, eyes closed, back straight, observing my consciousness and quieting my consciousness, watching processions of mental imagery. Someone who isn't into that kind of meditation might find it an unknown territory to go into, chaos, and see it as much too chaotic to get involved with.

NYQ: You once wrote, "I won't write my poem until I'm in my right mind."

AG: Yes. Of course, *that* poem ("America") is like a series of one-line jokes, so to speak. At the expense of the body politic, at the expense of the mass media Hallucination of Being entertained by the middle class.

NYQ: Does this refer to an attitude of yours about state of mind?

AG: I'm referring to a nervously comical attitude toward America. It ends "I'm putting my queer shoulder to the wheel." What I'm saying is, my poetry—this particular poem—my poetry in general—shows as such drivel because the United States is in such a state of apocalyptic drivelhood, that we're destroying the world, actually, and we're really destroying ourselves, and so I won't write my poems until I'm in my right mind. Until America gets out of its silly mood.

---

[89]*Evergreen Review* no. 5, Summer 1958 (Grove Press, NY: 1958).

NYQ: Would you discuss travel, when you're in different places do you find yourself affected by the prosody of the place?

AG: I try to learn what I can. I got involved with mantra chanting when I was in India and brought it back to America. I do a lot of mantra chanting here. Just because I was interested in it and it had something to do with poetics, I thought. It also had to do with vocalization in that it did relate to preoccupations that I was familiar with in Pound's dictum "Pay attention to the tone leading of vowels." Sanskrit prosody has great ancient rules involving vowels and a great consciousness of vowels or a consciousness of quantitative versification. Like Pound is conscious of that too, tried to bring that to the awareness of poets in the twentieth century, tried to make people more conscious of the tone leading of vowels and renounce hyper attention to accentual rhythm. Pound said that he thought the future American prosody would be "an approximation of classical quantity," he thought that would be a formal substitute for iambic count, stress count. The whole poetic movement of the century climaxed in what was known as Beat or San Francisco or Hippie or whatever Renaissance movement was finally a realization of a new form of prosody, a new basis for the prosody. Actually I've written a great deal about the subject. I don't know if you're familiar with much of it, but some poetics is covered in a *Paris Review* interview. The relationship between Poetics and Mantra is gossiped on in a *Playboy* interview. A closer analysis of stress prosody, that kind of craft, sits in a preface to my father's book[90]—where I referred—(in answer to an earlier question) to one of the books that influenced me when I was young called *American Poetry*, edited with Introduction, Notes, Questions, and Biographical Sketches by A. B. De Mille, Simmons College, Boston, Secretary of the New England Association of Teachers of English, Boston, Allyn and Bacon, 1923, Academy Classic Series. It was, like the high school anthology, for most older high school teachers who teach now, their education. It was the standard anthology of the early twenties and used around the schools. So, they say in this book . . . I read Dickinson and Poe and Archibald Rutledge and Whittier and Longfellow and Thoreau and Emerson and John Hay Whitney, all the bearded poets of the nineteenth century in that book. This book described accentual prosody as "particularly well adapted to the needs

---

[90]*Morning in Spring,* Louis Ginsberg (New York, William Morrow: 1970.)

of English poetry . . . definite rules, which have been carefully observed by all great poets from Homer to Tennyson and Longfellow." They gave as an example of accentual prosody in this book for teachers and students:

Thŏu tóo/săil ón,/Ŏ Shíp/ ŏf Státe./

Remember that line? They had it marked: as above. As you notice they had an unaccented mark for O and then an accented mark for Ship. When you read it you will realize that O is an exclamation, and, by definition, you *can't* have an unaccented mark for that and an accented mark for Ship. Which means that by the time 1923 had come teachers of English prosody had so perverted their own ears and everybody else's ears that they could actually write down "O" as unaccented. See, it was done like that. Well, what it means is that nobody could pronounce the line right. They were teaching people to mispronounce things. It would have to be many long vowels, but when you got up on the elementary or high school lecture platform, they used to say: Thŏu tóo/săil ón,/ Ŏ Shíp/ ŏf Státe." Hear? Another example they had in there was:

Whŏse héart/-strĭngs áre/ ă lúte

when it quite obviously is: Whŏse héart-stríngs ăre ă lúte. So, in other words, that's where "craft" degenerated. That's why I'm talking about how do we get out from under that. Because that was the Poetry Society of America's standard of poetics. And that's what Pound was fighting against. And replacing with a much more clear ear. And of course that's what Williams was working on, and that's what Creeley, Olson and Kerouac have always been compensating for. That's why I'm so mean about the use of the word "craft." Because I really wanted to make it clear that whatever people think craft is supposed to be, that what they've been taught at school, it's *not* that at all. One had better burn the word than abuse it as it has been abused, to confuse everybody.

NYQ: You have been giving readings with your father.

AG: We've done about four a year since 1965. We started at the PSA. But we don't do it often. It would get to be too much of an act or something.

Generally we do it when there's some sentimental or aesthetically inter-esting occasion. Like at the PSA, that was interesting aesthetically. At the Y the other night, that was interesting because it is the traditional place for "distinguished poets" to read. I do it because, partly, to live with my father, because he's not going to be here forever. Nor am I. As a poet I'm interested in living in the same universe with him, and working in the same universe with him. We both learn something from it, get a little bit into each other's souls, the world soul. A father can learn a son's soul and a son can learn a father's soul, it's pretty much knowing God's soul, finally. It's like a confrontation with my own soul which is sometimes difficult. But it usually winds up pleasurable. Sometimes I have to see things in myself or face things in my father that are quite hideous. Con-front them. So far this has turned out to reconcile us more and more.

NYQ: You read the "Wales Visitation" poem on the Buckley TV inter-view?

AG: Yes—"Wales Visitation."

NYQ: Is this your favorite poem?

AG: Of my most recent poems, this is, like an imitation of a perfect nature poem, and also it's a poem written on LSD which makes it exemplary for that particular modality of consciousness. It's probably useful to people as a guidance, mental guideline for people having bum trips because if they'll check through the poem they'll see an area which is a good trip. An ecologically attuned pantheistic nature trip. Also it's an example of the fact that art work can be done with the much maligned celebrated psychedelic substances.

NYQ: Didn't T. S. Eliot say that he didn't believe in that?

AG: Yeah, but Eliot was not a very experienced writer, he didn't write very much, he didn't write very much poetry. Anyhow there's a tremendous amount of evidence that good work can be done in all states of consciousness including drugs. Not that drugs are necessary. It's just that it's part of the *police* mythology that nothing can be done, that LSD leads only to confusion and chaos. That's nonsense.

NYQ: In nondrug states, do you ever work half asleep?

AG: Yes, as I said I keep a notebook at my bedside for half-conscious, preconscious, quasi-sleep notations. And I have a book out now called

Indian Journals which has such writing in it, including poems emerged out of dreams and remembered in half waking, long prose-poetry paragraphs, using double talk from a half sleep state.

NYQ: That seems a very relaxed and vulnerable kind of writing, as opposed to what you spoke of before, where you tried to get everything into the mind.

AG: They're both related to consciousness study. Take it as part of a tradition going back to Gertrude Stein who was a student of William James at Harvard, whose subjects were varieties of religious experience and alterations of consciousness. That was James' big subject— the pragmatic study of consciousness, the modalities of consciousness. She applied her Jamesian studies and her medical studies to the practice of composition and saw composition as an extension of her investigations into consciousness. That's the tradition that I would like to classify myself within, and I think that's a main legit tradition of poetics, the articulation of different modalities of consciousness, almost, you can't say *scientific*, but the *artful* investigation or articulation of extraordinary states of consciousness. All that rises out of my own preoccupation with higher states of consciousness on account of, as I said over and over, when I was young, twenty-four or so, some poems of Blake like "The Sunflower," "The Sick Rose" and "The Little Girl Lost" catalyzed in me an extraordinary state of mystical consciousness as well as auditory hallucinations of Blake's voice. I heard Blake's voice and also saw epiphanous illuminative visions of the rooftops of New York. While hearing Blake's voice. While reading the text of "The Sunflower," "The Sick Rose" and "The Little Girl Lost." This was described at great lengths on other occasions. But I want to go back to that just to reiterate that I see the function of poetry as a catalyst to visionary states of being and I use the word "visionary" only in these times of base materialistic media consciousness when we are so totally cut off from our own nature and nature around us that anything that teaches nature seems visionary.

NYQ: You once said you were a worry wart. Yet you have such a sense of joy and freedom, in reading, and in writing, too.

AG: Ideally, the ambition, my childhood desire, is to write during a prophetic illuminative seizure. That's the idea: to be in a state of such complete blissful consciousness that any language emanating from that state will strike a responsive chord of blissful consciousness from any other body into which the words enter and vibrate. So I try to write during those "naked moments" of epiphany the illumination that comes every day a little bit. Some moment every day, in the bathroom, in bed, in the middle of sex, in the middle of walking down the street, in my head, or not at all. So if it doesn't come at all, then that's the illumination. So then I try to write in that too. So that's like a rabbinical Jewish Hassidic trick that way. So I try to *pay attention all the time*. The writing itself, the sacred act of writing, when you do anything of this nature, is like prayer. The act of writing being done sacramentally, if pursued over a few minutes, becomes like a meditation exercise which brings on a recall of detailed consciousness that is an approximation of high consciousness. High epiphanous mind. So, in other words, writing is a yoga that invokes Lord mind. And if you get into a writing thing that will take you all day, you get deeper and deeper into your own central consciousness.

NYQ: And does this lead you to a greater reality?

AG: A greater attention. Not attention, more feeling emerging out of that. So you walk down the city streets in New York for a few blocks, you get this gargantuan feeling of buildings. You walk all day you'll be at the verge of tears. More detail, more attention to the significance of all that robotic detail that impinges on the mind and you realize through your own body's fears that you are surrounded by a giant robot machine which is crushing and separating people, removing them from nature and removing them from living and dying. But it takes walking around all day to get into that state. What I mean is if you write all day you will get into it, into your body, into your feelings, into your consciousness. I don't write enough, actually, in that way. "Howl," "Kaddish" and other things were written in that way: all-day-long attention.

ALISON COLBERT
Bronx, New York

Partisan Review, 1971

Alison Colbert: I want to ask you first about the recording, *Songs of Innocence and Experience*, which I listened to recently. I want to ask you when the idea for the music and the songs came to you. You said in the liner notes that you had a vision of sorts.

Allen Ginsberg: Many years ago, described at length in a bunch of interviews five, six years ago. Mainly, *Paris Review* No. 37 (see Tom Clark) . . . of, uh, Blake's voice, not a vision, but auditory illumination, but specific music to the voice on the way back from Los Gatos in a Greyhound bus to San Francisco coming from darshan with Neal Cassady's ashes. A couple of lines kept running through my head in musical tone.

> Fayette, Fayette, thou art bought and sold
> and sold is the happy morrow
> Thou gavest the tears of pity away
> In exchange for the tears of sorrow.

which is a fragment of Blake, not in *Songs of Innocence and Experience*, about the French Revolution, about Lafayette's relation to Tom Paine, and Lafayette's defense of the king and queen.

AC: He sold out, in other words.

AG: Yeah, Blake was saying, though Lafayette was like a big American revolutionary, he sold out for pity. That is, I gather.

AC: I'm a little rusty on history.

AG: So am I, so am I, but it keeps floating up out of my rusty unconscious. Anyway, this tune came with it, for the first time and then I had confused it with another poem called "The Grey Monk," which has the lines

> But vain the Sword & vain the Bow
> They never can work War's overthrow

so I got 'em mixed up and began humming a tune in my heart on the bus. Then when I got home I started working with a harmonium and making a tune for both poems, "The Grey Monk" and the poem that begins

> Let the Brothels of Paris be opened
> To the beautiful queen of France

which has a line about "Fayette, Fayette, thou art bought and sold/and sold is the happy morrow"

> Thou gavest the tears of pity away
> In exchange for the tears of sorrow

so it developed from that seed, then finished with working with, improvising with Daniel Moore of the Tibetan Floating Lotus Opera Company, who's a good musician and poet. . . . So we made a couple of tapes together with him on horns. And Don Cherry and Cyril Caster on trumpets and singing with it. So I worked out a melody for that one song and went from there home and sort of thought about it awhile, and then, at the Democratic Convention in '68 in Chicago, sang that song in Lincoln Park on Sunday of convention week with John Sinclair as M.C. for like a rock-political concert. So I was laying down "vain the Sword & vain the Bow / They never can work War's overthrow" and then sat under the trees and sang it to Phil Ochs also, who was there also at the time. Then when I got home from the convention after the police-state-shock/raw planet nerves exposed . . . upstate to a farm where I had a large organ, pump organ, nineteenth-century keyboard, foot pedal pump, no electric.

AC: Church organ.

AG: Yeah, I'd already had about five years experience singing mono-chordal—with one chord—C chord mostly, or C minor mantra chanting, with Hare Krishna mantra and with a lot of others. Tibetan ones. So I hadn't gone beyond one chord, but I'd gotten good at variation within one chord. So I began holding the chord and putting—like "The Grey Monk" is C-seventh, I found out later. So I began experimenting with other chords besides the C and improvising tunes on a Uher tape recorder that I had. I was holding the chord down and then improvising melodies for more of the Blake. I think the second one I did was "Piping Down the Valley Wild," the introduction, one night. And then I began thinking in terms of trying to write 'em down but I didn't know how. So Lee Crabtree, who was one of the early Fugs musicians, came up and stayed for a couple of months and taught me a little bit about chords and then Cyril Caster came up and stayed and taught me more about chords till I could work with two chords or three chords. But they started making chord notations for me and also transcribing it for me. And then I began learning how to write the melody line myself and then later someone taught me the basic principle of chords which is major chord is the first note then you count up four more, then you count up three more. . . .

AC: You never had piano lessons when you were a kid?

AG: Yeah, when I was nine.

AC: It didn't stick?

AG: That's right. I hit my mother over the breast finally, I was so angry. So, I slowly began learning notations so at this point, A.D. 1971, I can write lead sheets. Except for time; I can't put the time bars in, I haven't experimented with that. So, in other words, it begins with a vision, with a much earlier vision, years ago. With a realization of ashes of the body, ashes of sexuality, seeing Neal Cassady's ashes. With a realization of the ashes of the body of American democracy in Chicago. And then attempting to vocalize the wisest language sounds . . . from Blake.

AC: When you did the recording, were there rehearsals?

AG: I had it all written down, finally. I had the music written down and we actually rehearsed—a lot. I mean not a lot as professionals do, but like we had a couple evenings with Don Cherry and Bob Dorough and Cyril Caster and other people who are on the album. We had formal rehearsals with music stands and pianos at people's houses and we taped them—and filmed some of them even. At Robert Frank's house, the movie maker. And then some things were rehearsed at my house with Jon Sholle, the guitarist, who's a studio guitarist and is very brilliant and who immediately picked up—a kind of feminine intelligence. He was able to follow along and bring me on—musically.

AC: Did you make very many improvisations in rehearsal at all, things which sort of came out from working together?

AG: The whole thing is an improvisation in the sense that all the main lines are laid down, like the track with Elvin Jones is, I think, the second or third of three attempts done in the studio without rehearsal. The Elvin Jones was without rehearsal. Some of the people that were working on the record were, uh, unstable. And so surprises came in and we had to improvise whatever we could.

AC: Yeah. Cause that's one of the things I specially liked about it, especially like the *Songs of Innocence*, there's a very spontaneous sort of sound, which is really nice. Are you doing any more work on this? Are you going to put out another album?

AG: I did twenty-one out of the forty-five songs, so I'll make another album now and complete the whole cycle, for Volume II of that and that'll actually be monumental, because it'll be the complete *Songs of Innocence and Experience* from beginning to end, every one of them put to tune. I'll probably do that this spring. I've been writing the music down.

AC: As it comes to you?

AG: Working on one song at a time. I've been picking up on one song at a time as it rises, politically significant, like the "Schoolboy," "How can a bird that is born for joy / sit in a cage and sing?" I began working on around the time of Kent State,[91] last year. You notice that the poems have a particular reference, a little hermetic sometimes, to any time-space continuum, to any political event, and individual poems, like koans, like Zen koans, have a relevance right now either to—like it appeals to certain emotional softnesses and tendernesses and innocencies which are completely repressed in the revolutionary-counterrevolutionary battle, so they have to be recalled with things like the schoolboy poem. As I get into

[91]Kent State University is in Kent, Ohio. In May 1970, during a campus protest against the war in Vietnam, the Ohio National Guard fired on protesters, causing the deaths of four students. The event led to disorder at many American universities.

sitting meditation—which I've been doing a lot of lately, Blake's gnostic transcendental psychedelic inner glow comes on. . . .

AC: So it works both ways.

AG: Yeah.

AC: That's what's so incredible about it. You said somewhere—I read *Indian Journals* fairly recently—you said that Blake was sort of your guru.

AG: Yeah. Cause I'd had what would normally be considered a religious experience with Blake in 1948. When I went to India, running around looking for a guru in 1962-3, one funny thing that Srimata Krishnaji, a lady saint from Brindaban, recommended was "Take Blake for your guru."[92] Since guru, the teacher, is not necessarily a human or living human incarnation, it might be any influence, even a bird or the giant piteous cat-squeak of a whale recorded on a machine. Anything that will catalyze the total consciousness is the teacher. So she said, "Take Blake for your guru," and that put him in the context, oddly, of the Indian transcendental scene. In a personal way, though historically Blake always has been in that context, cause he's an eighteenth-century vehicle for Western gnostic tradition that historically you can trace back to the same roots, same cities, same geography, same mushrooms, that give rise to the Aryan, Zoroastrian, Manichaean pre-Hindu yogas.

AC: I didn't know that.

AG: Yeah, that's the whole point. If you're interested in the line of descent, I'll tell you—back from Blake through Paracelsus and Plotinus and Jacob Boehme all the way back to Pythagoras, and back from, back to in those days, the Eleusinian mysteries[93] and the Bacchic mysteries to the mystery cults, and back through the mystery cults to the Near East, back to the source of it all, you come to the same sources.

AC: And he was aware of all this?

---

[92]Today Brindaban is known as Vrindavan.

[93]In classical Greece, mysteries were secret ceremonies to be witnessed only by the initiates, who were sworn to never disclose their content. The most popular of these were the Mysteries of Demeter at Eleusis.

AG: Yes, Blake had a great collection of those texts—in fact, I understand he handled manuscripts.

AC: Yeah. He was a bookseller.

AG: A manuscript-seller. He handled some of the gnostic manuscripts. But one of the great people that influenced him was a guy named Thomas Taylor, the Platonist. Taylor is a very interesting figure, a very interesting psychedelic gnostic figure, in Europe and in America. Taylor was a great Latin-Greek scholar who translated all of the remaining gnostic fragments and made them available to eighteenth-century cats, so that Shelley, as well as Blake, Coleridge, other heads of that time, picked up on Thomas Taylor's translations.

AC: Yeah. And were enabled to take what they read further into their own thing.

AG: Yeah. And at the same time, Bronson Alcott, from America, of the Brook Farm commune, the early enlarged family experiments so typical of nineteenth-century America, went to England to get Thomas Taylor's books and translations—brought them back and loaned them to Emerson. So that the same sources Blake was using, tracing back to the Middle East, were also influential in the formation of the American individualistic transcendentalist tradition which is so influential now in, say, the antiwar revolutionary movement.

AC: The other thing I wanted to ask you about Blake was whether you'd been able to get into the *Prophetic Books* at all.

AG: No, I've been sort of hung-up in the simpler, earlier poems. I read in and out of the *Prophetic Books*.

AC: Because I tried at one point. It was pretty difficult.

AG: OK. The cosmic geography of the *Prophetic Books*, cosmology, that Blake outlines and changes around, year by year to the end, is parallel to the cosmography and cosmogony that you find outlined in the gnostic religions, pre-Christian, pre-Christ, including the names that Blake uses, like Ruha is the Mandaean[94] gnostic name for the evil

---

[94]A Gnostic sect still surviving in Mesopotamia. Aramaic equivalent of Greek *gnosticoi*, hence Gnostics.

female first principle, otherwise known as Sophia, wisdom, in other places, and in Blake's *Songs of Experience* known as Tirzah, the chick who created the whole cosmic chaos because she was reflecting the empty light of the abyss. She was a reflection of that light, in fact. But once she began reflecting the mirrors multiplied until the entire Indra's net of creation was established, like billions of mirrors shining on each other, creating a giant illusion of space and time. So, it all wound up in the Garden of Eden.

AC: Which is sort of what I expected. Is this the same as the Pandora figure?

AG: Probably they're parallel put-downs. Parallel anti-Women's Lib put-downs.

AC: Yeah. Because Kate Millett talks about this a good deal. She doesn't talk about the Indian. She talks mostly about the Western.

AG: The really interesting thing to realize is that the Indian and the Western converge in their origin. They come from the same Garden of Eden, or the same Middle Eastern source. The reason we don't know much about it is that when Constantine the Great accepted Christ as a Roman emperor accepting Christ, he also became pope. So it was a political deal. So the Christians had to make a deal to eliminate all the heresies, and that was the Council of Nicaea.

AC: The many, many cults.

AG: Yeah. These were the gnostic cults. So the books were all burnt around 313 A.D. The only records we have of them, which Taylor translated, is from the early Church fathers, arguing against these heresies, heresies like saying, I think Heraclitus said, "Everything we see when awake, is death; and when asleep, dream."[95] In other words, the Buddhist maya.[96] Everything we see when awake, is death, is the same thing as the Hindu or Buddhist proposition that all, or the Burrough-

---

[95]Perhaps Ginsberg had in mind the quotation from Heraclitus, "For when is death not within ourselves? . . . living and dead are the same, and so are awake and asleep, young and old," but he wrote a query on his own copy of the interview that the source was possibly Pythagoras.

[96]The word "maya" is Sanskrit for "deception, illusion, appearance," the continually changing impermanent phenomenal world of appearances, as well as forms of illusion or deception which the unenlightened mind takes as reality.

sian proposition, that all apparent sensory feelings, thoughts and impressions are illusory. So, in other words, what happened in the fourth century is that the basic Indian understanding that the apparent physical universe is only apparent, and really is a dream-structure in which we're trapped, because attached to a thing that's real—that was extirpated from Christ-doctrine, and also the books wiped out and burned, so that it took people like Paracelsus, Böhme, Blake, Shelley, Coleridge, Emerson, to perpetuate that memory out of their own intu-itions and glimmerings—and also checking out the hand-me-down legends and texts, the oral transmission. Poetry's carried it all along. Poetry's carried the dream-insight all along.

AC: I wanted to ask you about William Carlos Williams. I read the poem you wrote about his death, which I liked. I just had always had the impression that he was a very beautiful man, by all accounts. Did he, like when you were growing up in New Jersey, did he look at your manuscripts?

AG: Yeah. When I was grown up already. . . . First time I met him I was twenty-two. I had read his poetry when I was in college but college so perverted my brain about poetry that I still expected some sort of rhyme.

AC: Columbia English Department did that?

AG: Columbia English Department. And in those days, 1945–46, they were saying that Whitman was a jerk, that Shelley wasn't a real poet, that Williams was some kind of awkward crude provincial from New Jersey. That was sort of like the official doctrine, the party line of the English department, not only at Columbia, but at every university in the United States.

AC: Yeah. But in those days, Columbia was supposed to be more enlightened.

AG: No, no. The point was that in those days in the forties, there was nobody there teaching who wrote poetry, who wrote any kind of mod-ern poetry.

AC: So you went to Williams, I guess, partly because he did not have an academic attitude.

AG: I went to him because he baffled me. I'd picked up some of his books which were printed by the Cumington Press, which is like a little home press up in Vermont—and I couldn't understand why he was writing funny-paged writings, with the words scattered around on the page, why there were irregular lines, and I couldn't understand what he was talking about, really. I read it and I didn't understand it, literally—I didn't understand the literal meaning of what he was saying, if he was writing about wheelbarrows or whatever. So I went to interview him for a local labor newspaper in Paterson, as an excuse. I had the impression that he was some sort of stainless steel 1920s *moderne* visage out of Brancusi and Ezra Pound and Alfred Stieglitz's American Place Gallery, where he used to hang out. Instead I found a creaky-voiced, tender-toned, soft though sharp-eyed country doc, scratching his head, so I asked him, "Do you think of yourself as a poet, or as a doctor?" And he said, "As a doctor." Then I asked, "Why do you write almost-prose lines?" and he said, "Yesterday I heard a Polish laborer say, "I'll kick yuh eye." And he wrote it down on his prescription pad. "I'll kick 'y' 'u' 'h' 'e' 'y' 'e'," and he said, "How do you put that into iambic pentameter?" "I'll kick yuh eye—" it's a funny little rhythm all its own. So I suddenly realized he was hearing with raw ears. The sound, pure sound and rhythm—as it was spoken around him, and he was trying to adapt his poetry rhythms out of the actual talk-rhythms he heard in the place that he was, rather than metronome or sing-song archaic literary rhythms he would hear in a place inside his head from having read other writings. I suddenly realized he was inventing out of the actual ground of Rutherford, New Jersey, a different body-speech and that anything he said was absolutely natural, and didn't violate human being talk, didn't come from another era but came directly from the ground that he stood on. But I still didn't really understand the poems. I got the general idea—but then about a half year later I went to the Museum of Modern Art in New York and I heard him read, and he read one poem that I immediately understood was just somebody talking and making sense. And it made absolute sense.

AC: It has a rhythmic structure . . .

AG: Well, I'll kick yuh eye is a rhythmic twiddle.

AC: It doesn't have the bomp-and-bomp-and . . .

AG: It's not the iambic high school rhythmic structure. But it's the rhythmic structure of someone talking sincerely and earnestly, with the changes of mind that go on when somebody stops to figure out the word—and then begins again with another squaggle of phrasing—or stops in the middle of a sentence to go on to something else, because the mind changes. I heard him read another poem called "The Clouds," which is about people getting lost in their imaginations and forgetting about concrete earth-speech and fact things—and it was a poem that ended in talking about people's minds going off into the clouds, "plunging upon a moth, a butterfly, a pismire, uhh. . . ."

AC: Yeah.

AG: Not oh yeah. "Plunging upon a moth, a butterfly, a pismire, uhh, dot, dot, dot." Unquote. And when he read it, he read it almost impatiently, (fast) "Plunging upon a moth, a butterfly, a pismire, hupp. . . ." And then he waved his hands in the air—he just gave up. So I suddenly realized it was just like somebody talking in a bar, not finishing the sentence but just giving up with a gesture of impatience, and that it was a syntactical fact of speech that had never been written down before in poetry—and so I suddenly realized that his poetry was absolutely identical with speech, the highest speech, but absolutely identical, rhythmically and syntactically. And then I suddenly realized that if you began right where you are, with your own speech, then obviously you would have to create a whole new world of speech, that had never been written down before, which was what he was doing and what anybody could do. But it also meant that you had to listen, as he advised, very carefully, to your own sound, and to other people's talk-sound and talk-rhythms—and if you listened to "I'll kick yuh eye" or "plunging on a pismire, uhh. . . ."! you would arrive at all sorts of new unknown rhythms. And oddly enough just at this time I was talking with Kerouac quite a bit, who was, by his own nature, doing the same thing, and leaning for support on the new rhythms he was hearing in music at the time among the blacks around Minton's[97]—cause Kerouac at that time had been listening to the early Bird Parker, and Gillespie and Monk, and earlier a lot of Lester Young, and had been turning me on to all those musicians: and the key thing that he realized about them was that they had made the rhythm section into solo instruments with varying and variable rhythmic bases, just like Williams had variable

---

[97]Minton's Playhouse, a Harlem nightspot similar to the Cotton Club. Thelonious Monk played there often in the early 1940s.

rhythmic bases. So there was a parallel breakthrough going on in the music and in the poetry-speech.

AC: Could you like sort of work on this in your own head and try and get into it in your own writing?

AG: Yeah, immediately. I went over my prose writings, and I took out little four- or five-line fragments that were absolutely accurate to somebody's speak-talk-thinking and rearranged them in lines, according to the breath, according to how you'd break it up if you were actually to talk it out, and then I sent 'em over to Williams. He sent me back a note, almost immediately, and he said, "These are it! Do you have any more of these?" So apparently I'd caught on, to something so simple that the entire English department at Columbia University couldn't understand it, and never did get to understand it until perhaps two decades later—if at all, 'cause I don't think it's still understood. It's just a very simple basic principle that you listen to speech to hear rhythms and attempt to isolate the archetypal rhythms of actual speech and then remodel them in the poems. That's the whole basis, really, of what was later known as projective verse in Charles Olson and the whole gang of poets, friends and lovers of his. Black Mountain and elsewhere, San Francisco and New York. Oddly, you see, Williams influenced all the Black Mountain people, he influenced Gary Snyder and Philip Whalen and Lew Welch out in San Francisco and he influenced Robert Duncan and Mike McClure in San Francisco, he influenced Frank O'Hara and Kenneth Koch and John Ashbery here. The influence was that originality of taking the materials from your own existence rather than taking on hand-me-down poetic materials, speech units, rhythmic units and trying to adapt your life to them—you articulate your rhythm, your own rhythms. The concept of that led in the forties to Abstract Expressionist painting and (Willem) de Kooning and (Franz) Kline, it led in music to Ornette Coleman and uh, who was a teacher there? The guy who died two years ago. (John) Coltrane. It was the same American rediscovery of individual soul's impulse that led into Coltrane. It also influenced LeRoi Jones (Amiri Baraka) by quite a bit, Williams's practice. It brought Jones back to Newark, in a sense. If any literary influence had tended in that direction, Williams's influence tended to bring Jones back home to his own speech and to his own soul and to his own body and to his own color and to his own town.

AC: I've never been sure of the religious part of what's called the Beat philosophy, just what the source of this was.

AG: Goddard's *Buddhist Bible*, which has now been reissued, and available, which is like a great compilation of Buddhist texts, Lankavatara Sutra, Surangama Sutra, Diamond Sutra, Sutra of the Sixth Patriarch, Bodhisattva's Vows, Four Noble Truths, Theory of Interdependent Origination. Kerouac and Cassady and other people were reading in that in 1953 around San Jose, when Kerouac was living with Cassady. Cassady was at the time hung up on Edgar Cayce, ideas of reincarnation, so Kerouac, always impressed by Cassady's intuitions, went back to the sources, went back to the original Oriental texts, which he found in this one compilation, the Buddhist Bible, which he studied real hard. So that Kerouac then began writing on Buddhist matters, and he had a great unpublished thousand-page manuscript called *Some of the Dharma*,[98] which is Buddhist meditations, haikus, philosophical reflections, quotes from Surangama Sutra,[99] little poems all sizes. Then I had a sort of mystical experience in 1948, as I mentioned, relating to Blake, which specifically was an auditory illumination, you could call it an auditory epiphany, hearing Blake's voice—and at the same time a visual illumination of light permeated the universe, and a heart epiphany of the ancient of days, the father of man being present everywhere, in everybody always. So, in other words, I had a cosmic vibration breakthrough. Kerouac had one in the fifties, which he describes in his Sutra Scripture, *Scripture of the Golden Eternity*, a little pamphlet put out in Newark by Corinth Books, in which he describes sitting and doing breathing exercises and suddenly fainting in his back yard—with his eyes, sitting and looking at the sun, with a golden light coming through his eyelids, realizing that this golden eternity is the permeating suchness of the entire universe. Burroughs always has had a preoccupation with alteration of consciousness, from an early text of 1951, *Junkie*, pseudonym William Lee, the last pages discussing yage, a drug in South America, as perhaps being the final fix. So we already had "native moments," what Whitman calls "native moments," which we then began checking out with ancient moments and ancient texts. And

---

[98]*Some of the Dharma*, ed. David Stanford, Penguin, NY, 1998.
[99]Sutras are discourses of the Buddha and traditionally are attributed to Ananda, a student of the Buddha who recited them from memory at the first Buddhist council after the Buddha's death. The Surangama is one of the more important of these.

then, in the early fifties in San Francisco we ran into Gary Snyder, who was already studying Zen Buddhism and sitting, doing regular meditation, and Phil Whalen, who was a Buddhist poet from Reed College like Gary, both of whom had met Williams up in Reed College in 1948, the same year that I picked up on Williams.

AC: So it all sort of converged?

AG: Yeah. Now the Zen practice was paying complete absorbed attention to the immediate teapot and teacup in front of you and pouring the tea with complete absorption and intention, with the mind focused there, observing every wavelet and droplet coming out of the spout into the teacup and then serving it with complete presence to the person in front of you. Blake's proposition was that "concrete particulars" were the essence of poetry and consciousness observation to see eternity, uh, no, "to seek all Heaven in a grain of sand, and eternity in an hour." Williams's proposition from American roots was: "so much depends/ upon/ a red wheel/ barrow/ glazed with rain/ water/ beside the white/ chickens." When Williams said "so much depends," he means all human consciousness depends on direct observation of what's in front of you. So all those tendencies converged at once around '48 to '53 in our consciousness, and I think elsewhere, too. See, Berkeley at the time was going through what was called the Berkeley Renaissance,[100] where Harry Smith the great filmmaker was hanging around with Jack Spicer and Robert Duncan the poet, and Timothy Leary was on the scene and knew Robert Duncan. And that was called the Berkeley Renaissance, where Harry Smith was beginning to experiment with optical phenomena in relation to mandalas and Tibetan and American Indian psychic color effects, who built machinery for casting light-color shadows on the wall, who left his machinery to a guy named Jordan Belson in San Francisco who was a great filmmaker, who made a movie called *Mandala* in the fifties and was one of the first people in the mid-fifties, with Gerd Stern, to begin experimenting in large movie theaters with oils on water on plates with light casting on a movie screen, which led to light shows, which led a decade later to

---

[100]In the late 1940s Robin Blaser, Robert Duncan, and Jack Spicer envisioned a Berkeley Renaissance. The story of this is described in Duncan's "Medieval Scenes" and "Venice Poems" as well as in Spicer's early poems. See Michael Davidson, *San Francisco Renaissance: Poetics and Community at Mid-Century*, 1986.

large-scale use of that in Ken Kesey's experiments with Neal Cassady
on mixed-media feedback. Which then led to the trips festivals, elec-
tronic cut-up mixed-media decor which led into use of that
omnipresent rainbow of consciousness in the Fillmore.[101]

AC: So it all started in '48?

AG: I don't know if it all started then—it crested then. From gossip I
know, there was a lot of tip-of-the-iceberg showing, or the mountain
rose above sea level just about then. Synchronicity. It was inevitable,
cause it was the end of the war. A couple of years after the end of the
war, so everybody's mind was beginning to drift. The thing I wanted to
point out was that there was a breakthrough of, let us say, an enlarged
or cosmic consciousness or a big consciousness. There was that con-
sciousness breaking through the social norms and the social repres-
sion of consciousness and through the institutionalized repression of
consciousness, right then and there.

---

[101]See Joel Selvin's *Summer of Love: The Inside Story of LSD, Rock and Roll, Free Love, and
High Times in the Wild*. (New York: Plume Books, 1995); also *Electric Kool-Aid Acid Test*
by Tom Wolfe for the psychedelic cultural history of San Francisco and Ken Kesey
and Neal Cassady's friendship.

YVES LE PELLEC
August 1972,   New York City

Published as *"La Nouvelle Conscience"* in the "Beat Generation" issue
of *Entretiens*, 1975 and as "The New Consciousness" in the book
Composed on the Tongue (1980, Grey Fox)

Yves Le Pellec eventually became Allen Ginsberg's French translator.
The tape of the interview was transcribed by Le Pellec.

—DC

*This conversation was recorded in Manhattan on a hot, sultry, electric night
of August 1972 in Allen Ginsberg's small apartment on the Lower East Side.
The windows were open and on the tape one can hear background noises of
children laughing, dogs barking, and, at frequent intervals, the wailing of
police cars. Allen was just back from the Miami Republican Convention in the
course of which he had spent three days in jail with some Yippies and Zip-
pies.[102] He seemed a little tired, having caught a cold, and spoke in a deep,
slightly hoarse voice. But he was very courteous and, though I was meeting
him for the first time, willing to offer as much help and information as he
could. I was struck and charmed to see how thoroughly and scrupulously he
answered any question. He is a long-winded talker, picking his words care-
fully, I would say almost with delectation, as if they were ripe cherries, and
associating them in long sentences without losing the thread of his exposition.
Originally I had come to ask a few factual questions about Kerouac, but our
conversation expanded and when I left late in the night I realized I had just
been hearing a fairly detailed chronicle of the whole poetic movement he had
been associated with since the 40s. From the outset he asked me who, "among
our group," I had already visited, and as I had recently met Kenneth Rexroth
on the West Coast, we started speaking of his appearance in* The Dharma
Bums *under the pseudonym of Rheinhold Cacoethes.*

—YVES LE PELLEC

---

[102]The Zippies were a short-lived faction within the Yippies responsible for much
dissension during the 1972 demonstrations in Miami during the Democratic and
Republican conventions.

Allen Ginsberg: Cacoethes in English means bad habit, I think.

Yves Le Pellec: Yes, and I found a short note Jack (Kerouac) sent to Larry Ferlinghetti saying Rexroth had the habit of scribbling all the time, that he was an "inveterate scribbler."

AG: Well, Rexroth was an older person, so he wasn't among the younger comrades and he had at the time a great deal of pride built up over many years of literary isolation. His reaction to Kerouac was that of an elder who was very often surprised, shocked and sometimes irritated by what he interpreted to be bad manners, which were on occasion bad manners like a few times when Jack was drunk in his house—not really bad manners but sick manners—but Rexroth went through many modifications of his judgment. His judgment is generally benevolent and friendly now but at one time he thought Kerouac was a horrible writer and a very bad poet and he wrote a devastating attack on Kerouac in the *New York Times* on the appearance of the . . .

YLP: *Mexico City Blues?*

AG: *Mexico City Blues*, a very stupid review. *Mexico City Blues* is a great classic, I think. I mean it taught me poetics, and it taught Michael McClure a great deal and is one of the great presentations of Buddhism in American terms. It has many many virtues, it's a completely original book, with a self-invented poetics in it, it has a marvelous ear and marvelous rhythms. So that's a work of genius, an extended work of genius, 242 linked poems like a Shakespearean sonnet sequence, it's just a remarkable monolithic work of jazzing up the American language, as Céline said about Henri Barbusse's prose. He said he liked Barbusse because he jazzed up the language.

YLP: But Rexroth did a lot to help at the beginning: his first articles, the "San Francisco Letter" [in *Evergreen Review*, No. 2, 1957] . . .

AG: Yes, a great deal, a great deal. His first appreciations were accurate and very sympathetic and good-hearted. The difficulty came after our work was out in public and was very heavily attacked, particularly Kerouac's, by a sort of literary establishment that were pleased with the status quo in America, many of whom actually were unconsciously involved with the CIA through the Congress for Cultural Freedom, which published *Encounter*, a "front" and a magazine which was being funded by the CIA. Kerouac's completely unofficial version of America was very displeasing to them and their immediate reaction was that it was some sort of cult of violence, exhibitionism, rather than realizing it was a sincere heart-speech, a pacifistic heart-speech really. So these attacks delayed the acceptance of both the mode of speech, that is the style of speech and the physiology, the rhythms of such speech, as well as the general ideas and the intellectual background, which was gnostic rather than rationalistic. This stunted the development of American culture for a long time. So that when the style, the stylistic attributes, were picked up by later generations—that is to say spontaneous speech, the use of drugs on the side, marijuana and psychedelics, the preoccupation with American Indians, the second religiousness that Kerouac spoke of (the phrase being Spengler's), the interest in black music or the sense of a Fellaheen[103] subterranean underground nonofficial existence—the intellectual background and the rationalization for it, the literary background for it, was not transmitted rapidly enough. The intellectuals said that it came out of no tradition, they were not smart enough or learned enough to know the tradition, or sensitive enough to understand it. The tradition for instance in Kerouac is the black blues. It wasn't until the 60s—'62 or '63—with Amiri Baraka (LeRoi Jones) leading an attack on white culture and presenting the black culture and black blues as a major art form in America, that it got to be accepted as a major art form with the intellectual distinction that it has. Well, Kerouac was dealing with those forms earlier but none of the academic critics that attacked him in the 50s had any idea of the body of poetics that you find here—[showing a book] *The Blues Line* by Eric Sackheim, 1969—the body of poetics which Kerouac was drawing on for his *Mexico City Blues*. Most of his critics had no idea of this as literature, the true literature in America, and it was. Also in *Mexico City Blues* or in other writings of '53, '54, '55, Kerouac wrote with great sophistication on Buddhism and the basic tenets of Buddhism being that existence is *Dukha*, or suffering—but most of his critics were not familiar with that. So when he cited, say, the *Prajnaparamita Sutra* or *Diamond Sutra* or *Lankavatara Sutra*, probably many of his critics thought he was talking gibberish, without realizing he was citing very ancient and honorable texts. So that Kerouac was met with a barrage of

---

[103]Plural of fellah, a peasant or laboring man in Egypt and Syria.

enormous stupidity both in America and France, as a reaction to his presentation and transmission of wisdom texts. So when Rexroth—who was the older critic who took everybody seriously, read all the works and knew about Eastern thought—read *Mexico City Blues* so improperly and thought that it was some childish scatological babblings, it was a great blow to the advancement of the general culture in America, it just killed maybe ten years' time.

YLP: Some people have discussed Kerouac's attitude towards jazz. I think he has been reproached for having a "white man's" attitude, liking musicians like George Shearing for instance rather than more important musicians, and not being really into the soul of blues.

AG: Well, I think that's a very superficial *examen de texte*. If you look at Kerouac's texts, all of them, his approach was primarily oriented to Charlie Parker. He didn't write about Charlie Parker, he participated in those situations; he was in Harlem, at Minton's when Parker and Thelonious Monk and Dizzie Gillespie were there playing, back in the early 40s. He was there on the spot, he was there participating in some of the first recordings made there, including recordings of Charlie Christian. I remember that from 1944 he was introducing me to sound, to Parker. That was his main interest. Then he was interested in the later developments. By the 50s the great bop revolution had reached its peak and it was time for another wave, and there was another wave of exfoliation, or spreading out and development of the rhythmic changes that had been taking place in bop, and a passing of these new musical techniques onto the white culture. People like Lennie Tristano and George Shearing picked them up and so Kerouac noted that and wrote about that. He followed it closely, like a baseball player interested in the progress of the teams throughout the season and who all the players were. So he wrote about the American jazz milieu. Very early, at a time when very few people were even paying attention to this music, Kerouac made a little record with Zoot Sims and Al Cohn—do you know that recording of haiku?[104] That was the only recording of jazz and poetry that was original and recorded in

---

[104]*Blues and Haikus*, Hanover Records HM 5006 (1959). The texts are included in *Scattered Poems* (San Francisco: City Lights, 1971).

the studio spontaneously. See, all the rest was sort of like background music, or prepared music or music that had nothing to do really with the poetry. Kerouac would pronounce a little haiku like "The bottom of my shoes are wet, from walking in the rain" and Al Cohn or Zoot Sims would take the rhythmic thing (*sings*) "ta ba da ba ta ta, ta ta, ta da da ta ta ta," a little haiku of music, then Kerouac would go, "In my medicine cabinet the winter fly has died of old age," then Al Cohn would go "Ta Ra ta ta da ta ta, ta ra ta ta, ta ra da da da." So they would make rhythmic structures mirroring each other. Kerouac actually was a musician, or good enough to work with musicians. He had a sophistication, a prosody, a musical ear much greater than any of his critics were capable of . . . So I think that the criticism of his interest in the younger white musicians is out of the context of the development of jazz in America. For at that time a great deal of the notoriety and energy of the original bop changes went over to Shearing, Tristano, Lee Konitz and others. The next great round of creation came in the late 50s with the public emergence, not onto the commercial scene but onto the social scene in New York, of Ornette Coleman, Don Cherry, Cecil Taylor, Albert Ayler . . . Kerouac was then very much on the scene also for it was associated with his mood at the time. One of the people that he influenced a great deal, both in his appreciation of jazz and his appreciation of black culture, was LeRoi Jones. Around 1958 Jones became a sort of intellectual arbiter and energy setter in New York City. Just about a few blocks from here he had a big house, gave parties in his floor-thru and had his magazine *Yugen*. Kerouac was still in the city and sociable and saw Jones quite a bit, so he was still around when the next great phase of black music arrived. Beyond all that, Kerouac's ultimate forte was with alcoholic street musicians, the real blues, the ultimate blues of people that don't even have an instrument but just stand in the doorway and say (*sings*):

> I'm standing in the doorway
> Got no place to go
> Standing in the doorway
> Got no place to go
> Jack Kerouac gave me fifty cents
> And I'll go out and blow.

Kerouac could improvise blues, which is *the* tradition, the great classic tradition. In that sense he was himself black. Partly what destroyed him also was the alcoholic bum he was, the bum of the Third Street in San Francisco or Bowery here. He wrote *Third Street Blues* and *Bowery Blues* which I don't think have been completely published.

YLP: Aren't they in *Scattered Poems?*

AG: Fragments of them. So Kerouac spent a lot of time drinking in doorways with blues singers and was very close to the actuality of blues, universal blues, neither black nor white. I think that the sort of criticism that's proposed here is . . . "distant folly." You may not like Shearing but I think he is a good musician. The contribution Shearing made, he made then, that is to say in the late 40s, 22 years ago. We are talking about an era when musicians like Shearing, Konitz, Tristano were beginning to, like, make a scene of their own, which I don't think was an important scene, historically, and possibly Kerouac in his enthusiasm overpraised them but I think the time capsule that he gives is perfect, perfect for that time.

YLP: Could we speak of the early days in New York, the mid-40s, when you were at Columbia. From what I've read and heard I found that many of Kerouac's friends considered him as the intellectual, almost the scholar. Neal, for instance, regarded him as a man who had a university culture. Yet he had spent only one or two years at Columbia?

AG: And then in '49 he went to the New School for Social Research. He had already written *The Town and the City*. By his standards that was a very great piece of prose, traditional prose, but it was like the last possible gasp of such orchestral bildungsroman prose, as in Thomas Wolfe. It was a great contribution to that actually. It's still a very valuable book and there are great prose passages in it. I think the description of Times Square is prophetic. It leads on to the whole hippy movement twenty years later.

Kerouac was always sort of an exile in the university community. He was a football player but when he quit I think the football bureaucracy got mad at him and took away his scholarship and he was not able to pay for his housing there. He was very brilliant in class, he got good marks from Mark Van Doren, who was a Shakespeare scholar and a poet.

YLP : He was your professor too?

AG: I had him for one term, in fact I had six or seven professors each term. The one professor who had good relations with me and Jack and who respected Jack's prose then, the only one was a man named Raymond Weaver. He was the first biographer of Melville and the discoverer of the manuscript of *Billy Budd*. He wrote a book called *Herman Melville, Mariner and Mystic*. At Columbia he was a considerable intellectual presence, like an old-fashioned scholar and at the same time a very modern soulful intellectual. He was sort of mystic, gnostic, he had lived in Japan and used some Zen for teaching. So Kerouac had written a manuscript the name of which I don't remember . . . It was before *The Town and the City* . . .

YLP : Wasn't it "The Sea is My Brother"?

AG: I think it's in between, it was another little thing, a little romance about angels coming down fire escapes [laughs]. He took it to Weaver and Weaver recommended that Jack read the Egyptian Gnostics, Jacob Boehme, Blake, *The Egyptian Book of the Dead*, perhaps *The Tibetan Book of the Dead* and I think some Zen classics. So Weaver perceived immediately the magical aspect of Kerouac's character and his mystical potential. Everybody else around that scene was very materialistic in the sense of "If you write a story it should have a middle, a beginning, and an end, and you shouldn't have too many fancy words because you know it's not in the tradition of realism that might grow out of the older proletarian novel and make sense in the new America facing the post-war future . . ." Everybody was writing sort of rationalistic discourses putting down the communists and very heavily political in a very negative way, in a very status quo way, and most of them were writing about manners, and good manners were Henry James and Jane Austen in those days. Whereas Kerouac was writing about the descent of angels in workman's overalls, which was basically the really great American tradition from Thoreau through Whitman . . .

YLP : Was Whitman taught at Columbia?

AG: He was taught but he was much insulted. I remember, around the time of the writing of *On the Road*, a young favored instructor at Columbia College told me that Whitman was not a serious writer because he had no discipline and William Carlos Williams was an awkward provincial, no craft, and Shelley was a sort of silly fool! So there was no genuine professional poetics taught at Columbia, there was a complete obliteration and amnesia of the entire great mind of gnostic western philosophy or Hindu Buddhist eastern philosophy, no acceptance or conception of a possibility of a cosmic conscious- ness as a day to day experience or motivation or even once in a life- time experience. It was all considered as some sort of cranky pathology. So Whitman was put down as "a negativist crude yea-sayer who probably had a frustrated homosexual libido and so was general- izing his pathology into oceanic consciousness of a morbid nature which had nothing to do with the real task of real men in a real world surrounded by dangerous communist enemies" *(laughs)* or some- thing like that . . .

YLP: The same things were said about you and Jack in the first reviews you got. The term "narcissistic" recurs very often.

AG: Sure, it's just typical of the Pentagon to call anybody that criticizes their policy narcissistic and unrealistic. You know, these people are depending on sixty billion dollars a year, that's a lot of heavy metal and appears real to them! And anybody who thinks that their heavy metal is the by-product of a mass hallucination would be considered mad by War Secretary (Melvin) Laird. The crucial point, I think, is that the original criticism of the texts that came out—McClure's, Snyder's, my own, Kerouac's—was based on a very narrow view of human nature, a definitely prepsychedelic experience of human nature, at a time when there was a definite shrinkage of sensitization, of sensory experience, and a definite mechanical disorder of mentality that led to the cold war and to the present genocidal ecocidal mass air-war in Indochina. The desensitization had begun, the compartmentalization of mind and heart, the cutting off of the head from the rest of the body, the roboti- zation of mentality that could lead Harvard and Columbia intellectuals like Kissinger and (Arthur) Schlesinger, all those supposedly realistic, mature, ripeminded academics to pursue a 1984-style cybernetic war- fare with all the moral rationalizations of self-righteous self-interest that you found in France when you had people asserting all sorts of moral principles to justify the war in Indochina or the war in Alge- ria . . . Which by hindsight are now considered to have been madness and cruelty and in fact criminality. At the same time we were being attacked by the university academic intellectuals because we were opening up an area of another consciousness, a planetary ecological consciousness, in a sense. Consciousness within the academy was narrowing down, becoming more anxious and rigid,[105] and it was the initiation of the cold war theoretics for them, the beginning of that grand international paranoia.

YLP: But did you have the feeling that you were starting something, that a "new generation" was appearing? This spiritual awakening, which is obvious today, did you feel it just after the second World War?

AG: Not instantly, no. Actually the first perceptions that we were having, the first perceptions that we were separate from the official vision of history and reality, began around '45, '46, '47. We realized that there was a difference between the way we talked—Jack, Burroughs, myself, as comrades among ourselves in order to get information and give each other our best stories, just like between Neal and Jack, in order to share experience and find our ultimate heart or vision—and what we heard on the radio if any president or congressman or even literary person began talking officialese. The air was filled with pompous personages orating and not saying anything spontaneous or real from their own minds, they were only talking stereotypes. I remember Burroughs saying during one presidential campaign, I think when Truman was running for president, that if an elephant had walked up in front of all those candidates in the middle of a speech and shat on the ground and walked away, the candidate would have ignored it. Consciousness wasn't present there on the occasion when they were talking, consciousness was an abstract, theoretical state. A theoretical nation, the actual nation was not there. Which is basically the same thing that Ezra Pound and William Carlos Williams and Sherwood Anderson had been saying all along. So we saw the difference between our own speech, our own company, and the national company full of Ionescoesque hallucinations of language.

Then we began running into people on Times Square, Huncke,

---

[105]Anxious about youthful emotions, guilt by association, homosexuality, political & social radicalism, anxious about the smell of self-McCarthyism, Rosenberg executions, Chambers-Hiss case. *Time* magazine's frigidities. Cigarette ads featured "men of distinction" that looked like impeccably mustached CIA agents. The code word was "Serious."—AG

whom Kerouac describes already in '48, '49 in *The Town and the City*. Huncke was addicted to morphine, and in observing him we saw the difference between authoritarian law as imposed on Huncke's suffering as a sick junkie and what we saw in person: he was just sick and needed help and there was no reason why he shouldn't get his maintenance therapy, as it is called now, maintenance supply of drugs from a doctor. Huncke told us that the narcotics police themselves were peddling dope under the table just as in those very days General Raoul Salan and the entire French intelligence in Indochina were organizing the dope trade for their own benefits or for their own paramilitary fundings—I've just been reading a book that goes over all that, *The Politics of Heroin in Southeast Asia*[106]—and all that was presented in France as some sort of higher morality. *(laughs)*

YLP: Heroic defense!

AG: Yes! So we saw that the police were making big moralistic speeches to the New York *Daily News* appealing to the most lower guttersnipe emotions accusing people of being dope fiends. Our friend Huncke was being called a Fiend! A new category of human being, namely the "human Fiend," had been created by the police at the same time as the police themselves were peddling dope and being brutal and violating the law and creating a police bureaucracy in this area. And the newspapers and *Time* magazine, the *New York Times*, the *Daily News* were of no help at all, it was all the police-bureaucracy line. So we realized there was simply a separation between our thought and public thought, between private consciousness and public consciousness. But Whitman all along had said that private consciousness is public consciousness, that the State doesn't exist (as a living Person), only people exist through their own private consciousness. So we realized that we were in the midst of a vast American hallucination, that a hallucinatory public consciousness was being constructed in the air waves and television and radio and newspapers, even in literature. By hindsight we saw that part of that consciousness was being paid for by the CIA to the extent that they invaded the literary field, and the students' world later on

[106]By Alfred W. McCoy (New York: Harper & Row, 1973). See also *Allen Verbatim: Lectures on Poetry, Politics & Consciousness*, ed. by Gordon Ball (New York: McGraw Hill, 1975).

with the National Students Association suborned and paid for by the CIA. But we didn't know all this in political terms, all we knew was that like we were making sense to each other, you know, talking from heart to heart, and that everybody else around us was talking like some kind of strange lunar robots in business suits. Everybody sounded like the police in some funny way, even the professors at the university. I remember in 1946, as early as that, I was considered unsavory on the campus. My deportment was proper and I was editing the campus newspaper, but I had a beard, I didn't shave every day, I shaved every five days, and I had very little money so I walked around in secondhand clothes and I had a somewhat Chaplinesque look. And Kerouac was lit-erally banned from physical presence on the campus because he quit the football team and had strange Dostoyevskian friends! It takes a Russian police state to conceive of such a stupid social situation. Because we all knew everybody, it was a small campus, four hundred students, and everybody knew everybody. And you banned somebody who was a writer, a poet! So, I remember, Kerouac came and stayed with me one night in 1946. He had spent the evening with Burroughs talking, Burroughs had warned him against his mother, he thought Kerouac would never be able to get away from her if he didn't make a break, and Jack was all disturbed so he came to my room in the resi-dence hall, Livingston Hall I think, and said "I have been talking to Burroughs and he said the most interesting thing," so we started talk-ing about that. I had written a long poem modeled on Rimbaud's "Bateau Ivre" and the Baudelaire voyage theme, it was called "Le Dernier Voyage," a very stupid poem, fifteen pages of rhymed couplets or something, and I read it to him. Then we went to bed, in the same bed, in underwear; at the time I was a virgin though I was in love with Kerouac, but I was very afraid to touch him . . .

YLP: And what about him?

AG: Well, let me finish the story. So we slept very peacefully. But I had written on the windowpane, which was very dirty, like this window out here (rubbing dust off the windowpane) lot of dust here, "Butler has no balls"—Butler was the president of the University—and "Fuck the Jews" and drawn a skull and crossbones, thinking that the chambermaid would look at it and wash it off and clean the window. But instead she looked at it and reported it to the dean. So about eight in the morning the assistant dean of the student faculty relations burst into the room

and saw me and Kerouac in bed. Now it turned out this man was for-
merly the football coach that Kerouac had worked with. So naturally
the football coach assumed the worst. And Kerouac saw the situation
and did something very characteristic of him: he jumped out of bed,
ran into the next room, jumped into my roommate's bed, pulled the
covers over his head and went back to sleep (*laughs*) and left me to face
the situation. The assistant dean said, "Wipe that off the window," so I
wiped it off the window, and when I went downstairs after about an
hour I found a note in my box charging me 2 dollars and 35 cents for
having an unauthorized visitor overnight and a note saying the dean of
the college wanted to see me. So I went to see Dean McKnight an hour
later. I sat down in his office and he looked at me very seriously and
said: "Mr. Ginsberg, I hope you understand the enormity of what you
have done." These were his opening words. Burroughs had just given
us *Voyage au Bout de la Nuit*. Do you remember that scene when Céline
is in the middle of the battlefield and realizes he is in a place sur-
rounded by dangerous madmen? So I looked at the dean and remem-
bered that phrase and thought "Watch out, he's a dangerous madman."
(*laughs*) The "*enormity*"! The word itself is incredible! So I said, "Yes I
do, sir, I do!" cringing and crawling, "I do," thinking "what can I do to
get out of this situation, how can I apologize?" The dean was mad and
Columbia College was mad! Well, what I am trying to point out is the
difference between the private consciousness that we had and the offi-
cial public consciousness. The private consciousness was the cama-
raderie and common sense of talking very late at night, showing
poems, sleeping, having the personal subjective relations we had, the
public sense was "Mr. Ginsberg, I hope you realize the enormity of
what you have done!" I mean, to have banned Kerouac from the cam-
pus to begin with was an act of such great hysteria and stupidity and
desensitivity and, I think, so unacademic! I mean, imagine Socrates
trying to ban Alcibiades from conversation! It just wasn't proper, it
just wasn't classical. And those people were posing as the inheritors of
tradition and the guardians of learning and wisdom. In fact just at that
point they were putting themselves in the service of the military and
were building an atom bomb, secretly making the biggest political
decision of the century without consulting the democratic populace,
and by later examination one saw that the entire university had been
turned over to capitalist vocational training.

YLP: Did this awareness lead you to a political expression?

AG: I personally, and I think everybody around, immediately questioned the whole structure of Law and immediately apprehended the basic principles of philosophical anarchism. Kerouac had been a communist already, I even think he had been a member of the party. As a member of the National Maritime Union, before it had been taken over by right wing types and the CIA so to speak and the government—he was very overtly communistic for several years, from '39 to '41, '42.

YLP: I didn't know that. I'm surprised. From what he wrote I would never have imagined him as a Marxist! And he had read Marx?

AG: Certainly, sure, Kerouac was very learned, you know, he was always very learned. I don't think he read it with any formal scholarship but I'm sure he read in and out of *Das Kapital* and read through *The Communist Manifesto* and maybe a few other things and he read the *Daily Worker*. It was not a phase that lasted very long, it was only two or three years. When was he born, in 1924?

YLP: 1922.

AG: Well, it was 1940, '41 when he was a communist. He was 19, 20 years old, he was still like a young vigorous seaman. It's proper. Like any young man now. Just normal, quite normal. I think he got to dislike communist ideology later on because the Marxists' reception of his prose, my poetry, and Burroughs' prose was very stupid at first. I think the Marxists in general now feel that we were some sort of helpful, hopeful, useful, prerevolutionary something, they've fitted us in somehow. But at the time there was a large attack by the left against the idea of revolution of consciousness, sexual revolution particularly, and psychedelic revolution involving chemicals and dope, even involving marijuana which is after all an old folk culture tribal totem. You would have thought they were smarter than that but they had very little anthropological training. There were two aspects that Kerouac objected to. First the tendency among the Marxists to deplore our bohemianism as some sort of petit bourgeois angelism, archangelistic tendencies, and to deny the existence of God, to deny the existence of the great empty universal consciousness. And also the left attempted to make the cultural revolution we were involved in, which was a purely

personal thing, into a lesser political, mere revolt against the tempo-
rary politicians, and to lead the energy away from a transformation of
consciousness to the materialistic level of political rationalism. But
the Marxist rational interpretation of the psychological situation we
saw in America was not sufficiently understanding, delicate, tender, to
really apprehend the full evil of American society as far as its psychic
effects on ourselves and, say, Dean McKnight were concerned. It was
too linear, as they say now, an interpretation of the economic causa-
tion of the mass stupidity. Around 1948 we began having definite
visionary experiences. We had been reading Rimbaud by the time, his
letter to Georges Izambard about *"un long, immense et raisonné dérègle-
ment de tous les sens."*[107] Under the influence of Burroughs we already
had had the experience of some of the opiates, on and off, and by 1952
we already had had the experience of peyote partly as a result of trans-
lations of Artaud's *Voyage au Pays des Tarahumaras*, which appeared in
Transition magazine in the 40s, and (Aldous) Huxley's *Doors of Percep-
tion*. Some of Kerouac's writings of '52, particularly his *Visions of Cody*,
are some of the most brilliant texts written about the psychedelic
experience, especially the description of him and Neal on peyotl. So I
am talking about the development of a new consciousness, as they say
now. I think the phrase is used in *The Greening of America* . . .

YLP: Yes it is and I think Huxley already spoke of "expanding con-
sciousness" in *The Doors of Perception*.

AG: In fact the phrase "New Consciousness" was already being used
way back in the 50s. I think there was a little interview essay[108] with
Burroughs and Corso and myself which uses the phrase with capital
letters too . . . As it proved, in America, it was necessary to go through
a long period of change of consciousness before people could be liber-
ated from the hypnotic hallucination that they'd been locked in. It
would have been premature to speak in political terms in those times.
In fact we were definitely thinking in non-political terms, apolitical
terms. The first necessity was to get back to Person, from public to

---

[107]*Je dis qu'il faut être voyant, se faire voyant. Le Poète se fait voyant par un long, immense
et raisonné dérèglement de tous les sens."* "I say one must be a seer, make oneself a seer.
The poet makes himself a seer by a long, gigantic, and rational derangement of all
the senses." See "Letter to Paul Demeny," May 15, 1871, *Arthur Rimbaud: Complete
Works*, trans., Paul Schmidt, New York: HarperCollins, (1976).

[108]Published in *City Lights Journal*, no. 1 (1963).

person. Before determining a new public, you had to find out who you are, who is your person. Which meant finding out different modalities of consciousness, different modalities of sexuality, different approaches to basic identity, examination of the nature of consciousness itself finally—on a very serious level, meaning not only psychoanalysis and drugs but also meditation and ascetic experience, isolation and solitary experience, and *shabda* yoga[109] and jazz and sexual exploration. And recovery of natural tongue, of speech forms that are real rather than literary forms, and recovery of body movement and song and dance, in those days catalyzed primarily by rhythm and blues, the precursor of rock 'n' roll. Like Neal and Jack driving around listening to black rhythm and blues.

YLP: But later on you move towards a more "political" consciousness, in "Wichita Vortex Sutra" for instance.

AG: Yes, that's in the mid-6os.

YLP: Whereas Jack sort of went away from it.

AG: Yes and no. There's a lot of rot on this. For one thing, it wasn't so much that I evolved as that there were finally enough people conscious to make a group, not just a few—ten, twenty people corresponding. We didn't have a post-Beatles mass, a whole generation of longhaired kids who had the experience of a second religiousness and of the recovery of their own body and of a sexual revolution. Then you can begin thinking how do we get together and how do we change things? What do we want and what do we formulate? What do we really want with America? What percent of the American energy supply does the aluminum industry take? Actually one third. And the Pentagon? Actually forty percent. Now we can have actual research and it can be disseminated. So it was not possible really to be political until you had enough people to make a new nation. Also by that time Kerouac had suffered so much attack and abuse from all sides, left and right, particularly left in terms of the venomousness of it, and had become so entangled in personal problems with his mother and most of all had become so ill physically with alcoholism that he was not in a position to go out in the world very much. From 1960 on, every move he made outside of his home was dangerous for him because

---

[109]A version of yoga using sound as the means for transforming consciousness, as practiced in the Sikh religion.

he would always drink himself so ill and get in trouble, people beating him up actually, left-wing literary critics beating him up. (laughs) Once he came drunk into a bar saying "I'm Jack Kerouac," and some radical goon, from the longshoremen's union, I think, beat his head on the pavement. It wasn't political quite, it was just some sort of macho ego thing. And Kerouac was very open, totally helpless. Then there was the tendency to vulgarize the renaissant spirituality of what he had proposed. One built-in stereotype which still exists and is poisoning the left here insists on "hatred" as a "revolutionary weapon," an old-fashioned prepsychedelic nineteenth-century hatred, father and mother hatred actually, which was contrary to his nature as it is contrary to mine. This hatred is at the root of most radical consciousness in America as we saw in the last four years when the entire left went into a completely masturbatory period of social violence, calling everybody pigs, with self-righteousness and self-isolation which finally led to the election of Nixon. That gross element in the left repelled Kerouac, who felt that it was a betrayal of what he had prophesied. He prophesied a spiritual angelic generation that would ultimately take over with long hair and exquisite manners, you know, "wise as serpents and harmless as doves." Instead they were, like, greedy as pigs and harmful as dogs. It's still a problem, the left being poisoned by its own anger. Also his thing was very wise in that he was basically himself a populist redneck and his mother was like a French Canadian peasant, narrow-minded, selfish, naive, hard-hearted, family-oriented lady. She wanted to keep Jack to herself and needed Jack, and he was tied up with her in the sense that he said he didn't want to throw her to the "Dogs of Eternity," as he thought I had done, putting my mother in a mental hospital. So he felt bound to take care of her and, having to live with her, he had to put up with her opinions. In that sense Jack was always an "Americanist," always interested in American archetypes and his mother was like a George Wallace archetype so to speak. Like Céline, like Dostoevsky in old age, like Pound in some respect, like any Tolstoy anywhere, he had an odd cranky appreciation for right-wing archetypes that most left-wing writers are not subtle enough to appreciate. And so in a sense he fulfilled an interesting role there in poetizing that type. Harmless I would say, because it served to curb violent left excesses in myself and in other people. I mean, I always had Kerouac in mind when I got on a peace march and I always made sure it was like really straight, pure, surrealist, lamblike, nonviolent, magical, mantric, spiritual politics rather than just marching up and down the street screaming hatred at the president. In a funny way he didn't have a position, he was just himself, his own character, reacting. He was against the war actually, in a redneck way. On a TV program with pro-war-scholar William F. Buckley, Jr., in 1968, he said of the South Vietnamese politicians "All those guys, all they're trying to do is steal our jeeps." That's a very archetypal proposition and it's really true. He put the whole thing in a very intelligent way that could be universally understood, unlike New York dialectical doubletalk.

YLP: Well, you could hear that sort of thing in any Café du Commerce in Bretagne.

AG: Yeah. If you read his essays like *The Lonesome Traveler* they were really attacks on the police state. Always. The whole thrust of his work was towards individualism and freedom, the only thing is he very definitely took a stand on communist brainwashing. He designated it with the name of "Arapatienz," I don't know what his source was, in the encyclopedia I think, from the name of the Russian who invented mind-conditioning. That he felt was the great evil, which he ascribed to Russian communism as well as to the American *Time-Life* network. So his preoccupation was with individualism. Later he never got into a communal effort, possibly that was because there was no commune sufficiently mature and sweet to be able to take care of him and his mother. And above all there was the problem of his physical illness. When he died his body was in a terrible condition: he had a broken arm, a hernia in his belly-button area that he refused to have fixed, and apparently his liver was gone. I believe it was the night he finished the last chapter of *Pic*, his last novel, that he had the hemorrhage.

YLP: He had started *Pic* in '51, '52, hadn't he?

AG: I don't remember. It's just a little thing that he did long ago. This is just the last chapter. Actually there is another last chapter that he didn't add, when Pic meets Dean Moriarty on his hitchhiking North. He wrote that chapter but I think his mother or his wife didn't like it. When that suppressed chapter is published ultimately, you'll see it's all tied back to the *On the Road* themes.

YLP: Yeah, and the episode of the Ghost of the Susquehanna recurs in *Pic*. Speaking of that period, I found a letter from you to Neal saying you didn't like Jack's new style when he started writing *On the Road*.

And John Clellon Holmes also says in a letter of that time that he thought Jack was on the wrong way.

AG: I changed my mind very soon after that. It was a sort of superficial egotism on my part not to understand what he was doing. I was just a stupid kid. What did the letter say, I don't remember, that his new writing was all crazy or something?

YLP: Yes and that it could be interesting only for somebody that had been blowing Jack for years. (*laughs*)

AG: Oh, what a stupid thing to say! You know, I was very naive, he taught me everything I knew about writing. It took a long time, a couple of years I think, for me to appreciate his ability there and even a longer time for me to begin practicing in spontaneous composition. But my stupidity about his prose couldn't have lasted too long because pretty soon after I was running around New York with his manuscript trying to get people to publish it. At that time I was still writing very laborious square rhymed verse and revising, revising and revising. He was on my neck to improvise more and not to get hung up but I resisted that for a long, long time. All my conceptions of literature, everything I was taught at Columbia, would fall down if I followed him on that scary road! So it took me a long time to realize the enormous amount of freedom and intuition that he was opening up in composition.

YLP: Was it with "The Green Automobile?"

AG: No, that was revised but I was getting close. Actually it wasn't till I went out to his house, when he was living in Northport I believe, so this was '53, '54 really . . . I sat down at his typewriter and just typed what was in my head and came up with a funny poem about the Statue of Liberty that was about three pages, a very sloppy poem and I never published it because it was inferior. But he looked at it and pointed out all the interesting images and he said, "See, you can do it too." It was just that I was afraid to try, afraid to throw myself out into the sea of language, afraid to swim.

YLP: In a June '53 letter to Neal, you said that you are "trying to build a modern contemporary metaphorical yak poem using the kind of weaving original rhythm that Jack does in his prose." And it comes out with "Howl."

AG: Yes about two years later. "Howl" is very definitely influenced by Jack's spontaneous method of composition. So I always found Jack extremely right, like a Zen master, and completely alone in his originality, and because of that I always hesitated to question his judgment thereafter, he always had a depth of character and appreciation that I found later on to be prophetic and useful. Same with Burroughs. I developed after a while almost too much respect perhaps, in the sense that usually I found it best to listen and absorb and learn rather than assert myself egoistically, even in respect to Jack's later political ignorance.

YLP: He often referred to you as being a sort of devil.

AG: Well, I guess it's because of my devilish jealousy, or rather ignorance, in relation to his discovery of his own prose style.

YLP: What do you think Burroughs brought to both of you, since he was older and was considered as a sort of teacher by the two of you?

AG: Well, first of all, the contact with an older European tradition. Burroughs had been in Europe and particularly at that crucial time described by Isherwood in *Prater Violet* and *I am a Camera*: the Berlin of Brecht and George Grosz, the glorious artistic time of the Weimar Republic in which, despite, or perhaps because of, the obvious cruelty of the police state that was emerging, it was clear to more and more liberated minds how true, free, tolerant, a bohemian culture might be. Burroughs had lived in that aura and brought that over to New York. And also on the first formal visit Jack and I paid to Burroughs really to find out who he was, if he was really evil or like some sort of extraordinary melancholy blue child. He was reading a lot of books that we didn't know about and so we took our reading from him. He had Kafka's *Trial*, Cocteau's *Opium*, he had (Oswald) Spengler's *Decline of the West* which influenced Kerouac enormously in his prose as well as his conception of Fellaheen, he had Korzybski's *Science and Sanity*, so that was like a preliminary western version of the later oriental teaching of the difference between concept and suchness, between word language and actual event—he had Rimbaud's *Season in Hell*, Blake which I picked up on, he had *A Vision* by William Yeats, a sort of gnostic analysis of history and character, he had Céline's *Voyage au Bout de la Nuit* . . . If you take all those books, it takes one year or two to read them through seriously and get them all together. Burroughs had studied English and archeology at Harvard and his preoccupations were anthropological. He was interested in Kwakiutl Indian potlatch ceremonies, which I had never heard of before; in the berdache, American Indian shamanistic transvestite figure; in the psychology of apes; in primitive mind; he was interested in the psychopath as R. D. Laing is now interested; in the crude sense that the psychopath has a certain freedom of mental conception that the so-called normal person doesn't have. So Burroughs was primarily a master of gnostic curiosities and in his approach to the mind he had the same yankee practicality and inquisitiveness as his grandfather who had invented an adding machine.

YLP: He had also this orgone machine . . .

AG: He was exploring the Reichian orgone therapy.[110] I went to a Reichian around that time myself. Burroughs was at that time being psychoanalyzed by a Doctor Federn, who had been analyzed by Freud, so he had in a sense a direct transmission to the source of psychoanalysis. And he psychoanalyzed us. I spent one year talking, free-associating on the couch every day while Burroughs sat and listened. I really did explore a great deal of my mind and then began exploring some of my emotions. I remember bursting out in tears one day toward the end and saying "Nobody loves me!" It took a great deal of patience of Bill to sit there for a year until I bared very frail sensitive private fears. He was a very delicate and generous teacher in that way. Jack spent a good deal of time in that same relationship to Bill, being psychoanalyzed or psycho-therapized, whatever you want to call it, we didn't have to use that category, simply Bill sat and listened to us for a year. Burroughs very accurately predicted that Kerouac would move in concentric circles around his mother's apron strings until he wasn't able to go away ten feet from the house. At that time we were all living together in a house near Columbia, it was probably 1945.

---

[110]Orgone theory is central to Wilhelm Reich's cosmology, and he wrote extensively on it. See: Wilhelm Reich, *The Function of the Orgasm: Sex-Economic Problems of Biological Energy (Discovery of the Orgone, Vol. 1)*, April 1986, Noonday Press. For William Burroughs's discussion see *The Job: Interviews with William Burroughs* by Daniel Odier (New York: Grove Press, 1970).

YLP: What about his sense of humor?

AG: Oh, exquisite. Like in *Naked Lunch*. He would tell stories and laugh. *(cackling to imitate Burroughs's laugh)* He and Kerouac wrote a book together by the way, around that time, called "And the Hippos Were Boiled in Their Tanks" after a news story that they heard on the radio. It was about a fire in, I think, the Saint Louis Zoo, which the announcer ended: "The fire consumed two buildings and three acres of forestland and the hippos were boiled in their tanks." Burroughs thought that this deadpan yankee bizarre image was characteristic of the most blatantly desensitized mad humor in America. Like saying "And the Vietnamese were burned alive in their huts," so to speak. So that was the title and Jack and he each wrote a chapter. It was written in the style of Raymond Chandler, hardboiled. That was very early, before *On the Road*. I think Sterling Lord has that manuscript.

YLP: And Neal Cassady? He must have been very different from all of you when he arrived in New York in '46.

AG: Not much different from Jack. We were all lonely, we were solitarios, isolattoes, as Whitman said. He was different in that he just wasn't around Columbia, I suppose, but he had a lot of the knowledges that Jack was interested in: the knowledge of America particularly. He was different in the sense that we were eastern and he was from the west Rockies so he represented America to us, a child of the rainbow of the Far West, probably the same mythology that he might have represented to a Frenchman with *le mythe du Far West*. I think Neal though had a good deal more sexual experience than any of us and was much more open about it and took it as a sort of joyful yoga, and transformed it into a spiritual social thing as well as a matter of esthetic prowess. There was an element of esthetic prowess but there was also an element of faith in sexual intercourse and intimacy as an ultimate exchange of soul. That's certainly why I fell in love with him and that's why he responded to me. Because though he was primarily heterosexual, except when he was sixteen or seventeen and slept with some older people around Denver, I think he saw that I needed someone so much, because I was almost virginal and was so locked up in myself, that he sort of opened up and left himself naked and put his arms around me. He was a very beautiful open soul, very American-Whitmanic, universal in that sense. That aspect of emotional generosity and "adhesiveness" as Whitman called it was precisely the portion of Adam which had been extirpated from American public life, and even from private consciousness, in the years between Thoreau and Whitman and the post-war generations, through the development of a competitive macho capitalist selfish ethic. So Cassady was recovering a tradition of generosity of emotion and magnanimity of body and soul that were praised by Whitman and Sherwood Anderson and Hart Crane. That was really the ideal stuff of the American Adam, the image that Europe entertained, the Billy Budd that Melville had proposed, the large magnanimous citizen that Whitman proposed as necessary. It was a restoration of tenderness through the person of a definitely macho and masculine ideal western tough kid. It was a very early exhibition of that sexual democracy that's spread now and accepted by the entire psychedelic unisex generation.

So that was Neal's particular heroism, and the interesting thing about *On the Road* and *Visions of Cody* is Kerouac's approach to Neal as a Whitmanic lover to another man and the books are real love songs of a very ancient nature. They are not to be categorized as homosexual because Jack and Neal never made love genitally, they never had sexual relations. But they had a very noble, thrilling, love tenderness, heart palpitations for each other, which is characteristic of normal masculine relationships and, as I keep saying, is almost obliterated in modern culture, probably in France as much as America now, because the machine state and the military police bureaucracy state has cultivated a paranoia between men and a suspicion and fear of body contact and soul contact in order to keep people separate, in order to divide and conquer. There is a very brilliant passage in Whitman's preface to *Democratic Vistas* that studies comradeship in political terms. [He reads a long passage from Whitman.][111] I think that is at the interesting heart of the relationship between Jack and Neal and myself and Burroughs, sometimes, as in my case and Burroughs', overtly homosexual, and in Neal's and Jack's case it's more fitting to the Adamic tradition that Whitman is proposing. But the point is that in our private relationship we found the whole spectrum of love if not convenient at least tolerable and charming. And that was a world of private sociability and discourse which was the inverse of the lack of adhesiveness and the lack of recognition of Person, the objectifica-

---

[111]See dedication page of *Fall of America* (San Francisco: City Lights, 1972).

tion, reification, depersonalization, mechanicalization of Person that made it possible for, say, the Germans to produce an Eichmann, merely following orders, or for the Americans to produce a whole generation of people who can watch U.S. wars' mass murder on television and not recognize it as personal to them. So I think *On the Road* and *Visions of Cody* and all the other books have a basic political prophetic value, not merely in discovering the body of the land, not merely in noticing the minority black superculture transcending the white superficial culture, but as the presentation for the first time in a long time of unabashed emotion between fellow citizens. And I think that sweetness of emotion is perhaps Jack's major contribution to both the literary prophetic scene in traditional American letters as well as to the generational development that's gone on since his time. And that's why his later grumpiness at left-wing anger and hostility seems somewhat relevant and in place justified, or at least understandable.

YLP: Allen, there is another thing I'd like you to explain. Why did you make your breakthrough in San Francisco?

AG: Well, there was a rigidity in the N.Y. literary world here which we've described at great length already. There wasn't an admission that there was another breath in America till the early 60s. So from 1950 to 1960 the town that was most perceptive was San Francisco. It had a tradition of philosophical anarchism with the anarchist club that Rexroth belonged to, a tradition receptive to person rather than officialdom or officiousness. There already had been a sort of Berkeley Renaissance in 1948 with Jack Spicer poet, Robert Duncan poet, Robin Blaser poet, Timothy Leary psychologist, Harry Smith, great underground filmmaker, one of the people who originated the mixed media lightshows . . . They were there all around the same time, they didn't know each other well but they were passing in the street, you know. And specifically there were little magazines like *Circle* magazine, George Leite was there as an editor, there was a tradition that didn't exist in the more money-success-*Time*-magazine-oriented New York scene. There was also a gnostic tradition quite well developed in San Francisco and peyotl was understood in its place as part of the American Indian tradition. And there were a lot of interesting people: (Philip) Lamantia, who had this sophisticated connection with the Surrealists— very few people in the New York professional literary world had his experience of letters, even the older people were naive compared with Lamantia—Bern Porter was there, he was an atomic physicist who quit and became a literary friend of Henry Miller, and Henry Miller was in Big Sur, Ferlinghetti was there, with the first pocket paperback bookshop, so there was a whole underground tradition so to speak. That's why we had a place to publish there. For instance I took my poetry and Burroughs' *Junkie* and Jack's *On the Road* and other manuscripts around all over New York and nobody would publish them. I went to the big publishers, some of whom were friends of mine from Columbia by then, some of whom were even poets, and they rejected our books saying "The prose is bad" or, about *Junkie*, "Well that would be interesting if it were written by somebody famous like Winston Churchill." *(laughs)* So I took all the manuscripts to San Francisco and began circulating them there and found a much warmer reception. Rexroth and Duncan saw them and immediately understood them as being, you know, good writing, classic. I remember Duncan's reaction to *Visions of Cody*. There's a great passage in which Kerouac goes on for pages sketching all the reflections on a curved shiny fender which was reflecting images from a plate glass window. Duncan saw that and said that anybody who could write five, or fifty pages I forget, of description of reflections in an automobile fender obviously had original genius. Whereas the reaction in New York was "What kind of mongolian idiot would even be interested in that?" or the later reaction of stupidity by Truman Capote "That's not writing, that's typewriting," superficial gaga putdown. So that's why. San Francisco was not involved with the coldest aspects of American frozen consciousness.

YLP: I found that most critics tended to regard your works as manifestoes rather than as works of art. Don't you think there were some misconstructions and misunderstandings there?

AG: The works produced as literary works should be looked at as such, so that whether a consciousness revolution is a by-product of that or is a concomitant potential of that, it doesn't come as too great a shock. They offer some sort of material base to it, being a series of verbal classics. "Howl" is not a bad poem, it's in the tradition of strong rhythmical panegyrics like Poe's "Bells" or Vachel Lindsay's "Congo" or Hart Crane's "Atlantis" or even Shelley's "Adonais." *On the Road* is a great and beautifully prosed picaresque novel of a traditional nature, *Naked Lunch* is an extension of a style already introduced and accepted as literary style by Jonathan Swift in *Gulliver's Travels* or *A Modest Proposal*. Very definitely all of our work is built on a very firm base of connaissance of twentieth-century writing from Gertrude Stein through the French Surrealists with much of their knowledge appreciated and genuflected to and then developed further in an American context for an American tongue, slightly later in the century when there was a slightly more obvious opening of a new consciousness. So it's all within the context of a known literary tradition for the French and I was surprised that it was not immediately apparent to French writers back in the 50s that what we were doing was working a further and perhaps welcome extension of their own work. Part of the dryness the French always complained about in their own work of the 50s and 60s comes from that sort of lack of comradely generosity. There were nice people around; Michel Mohrt arranged for me and Burroughs to go and visit Céline in 1958. We did want to touch home-base, we did want to visit our heroes and receive their blessing and we did do that. Burroughs and Céline were like two cousins literarily and the conversation was interesting and very straight. I wrote a little bit about it in a poem called "Ignu" . . . I've always thought that we were recognizable so I wished that just the primary attention would be paid to the texts as literary texts, just so they could be inserted in the curriculum as they apparently now are, just so they would affect the consciousness of the younger generation. The longer the texts are considered as some sort of strange unicornlike objects that are not literature and not not literature, the slower the development of consciousness will be. These texts should have been around in France a long time ago.

YLP: Well, they are widely read now and are published in paperback. *On the Road* for instance is being published in a new Gallimard paperback edition.

AG: That's good. See, it was the best we had in America in terms of trying to explore new territories of consciousness at a time when France, because of preoccupation with the war and perhaps exhaustion after an enormous effort earlier in the century, had come to a fixed classical overrationalistic mentality, and it was necessary for another bath of emotion and comradeship and trust.

YLP: Which appeared maybe in May '68.[112]

AG: Yes. I wish that when it appeared in May '68 it had been more informed by adhesive tenderness between men as an overt understanding, and greater experience in psychedelic modes, and some of the mellowness of character that Kerouac displays in *Doctor Sax*, the sense of Buddhist time illusion. I doubt that the revolution would have succeeded any more than it did but perhaps the effects might have been more durable if more consciousness had been reclaimed as well as more matter. But the French did astoundingly well anyway I suppose in that situation, they inspired everybody else really . . . But the conditions of revolution in late twentieth century are conditions unforeseen by any other civilization. We are going to the moon, we have drugs that go to the moon inside, we've recovered the archaic knowledges of the Australian aborigines, most primitive societies are available to us if we take the effort, many different forms of magic warfare or peacefare are available, apocalypse, classical Armageddon, destruction of the planet, millennium, all this is a possibility. So it's now unlike it ever has been in history, this should mellow everybody out and make it possible for everybody to work together, to create a revolution that had no enemies, a revolution by apokatastasis.

YLP: What do you mean exactly?

AG: Apokatastasis is the transformation of satanic energy to celestial energy. In that sense the former notions are irrelevant to the revolutionary steps of consciousness that we're forced to take in order to survive without tearing the earth asunder. It means that Americans have to stop calling policemen pigs and use a different mantra on these human souls in order to transform them into helpers of the gods rather than enemies of the gods. That's the direction.

YLP: Love?

AG: No, that's too much of a used word and too vague. It's more emptiness than love, detachment rather than attachment. Certainly not sentimental love. It would mean flower power, definitely ecologically, the

---

[112]Maoist-inspired students calling for the total transformation of society sparked a nationwide strike in France from May to June 1968, affecting ten million workers, but electorate reaction barred revolutionary change.

power of greenery to regenerate the earth. Many people have not taken into account this aspect of flower power. They think it just means sticking a blossom into the muzzle of a rifle in front of the Pentagon, but also it is the preservation of the Amazon jungle as one of the great lungs of the planet.

YLP: Do you spend much time in the country now?

AG: Yeah, the last four, five years. We live on a commune piece of land without electricity. We do organic farming and grow our own food. All the extra money I get I put into that, to grow food instead of spending it on plastic beer cans in the city. I don't want to be just a consumer. I consume enough paper as it is, enough electricity, so I should produce something in return, like vegetables. *(he goes to the kitchen and comes back holding a jar of tomatoes)* Paul Goodman showed a long time ago that the kind of industrialized farming we have is not only inhuman but is also destructive of the soil, we already saw that with the Dustbowl in the 3os. So as Goodman proposed, it would probably be useful to have decentralized specialized organic farming and that's what we are doing. It's Peter Orlovsky's work, organic farming, that's his specialty.

YLP: I was surprised to see the importance young Americans grant now to the Do it Yourself thing.

AG: It's the old Thoreau tradition. The reason for that is that if you don't do it yourself you are a prisoner of the robot state, the electric company, the transportation company, the food monopolies and the chain stores. You live in a suspended state where you don't even know where your power comes from, you leave the faucets running and the lights on all night just because you don't even know that the water supplies are slowly diminishing and maybe we have only another twenty or thirty years of clean water before it all goes away. You live in a situation where you let people dump your garbage out in the Atlantic Ocean so that in the last twenty years 40 percent of the life of the oceans has been destroyed.

YLP: Do you think that moving out of all this is sufficient in the long run?

AG: Well, probably not but on the other hand the direction of mono-lithic conspicuous consumption of power and piglike abuse of natural resources both in America and Russia can only lead to planetary disaster. So now is the time for experimenting in decentralized nonpollutive power sources. And that means to some extent going back and reexamining nineteenth-century technology, miniaturizing it and making use of it in twentieth-century terms, adding a little of twentieth-century finesse and cybernetization perhaps. Nobody knows but nobody is experimenting in that area.

YLP: Except in China.

AG: In China certainly. The intuitive move of many young Americans into country communes is very similar to the cultural revolution in China, when everybody was told to get out of the cities and go back to work with the workers in the fields for a while, and see what life was really like at the source of production. It's impossible for French Parisian kids or New York kids to conceive of a blueprint for a new society if they don't even know where water comes from, if they've never seen a tomato grow, if they've never milked a cow, if they don't know how to dispose of their shit, how can they possibly program a human future? It would be all abstract in their heads, like a mathematical equation, and would produce a monstrosity. Going back to the primary sources would also lead back to the people, as the Chinese said. It would also lead to a recovery of natural Adamic consciousness. It would mean the end of the noxious aspects of elitism that everybody is always complaining about, it would lead to an appreciation of the sacred character of nature that we lost, an appreciation of the American Indian virtues and earth ceremonies. Also to an appreciation of the fact that, as Gary Snyder pointed out, that in our times the exploited masses are not merely the Third World populations whose resources and labor are being stolen by the Americans who consume half of the world's raw materials with only 6 percent of the population, but the exploited masses also are the green grass, the trees, the soil, the birds, the whales and the fish, nonhuman sentient beings who are being destroyed by human greed. A funny thing to insert into Marxian ratiocination. *(laughs)* I think the psychedelic Marxists would understand it or it will maybe take ten years for them to realize that it's simple common sense. In 1951 Kerouac said "The earth is an Indian thing." It's very interesting that Kerouac's original preoccupation was the town and the city, the town versus the city, the monstrosity of the city and the humanity of town, and the destruction of town by mass production capitalism and hard-hearted exploitation.

YLP: I found a letter from Bill Burroughs to Jack saying he knew that Jack expected to find Lowell in Mexico, which is quite in keeping with what you've just said . . .

AG: That's interesting. Cities are too horrible and are forcing people out but the main thing to do is not to see it as a mechanical exodus from an overcrowded urban madhouse but to take the opportunity then to spiritualize the country scene, get the vibrations of the trees and then come back to the city and turn the buildings into trees.

YLP: Unfortunately the people who make the buildings are not those who go get the vibrations of the trees.

AG: No, people who had the power in the city always had country estates, all the way back to Louis XIV. Look at Versailles. The exploiters always dug the country.

YLP: Note that in Versailles they planted orange trees to overpower the smell of their shit.

AG: That was a good idea! You don't realize the helplessness of the fat-assed, fat-bellied American Jewish boy in New York who really doesn't know how to walk and doesn't know how to climb, is short of breath and smokes too many cigarettes and eats too much meat and doesn't know how to do without electricity and doesn't know where the water comes from and thinks that his shit should go in the toilet and be flushed out into the ocean and that takes care of that. A completely hopeless creature. And it's precisely this sort of intellect, this sort of dependent, subservient, overweight mind separated from the body of the earth that was the primary critic of Kerouac when he first came on the scene with *On the Road*. They resented the holiness and whole-someness of his body. Of course there was a lot of dewy lyricism in the book, and a lot of inexperienced joyfulness and foolish historical puns but the essential rightness and healthiness were there. And yet his work was seen by city intellectuals as being barbarian, ignorant, anti-social, anti-intellectual, mindless! Everybody was seen inside out. But the transmission of consciousness and ideas through time was already a heavy element in the original literary activity we were concerned with, in that Gary Snyder and Philip Whalen met William Carlos Williams in 1950 and learned directly from him, as I learned directly from him, as he cooperated with me, wrote prefaces to my early work and incorporated my letters to him in his Paterson text, as Robert Duncan and Charles Olson for many years corresponded with Pound and Williams, as Robert Creeley wrote back and forth to Pound for instructions on running a magazine when Creeley was running *Black Mountain Review* as far back as 1945, as Philip Lamantia was in connection with (André) Breton and the Surrealists during the war in New York . . . We were carrying on a tradition, rather than being rebels. It was completely misinterpreted. We were rebelling against the academic abuse of letters but we went to the living masters ourselves for technique, information and inspiration. The academic people were ignoring Williams and ignoring Pound and Louis Zukofsky and Mina Loy and Basil Bunting and most of the major rough writers of the Whitmanic, open form tradition in America. But we had that historical continuity, from person to person. There is no gap. And since that time, this poetics has moved on into a larger democratic field, even returned true lyric to pop music, through Dylan.

So there's been a comradeship, a Whitmanic adhesiveness from generation to generation, from the older generation to ourselves and from ourselves to a younger group of geniuses who are reflecting our own explorations back on us and teaching us how to go further. There was a necessary democratic revolution of consciousness that I find charming and hopeful and exemplary. That sense of comradeship is what I find exemplary, and that is basically the key to what was discovered between us as a smaller group of people back in 1945.

ALLEN YOUNG
September 25, 1972,   Cherry Valley, New York

From the book *Allen Ginsberg: Gay Sunshine Interview*
(1974, Grey Fox)

Young wrote in his April 11, 1973 introduction to the book version of the interview: "Around the same time [that Young read *The Dharma Bums*], I discovered that Ginsberg was staying on a farm . . . located in Cherry Valley, N.Y., only 20 miles away. . . . I wrote Ginsberg a letter, inquiring about Kerouac's characterization of him in *The Dharma Bums* . . . and asking him if he would be willing to be interviewed for a gay liberation newspaper. We didn't manage to get together that summer, but Ginsberg's response was very favorable, and, finally, in late September of 1972, a full year later, I arrived at the Cherry Valley farm. The goldenrod, a tiny yellow wildflower people seem to prefer to call a weed, was in full bloom.

"Several of Allen's friends, who shared the house with him, greeted me when I arrived, and told me that Allen was waiting in the kitchen of the old farmhouse. Allen was very friendly, and immediately we began a conversation in which he expressed as much interest in me and my life as I did in his. For a few moments, he interrupted the conversation to show me around the house. As we passed an ancient pedal organ, he sat down and played chords while intoning a spontaneous poem about my visit and our shared gay consciousness. Sadly, my tape recorder was not yet going, so that particular bit of Ginsberg creativity has not been preserved. Then we spent more than an hour walking around the land, with Allen indicating the various features of the ecologically sound farm, including the absence of electricity and an ingenious water-pumping device known as a hydraulic ram. We picked some cherry tomatoes from the garden, shared reminiscences about the 1958 Columbia poetry reading . . . and began to exchange viewpoints and experiences from our respective political and cultural backgrounds.

"I soon became frustrated and nervous about not having the tape recorder with me, so we returned to the kitchen and began the interview."

After Young transcribed the tapes, he submitted the manuscript to Winston Leyland, the editor of the gay literary magazine, *Gay Sunshine*. Leyland published the interview in issue no. 16, January 1973.
—DC

Allen Young: One of the things that provoked this whole conversation between us was my reading of *The Dharma Bums* last summer. In that book the character Alvah, who is quite obviously you, is portrayed by Kerouac as heterosexual. There are a number of sexual encounters and there isn't any indication that there was any kind of homosexuality in this group of people.

Allen Ginsberg: That was Kerouac's particular shyness. You know, I made it with Kerouac quite often. And Neal (Cassady), his hero, and I were lovers, also, for many years, from 1946 on, on and off, at least I wanted to be, and we got to bed quite often, didn't really fully . . . finally he didn't want any more sex with me, he rejected me! That's what he did! But we were still making it in the mid-1960s after having known each other in the mid-forties, so that's a pretty long, close friendship—Neal and Jack, for that matter.

AY: Did Jack Kerouac identify himself as being a gay person?

AG: No, he didn't. A lot of that took place in the cottage we all held together, and then I had been living with Peter (Orlovsky) for several years. Peter, Jack, Gary [Snyder] and I and various other people were all sleeping with one or two girls that were around. Jack saw me screwing and was astounded at my virility. I guess he decided to write a novel in which I was a big, virile hero instead of a Jewish Communist fag.

AY: What was your reaction to that? Did you feel that he was hiding?

AG: I didn't notice. *On the Road* has one scene in the original manuscript in a motel where Dean Moriarty screws a traveling salesman with whom they ride to Chicago in a big Cadillac; and there's a two-line description of it which fills out Cassady's character and gives it dimension. That was eliminated from the book by Malcolm Cowley in the mid-fifties, and Jack consented to that. So Jack actually did talk about it a little in his writing.

In a book that's being published now, *Visions of Cody*, there's a longer description of the same scene. It was written in 1950–51 by Kerouac and was his first book after *On the Road*, a sequel to it. It was a great experimental book, including a couple of hundred pages of taped, transcribed conversation between him and Neal, over grass at midnight in Los Gatos or San Jose, talking about life to each other, the first times they got laid, and jacking off, and running around Denver.

AY: Why is it first coming out now?

AG: Kerouac always wanted it published. But the commercial publishing world wasn't ready for a book of such great looseness and strange genius and odd construction. It's more like a Gertrude Stein *Making of Americans* than it is speedy Kerouac.

AY: Was it a fight for Kerouac to get his stuff published?

AG: Oh, yeah. *On the Road* was written in 1950 and was never published till '57, even though he had previously published his great book *The Town and the City*. The commercial insistency was that he write something nice and simple so everybody could understand it, to explain what the beat generation was all about. So he wrote *The Dharma Bums*, to order, for his publisher, a sort of exercise in virtuosity and bodhisattva magnanimity. He wrote in short sentences that everybody could understand, describing the spiritual revolution as he saw it, using as a hero Gary Snyder; actually, "Japhy Ryder" is Gary Snyder.

AY: So then your portrayal as a heterosexual doesn't have anything to do with being in the closet.

AG: No. I came out of the closet at Columbia in 1946. The first person I told about it was Kerouac, 'cause I was in love with him. He was staying in my room up in the bed, and I was sleeping on a pallet on the floor. I said, "Jack, you know, I love you, and I want to sleep with you, and I really like men." And he said, "Oooooh, no . . ." We'd known each other maybe a year, and I hadn't said anything.

At that time Kerouac was very handsome, very beautiful, and mellow—mellow in the sense of infinitely tolerant, like Shakespeare or Tolstoy or Dostoevsky, infinitely understanding. So in a sense—there's a term that I heard Robert Duncan use for poetry and I've heard others use for relation between guru and disciple—as a slightly older person and someone who I felt had more authority, his tolerance gave me *permission* to open up and talk, you know 'cause I felt there was space for me to talk, where he was. He wasn't going to hit me. He wasn't going to reject me, really, he was going to accept my soul with all its throbbings and sweetness and worries and dark woes and sorrows and heartaches and joys and glees and mad understandings of mortality, 'cause that was the same thing he had. And actually we wound up sleeping together maybe within a year, a couple of times. I blew him, I guess. He once blew me, years later. It was sort of sweet, peaceful.

AY: Did you experience any kind of a split between your hipster circle and getting involved with other gay people as you were coming out?

AG: It's in a poem that I read at the Chicago Seven trial. This is a dream I had in 1947 while I was at Columbia: [*Ginsberg reads "In Society"; see pp. 238–239 above.*]

There were a whole group of queens around Columbia at that time who were doing things like going down to hear Edith Piaf sing at the Plaza Hotel and interested in status and money. They had cultural interests that went back to Lotte Lenya and things like that, but at the same time it was an overly aristocratic, elitist thing.

AY: Did you associate that with the faculty at Columbia also?

AG: There were a couple of guys on the faculty at Columbia that participated in that sense of things rather than in an open democratic Whitmanic gaiety, because to be open and democratic and Whitmanic meant, like, open, friendly, kissing the football players! In public, no less. Whereas the closet queen gaggle could get together and go down to the Plaza . . .

AY: Well, was kissing the football players in any sense a reality, or just a Whitmanesque fantasy?

AG: I was kissing Jack Kerouac, who was on the Columbia varsity team in those years. It was a Whitmanesque fantasy, which, like all Whitmanesque fantasies, were practical realities. Of course, then a faculty guy probably couldn't do it, but nowadays a faculty guy could.

I was silent about it [homosexuality] at Columbia the first year I was there, between the ages of sixteen and seventeen. At seventeen something shook me loose from the authoritarianism of the culture and from the authority of Columbia. I think it was the jailing of a friend, who I loved, who knew Jack well. And then also I was interested in Rimbaud and Whitman, and I had met Burroughs by then. I was getting teaching from Burroughs that included Blake and Spengler (*The Decline of the West*); and semantics was important, separating words from the objects they represent, not getting confused by labels, like gay or queer, in those days.

So it was just a whole change, growing up out of high school and puberty and closed-in-ness. It wasn't closet; it didn't have that much style about it. It was just timidity and fear of rejection. All through high school I was secretly in love with all sorts of boys—particularly one boy from East Side High, Paterson, who I actually followed to Columbia.

AY: Whose name begins with "R." You mention him in one of your poems.

AG: Yeah. Very soon I was babbling at great length. The permission for that openness came from Burroughs and Kerouac, who I was living with. They were wide-brained, international, hip, Jack London, Doctor Mabuses,[113] all.

Kerouac was a very funny, strange, heroic figure, a seminal figure for many ideas and attitudes. He had a lot of trouble; he drank himself to death. And he ended, like many older writers, reactionary in a funny, interesting, characteristic way, a way that's teaching rather than negative. But the basic thing about him was Character, with a capital C, was an enormous mellow, trustful tolerance and sensitivity. And that's why he's such a great writer and observer. You know, he held everything dear, as a sensitive young fellow, even my fairy woes. In fact, we wound up in bed together.

AY: You're saying that this really wasn't where he was at sexually?

AG: Well, he was very mixed sexually. He had a lot of trouble with attachment to his mother and his mother's dependence on him. He was a football player, and he liked girls. He liked to eat girls and was really hung up on them. That's what really excited him: black panties! black stockings! He also appreciated beautiful boys and had a really

---

[113]Fritz Lang character and protagonist in three of his films.

novelistic, personal appreciation of older queens—which was like a sharing of common humanity, and a sharing of emotions, even a sharing of the erotic, except that he didn't feel it was right for him to participate in the erotic.

As a novelist, he opened himself to the art of gaiety and some of its attitudes and styles in writing. In some of his poems there's a lot of stuff about himself, where there's all sorts of high teacup bullshit. In those days "high teacup" was a lifted pinky infra-language. I'm saying he had mixed feelings at different times, but I think it would have been abusive of his character to point an accusing finger and say, "You're a fairy!" There was a certain tendency among gay people there to plaster labels over everybody, including themselves, instead of seeing the nameless love that everybody is. Just as there was a tendency among macho heterosexuals to plaster labels, so there was a counterbalancing tendency among homosexuals to overreact to that and camp too heavily, so that he was sensitive about being put down as a fairy, which he wasn't. *(Calling over to Peter Orlovsky on the other side of the room where he was not listening to the interview:)* Was Jack a fairy?

Peter Orlovsky: Was Jack a fairy? No . . . in a tiny sense of the word.

AG: Perfect, in a very tiny sense of the word. *(To Peter:)* We all made it with Jack once.

PO: [One time] he was so drunk he couldn't even get it up.

AG: *(laughs)* Yeah. Well, no, he came that time. We were at Clellon Holmes', remember? I blew him and you screwed me.

PO: What about on Second Street? Do you remember that? Jack was gallantly drunk, lying in one of the small, side rooms, and you tried to blow him. He couldn't get it up and he was talking about his little cock; it was so tiny, so small, shriveled up, sad.

AG: He was very apologetic. But ten years ago he was asking me to blow him all the time. In '64–'65 he said: "I'm old, ugly, red-faced, I'm beer-bellied and I'm a drunk and nobody loves me anymore. I can't get girls, come on and give me a blow job." There were times he'd get drunk and be really insistent on it. By that time he'd gotten beer-bellied, florid-faced, and I no longer saw him as the romantic, handsome, young glamor-beau of postwar, dark, doomed, maddened Spengler hippiedom.

So, I got freaked out at the whole idea of bodies and sex, in fact. That was one of my first lessons in chastity. There's a line in Yeats: "Old lovers yet may have all the time denied, grave is heaped on grave that they be satisfied." I found actually in the course of time that everybody I really loved and wanted to go to bed with, I finally did. It may have taken twenty or thirty years, and we may have both fallen into ruins and baldness and all our teeth fallen out, but desire always found its way, even if it took decades. There's a lesson there. Once you become a little detached, once you lose neurotic, obsessive attachment, then, when things are floating lightly, then you find love objects that you once worshipped drifting in on the tide, back to you, more than you can deal with; in fact, horrifyingly rottened up from the sea.

An element in the gay lib struggle and metaphysics that I don't think has yet been taken up is that of disillusionment with the body. I'm not trying to be provocative in that—just the age-old realization of over-forty, over-fifty, and over-sixty, and over-seventy and over-eighty, finally, the age-old grinning skeleton, with the spiritual lesson behind it, of detachment from neurotic desire. I think there's a genuine eros between men that isn't dependent on neurotic attachment and obsession, that's free and light and holy and lambent—which is more or less what we all get during our first fantasies, loves and devotions. Some of us are lucky enough to be able to act it out and receive back and forth. But it can only come in like the tide when you're free to float in it. If there's too much of a neurotic grasping to gaiety, to gayness, even to gay lib, then it makes everything too tense, and the lightness of the love is lost. So the gay lib movement will have to come to terms sooner or later with the limitations of sex.

If you consider sex from a Hindu, Buddhist, Hare Krishna, even Christian fundamentalist viewpoint—a warning about the body and a warning about attachment itself—it becomes interesting. Burroughs has actually written about it at length in a way which hip people and even radicals have found very interesting: the sex "habit"[114]—sex as another form of junk, a commodity, the consumption of which is encouraged by the state to keep people enslaved to their bodies. As long as they're enslaved to their bodies, they can be filled with fear

---

[114]Burroughs's use of the concepts of "addiction" and "junk" are as often metaphorical as they are literal, and he often exchanges one idea for some parallel idea or condition; "sex" in no small part fills this role.

and shock and pain and threat, so they can be kept in place. The road of that, he said, leads to the great palace of green goo, the garden of green goo, green goo trap, with everybody schlupping together in this green stuff.

I find, as I'm growing older, no less flutterings of delightful desire in my belly and abdomen. But also I'm becoming more tolerant of other resolutions between people besides sex. When I was in Australia, I had a crush on a beautiful young Dobro[115] player who traveled around with me. He sought me out and waited all day at my hotel and put himself at my service to play music with me. He wanted to play mantras and then turned out to be a great blues player, and he taught me blues. And he went to bed with me the first night, when I really got entranced by his . . . servility . . . and availability, and generosity, genius and sense of duty. And then he didn't want to go to bed with me after that, but he loved me. I was the first man he had ever been to bed with. How am I going to deal with somebody who really loves me but doesn't want to play with my cock and doesn't necessarily want me to blow him? But he didn't mind sleeping in bed naked with me, you know, because he loved me. There was a weird thing . . . but is that any more weird than my desires?

So I finally got into a scene which was like the old nineteenth-century thing recommended by Edward Carpenter, and Whitman— people sleeping together. It's called "carezza," a platonic friendship in which people sleep together naked, caressing each other, but don't come, saving their seed for yogic or other reasons. So I did that with this kid.

For the next couple of weeks we were running around Australia. I found the intensity of my devotion to him in the heart area—a warm, aching feeling in the heart—growing and growing and growing, and becoming more and more desirous and narcotic-like, and more and more satisfactory to carry around with me. And I found him responding in a similar way to me, and I realized that that same warmth was growing in his breast to me, and that what was building it was the naked chastity that we were practicing together. When we got on the stage and played together—I was singing mantras, blues and playing

---

[115]The trademark name of a type of acoustic guitar with steel resonating discs inside the body of the guitar, an abbreviation of the Dopera Brothers who invented the instrument in the 1950s.

harmonium and he was playing Dobro—the erotic communication between us got ecstatic and delirious. And it couldn't be withheld. We'd keep bursting out in song and eye glances which turned the audience on completely, and turned me on, and turned him on. So I was feeling another kind of very subtle, ethereal orgasm that seemed to occupy the upper portions of the body rather than the genital area.

Though I've always been prejudiced against that kind of sublimation, thinking of it as some sort of sublimation of primary, holy sex drives, the experience was so delicious that I can't really put it down for any moral reason at all. I recommend it; everyone should have that experience, too. You can get real close with people that you love who wouldn't otherwise want to sleep with you sexually. But you could have like a total relation. So there's all sorts, all forms. . . . "Smash sexism!"

I know lots of men who are thinking along those lines. They may not want to sleep naked together but there's a love thrill in the breast they have for each other, and yet are completely heterosexual. And I wouldn't be surprised if that's, among the mass of men, a universal experience, completely accepted, completely common, completely shared.

The idea of a buddy is just the thin, label, vulgarization of it. The tradition of comradeship, of companionship, spoken of in the Bible . . . between David and Jonathan . . . all the way up to the body relationships as we know them . . . all these probably are intense love relationships which the gay lib group, in its political phase, has not yet accepted and integrated as delightful manifestation of human communication, satisfactory to everybody. In other words, there's a lot of political and communal development open to the gay lib movement as it includes more and more varieties of love, besides genital, and it may be that the bridge between gay liberation and men's liberation may be in the mutual recognition of the masculine tenderness that was denied both groups for so long.

AY: In "Kaddish" you say something about the weight of your homosexuality: "Matterhorns of cock, Grand Canyons of asshole." Did you use those big metaphors because homosexuality was a heavy thing for you?

AG: When I was a sensitive, little kid, hiding, not able to touch anyone or speak my feelings out, little did I realize the enormous weight of love and numbers of lovers, the enormity of the scene I'd enter into, in which I finally wound up a public spokesman for homosexuality at one point. In that sense, "Matterhorns of cock, Grand Canyons of asshole." Taking off my clothes in public and getting myself listed in *Who's Who* as being married to Peter.

AY: In a number of poems your homosexuality flows very naturally. Did that really happen?

AG: About 1953 I wrote a big, long, beautiful love poem to Neal Cassady called "The Green Automobile." I made the love overt. I didn't make the genital part overt but I made every other aspect: tenderness, kneeling together, holding on, traveling together, and then ultimate separation.

The next poem that had some overt thing was a little poem in '53–'54, that mentioned the "culture of my generation, cocksucking and tears."

Living in Neal Cassady's house I wrote a little poem, from a line by Whitman, about lying down between the bride and the bridegroom. This was one of Whitman's great lines. In a fantasy I just wrote a description of what I would do, my love fantasy, between Neal and his wife, say, given permission by his tolerance.

The crucial moment of breakthrough in terms of statement came while writing "Howl": "let themselves be fucked in the ass by saintly motorcyclists, and screamed with joy." Usually the macho reaction to that image of being fucked in the ass would be just like in this new James Dickey film, *Deliverance*,[116] where it's supposed to be the worst thing in the world.

AY: You have a line somewhere: "who wants to get fucked up the ass, really?"

AG: That's in the book *Kaddish*, in the poem "Mescaline." On mescaline, who wants to exist in the universe to begin with? Who wants to have a name? Who wants to have an ego? And also who wants to be queer? Who wants the pain of being fucked in the ass at times when it is painful, when it occasionally is. That's part of the scene, too. Sometimes you never know it in advance. Things seem to be all right, and all of a sudden it turns out to be painful. So, who wants to be fucked in the ass that way, really?

---

[116]The James Dickey novel was made into a film directed by John Boorman in 1972 that features a male rape scene.

The outrageous presentation came with "Howl," where I suddenly realized how funny it would be in the middle of a long poem, if I said: "Who let themselves be fucked in the ass . . . and screamed with joy," instead of "and screamed with pain." That's what the contradiction is in that line. An American audience would expect it to say "pain," but instead you have "and screamed with joy"—which is really true, absolutely, one hundred percent.

And again I have a line like: "who blew and were blown by those human seraphim, the sailors, caresses of Atlantic and Caribbean love," referring to Hart Crane, actually. It was an acknowledgment of the basic reality of homosexual joy. That was a breakthrough in the sense of a public statement of feelings and emotions and attitudes that I would not have wanted my father or my family to see, and I even hesitated to make public. So that much was a breakthrough: literarily coming out of the closet.

AY: Did critical reaction to you ever focus on the fact that you were homosexual?

AG: Yes, Norman Podhoretz, in *Partisan Review*, made a big attack on all the beatnik literature, the "know-nothing bohemians." He said that though my poetry was not too bad, its chief force rested on this somewhat questionable insistent proclamation of being queer, homosexual all the time, which, if frank, was not that interesting socially. It was a put-down which acknowledged and at the same time dismissed, while it called Kerouac a "brute."

Walt Whitman is very important on male tenderness. He's never been brought forth as a totem or as a prophet by either gay lib or by the radical left for some very precise statements he made on the subject of men's lib, this is in *Democratic Vistas*, or prospects for democracy, in which he's talking about how possibly materialistic competition in America will turn it into the fabled "damned of nations"—which it now has become. It may be that "we are on the road to a destiny, a status, equivalent, in its real world, to that of the fabled damned." It says, "intense and loving comradeship, the personal and passionate attachment of man to man—which, hard to define, underlies the lessons and ideals of the profound saviours of every land and age, and which seems to promise when thoroughly develop'd, cultivated, and recognized in manners and literature, the most substantial hope and safety of the future of these states—will then be fully expressed."

Then in a footnote, he says:

"It is to the development, identification, and general prevalence of that fervid comradeship, (the adhesive love, at least rivaling the amative love hitherto possessing imaginative literature, if not going beyond it,) that I look for the counterbalance and offset of our materialistic and vulgar American democracy, and for the spiritualization thereof. Many will say it is a dream, and will not follow my inferences: but I confidently expect a time when there will be seen, running like a half-hid warp through all the myriad audible and visible worldly interests of America, threads of manly friendship, fond and loving, pure and sweet, strong and life-long, carried to degrees hitherto unknown—not only giving tone to individual character, and making it unprecedently emotional, muscular, heroic, and refined, but having the deepest relations to general politics. I say democracy infers such loving comradeship, as its most inevitable twin or counterpart, without which it will be incomplete, in vain, and incapable of perpetuating itself."

Then, in the preface to the 1876 edition of *Leaves of Grass*, he adds, in a long footnote:

"Something more may be added—for, while I am about it, I would make a full confession. I also sent out *Leaves of Grass* to arouse and set flowing in men's and women's hearts, young and old, (my present and future readers,) endless streams of living, pulsating love and friendship, directly from them to myself, now and ever. To this terrible irrepressible yearning, (surely more or less down underneath in most human souls,)—this never-satisfied appetite for sympathy, and this boundless offering of sympathy—this universal democratic comradeship—this old, eternal, yet ever-new interchange of adhesiveness, so fitly emblematic of America—I have given in that book, undisguisedly, declaredly, the openest expression. . . . Poetic literature has long been the formal and conventional tender of art and beauty merely, and of a narrow, constipated, special amativeness. I say, the subtlest, sweetest, surest tie between me and Him or Her, who, in the pages of *Calamus* and other pieces realizes me—though we never see each other, or though ages and ages hence—must, in this way, be personal affection. And those—be they few, or be they many—are at any rate my readers, in the sense that belongs not, and can never belong, to better, prouder poems.

"Besides, important as they are in my purpose as emotional expressions for humanity, the special meaning of the *Calamus* cluster of *Leaves of Grass*, (and more or less running through that book, and cropping out in *Drum-Taps*,) mainly resides in its Political significance. In my opinion it is by a fervent, accepted development of Comradeship, the beautiful and sane affection of man for man, latent in all the young fellows, North and South, East and West—it is by this, I say, and by what goes directly and indirectly along with it, that the United States of the future, (I cannot too often repeat,) are to be most effectually welded together, intercalated, anneal'd into a Living Union."

So, that's really the direction, I think, for gay lib, for men's lib, the whole thing, you know, the release of emotions, finally a release of tenderness and that's the thing being suppressed.

AY: Some people in the gay movement who call themselves "effeminists"[117] would say that this sort of romanticization of masculine love is anti-woman, that it's another expression of male supremacy along the lines of Greek love; that the Greek society which tolerated and nurtured homosexuality was at its root a male supremacist society.

AG: I don't know. I don't think that's so in the long run. I think it's too genuine a feeling. With Whitman it didn't seem to interfere with his relations with women, because he had women friends who felt the same as he and who were, I think, married householder lesbians.

Whitman was saying that emotional giving between men, acceptance between men, has not been developed in America. One would say nowadays that it's been repressed by the spirit of competition and rivalry characteristic of capitalist home economics. A concomitant potential of a communal fraternality would be brotherly tenderness at least. That tenderness has been denied to the Southern redneck and is responsible for his disrelation both with men and women. We don't

---

[117]The effeminists were a group of male radicals within the Gay Liberation Front in New York City who, according to Toby Marotta's valuable history of the early gay liberation movement, *The Politics of Homosexuality* (Boston: Houghton Mifflin, 1981), "felt that the widespread devaluation of traits and behavior presumed feminine made them outcasts even among other radical males [who] withdrew into groups of their own and resolved to do battle against sexism" (p. 312). One group of them "issued broadsides and journals in the name of a group called The Flaming Faggots" (p. 122, n.).

yet know what the result would be of men forming closer emotional ties, or of the making conscious of those emotional ties and the acceptancy of them as a political significance. What's the alternative? You can bring up the specter of Greek love and its anti-feminist con- comitant and point out aspects of that in behavior of the beatniks—a fear of women, at least with me. But you would also have to see it as a real, heartfelt, native development, out of the fear and restrictiveness of the situation that we were brought up with: distrust, hatred, para- noia and competition between men rather than cooperation; and the same also between men and women.

Whitman was most sensitive and conscious of that because of his blocked love for men, because he couldn't make it with men openly and publicly. He had to find a way of expressing his adhesiveness, as he calls it.

I think a liberation of emotion between men would also lead to a liberation or straightening out of relations between men and women, because men would no longer have to be men in relation to women in the sense of hard and conquistador. They might have a much more relaxed relationship in which they weren't continuously obliged to be sexualized but could be just friends, or fond. Men's non-sexual friendship with women is now considered unmanly. So the develop- ment of frankly emotional, non-genital friendships with men might mean also the development, the opening up of frankly emotional non-genital friendships with women.

What is the effeminist alternative position between men? In other words, what do they propose besides saying, "No, you shouldn't feel good with your fellow man; heterosexuals should not develop toward emotional relations with heterosexuals?" They're pointing out the danger of an exclusive club, but we've already had that exclusive club in another form with the Hemingway macho scene, or with the mili- tary muscular macho scene. I'm saying and Whitman is saying that the antidote to the Hemingway macho and military macho scene is the development of frank, emotional tenderness and an acknowledgment of tenderness as the basis of genital or non-genital emotion. It may resolve itself in more men friendships, a democratization of friend- ships, so that it's not exclusively friendships between men and women on a sexual basis. I think it would resolve a lot of the macho conflict and contradictions.

I think that's one of the definitions of gaiety, or homosexuality:

that there is a built-in conditioning, from very early times, in which both genital and emotional flow goes toward men more than, as is more usual, toward women. I thought the point of gay lib was to admit that variety of development as being viable, making a place for that. Otherwise, what is a homosexual? Unless you want to have a homosexual liberation front which proposes that men should develop out of homosexuality to a more equal and democratic relation with both men and women. But I think you could say: let the straight flower bespeak its purpose in straightness, which is to seek the light, and let the crooked flower bespeak its purpose in crookedness, which is to seek the light. The crooked flower has to go around the rock to seek the light. But the point was to get to the light of love, and the straight flower just grew up straight, right into the light of love. So you have either biological or conditional man-love and a gay lib movement which purports to release and make public those emotions. One thing that gay lib could do would be to break down the fear barrier that queens have against women. Breaking down the fear barrier between men and men would probably tend toward that.

Another point I'd like to take up is the traditional, effeminist possibly, objection to the "sexist" relations between older men and younger men. I saw some effeminist manifestos [on this point] in Berkeley. I took that question to Gavin Arthur, who died this year in San Francisco. He was a great gentleman, with beautiful manners, an astrologer, a teacher, a guru, and a grandson of President Chester Arthur. Neal Cassady slept with him occasionally, taking refuge in San Francisco from his travels with (Ken) Kesey, back and forth from the railroad; and Gavin Arthur had slept with Edward Carpenter, and Edward Carpenter had slept with Walt Whitman. So this is in a sense in the line of transmission . . . that's an interesting sort of thing to have as part of the mythology.[118] Kerouac's heterosexual hero who also slept with somebody who slept with somebody who slept with Whitman, and received the Whispered Transmission, capital W, capital T, of that love.

AY: Kerouac's heterosexual hero? Who would that be?

---

[118]A document that Gavin Arthur wrote for Allen, describing how Edward Carpenter made love to him in the manner in which Walt Whitman had made love to Edward Carpenter, is contained in the Grey Fox Press book version of this interview as well as in both *Gay Sunshine Interviews*, vol. 1, and *Gay Roots: Twenty Years of Gay Sunshine*.

AG: Neal Cassady, Dean Moriarty, who slept with Gavin Arthur, who slept with Edward Carpenter, who slept with Whitman. And I slept with Dean (Neal Cassady), so . . . So speaking from that line of transmission . . . what was whispered to me in that line of transmission by Gavin Arthur on the subject of older and young people making it: he says that's like an ancient thing, and it's very old and very charming for older and younger to make it—which you realize as you get old, too—and nothing to be ashamed of, defensive about, but something to be encouraged—a healthy relationship, not a sick neurotic dependency.

The main thing is communication. Older people have ken, experience, history, memory, information, data and also power, money and also worldly technology. Younger people have intelligence, enthusiasm, sexuality, energy, vitality, open mind, athletic activity—all the characteristics and sweet, dewy knowledges of youth; and both profit from the reciprocal exchange. It becomes more than a sexual relationship; it becomes an exchange of strengths, an exchange of gifts, and exchange of accomplishments, an exchange of nature-bounties. Older people gain vigor, refreshment, vitality, energy, hopefulness and cheerfulness from the attentions of the young; and the younger people gain gossip, experience, advice, aid, comfort, wisdom, knowledges and teachings from their relation with the old. So as in other relationships, the combination of old and young is functionally useful. It's far from sexist, in the sense that the interest of the younger person is not totally sexual; it's more in the relationship and the wisdom to be gained.

In Edward Carpenter's and Whitman's theory the older person made love to the younger person, blew the younger person, and there was the absorption of the younger person's electric, vital magnetism (according to a charming, theosophical, nineteenth-century theory). And it's something that somebody older like myself does experience as a natural fact. When you sleep with somebody younger you do gain a little vitality of breadth and bounce.

AY: You've referred to Whitman and Edward Carpenter, and in some of your poems you mention García Lorca. For me it was a very recent discovery that these famous writers were gay like myself, that I had this bond with them. I'm curious as to how you made this discovery?

AG: Lorca's "Ode to Walt Whitman" speaks of "the sun singing on the navels of boys playing baseball under the bridges," which is an image of such erotic beauty that immediately you realize that he understood, that he was there; that was an emotion he felt. Then, later on I met somebody in Chile who knew him and said that he'd slept with boys. In fact, some sort of argument about a boy may be the cause of the shooting of Lorca. I don't think there's any written biographical history.

This sex epiphany is all in Whitman's texts, his homoerotic rhapsody, including a description of the time he lay down with a friend— it's in part 5 of *Song of Myself*:

I mind how once we lay such a transparent summer morning,
How you settled your head athwart my hips and gently turn'd over
  upon me,
And parted the shirt from my bosom-bone, and plunged your tongue
  to my bare-stript heart,
And reach'd till you felt my beard, and reach'd till you held my feet.

AY: You don't get it in high school.

AG: But school is irrelevant to poetry and everything else anyway. I mean school is something from the nineteenth century. Poetry has gone back to 15,000 B.C. There's Whitman's "We Two Boys Together Clinging":

We two boys together clinging,
One the other never leaving,
Up and down the road going, North and South excursions
  making,
Power enjoying, elbows stretching, fingers clutching,
Arm'd and fearless, eating, drinking, sleeping, loving,
No law less than ourselves owning, sailing, soldiering, thieving,
  threatening,
Misers, menials, priests alarming, air breathing, water drinking,
  on the turf or the sea-beach dancing
Cities wrenching, ease scorning, statutes mocking, feebleness
  chasing,
Fulfilling our foray.

In "No Labor-Saving Machine" he writes:

> But a few carols vibrating through the air I leave,
> For comrades and lovers.

And Whitman says [in "A Glimpse"]:

> A glimpse through an interstice caught,
> Of a crowd of workmen and drivers in a bar-room around the
>   stove late of a winter night, and I unremark'd seated in
>   a corner,
> Of a youth who loves me and whom I love, silently approaching
>   and seating himself near, that he may hold me by the
>   hand,
> A long while amid the noises of coming and going, of drinking
>   and oath and smutty jest,
> There we two, content, happy in being together, speaking little,
>   perhaps not a word.

Perfect! And real, absolutely real! That is actually life. That's even heterosexual life. That's the undescribable reality of human relationship in America. It can't be called gay . . . it's . . . that's something in the line of what I was talking about before in terms of what needs to be . . . the adhesiveness that Whitman spoke of that is latent in all of us now, and that's ready to be opened, and God knows how many, in the last ten years, how many younger boys that I've run across that I just sat and held hands with anyway and felt love feelings toward them and felt love feelings toward me, in situations like this which were maybe in a college, or somewhere, which had nothing to do with quote queer unquote, or even gay. Gay is too much of a category!

AY: From what you said before, this existed to a certain extent with the bohemians or the hipsters . . .

AG: Oh, it existed back to Cro-Magnon man!

AY: I think definitely a tension exists today between gay freaks and straight gays. There are some people in gay liberation who say, "I have more in common with a heterosexual freak than with a gay person who's into very short hair and alcohol." And then there are other gay people who say, "My loyalty is to other gay people, and the freak culture is very macho."

AG: The form I felt it in was between the heartfelt, populist, humanist, quasi-heterosexual, Whitmanic, bohemian, free-love, homosexual tradition, as you find it in Sherwood Anderson, Whitman, or maybe Genet, a little, versus the privileged, exaggeratedly effeminate, gossipy, moneyed, money-style-clothing-conscious, near-hysterical queen. Of course, there's nothing more ancient or/and in a sense honorable than the old shamanistic transvestite that we see running up and down Greenwich Avenue or, among the American Indians, a shaman who dresses himself up like a woman and even takes a husband. The screaming young queen—there's something very ancient and charming about that; great company, total individuality and expressiveness. Sometimes you fear it's the screaming, hysterical outside of somebody who's going to have a nervous breakdown and wind up in the church, or something. But then there's also the pettish, spiteful, anal retentive, disciplinarian.

But when I was younger the split was more between the grubby, beatnik, open-hearted . . . I couldn't call myself a fairy exactly . . . queer? . . . I have used that, but I've never found the right word . . . the nameless lovers, the nameless gnostic lovers . . . and the monopolistic queens, say, who had privilege and money. So that was the distinction. It was more between the cold-hearted and the warm-hearted.

AY: In the gay bars of New York did you find both?

AG: Oh, I found both definitely. There were lots of outspoken, funny old sailor queens from the twenties; and then there were all sorts of prissy-mouthed, paranoiac, fearful, conservative-reactionary, short-hair, worried advertising martinets. And everything in between. There is a manneristic fairydom that depends on money, chic, privilege and exclusive, monopolistic high style, and I would say that it is usually accompanied by bitchiness and bad manners and faithless love, too. I like homosexuality where the lovers are friends all their lives, and there are many lovers and many friends.

AY: Could you say something about your relationship with Peter Orlovsky?

AG: We met in San Francisco. He was living with a painter named Robert LaVigne in '54. I was having a very straight life, just trying it out, working in an advertising company, wearing suits, living up on Nob Hill in a nice big apartment with Sheila, who was a jazz singer and worked in advertising. Things were somewhat unsatisfactory between us. We'd been taking a little peyote, so we were into a psychedelic scene, too.

We got into an argument, so I wandered down one night into an area of San Francisco I'd never noticed then called Polk Gulch, now known as a notorious gay area with lots of gay bars. It was then more of a bohemian section, somewhat gay, artistic. Hotel Wentley was there, right on the corner of Sutter and Polk, and a Foster's cafeteria. I went and sat in the Foster's, late at night. I ran into Robert LaVigne and got into a big, interesting, artistic conversation about the New York painters I knew—Larry Rivers, (Willem) de Kooning, and (Franz) Kline. LaVigne was a provincial San Francisco painter, so I was bringing all sorts of fresh poetry, art news from New York.

He took me up to see his place and his paintings, about four blocks away on Gough Street in an apartment that I subsequently lived in for many seasons and still use now. I walked into the apartment and there was this enormous, beautiful, lyrical, seven-by-seven-foot-square painting of a naked boy with his legs spread, and some onions at his feet, with a little Greek embroider on the couch. He had a nice, clean-looking pecker, yellow hair, a youthful teeny little face, and a beautiful frank expression looking right out of the canvas at me. And I felt a heart throb immediately. So I asked who that was, and Robert said, "Oh, that's Peter; he's here, he's home." And then Peter walked in the room with the same look on his face, a little shyer.

Within a week Robert said that he was going out of town or breaking up with Peter, or Peter was breaking up with him. He asked me if I was interested in Peter, and he'd see what he could arrange. I said, "Ooh, don't mock me." I'd already given up. I already had had a historic love affair with Neal Cassady a decade earlier. So I was already a tired old dog, in the sense of the defeats of love, not having made it, not having found a permanent life companion. And, in 1955, I was already twenty-nine. I wasn't a twenty-year-old kid with romantic

notions. That night we were in Vesuvio's bar. Robert had a big conver-
sation with Peter, asking Peter if he was interested, sort of like a *shad-chan*, a matrimonial arranger.

Then I went home one night. I went to Peter's room. We were to
sleep together that night on a huge mattress he had on the floor. I took
off my clothes and got into bed. I hadn't slept with too many people.
Never openly, completely giving and taking. With Jack or Neal, with
people who were primarily heterosexual and who didn't fully accept
the sexualization of our tenderness, I felt I was forcing it on them; so I
was always timid about them making love back to me, and they very
rarely did very much. When they did, it was like blessings from
heaven. If you get into it, there's a funny kind of pleasure/pain,
absolute loss/hope. When you blow someone like that and they come,
it's great! And if they touch you once, it's enough to melt the entire
life structure, as well as the heart, the genitals and the earth. And it'll
make you cry.

So . . . Peter turned around (he was in his big Japanese robe),
opened up the bathrobe—he was naked—and put it around me and
pulled me into him; and we got close, belly to belly, face to face. That
was so frank, so free and so open that I think it was one of the first
times that I felt open with a boy. Then, emboldened, I screwed Peter.
He wept afterwards, and I got frightened, not knowing what I'd done
to make him cry, but completely moved by the fact that he was so
involved as to weep. At the same time the domineering, sadism part of
me was flattered and erotically aroused.

The reason he wept was that he realized how much he was giving
me and how much I was demanding, asking and taking. I think he
wept looking at himself in that position not knowing how he'd gotten
there; not feeling it was wrong, but wondering at the strangeness of it:
The most raw meat of reasons for weeping.

Then Robert hearing, seeing the situation, came in to comfort
Peter a little bit. I was very possessive and I pushed Robert away. That
got me and Robert into a funny kind of distrust that lasted for a year or
two before our karmas finally resolved. He then realized he was well
off on his own; and I was burdened with the karma of love.

Peter was primarily heterosexual and always was. I guess that was
another reason he was shocked—the heaviness of my sadistic posses-
siveness in screwing him. For the first time in my life I really had an
opportunity to screw somebody else! I think that wounded him and

thrilled me a little bit. So we still had to work out all that in our rela-
tionship over many, many years. It's painful sometimes.

We slept together perhaps one more time. Then I had to go to New
York for my brother's wedding at Christmas, '54. I came back and
moved into that apartment where they were living, at their invitation.
And then there was a triangle of Robert, me and Peter. Peter had not
made up his mind whether or not he wanted to make a more perma-
nent relationship with me. I had my eyes on Peter for life-long love; [I
was] completely enamored and intoxicated—just the right person for
me, I thought. Robert was not sure he hadn't made a mistake, seeing
the flow and the vitality that was rising up in both me and Peter. And
Peter began withdrawing. He was caught in this rivalry between me
and Robert, and, at the same time, there was his unsurety of me and
his relation to me. Basically he liked girls anyway, so what was he
doing lying there being screwed by me?

So I moved across from the Hotel Wentley and got a room. I was
working in a market research job. I had the brilliant inspiration that
all the categorizing and market research I was doing could be fed into
a machine, and I wouldn't have to add all those columns anymore. So
I supervised the transfer for the company, and that left me out of a
job just nicely, like a seamless occlusion. Then I got unemployment
compensation.

I was being psychoanalyzed at Langley Porter Clinic, an elite
extension of U.C. Berkeley Medical School. He was a very good doctor,
and I said: "You know, I'm very hesitant to get into a deep thing with
Peter, because where can it ever lead. Maybe I'll grow old and then
Peter probably won't love me—just a transient relationship. Besides,
shouldn't I be heterosexual?" He said, "Why don't you do what you
want. What would you like to do?" And I answered, "Well, I really
would just love to get an apartment on Montgomery Street, stop work-
ing and live with Peter and write poems!" He said, "Why don't you do
that?" So I said, "What happens if I get old or something?" And he
replied, "Oh, you're a nice person; there's always people who will like
you"—which really amazed me. So, in a sense he gave me permission
to be free, not to worry about consequences.

So then I waited for Peter, and Peter stayed up at the Gough Street
apartment and went to school. I got this room and started writing a lot
and waited and waited for Peter. Neal Cassady came by a couple of

times. I made it with Neal. I can remember one of the last really wild times I made it with him, because I had a room of my own and there was privacy, finally. He was lying there naked, and I was sitting on his cock, jumping up and down trying to make him come.

And I just waited and waited [for Peter]. There was nothing I could run after or pursue, because I couldn't claim anything by force. Things got too difficult where Peter was living, so he got a room himself in the Wentley, across the street from where I was. And there was embarrassment, coldness—not knowing where each other was, what we would do. I was waiting for him to make some sort of decision. A couple of times we drank a little to see if we could get over the low. We didn't sleep together at all, though I was longing to.

Then one day he was lying in bed, and he started crying again. He said, "Come on and take me." I was too overwhelmed and frightened to even get a hard-on. I didn't know what to do. We both had our clothes on. I was afraid he was interpreting it as me screwing him again, rather than really just having each other. But that soon got resolved, and we moved in together, into an apartment in North Beach. We found an apartment, and it had a room for him, a room for me, and a hall between us; and a kitchen together. So that gave us both a little privacy, and, at the same time, we could make it when we wanted.

He was very moody, very sweet, tender, gentle and open. But every month or two months he'd go into a very dark, Russian, Dostoevskian black mood and lock himself in his room and weep for days; and then he'd come out totally cheerful and friendly. I found after a while it was best not to interrupt him, not to hang round like a vulture; let him go through his own yoga.

The key thing was when we decided on the terms of our marriage— I think it was in Foster's cafeteria downtown about three in the morning. We were sitting and talking about each other, with each other, trying to figure out what we were going to do, who we were to each other, and what we wanted out of each other, how much I loved him, and how much did he love me. We arrived at what we both really desired.

I'd already had visionary experience: an illumined audition of Blake's voice and a sense of epiphany about the universe. He had had an experience, weeping and lonesome, walking up the hill to his college, and having a sense of an apparition of the trees bowing to him. So we both had some kind of psychedelic, transcendental, mystical image in our brains and hearts.

We made a vow to each other that he could own me, my mind and everything I knew, and my body, and I could own him and all he knew and all his body; and that we would give each other ourselves, so that we possessed each other as property, to do everything we wanted to, sexually or intellectually, and in a sense explore each other until we reached the mystical "X" together, emerging two merged souls. We had the understanding that when our (my particularly) erotic desire was ultimately satisfied by being satiated (rather than denied), there would be a lessening of desire, grasp, holding on, craving and attachment; and that ultimately we would both be delivered free in heaven together. And so the vow was that neither of us would go into heaven unless we could get the other one in—like a mutual Bodhisattva's vow.

Well, this is like a limited version of that, almost intuitive, the vow to stay with each other to whatever eternal consciousness: him with his trees bowing, me with Blake eternity vision. I was more intellectual, so I was offering my mind, my intellect; he was more athletic and physical and was offering his body. So we held hands, took a vow: I do, I do, you promise? yes, I do. At that instant we looked in each other's eyes and there was a kind of celestial fire that crept over us and blazed up and illuminated the entire cafeteria and made it an eternal place.

I found somebody who'd accept my devotion, and he found somebody who'd accept his devotion and who was devoted to him. It was really a fulfillment of fantasy, to a point where fantasy and reality finally merged. Desire illuminated the room, because it was a fulfillment of all my fantasies since I was nine, when I began to have erotic love fantasies. And that vow has stuck as the primary core of our relationship. That's the mutual consciousness; it's the celestial social contract, valid because it was an expression of the desire of that time, and it was workable. It's really the basic human relationship—you give yourself to each other, help each other and don't go to heaven without each other.

There's this mythology of Arjuna, from the *Bhagavad Gita*, getting to the door of heaven. He's got this little dog following him, and they say, you can come in but you can't bring your dog. And he says, well, no, if I can't go in with my dog, I won't go. And then they say, Oh, come on, you can go in, just leave him behind, it's only a dog. And he says, no, I love my dog, and I trust that love, and if I can't bring that trust in, then what kind of heaven is this? And the third time, he says, no, no, no, I'll stay out and put the dog in heaven but I won't go in without the dog. I vowed to tears with my dog, I can't leave my dog alone. And so, finally, after the third time, the dog turns out to be Krishna, the supreme lord of the universe and heaven itself. He was only trying to get heaven into heaven. And his instinct was right. And our instinct was right. It was enough to bring us through very difficult times—all through the change of status, beat generation and fame, the alteration of social identity that fame entails.

Our relationship has lasted from 1954 on. The terms have changed tremendously. Peter's gone through a lot of changes, and we've separated for a year at a time. And always come back. We've gone through a lot of phases of sleeping with people together, doing orgies together, sleeping alone together. Now Peter sleeps with a girl. I very rarely sleep with him. But the origin of our relationship is a fond affection. I wouldn't want to go to heaven and leave Peter alone on earth; and he wouldn't leave me alone if I was sick in bed, dying, gray-haired, wormy, rheumatic. He'd have pity on me. We've maintained our relationship so long that at this point we could separate and it would be all right. I think the karma has resolved and worn out in a sense.

The original premise was to have each other and possess each other until the karma was worn out, until the desire, the neurotic attachment, was satisfied by satiation. And there's been satiation, disappointment and madness, because he went through a long period of speed freakery in the mid-sixties which really strained things. We had times of hostile screaming at each other such as happens in the worst of homo- and heterosexual marriages, where people have murder in their hearts toward each other. That burned out a lot of the false emotion of youth, and the unrealistic graspings, cravings, attachments and dependencies. So he's now independent, and I'm independent of him. And yet there's an independent curiosity between us.

AY: There were some vague stories going around about your visit to Cuba in 1965 and departure. I'd like to know more about what you did in Cuba and what you said that eventually got you deported.

AG: Well, the worst thing I said was that I'd heard, by rumor, that Raul Castro was gay. And the second worst thing I said was that Che Guevara was cute. The most substantial thing was that I went around wondering why their marijuana policy, as of 1965, was so down and unscientific. I didn't accept the answer I got which was that the Batista soldiers used to get high and shoot at them, because I didn't think that was true. By hindsight, it doesn't seem really relevant to their needs, but at the same time, the denial of marijuana doesn't seem relevant to their needs, either.

There was persecution of homosexuals in the primarily gay-oriented theater group at the time. Instead of finding a place for that, they tried to break it up and sent everybody out to the sugar-cane fields to work. This was an attempt to humiliate them, to use sugar cane for humiliation rather than community. And it wasn't in the newspapers. It was a secret campaign, with all the Young Communist League party-hack, flag-waving kids, like the Nixonettes, so to speak, accusing everybody they didn't like of being faggots.

It was considered bad form to wear beards and long hair, even though that was the characteristic style of Castro and the liberators up on the main drag, La Rampa. People were being stopped by the police and busted for having long hair, accused of being existentialists and degenerates. A bunch of young kids belonging to a poetry group I knew, El Puente (The Bridge), were being bugged by the police, not allowed to publish, and were called fairies. One evening the whole group of Escritores del Encuentro Inter-Americana, sponsored by Casa de las Américas, went to the theater to hear a concert of "feeling" music. We were joined there by a whole bunch of young poet kids. When we left the theater, they were all stopped by the police, arrested and told to stop hanging around with foreigners. Some of the young poet kids were translating my work.

So there was this police bureaucracy in Cuba that was very heavy and was coming down heavy on culture, in terms of beards, sexual-revolution tendencies, sociability, and homosexuality. In other words, there was no real cultural revolution; it was still basically a Catholic mentality. As in many Communist countries, the police bureaucrat party hacks were like Mayor Daley ward-healers: flag-waving, fat-assed square types. Self-seeking squares, not at all spiritually communist, were getting control of the police and emigration bureaucracies and setting themselves at odds against the people who screw with their eyes open, listen to the Beatles and read interesting books like Genet, and *fought* at the Bay of Pigs against the Americans. Even people who had been up in the mountains with Castro were very secretive about smoking grass. The press was monolithically controlled and boring, and the newspaper reporters for the press reminded me very much of the self-righteous newspaper reporters from the *Daily News* as far as their opinionation and argumentativeness.

I just continued talking there as I would talk here in terms of being antiauthoritarian. But my basic feeling there was sympathetic to the revolution. I had friends living there, was invited there as a guest, and I took part as a judge in a literary contest, and I just shot my mouth off! The worst thing was the talk about homosexuality and the challenge to the official position about it. Castro had taken an official position in a speech at the university in which he had attacked homosexuality. He called it degenerate or abnormal, saw it as a cabal, perhaps, a conspiracy. I think he praised the Young Communist League for turning in fairies.

I suggested to Haydee Santamaria that they invite the Beatles and got the answer: "They have no ideology; we are trying to build a revolution with an ideology." Well, that's true, but what was the ideology they were proposing? A police bureaucracy that persecutes fairies? I mean, they're wasting enormous energy on that. Some of those "fairies" were the best revolutionaries—people that fought at the Bay of Pigs, Playa Girón.

I slept with one young poet, secretively. I took one stick of grass one day, walking along a shady street with a bearded fellow who said he'd been up in the mountains with Castro and that they had smoked up there. But that was the extent of my "criminal behavior."

I was in my hotel room one morning toward the end of my stay in Cuba when three uniformed, olive-clothed, mute soldiers came in with an officer. He said he was head of immigration, that I had to pack my bags, and that I was being deported on the next plane out, to Prague. I asked if they had informed the Casa de las Américas, and the answer was, no; there will be time enough later. They wouldn't let me make a phone call to the Casa, which was my host, and they took me downstairs. I shouted in the lobby to Nicanor Parra that I was being deported and they should get in touch with the Casa de las Américas and warn them. I was driven out to the airport. On the way I asked why I was being deported. The officer said, "For breaking the laws of

Cuba." And I said, "What laws?" He responded, "You'll have to ask yourself that." And that answer, I thought, was like the answer I got from Dean McKnight at Columbia University when I got kicked out for staying overnight in my room with Jack Kerouac. And we hadn't made love at all. We just slept there because Kerouac had no place to sleep that night.

I didn't go round screaming to *Time* magazine that I'd been unjustly kicked out of Cuba. I just gave them the benefit of the doubt, understanding that I was like a pawn. It was a fight between the liberal groups and the military bureaucracy groups. I realized also that the more the United States put pressure on Cuba, the more power the right-wing military, police bureaucracy and party hacks would get. The real problem was to relieve the pressure in America, to end the blockade rather than to "blame" the Cuban Revolution, Castro, or Marxism—although I don't think Castro was very tactful on the question of homosexuality. There was an excessively macho thoughtlessness on his part, and insensitivity.

I thought one of the most brilliant and interesting results of gay liberation was the confrontation with the repressive, conservative police bureaucracy in Cuba. I think the confrontation between the Venceremos Brigade[119] and gay lib showing the Cuban mental block on the subject of homosexuality was one of the most useful things that gay lib did on an international scale. At least it brought the question to front-brain consciousness. Gay lib people went there to offer themselves and, I think, less to confront the Cubans than to find out what the scene was. They were, obviously, faithful in terms of change and sympathy with the revolution. Since it was a gay lib group, the right-wing, capitalist press couldn't take advantage of the confrontation to put shame on Cuba, because otherwise they'd have to defend gay lib! So, it was gay lib taking the bull by the horns, within the context of brotherhood, challenging the Cuban macho, repressive mentality in a constructive way. I don't think the Communist Party there reacted very well. What was the result?

---

[119]The Venceremos Brigades were groups of young Americans sympathetic to the Cuban revolution who went by ship to help the Cubans harvest the sugar crop in 1969 and 1970. Allen Young had helped set up the initial Venceremos Brigades. See Toby Marotta, *The Politics of Homosexuality* (Boston, Houghton Mifflin, 1981), and Young's own *Gays Under the Cuban Revolution* (Grey Fox, 1981).

AY: In the interim period the brigade has adopted a policy of excluding gay liberation people. There was a fifth brigade that did not have gay liberation people on it. The Cubans have since come up with a detailed, rather specific policy statement on homosexuality, declaring it to be a "social pathology." The pro-Cuban Venceremos Brigade people have related with hostility to the radical gay lib movement. Large numbers of New Left people who formerly were very sympathetic to Cuba have reduced their expression of sympathy for Cuba because of the gay question. The Cubans, basically, have forced a lot of people to choose between the Cuban revolution and gay liberation, and they're quite surprised to find people choosing the gays.

AG: When Castro originally had his revolution, he said it's a Marxist revolution but it's still a humanist revolution. If it's a humanist revolution, they cannot put down gays. Otherwise, it's doubletalk. I think it's important to support any separation out from American imperialism and conspicuous consumption, and any sort of independence from American psychological domination. But, on the other hand, the reason for doing so is to become human again and independent.

If the definition of human and independent means sustaining an old, authoritarian viewpoint toward sexuality—the monotheistic, Catholic viewpoint—then it would be better that American radicals at least realized that they're dealing with human beings in the Cuban situation rather than with divine authorities. I am willing to accept the fact that the Cuban revolution is a genuine relief from Mafia capitalist domination, the previously corrupt society of Cuba, and a release from America.

In other words, I feel the Cuban revolution is important and should be supported. They'll learn, soon enough. They're gonna see the end of the world anyway and end up with long hair and pansexualities. They're going to have to take it as state policy before they're over, just to relieve their population problem. I think the gays are dealing in the long run from a position of great strength, because their position is founded in ancient rules of mammal behavior and ecological necessity as far as the future and the recognition of common humanity. So I think gays can afford to say, "Ahhh."

AY: When I was there in '71 at the journalists' conference, there was a reception, at the side of a big swimming pool. Everybody was crowding around Fidel. He was loving it and getting involved in lively conversation with different people. I was feeling very out of it. I was the only male that didn't have short hair, a suit and a tie, except for some Africans in African dress. The whole idea of pushing into a crowd to talk to a famous man was something I wasn't exactly into.

I decided to get involved in conversation with some other people. I spoke with a very important comandante, a black guy, who had fought with Fidel in the hills and was on the central committee. Karen Wald, an American who was with us, asked him what he thought about machismo. And he said, "Oh, man, that's good!" I can't figure out to this day whether he was putting her on or whether he was simply expressing his very gut reaction—which is that machismo is an important thing for a Cuban man to appreciate.

AG: The question does finally boil down to machismo, both here in the United States, and there, in terms of revolutionary tactics. Gay lib, in a sense, has a good approach to straight people with smug, middle-class ideas about power coming out of the barrel of anything, actually.

AY: I think there's been a certain schizophrenia in the radical section of gay liberation. People have said they're against power. In fact, most of the people I know in the radical wing of gay liberation don't even like and don't use the slogan "gay power" because of the word "power."

AG: Gregory Corso has a great poem called "Power"[120] which I invite you to check out. It was written in 1959: "Standing on a street corner waiting for no one is Power" . . . "A thirst for Power is drinking sand."

AY: On the one hand people were attacking the whole notion of power, trying to do away with power in personal relationships. On the other hand, there was this desire to be a part of the left, a desire best epitomized by the slogan, "Go left, go gay, go pick up the gun"—a variation of the Panther slogan.

AG: Though it may serve as a vehicle for machismo among gays, it also serves as a deflating slogan for the pompousness of black- or white-power slogans that are actually a bit ridiculous sometimes.

The slogan "Power comes out of the barrel of a gun"[121] was irrele-

---

[120]See *Happy Birthday of Death*, New Directions, 1960.
[121]The full quote is "Every Communist must grasp the truth; 'Political power grows out of the barrel of a gun.'" *Selected Works of Mao Tse-Tung* (Oxford, New York: Pergamon Press, 1961).

vant in the American situation all along. There wasn't enough imagi-
nation in terms of tactics and poetry. How do you transform and con-
vert America? It was a sign of the poverty of imagination that finally
people fell into violence, when all along the whole problem had been
mental violence, blindness and rage. Gay lib really did turn all of the
machismo of the left inside out.

AY: Do you feel that gay liberation has influenced you personally in
any way?

AG: I use the word "gay" now which I never did before. And that's
important when you change somebody's language. I find myself drift-
ing toward the gay lib group whenever there's a big parade, because it's
generally so sincere and interesting. The ideology there at least is per-
sonal. Gay lib has influenced my thinking on a few other things—like
junkie liberation.

If you can have gay liberation from the oppression of the macho
oppressors, then you can have junkie liberation from the oppression
of the macho Mafia CIA, fuzz AMA, the Truman-Nixon oppression of
punitive treatment of junkie illness rather than medical treatment.
There should be a Junkie Liberation Front. They're the most
oppressed group in America, in the sense that they're hunted down
like dogs by people with guns. They're always under the threat of jail.
They're sick. They've got a legitimate illness, and they're not being
treated with legitimate medical means. But they are thrown over into
the hands of the most corrupt police agents in America—narcs who
have relations with the Mafia and peddle—as proved by the Knapp
Commission[122] and various other documented analyses. They've suf-
fered the greatest image distortion of any group in America. There
was never a category of human being in America that was invented as
low as the fiend category for heroin addicts. They didn't even say
liquor fiends.

And they are the victim of slander. They're called a criminal class,
violent murderers when they're not; when it's the alcoholics who are
really out of control.

AY: Half of the Pentagon generals are alcoholics, too.

---

[122]See *The Knapp Commission Report on Police Corruption* (New York: G. Braziller,
1973).

AG: So, I mean, there should be Junkie Liberation. The idea of the gay liberation front turned me on to the terminology Junkie Liberation Front. It added a little fillip to the relationship that Peter and I have.

AY: One of the quotations that floats around gay liberation ascribed to you is your reaction to the Stonewall riot: "The fags have lost that wounded look." What were the circumstances?[123]

AG: It was an interview in the *Village Voice*. I wasn't there at the riot. I heard about it, and I went down the next night to the Stonewall to show the colors. A crowd was there, and the place was open. So I said [to myself], the best thing I can do is to go in; the worst that can happen is I'll calm the scene. They're not going to attack them when I'm there. I'll just start a big Om.

  I didn't relate to the violent part [of Stonewall]. The trashing part I thought was bitchy, unnecessary, hysterical. But, on the other hand, there was this image that everybody wanted to make that they could beat up the police, which apparently they managed to do. It was so funny as an image that it was hard to disapprove of, even though it involved a little violence.

AY: Did you at that moment anticipate that this might lead to something called gay liberation with organizations, publications and so forth?

AG: It seemed to have been there all along, somehow. There was already that in rudimentary form with the Mattachine Society and One.[124] They were more sedate, but they did some interesting things in their time.

---

[123]The Stonewall riots, a six-day series of demonstrations and street battles with the police that began when the police raided the Stonewall Inn, a gay bar in Greenwich Village, took place over the end of June and the beginning of July 1969. Out of the uprising came the Gay Liberation Front (GLF) and later the Gay Activists Alliance (GAA), ushering in what is known as the "gay liberation" phase of the modern gay and lesbian rights movement. It is in honor of the Stonewall rebellion that gay pride marches and other celebrations are held worldwide at the end of June. Allen made his famous comment about the riots to *Village Voice* reporter Lucian Truscott after Allen had been dancing inside the Stonewall Inn on the night of Sunday, June 29: "You know, the guys there were so beautiful—they've lost that wounded look that fags all had 10 years ago."
[124]Two early homophile organizations. In 1951 Harry Hay, Chuck Rowland, and a handful of other men started the Mattachine Society in Los Angeles, one of the first

I published the poem to Neal, "The Green Automobile," in 1959 in the *Mattachine Review*. Because it was a frank love poem, approving of the gay love relationship, it brought forth a rebuke from psychiatrist Karl Menninger of Topeka, Kansas. He wrote a strange letter to the Mattachine Society, denouncing the poem and saying they were trying to cure everybody, and here was this terrible poem boasting of these perverted feelings!

I went to a few of Mattachine's meetings and gave a little poetry reading there in San Francisco. But I never got involved politically with them, just literarily. Of course, San Francisco was always more advanced than New York in terms of the acceptance of homosexuality. It's like a Parisian city. There was a historic, famous bar [the Black Cat] in San Francisco's North Beach, near what was called the Monkey Block, which was maybe the greatest gay bar in America. It was really totally open, bohemian, San Francisco, Viennese; and everybody went there, heterosexual and homosexual. It was lit up, there was a honky-tonk piano; it was enormous. All the gay screaming queens would come, the heterosexual gray-flannel-suit types, longshoremen. All the poets went there.

AY: Martha Shelley has a great first line in "Gay Is Good," one of the first gay liberation articles: "Look out straights, here comes the Gay Liberation Front, springing up like warts all over the bland face of Amerika, causing shudders of indigestion in the delicately balanced bowels of the Movement." At the end of the article she says, "You will never be rid of us because we reproduce ourselves out of your bodies."

AG: There's too much of a conflict there. The point is that nobody's straight. It's like calling someone a pig. Everybody has dreams that have some homoerotic content. So the problem is to make it safe for "straights" to feel the whole spectrum of feelings instead of single-level feelings, just as it's important for gay people to feel a whole spectrum of feelings. The politics of challenge in that sense doesn't seem to make too much sense. You don't woo somebody by challenging them. You woo them by giving them a place where they're comfortable, making it safe for them to get a hard-on . . .

gay organizations in the United States and a forerunner of the current lesbian and gay rights movement. ONE Institute, a pioneer in what would today be called lesbian and gay studies, also was founded in Los Angeles, in 1956, as an educational institution.

AY: I think there are definitely tensions in the movement now between the people who say having lots of blow jobs is liberation, and those who say that we are trapped in a meat-meets-meat approach and have to get out of that and relate to each other as people.

AG: It's an important human experience to relate to yourself and others as a hunk of meat sometimes. That's one way of losing ego, one holy divine yoga of losing ego: getting involved in an orgy and being reduced to an anonymous piece of meat, coming, and recognizing your own orgiastic anonymity. It's not a place where you want to live all your life, but it's certainly a place where you want to see and experience as a lesson, experience of consciousness that's valid for a certain level, and experience of great, divine beast consciousness. That's what they used to have the Dionysian orgy for; it's an ancient ritual; I don't think there's anything wrong with relating to people on the level of pure meat, as long as you don't get trapped into that all the time as a single level of consciousness—as some queens do.

The gay lib answer is obviously not going to be just simply lyrical sexuality. The use of sex as a banner to *épater le bourgeois*, to shock, show resentment or to challenge, is not sufficiently interesting to maintain for more than ten minutes; it's not enough to sustain a program that will carry love through the deathbed or help out Indochina. Or even get laid, finally. You have to have something more. You have to relate to people and their problems, too.

I dig baths and orgies. I think orgies should be institutionalized: impersonal meat orgies, with no question of personality or character or relating to people as people.

Anyone who insults Dionysus had better watch out! The leopards come and get them, or else they get turned into vine leaves, in Ezra Pound [Canto 2], when they practice god-slight, the insult to Dionysus.

AY: The problem with that approach is that as long as your meat is young and attractive, you're doing o.k. But if it doesn't meet the standard . . .

AG: When you get to be my age, that's when you really appreciate orgies, in the dark when nobody sees anybody and doesn't give a shit who they're being screwed by. The paranoia in Turkish baths, are you acceptable or not, is another problem. But orgy is one way for people to equalize—for fat people, thin people, handsome people and ugly people, hunchbacks and one-legged people and rachitics all together in the dark.

Peter and I used to get into scenes in San Francisco with girls and boys together, very nice. He liked girls, and that situation would set up a nice vibration when other men would come in. Since Peter and I were already close and making it, that opened the door to anybody. He'd make out with girls and I'd make out with boys. Sometimes I'd make out with girls too. Or we'd make out with each other. We had a two-year period in San Francisco where almost every party we went to we took off our clothes and wound up in bed with one or two people. We didn't try to start orgies; we just took off our clothes, wandered around the party, had a good time and didn't make a big scene out of it.

AY: I remember hearing that around the time of the original excitement about LSD, Timothy Leary made some statement saying that he cured homosexuality. I recall you said you had a heterosexual experience under the influence of LSD.

AG: [I had] an emotionally heterosexual fantasy experience in relation to my mother and girls. But everybody has that on LSD. It was a breakthrough of heterosexual feeling/emotion in relation to my mother, and there were so many girls that I'd rejected. When Leary was looking around for information and rationalization on LSD, I told him that it probably would loosen up some of the blocks in homosexuality. The reverse is true, too: it would probably loosen up some of the blocks in heterosexuality, which it's notorious for doing. Leary or someone else carried the ball a little too far on that one, to say that I experienced heterosexual breakthrough for the first time in my life. I've got a venerable heterosexual or bisexual history to begin with.

In the context of the arguments about LSD I gave congressional testimony:

> "One effect I experienced in Peru I would like to explain. From childhood on I had been mainly shut off from relationships with women, possibly due to the fact that my own mother was, from my early childhood, in a state of great suffering, frightening to me, and [she] finally died in a mental hospital. In a trance state I experienced in a *curandero's*[125] hut [in Peru] a very poignant memory of my mother's self, and how much I had lost in my distance from her and my distance from other later

---

[125]Spanish for healer, medicine man.

friendly girls. For I had denied most of my feelings for them out of old fear. And this tearful knowledge that had come up while my mind was opened through the native vine (some yage) did make some change toward a greater trust and closeness with all women thereafter. The human universe became more complete for me, and my own feelings more complete. . . ."[126]

AY: In your development as a yoga person, have you come across the somewhat anti-homosexual writings of certain yoga masters?

AG: No, I've never seen any of those. A couple of months ago I got into a conversation with a teacher I'm working with now, Chögyam Trungpa Tulku, incarnate lama, and asked him what he thought about homosexuality. He said he thought it was interesting. I asked him if he thought it was negative or bad. He said, "No, it doesn't make any difference what forms the bodies are; the important thing is the communication." This is very sensible, clear and really important. With communication overt, homosexual lovemaking is obviously terrific and charming. Without communication it's a drag, and heterosexuals the same.

AY: I've always felt there was something particularly mystical about two men doing sixty-nine, something in the configuration of the bodies.

AG: Yes. There's a mysticism when you screw somebody in the ass, or in being screwed. There's a mysticism in being screwed and accepting the new lord divine coming into your bowels—"Please Master." There's a great mysticism in a girl being screwed, or climbing on a guy. Any position is mystical.

The official answer of my Tibetan teacher was interesting: the quality of the emotion is important; communication is important. And the form is not, obviously.

There's a tradition against marriage in the Gelugpa, a yellow-hat sect of Tibetan Buddhism. It slowly evolved historically that the monks all make it with each other. But basically the bias, if any, in yoga is toward chastity, retention of sperm. Sperm is art, poetry, music, yoga. Sperm is *Kundalini*, serpent power: a shivering tingling that runs and takes over the top of your head and spreads throughout

---

[126]Hearings of the Special Subcommittee on the Judiciary, U.S. Senate, Senate Resolution 199, "LSD and Marijuana Use on College Campuses," June 14, 1966. (See *Deliberate Prose*, pp. 67–82.)

the whole body. Retention of sperm is apparently one of the basic understandings of some forms of yoga.

So it's not really homosexuality or heterosexuality that would be disapproved. It would be attachment to any kind of "pleasure," as a neurotic attachment. As Burroughs might say, an attachment to the green goo factory, an attachment to body. The body itself may be the by-product of a large-scale conspiracy by certain forces as Burroughs says, trying to keep people prisoners in a prison universe made out of parent matter, subjected to appearances and apparent physical conditions defining their limitations. As Blake, the Buddhists, and Burroughs would say, the real world is a world of complete, blissful, empty silence. In other words, the anti-body yoga position is not anti-homosexual; it's pro-empty or pro-transcendental. We are so free of our bodies that we are able to stay in them; and it's all right to be in them and use them. That's the Buddhist position. You're so free of the body you don't have to be afraid of it.

We have the question of what is sex, which William Burroughs has addressed himself to. He's one of the few gay lib "heroes," one of the few homosexual theorists who has theorized up to the point of outside-of-the-body, and detachment from sexuality. In fact, the cut-ups were originally designed to rehearse and repeat his obsession with sexual images over and over again, like a movie repeating over and over and over again, and then re-combined and cut up and mixed in; so that finally the obsessive attachment, compulsion and preoccupation empty out and drain from the image. In other words, rehearsing and repeating it over and over, and looking at it over and over, often enough. Finally, the hypnotic attachment, the image, becomes demystified. His particular sexual thing is being screwed, because Burroughs can come when he is screwed; he's one of the few men that can.

AY: You mean without manipulating himself?

AG: Yeah . . . look, no hands!

AY: I remember he has that image in *Naked Lunch*, but I couldn't believe it.

AG: Burroughs and I made it a lot over many years, back in '53, so I know his body. And there's the image of the hanged man in Burroughs, the guy that's being hanged and coming involuntarily. At the end of "The Blue Movie" [episode in *Naked Lunch*], there is a sequence of rehearsals over and over again, like cut-ups of the same scene and the same characters over and over: getting hung, and spurting, and the hangman coming up and sucking it, and Mark coming up, and Mary coming up and gobbling at his crotch just as his neck is being snapped, and he's involuntarily coming—just as Burroughs involuntarily comes when being screwed. At the end of that "Blue Movie," they all appear on the screen, bowing, tired, a trickle of saliva, or come, on someone's lower lip, a rope around Johnny's neck, Mary completely worn out and tired. The image is completely washed out through experience.

That's very early Burroughs—'58 or '59—and leads to more and more cut-ups of his most favorite and tenderly sentimental images. He can finally look at it at the end of the spool; he can look at his most tender, personal, romantic images objectively, and no longer be attached to them. And that's the purpose of the cut-ups:[127] to cut out of habit reactions, to cut through rehearsed habit, to cut through conditioned reflex, to cut out into open space, into endless blue space where there is room for freedom and no obligation to repeat the same image over and over again, to come the same way over and over again.

So Burroughs is one of the very few gay liberation minds who is thinking in ultimate philosophical terms about sexuality, about the nature of "apparent sensory phenomena" (that's his phrase). He's one of the few that has actually questioned sex at the root—not merely rebelled from heterosexual conditioning or heterosexual, social/moral fixed formations—to explore love between men as he has experienced it. He's seen it inside and outside, divine and degraded. But also to go beyond that and look at it through the eyes of a Sufi or a Zen master, or a sufud adept Tibetan monk saying, "Ah."

Burroughs has contributed a great deal of space which a new generation hasn't yet caught up with. His style was picked up by younger people. The whole cutup-collage thing did influence even underground press writing quite a bit. The further philosophical, practical Yankee examination of sensory phenomena perception that Burroughs had gone into still awaits discovery by the gay lib left. Other-

---

[127]On cut-ups, see William Burroughs, *The Ticket that Exploded* (New York: Grove Press, 1992), and Brion Gysin and William Burroughs, *The Third Mind* (New York: Viking, 1978).

wise, you just get into some kind of a funky scene. There are a lot of young kids who carry sex banners and march around saying sex, sex, sex, great, great, great—doing it humorlessly, reacting to just the initial superficial attraction.

Let's see, what haven't we covered? I asked Swami Shivananda where I could find a guru, and he told me, "Your own heart is your guru." The main slogan, instruction, teaching, compass and fidelity of the whole love situation is the heart which must always be followed because there's no other place to go. And that will dissolve perplexities of ideology, or complexities of the political fix we're into. Following the heart a little more—there's a way of avoiding the pitfalls of hyper-intellectual, ideological dead ends, which both homosexuals and radicals have gotten into.

Rely on your feelings and trust your feelings. I think a lot of homosexual conflict comes from internalizing society's distrust of your loves, finally doubting your own loves, and therefore not being able to act on them. The other thing is I think it's important to accept rejection because the more you learn to accept rejection, the more you leave yourself vulnerable to be rejected, the more you have a chance of getting laid, of scoring, both for heart and for cock. The more you open yourself up and give yourself, continuously without rancor, and accept rejection from people who are either too timid or are afraid socially, or who just don't want you . . . the more open you'll be to your feelings, the more you'll communicate, the more likely you'll just connect.

One of the greatest difficulties, especially for the younger sexualists of all kinds, is the fear of making a move, because they're afraid of being rejected. So, the only thing is frank revelation of the heart: that applies politically, subjectively, personally. . . . It's the lack of trust in the heart that's messed up radical mentality as well as sexual mentality in America. If we don't interest ourselves in our hearts and accept our hearts, if we don't find our reality in our hearts, then the rest is a perpetual void of intellect, Urizen, Blake's Urizen, your-reason, rationalization, common error, and, ultimately, the heart becomes brightly empty. Thinking of that in terms of making judgments politically: does Tom Forcade have a brightly empty heart, or does he have a heart loaded with silver shit? Is Mark Rudd's heart brightly empty this day as he goes to the anonymous cafeteria for his oatmeal? Is Mother Machree's heart brightly empty as she comes out of the Capri bar at

3 A.M., leaving the most beautiful boy in San Francisco with long curls flowing down his lion-like shoulders with a smile in his eye and pearly teeth showing in the moonlight?

> Beauty is but a flowre,
> Which wrinckles will devoure,
> Brightnesse falls from the ayre,
> Queenes have died yong and faire,
> Dust hath closde *Helens* eye.
> I am sick, I must dye:
> Lord, have mercy on us.

This is "Time of Pestilence" (1593) by Thomas Nashe. It's maybe the great poem in the English language, and the greatest line is "Brightnesse falls from the ayre."

JOHN DURHAM
November 1–2, 1972,   St. Louis, Missouri

"The Death of Ezra Pound"
From the book *Allen Verbatim* (1974, McGraw-Hill)

During a visit to Webster College, John Durham of station KDNA was broadcasting a talk Allen was giving to a crowd of about forty people when news arrived of the death of Ezra Pound. An article in the December 1972 issue of *Fat Chance*, apparently by Durham, describes the setting: "At the Webster College guesthouse we took the air live and Allen answered questions from KDNA's phones as well as from Webster students and others, . . . taught basics of mantra chanting, and showed some of his enthusiasm (see 'Wichita Vortex Sutra') for the act of radio not as mere language language but as an art form and real communication medium, imagining, in the rain, those hundreds of dark, wet blocks in St. Louis, a place where poets are born (Eliot, Burroughs, Moore, Williams, etc.) but seldom are made."[128]

In the spring of 1971 Allen had invited Gordon Ball, who had lived for three years with Allen and other friends on Ginsberg's Cherry Valley farm as the farm's manager, to accompany him on a cross-country college reading tour. On the tour Ball taped many of Allen's readings and later, at Allen's suggestion, turned the tapes into the book *Allen Verbatim: Lectures on Poetry, Politics, Consciousness.* In

---

[128]The date has been extended from November 1 as given in *Allen Verbatim* to include November 2 for, according to Michelle P. Kraus's *Allen Ginsberg: An Annotated Bibliography, 1969–1977*, p. 41, entry no. 333, the news of Pound's death came around midnight on November 1. She quotes M. Castro's description of the scene in the November 17–December 14, 1972, issue of *Outlaw*: " 'November 1st, the night of Allen Ginsberg's reading was also the night in which Ezra Pound passed on. Ginsberg didn't learn of Pound's death until an hour and a half after the reading around midnight[,] when he was gathered with a smaller group of students interested in politics at Kirk House on the Webster Campus. . . . A discussion ensued about demon worship and the like, and the importance of not clinging, when Harry Cargas (head of Webster College's English Department) entered' with the news."

December of 1972 John Durham sent the tape of the KDNA broadcast to Ginsberg, who then gave it to Ball for inclusion in the book he was editing. Ball later became the editor of two volumes of Ginsberg's published journals.

Here the mantra Ah is rendered "Ah" when it is being discussed but as "Ahh!" when chanted, since when Allen chanted the mantra he did so with feeling, drawing the syllable out.

The mantra Ah was a favorite one of Allen's. Speaking about the mantra to Ekbert Faas, Ginsberg explained that in Buddhism, "the word 'space,' [is] the classical yogic and Buddhist term for appreciation of the actual space, and the emptiness of space, the infiniteness and the emptiness. That's what the mantra 'Ah' is. 'Ah' before the place or space, that we breathe into, 'Ah.' So it's almost literal appreciation and use of space. And traditionally it is the seed syllable mantra for purification of speech, as well as the appreciation of space like the color blue of the sky."[129]

The tape was transcribed by Gordon Ball and the present editor.
—DC

Allen Ginsberg and audience: Ahh! Ahh! Ahh!

AG: Now everybody outside who's listening will go: Ahh! Ahh! Ahh! Ahh! Ahh! Ahh!

Q: What's the function of a mantra?

AG: You keep your back straight, focus the mind on a single sound in the body connected with the breath or fill in the breath. It's to become conscious of breathing and to direct your attention in one spot. You then—the circumference of the body—actually the heart area as with the word "Ah" or the syllable "Ah," which is part of a triad: Om, from the head, Ah from the throat to purify speech, Hum from the heart area, the location of your intellect, for body speech and mind: Om, Ah, Hum. So the Ah is purification of speech and appreciation of limitless space. For instance, as I become aware of the neutral and interesting space between my voice and

---

[129]From *Towards a New American Poetics* (1978, Black Sparrow). See also an excerpt from Faas's interview with Ginsberg in the present volume.

the microphone, you on the other end of the electronic universe with me in your home can become appreciative, Ahh!, of the space between your own bodies and the radio that you're listening to. So Ah is appreciation of space that we all have lots of room to move around in, appreciation of the enormousness of the space outside of the house, appreciation of panoramic consciousness which includes the similar thought, say, of Ah emanating hundreds of square blocks of St. Louis as a possible unified field of body-mind consciousness, a single consciousness that unifies ourselves here in the room, a space where we can be together, a psychological space, as well as the situation here in the room where we can all do one thing together: Ahh! *(Silence from audience.)* Come on!

AG and audience: Ahh!

AG: It's one, in [that] this intersecting modality of sound and breath also goes outside and the radio can be one with us and we can all do one thing together. It's a hard corner of space to find but connected by sounds we can all get together, do the same breath, try singing it that way. Let's do eleven Ahs in succession and those who are listening try and do it with us, and with everybody connected, we'll be having the same breath. And locate the sound in your breath as a sigh. That makes it vibrate in the breast, makes you feel good. Indeed that's the heart feeling, so it is a feeling: Ahh!

AG and audience: Ahh! Ahh! Ahh! Ahh! Ahh! Ahh! Ahh! Ahh! Ahh! Ahh! Ahh!

AG: So the function of mantra is to unify. At times it can cause an amount of buzzing or tingling in the body and the skull too, which is hyperventilation. *(laughter from audience)*

(Discussion moves from mantra to various other subjects.)

AG: Has anyone ever tried like a moment of total blissful conscious silence over the radio? Should we try that? Would everybody like . . . ?

Radio Station Personnel: I'd have to turn off your mike. *(audience laughter)*

AG: I gotta explain the silence first! *[audience laughter]* Silence with back straight, for those listening and for those present. To get the breathing correct, we'll do some Ah first and then just have one minute of air silence while you do the Ah subvocally. So, beginning: back straight, ears lined up with the shoulders, spine straight, the back of the head upholding heaven as Roshi Suzuki says in his book *Beginner's Mind, Zen Mind.* Put your hands anywhere you want, but restful mind posture on the knees is easy. But main thing, just so you can breathe, pumping out your chest like a powder pigeon, and arching your inner back so your belly falls out like a baby's belly.

AG and audience: Ahh! Ahh! Ahh! *(Some seconds of silence, then:)* Ahh! Ahh! Ahh!

AG: The breathing of the great beast. So we had one minute of silence with straight back followed by three Ahs appreciative of limitless space both inside body and outside body. Anybody else got any ideas?

[Talk moves to other subjects. Harry J. Cargas, head of Webster College's English Department enters the room.]

Harry Cargas: Have you heard the news?

AG: No.

HC: Ezra Pound died.

AG: Ahh! . . . for Ezra Pound. When did he die?

HC: I don't know. Just heard it on the news—apparently within the last couple of hours.

AG: Rest in peace, Ezra. Beautiful man—he died very peaceful—I'm sure—he died in a state of bodily thinness and grace mentally, I think. He was like Prospero—wise, and a great teacher—and a great guru, and a great silent man at the end. The same silences we had here tonight for a minute he had for days and hours on end, and he wouldn't open his mouth and say anything for the last ten years unless he had sumpin' to say, something sensible and sharp. And then when he opened his mouth he always had something wry, factual.

Q: There's been some dispute recently about an award he received from the American Academy of Arts and Sciences.

AG: It's a bunch of neofascist scholastics attacking Pound—the neo-reactionary CIA-mongers attacking Ezra Pound. Irving Kristol, who's a former editor of *Encounter* magazine when it was being funded by the

CIA, and a great advocate of academic law and order, who wrote Harper and Row that they should show their book revealing the CIA involvement with opium traffic in Indochina to the CIA for prior criticism (which must have led to prior restraint before they published it), Irving Kristol had the nerve to write the *New York Times* that Pound didn't deserve a prize because he was morally corrupt because he supported fascism. Irving Kristol who just signed an ad that he was going to vote for mass murderer Nixon! It's like an immediate karma shot!

Q: What do you think about giving an award to somebody who held the positions that Pound held?

AG: It's irrelevant! If the award is lucky enough to find him, God bless the award. The award needs him—he doesn't need the award. Certainly, give him *all* the awards. It's a shame he didn't get the Nobel and all the other awards at once—he was the greatest poet of the age! Greatest poet of the age. . . . Means a great blessing tonight, that he's dead.

Q: What does that mean—greatest poet of the age?

AG: The one poet who heard speech as spoken from the actual body and began to measure it to lines that could be chanted rhythmically without violating human common sense, without going into hysterical fantasy or robotic metronomic repeat, stale-emotioned echo of an earlier culture's forms, the first poet to open up fresh new forms in America after Walt Whitman—certainly the greatest poet since Walt Whitman . . . the man who discovered the manuscripts of Monteverdi in Venetian libraries and brought them out in the twentieth century for us to hear; the man who in his supreme savant investigations of vowels went back to the great musicians of Renaissance times to hear how they heard vowels and set them to music syllable by syllable and so came on the works of Vivaldi also, and brought him forth to public light—

Q: Wrote a book on harmony.

AG: Wrote a book on harmony, wrote an opera on Villon. The book on harmony I don't know.

Q: Well, how about—some Jewish people have reacted very strongly to some of the negative things he said about Jewish people.

AG: Pound told me he felt that the Cantos were "stupidity and ignorance all the way through," and were a failure and a "mess," and that his "greatest stupidity was stupid suburban anti-Semitic prejudice," he thought—as of 1967, when I talked to him. So I told him I thought that since the Cantos were for the first time a single person registering over the course of a lifetime all of his major obsessions and thoughts and the entire rainbow arc of his images and clingings and attach- ments and discoveries and perceptions, that they were an accurate representation of his mind and so couldn't be thought of in terms of success or failure, but only in terms of the actuality of their represen- tation, and that since for the first time a human being had taken the whole spiritual world of thought through fifty years and followed the thoughts out to the end—so that he built a model of his consciousness over a fifty-year time span—that they were a great human achievement. Mistakes and all, naturally.

Q: Like *Leaves of Grass*, again, isn't it? Like Whitman again, where he did that in making *Leaves of Grass*.

AG: Yes, he did.

Pound also unmasked or demystified the nature of banks and money and currency; his ideas on the banks and on the hallucina- tory role of the banks in distracting everybody from the fact that the money they're issuing as credit really comes from the govern- ment because it's backed by the government and the banks couldn't issue it as real credit, paper money, checks—like the gov- ernment couldn't borrow money from the banks unless the gov- ernment backed the banks, so the government is borrowing money from its own authority, 'cause the only authority the banks have for lending out money, including money lent to the government, is the fact that the federal government is backing the banks. So Pound demystified that very simple point which has hallucinated people for the last hundred years in America, leading to Abbie Hoffman standing on the balcony overlooking Wall Street stock market and showering free money down to all these money junkies, matter-habit hookers.

So Pound I think affected many people in many ways, mostly in very revolutionary and charming ways, like Imamu Amiri Baraka probably came a great deal out of Pound with the particularism of his Black revolution. So the net substance and sum of Pound's energy finally comes to mean the liberation of the voice, liberation of the vowel, liberation of conscious attention to language, purification of language, so finally you could say as to Pound's karma, Ahh!

Q: Do you personally ignore Pound's involvement with fascism, or do you just accept it?

AG: No, I see it as part of character and humour, h-u-m-o-u-r, which is changeable. I think he was, as he pleaded, mentally ill for a while—if you listen to his records, the phonography records made in St. Elizabeth's, there's a splenetic, irritable voice. Whereas if you listen to the records made by 1958—in Milan, a very rare copy of "With Usura"—and later records at Spoleto in '66, you hear the voice of Prospero himself, whose every third thought is his grave: the fine old man with beautiful manners with the whispering paper-thin voice pronouncing syllable by syllable with great intensity and meaning each thought of the earlier younger man. So he'd come to a resolution of his woes, a rue; like Prospero, he drowned his books and plunged "deeper than did ever plummet sound" his magic wand of Pride, and took unto his counsel silence, broken only by good-humored advisements on rare sensible occasion as when he told me, "Stupidity and ignorance all the way through."

Q: Who was it who said that he was the last American to have lived through the tragedy of Europe?

AG: It sounds like him 'cause after all he died at the age of eighty-six— eighty-seven? And he'd gone to London before World War I and seen the most beautiful men of his generation destroyed by war madness, wrote many elegies about the "Charm, smiling at the good mouth, / Quick eyes gone under earth's lid," so he did remember World War I and he knew stories of the Civil War from his grandfathers in Idaho and Philadelphia, and he was also in Europe during the Second World War and in between the wars, and then all through the postwar generation.

I had the honor of smoking a stick of grass at his house on his eighty-second birthday and singing Hare Krishna to him with that harmonium I used tonight, and bringing him the first album of *Sergeant Pepper* by the Beatles that he'd ever heard, and bringing him Dylan—I spent an afternoon playing Dylan and *Sergeant Pepper* and Donovan's "Sunshine Superman."

Q: How'd he like it?

AG: Well, he didn't say anything, he just sat there with a wry smile occasionally playing around his lips. But I asked Olga Rudge what he thought—you know, did he like it—and she said, "Well, if he didn't like it he would have gotten up and left the room!" So he sat through about two hours of Dylan and the Beatles, so he heard that at least. That was nice. Patient man.

I first met him in a theater at Spoleto, at an opera. I came in and he was downstairs seated in the orchestra and I was up in a box in the balcony, so I went downstairs before the Magic Flute began and stood by his seat and took his hand and he stood up and stared at me and I stared at him and then we just stood there and stared at each other for a long time, very quietly and calmly, neither of us having to say anything. And then he was still standing there—just standing there—so I put my hand on his neck to ease him down. 'Cause I didn't know if he knew whether to sit down or not and I was unsure of myself so I thought I'd make a gesture, touch him, let him get off his feet again.

Then I went to see him in Rapallo where he'd lived in the twenties with Yeats and came up with the harmonium to a high hilltop where he had a small modest house which overlooked great blasts of blue Mediterranean spaciness and shore, and sat under a tree outside in a wooden chair and sang Hare Krishna mantra to him and Hari Om Namo Shivaye and Gopala Gopala Devaki Nandana Gopala mantras and then we went in to lunch and Olga Rudge—he hadn't said any-thing—said, "Ezra, why don't you ask Mr. Ginsberg if he wants to wash his hands," so he said, "She says, would you like to wash your hands?" And then he didn't say anything more that whole afternoon, except one time, when I talked about a visit with Burroughs in 1958 to Louis Ferdinand Céline, who I thought was the greatest French prose writer. And I'd asked Céline whom he liked among French prosateurs and he said C. F. Ramuz, Swiss writer, and Henri Barbusse, who wrote *Le Feu* (*Under Fire*, World War I) and Barbusse, he said, had jazzed up the French language some and so had Paul Morand, whose book name I had forgotten in telling Pound the story, and Olga said, "Ezra, what was the name of that book by Morand you liked so much—you didn't like his later work but you liked that." And he'd been spooning up spaghetti and he lifted his face and said, *"Ouvert la nuit"* and he finished eating his spaghetti. That book was from 1928 or so—so his mind was very crisp, and funny. And then he didn't say anything more all day. We went off to Portofino and just sat in the sunlight at a café table at the wharf by docks and fisherman's nets and I babbled on and on, whatever I had to say, whenever I thought of sumpin', and then shut up for a while, just keeping silence like meditation silence, and I kept thinking it was like being with Prospero, it was a pleasure. There was no weight in the silence—there was lots of room, there was no need for anxiety about it. It was like the silence of the Indian mouni, holy men who have taken a vow of silence, like Meher Baba did—there was a Baba-esque quality to his silence.

And then several months later visited in Venice in Calle San Gregorio where he stayed with Olga Rudge next to the Pensione Cici where he ate maybe every other day and picked up his mail and saw some few people. So I spent three weeks there reading through all the Cantos and a very good book on Pound called *Ideas Into Action* by Clark Emery, University of Miami Press, Coral Gables, Florida—the best description of the Cantos that I know. Best analysis, explanations—makes it easy to understand his theories of history and banking.

I used the Pisan Cantos as a guidebook to Venice: "and in the font to the right as you enter / are all the gold domes of San Marco." So I went to San Marco Church and looked for a font to the right which I found, but there was no water in it and the gold domes of San Marco didn't shine therein. So I went back to him after several days in which he didn't talk at all and said, "I went and couldn't find the gold domes shining in the font to the right." And he suddenly opened his mouth and said, "Ah, that was many years ago. Since then they've put a copper runnel around the baptismal fount and filled that with water so there's no longer water in the center to reflect the gold domes." And he didn't say anything more—until I had another specific question: "Where was Salviati's?" "Up the street, the glass manufacturing place." And then the conversation about stupidity and ignorance all the way through and the Cantos being a mess. So actually quite a humble person by then. I told him about acid, and a little bit about grass, and asked him if I was making sense to him and he said, "Well, you seem to know what you're talking about." I asked him if he'd received Basil Bunting's *Briggflats* and he didn't say anything but a great broad smile passed over his face—he really smiled, like a Crumb cartoon character smiling—and nodded. Basil Bunting being a great fellow poet who lived with him in Rapallo and with Yeats in the twenties and who'd come out of retirement with a great epic long poem, thirty pages, *Briggflats*.

Q: He's a British poet from the north—

AG: Yeah, from Newcastle, Northumbria, who taught Pound the great lesson "Dichten = condensare,"—poetics is condensation, compression of thought. Pound said "Bunting once told me my poetry referred too much and presented too little." So I said, "Well, I saw Bunting last month in New York and he said, 'Read Pound if you want hard, condensed exact precision in language.'"

The difference between presentation and reference is an interesting distinction. Reference—which he does do a good deal—is like referring to the situation without describing it, without presenting the facts—so that a later generation won't be able to figure out what you're talking about. Like, "That man in the White House" or sumpin'—all the abstract hatred laid down on Nixon as on Johnson before him, without the particular form that it attends.

Q: Which of the Cantos stands out in your mind right now?

AG: The last great Canto—

> The scientists are in terror
> and the European mind stops
> Wyndham Lewis chose blindness
> rather than have his mind stop. . . .
> When one's friends hate each other
> how can there be peace in the world?
> Their asperities diverted me in my green time. . . .
> Time, space,
> neither life nor death is the answer
> And of man seeking good,
> doing evil.
> In meiner Heimat
> where the dead walked
> and the living were made of cardboard.

"In meiner Heimat," in my homeland—America—"where the dead walked / and the living were made of cardboard." The most beautiful piece of his writing, and done in his old age. As he was being constantly attacked.

Q: Beautiful thing, it really is.

AG: Great despairing end, yes. He wrote that I think around 1955 or '60, perhaps. It was published in drafts and fragments of the final Cantos. It's so elegant, though, when he says, "When one's friends hate each other / how can there be peace in the world?" Very Chinese, probably a paraphrase of Confucius, really. And then there's a very funny leap there for an old man—it's a very abstract line: "Time, space, / neither life nor death is the answer." So death is not the answer either.

Q: He sounds like Lear, almost.

AG: Well, he sounds like Chuang-Tzu, he sounds like Buddha. He sounds like some Chinese philosopher of the void.

Q: But he constantly rejected the Taoists.

AG: No, he took a Taoist position, basically, or Confucian position. He accepted certain Taoist elements, but he was against religious flattery and incense-burning, I think, against dead ritual. And the incense-burning cult of the eleventh century or so was I think one of his favorite whipping boys. But he was a pragmatic mysticist—he was interested in Eregina, apparently, and Dun Scotus and people like that—Plotinus, I guess. There's a line to Eliot:

> "mi-hine eyes hev"
>   well yes they *have*
>   seen a good deal of it
>   there is a good deal to be seen
>   fairly tough and unblastable

It's in the Cantos.

See if I can remember any other incidents of that conversation with Pound. I sang him the St. Francis Canticle of All Creatures which I was trying to figure out how to vocalize 'cause it was a song and St. Francis probably sang it. So I sang it to a C chord, monochordal.

I asked him for a blessing for Sheri Martinelli who was an old girlfriend of his when he was in St. Elizabeth's and had suffered many years his absence. And so he finally smiled and nodded, "Yes," which brought tears to Sheri Martinelli's eyes on the Pacific Ocean edge a year later, '68.

Told him that the younger poets all learned from him, all derived from him, and were in a sense developing forms that he opened up. And he said, "That's very flattering, but would be very difficult to prove or substantiate."

EKBERT FAAS
March 27, 1974,   New York City

From the book *Towards a New American Poetics*
(1978, Black Sparrow)

Although Allen Ginsberg's interest in Buddhism went back to his youth, his interest in the religion deepened after meeting the Tibetan Buddhist lama Chögyam Trungpa in 1970. After their initial meeting, Allen attended a seminar with Trungpa and Gary Snyder in 1972. He took formal Buddhist vows that same year. In 1973 Trungpa invited Allen to help design a poetry course for what was to eventually become, with much assistance from Allen, the first accredited Buddhist college in the Western world, initially known as the Naropa Institute, to which the Jack Kerouac School of Disembodied Poetics was later added to train young poets.

As Allen's involvement with Trungpa deepened, a number of interviewers in the mid-1970s sought to understand and clarify what they correctly perceived to be a significant development in the poet's life.

The interview excerpted here took place first by phone on March 27, 1974, and then continued a few days later in Allen's East Tenth Street apartment after the two men discovered that they had been only five blocks apart during the entire phone conversation.

—DC

Ekbert Faas: What criteria do you have for evaluating your own work? Somewhere, you say, for instance, that in a poem each syllable should have a place and a purpose; or you claim that each poem should have an emotional feeling center.

Allen Ginsberg: Well, I write a great deal, and my habit of writing is to write in journals and notebooks, generally, not all the time, but that's the general habit. And then I cull from the notebooks things that seem to be complete poems.

EF: And what's a complete poem?

AG: Well, something that seems to have a beginning, a middle and end.
(*Laughing*) Or something that embarrasses me because of the truthful-
ness, or something that seems quite clear and hard, the images seem
hard in the sense of precise and not abstract, not generalized, but
composed of "minute particulars"; or if the sound is coherent, the
vowel structure is clear. Since each poem has a different form more or
less, there is no way of knowing it in advance. But by hindsight I can
look at a piece of writing and say: That's interesting. But for criteria
very specifically, almost the highest criterion I know is, if I weep while
writing it's generally good writing—weep for the truthfulness of what
I'm writing. And if I'm embarrassed by it, it very often turns out to be a
really good poem, in which I'm reclaiming areas of my awareness
which I hadn't admitted to myself or others.

EF: You spoke of your poems as having a beginning, a middle and an end.

AG: Yes, where the thought process begins, where there's a middle,
and where (*laughing*) I don't know, well, beginning, middle and end.
I'm thinking of a poem I wrote recently,[130] a description of breathing,
and I have the breath go over the Grand Teton Mountains to Northern
San Francisco into Hawaii and Australia and Saigon through the Indian
Ocean and Africa and Marseilles and on top of the Eiffel Tower and
then through the Atlantic to New York to Paterson, N.J. onto the Mid-
west finally arriving back in Teton village where it first began breath-
ing. So (*laughing*) it had a beginning, a middle and an end. And that's a
pretty good poem because it had that much clarity. But it's just the
beginning of a series of thoughts and an end of a series of thoughts, a
definite emotional end I guess and conclusive.

EF: So it's really the beginning and end of a thought process.

AG: Right.

EF: William Stafford says that his poems usually come to an end when
his powers to homogenize an experience come to an end.

AG: I don't quite know what that means. I don't mind it actually, if I
knew what it meant.

---

[130]The poem is "Mind Breaths." See *Selected Poems 1947–1996* (New York: Harper-
Collins, 1996), or *Mind Breaths* (San Francisco: City Lights, 1978).

EF: I think it means to impose some kind of order upon an experience or reality, to make an experience shapely.

AG: But then I don't know what that means either. In other words the experience one has while writing is the experience of one's own mind and to *put* order into it wouldn't that mean to rearrange one's normal sequence of thought-reflection?

EF: Anyway, I didn't mean to defend that statement.

AG: No, but I don't even know what it means because, well, I'm saying quite literally, I don't know what it means. I would—I also put order into my thoughts if I knew a way of doing that without violating the nature, the natural structure of the thought. I do feel that thought has a natural order of its own, so it's a question of transcribing the thought in its own order, and in that sense a beginning, a middle and an end.

EF: But of course it would be very difficult to describe that kind of . . .

AG: Not too hard. For instance, you know my book called *Empty Mirror*. There is this little poem that begins "How sick I am!" called "Marijuana Notations." So I got high and thought ahah at the beginning of a thought feeling sensation, and then made several ruminations about it, "does everybody feel this way?," and then suddenly switched the mind from that kind of introspective self-persecution to: "It is Christmas almost and people are singing Christmas carols down the block on Fourteenth Street." So that was the end, shifting the mind completely to something more real. It was a lot of subjective thought about my being sick or not being sick, very vague really, representative of a twenty-year-old or a twenty-four-year-old talking to himself. But then suddenly the mind comes down, suddenly I realize, you know, it's hopeless, and switched the thought like a cut-up almost, or like you switch back to the breath attention in mindfulness meditation. Suddenly I switched back to observing something real at the outside of myself, outside of my subjective babbling. So in that sense it has a beginning, a middle and an end. In almost like ah . . .

EF: But always in terms of actual experience.

AG: Well, in terms of the experience of the mind.

EF: Of actual thought process.

AG: Yeah, I would think that would be the criterion. Like one poem that does not have a beginning and an end is at the beginning of *Planet News*, a long poem called "Journal Night Thoughts," which is just sort of discursive free association. It's almost not a poem, but I included it because it's a specimen of that poetical mode, I mean it's very intense language, but at the same time it doesn't have any subject at all except the thoughts of the moment of that day, the thoughts at the moment of writing, recollecting all the thoughts of that day.

EF: It seems that you avoid the term automatic writing, preferring to speak of spontaneous writing.

AG: Well, automatic writing has to me the association that you are not completely conscious of what you are writing.

EF: Not completely conscious of what you are writing?

AG: Yes, just the hand moving and just the unconscious coming out as if dictated by voices or from the unconscious but, you know, the hand moving and creating words that the mind is not aware of.

EF: So graphing the movement of the mind, as you understand it, would imply a high degree of consciousness.

AG: Yes, depending . . . the process would be observing what is going on in the mind and writing it down, as simple as that, rather than to attempt to separate yourself from awareness of what your thought forms were. I would speak of spontaneous writing as writing down your thought forms as they occur. Automatic writing, I think, is a technical term which implies that you don't know what you're writing at the time. You really don't know, it might be ghost voices, and I think certain Surrealist practices were attempting to approach that.

EF: With the assumption that if you have a poetic genius whatever comes out might be a great poem, I suppose.

AG: I would have the same assumption about recording the thought forms of spontaneous mind[131] and really if you are a poetic genius, if

[131]In the Buddhist sense, "spontaneous mind" refers to "ordinary mind." Free from concepts. Also referred to as "first thought" and "pure perception." Ginsberg uses this sense interchangeably with the simple sense meaning "composed on the tongue."

your mind is shapely, then what comes out will be shapely. In other words if you see clearly into your mind, if you remember and recollect clearly what you were thinking, it would be thoughts in minute detail, thoughts of minute detail, it will have objects and it will have real describable things in it. It won't be sort of vague moods, it will be actual buses passing by, yellow street signs, those of Tenth Street.

EF: I am puzzled by the fact that you frequently mention Gertrude Stein as a model for your technique of graphing the mind-movement, but I don't think she graphed the movement of her thoughts the way you do.

AG: No, she wasn't interested in memory, she wasn't interested in the future, she had a specific subject which was a few sentences like "Napoleon ate ice cream on Elba" and then she would recombine the words over and over again in the normal order that they came to her. Actually "the graph of the mind-movement," that poetry is a "graph of a mind moving" is not my phrase, it's Philip Whalen's from his poetics statement at the end of Don Allen's anthology.[132] Philip Whalen once read Stein's "Composition as Explanation" to me aloud about ten years ago and it seemed exactly on the line of what I was interested in. Gertrude Stein made a lot of different kinds of art experiments, but her experiments were all with the present consciousness during the time of composition.

EF: When does a line break? It wasn't clear to me what you say about that in your *Improvised Poetics*. Sometimes you say it's determined by the page . . .

AG: That's Creeley's method . . . Basically breath-stop.

EF: . . . or that the thought breaks.

AG: Well, when you talk and then after a while like you run out of thought and words, and then you're gonna stop and take a breath and continue. So that's when you have your breath-stop. Breath-stop and the thought-division could be the same.

EF: Which would imply a new kind of thinking, wouldn't it?

AG: No, only an *observation* of your thinking. Same thing as usual, except you're observing it, and see where it stops, and there's a gap,

---

[132]*The New American Poetry 1945–1960*, Don Allen, ed. (New York: Grove Press, 1960).

and where a new thought begins. Because if you observe thought, it actually does stop and there are gaps in between. It's like a movie, it looks continuous, but actually it's a lot of different frames. And thought is the same way.

EF: Graphing the movement of the mind, while, I suppose, standing back from your own stream of consciousness.

AG: Observing it or recollecting it.

EF: I just don't have any personal experience of graphing the movement of my mind, so it's difficult to understand.

AG: Well, I'll show you. Really, it's very simple. Sit up with your back straight. Now, the primary attention could be on the breath coming out of the nostrils, so that—keep your mouth closed so there's no air pocket, and instead of thinking about the conversation or about me, or looking at me or anything like that—look toward *there* maybe (*pointing to the kitchen wall*)—so we don't get tangled in some kind of subjective thing, pay attention to the flow of breath out through the nostrils for maybe five breaths. Can you do it for five breaths just on the out-breaths? Forget about what's going on inside you on the in-breaths or any other time. And really sit up straight—Yeah, now—Okay. (*Pause*) Now are you able to put your complete attention on your breath going out . . . or do thoughts interrupt?

EF: I have an image of the hair in my nose.

AG: Okay, so pay attention to breath after it leaves your nose and the space in front of your face, into which the breath flows, for five breaths.

EF: Pay attention to the air coming out? And flowing into space?

AG: Yeah, into space and dissolving in space, or identify with the breath going out and dissolving. (*Pause*) Okay now, you're paying some attention to that.

EF: Yes.

AG: What else were you thinking about?

EF: Sort of an image of a balloon coming out of my nose as in comic books, and then I was also thinking about telling you about it.

AG: Okay, so there's three separate things going on. Now can you pay attention to the invisible space not in the form of a balloon. (*Pause*) I don't mean invisible but transparent. (*Pause*) Now during the last few breaths, what other thoughts came into mind? Anything about Frankfurt or anything like that?

EF: No, you see, when I meditate I usually try to concentrate on one image: a clock outside the town gate of Eibelstadt, a German medieval town I used to live in, and that sort of came into my mind.

AG: Okay, fine. Well, try one more time, and when the clock rises, switch your attention to your breath again. (*Pause*) So your thought is separated between the clock and the breath, and it's really discontinuous. There is a gap in between.

EF: I see. (*Both laughing*)

AG: Just literally speaking, there is a moment of shift, that's almost a fraction of a second. So if you do meditate and continue to practice attention to breath, you never can stop your mind from thinking and wandering and travelling in space, but you keep interrupting it, bring it back to the breath and the space in front of you.

EF: I suppose that's what I'm trying to do with the clock.

AG: Yeah.

EF: I mean I can never stay with that image for any longer than say thirty seconds.

AG: Okay, so that in that sense thought is discontinuous, I mean there are jumps and gaps in between the thoughts, and there are actually complete gaps in between for only a micro-second, but . . .

EF: And that's where you talk about śūnyata.

AG: Well, a gap is—that is Chögyam Trungpa's terminology—that gap in which we are just sitting in space or in just empty space. Now, if you practice that kind of thing either for meditation or just trying to recollect what you were just thinking about when you are writing you notice you get a lot of different thoughts with gaps in between, and in that sense thought is discontinuous, like movies, in the sense of the cinema, you know with twenty-four or fifteen frames a second. So you might have like three or four thoughts a second, or maybe one thought a second, but in between—and the thoughts jump around, so obviously you were thinking of a clock, but then you might think from a clock to the . . . Beats, your paper or your nostril. So that's all I mean when I say remembering or recollecting. So the process of writing for me is remembering and recollecting the thoughts I was just having while writing.

EF: And the writing is a kind of stepping back from that.

AG: Well, the mind steps back, and the writing registers it fast, you just scribble it fast. The first words that come to your mind, the first flashes of what it was, your clock, a hand, a gingerbread hand, iron letters, roman numerals, roof, grey tiles, those were the first things . . . So that's "clock roman numerals iron letters roof grey tiles," and you've got a line of Ginsberg poetry. That's exactly it, you see. It's the first thoughts, whatever you can get. I have a whole scene of clocks and roofs and whatever I remember from Munich or wherever I saw a clock, just picking up, you know, like when you sketch very fast.

EF: Yet these are still only fragments. How do you put them together? The way they come out?

AG: Like a haiku: clocks iron letters.

EF: No, I mean once you have written a line.

AG: You move on to the next.

EF: Add them up, just in the order in which they came out?

AG: Move on to . . . yes.

EF: And never shift them around.

AG: No, no, no, no, no. No, you couldn't—then you'd be stopping the whole process.

"Squawks Mid-Afternoon"

Michael Goodwin was the senior editor for *City* magazine in 1974 when Ginsberg came to San Francisco for a reading at the De Young Museum. To interview Allen he enlisted the help of two friends, *City*'s book reviewer, Ed Ward, and artist Richard Hyatt, who, Goodwin wrote in 1998, "brought an astonishing range of expertise, everything from Buddhist sutras and yoga practice to Chinese martial arts." The interview took place upstairs at Lawrence Ferlinghetti's City Lights Bookstore, where Allen was staying.

After the three-hour interview, Goodwin transcribed the tapes into a thirty-two-page, single-spaced document, and then proceeded to edit it to fit into a four-page magazine story. Goodwin explained that "I culled out what I thought would be the most exciting material for a general readership and then, looking to make the story as tight and punchy as possible, tightened the quotes even further by carefully removing less-interesting comments and delicately 'improving' some of the text. I was very careful, as I always am, not to change even a nuance of meaning.

"A week after the story was published I called Ginsberg to find out if he'd liked it. He hadn't. He was annoyed that I'd edited him. I pointed out that I'd had no choice: I was editing to a tight page-count. Ginsberg replied that he had no problem with my dropping whole sections of text, but he objected strongly to my editing his comments in the sections I did include. I got it. I understood that his answers had been spontaneous and therefore flawless. My meddling had marred their perfection."

For this collection of Allen's interviews, Goodwin kindly located his original transcription and retyped it in its entirety. I then restored the portions of the original interview that he had edited to their verbatim form and then lightly edited those in the manner

Allen himself explains in the interview, merely "blue-penciling and elimination . . . sometimes tying threads together that are separated by inattention and inaction," so that this new version of the interview retains much more of Allen's original speech.[133]

—DC

Allen Ginsberg: So now that we've got the machine, what do we want to . . . ? I've got lots of time. Have you got lots of tape?

Ed Ward: Yes, two 90-minute cassettes.

Michael Goodwin: I normally think of a poem as something that's perfectly polished, and yet, just before, you were talking about writing poetry by reciting it into a tape recorder . . .

AG: Well, that's talking, vocalizing it directly.

MG: And last night we were reading "Wichita Vortex Sutra" . . .

AG: That's tape recorded . . .

MG: Reading it out loud, 'cause I think poetry has to be read aloud, and the cadences in it reminded us of a preacher preaching a sermon. So I was thinking about rhythm, and I was wondering do you sit down and polish poems until they get perfect rhythm, perfect images, or do you just whip 'em off?

AG: You know, I try and sit down and polish my mind until it gets perfect.

EW: Polish it in your mind?

AG: Polish my mind.

EW: Polish your mind?

AG: Yeah, start with the mind, start with getting some more direct contact with the actual sequence of flashes of thought, of the thought process, or the vocal thought process. So sitting meditation—which I do now and then, and which I've done for prolonged periods of ten

---

[133]Readers curious to see the original published version of this interview will find it in the November 13–26, 1974, issue of *City*.

hours a day for several weeks at a time—is like taking an inventory of the sequences of thought and the recurrences of thought and the weights of certain thoughts as they come back over and over, and after a while you get familiar with your mind. You know what you're thinking about, you know what's bothering you, and you know what comes to interrupt your breath if you're paying attention to your breath when you're standing or sitting or just walking around, and after a while you become familiar with your consciousness.

So you get to familiarize yourself with that and you begin to recognize it when it comes to you like a fish coming down the stream of the whole phenomenon of consciousness. You just take out a hook and let it down. So it's just a question of getting it while it's going by, and doing it rapidly, in the sense that it's a special event, like somebody jumping out of a window. There's an old story about Flaubert teaching de Maupassant how to write, and Flaubert says somebody is jumping out a window, and in your mind, sketch a description of it so swiftly that you've got it down before the guy hits the ground. *(laughter)*

And that ties up with Williams's introduction to *Agora In Hell*, which is that if you want to describe a thing like a tree you have to pick out one particular thing that makes it different from all the other objects in its class, like a special weird branch that sticks out of the top like horns out of the top of the tree, which is what Williams does in his own poetry. That would describe the whole tree for you. Or a person, a passing guy with a bent, blackened nose, and already you've got a whole face suggested. So it would be a question of, with your thoughts, picking out that detail of your thought which is like the fin of the fish—because you can't get the whole fish and you can't get the whole stream, because everything's going too swiftly, and there's too much simultaneous perception, and thoughts going on at once and they all follow in sequence.

Also, if you do meditation you know there are gaps in between the thoughts. It's like movie frames, and there are actually gaps in between thoughts. If you slow down enough to sit and observe your thought processes there are sequences of thoughts with little blank spots in between them, dharmakaya thoughts, unborn awareness. Like Shakespeare says, "Every third thought will be my grave." Prospero says when he goes home to Milan after resolving the problems and calming his mind, "Back to Milan, where every third thought will be my grave." There's always been an awareness of thought as a discrete, solid thing that you can actually get onto instead of making up a cloud of babble.

So that's what I mean by "polish the mind," in that you actually do get an increasing awareness either through meditation or poetry, which is another yoga, of the actual stuff, *ichita*.[134] And then it becomes a matter of being a very faithful secretary. You can't get everything, so you get as much as you can so you have something solid to work with. In other words you're not doing something arbitrary, romantic, babble, bullshit, you're actually dealing with your mind-stuff just like a painter's working with an actual landscape. Solid in the sense that it's real, it's objective, it isn't even your subjectivity any more, you're just objectively watching something move. So there's no longer any question of egotism or self-expression or personal expression. All those theoretical things are like nonpracticing questions. But if you're actually practicing there's a real thing to work with, which is your thought-forms.

Then there are techniques of swiftness and deftness of attack and trueness and jiu-jitsu for grabbing a fin or a butterfly or a flower's knees or a cloud's wriggle or a foot's skip. Everything you flash on has some characteristic individual oddity, so it's a question of noting the oddities. The basic principle is something that Chögyam Trungpa formulated. The last couple of years we've been talking a lot about poetry, because he's a poet, and we've done a lot of chain-poems together. His basic principle is "First thought, best thought." The first thought is the best thought. And that was Kerouac's basic principle for his spontaneous writing, for the same Buddhist reasons of practical inquiry into the operation of the mind. Both Kerouac and Trungpa realized—and teach—a very simple thing, which is that the first way that you flash on a thing is the unselfconscious, naked, real first-mind way, which is totally private and odd, eccentric to you, but is so direct that anybody *can* understand it. Naturally that was your first take on a pretty face or a bus roaring down Kearny Street, a fleurette on a piece of linoleum, which is just below the threshold of your normal, discursive consciousness. If you can become conscious of those things—the first inkling of thought—that's the material, that's the rawest and freshest and brightest and most impersonal almost. It isn't

[134]Ginsberg apparently said *ichita* but probably intended *cita*, which means consciousness. The tape recordings of the interview are no longer extant.

you—it's the whole mind doing that, the mind reacting, mirroring. So it's actually mirroring what you see.

Richard Hyatt: Is your technique more that of a *sumi* painter, do you try to make the expression of your poetry as direct as your perception? Or do you work like, say, Michelangelo or DaVinci, who used chiaroscuro, building up layer upon layer?

AG: More *sumi*. As direct as possible. Because what's interesting— Philip Whalen has a phrase, "My poetry is a graph of the mind moving," talking about the same area. So if the mind is moving, you don't have much time to work on enriching the verbal surface. Yet on the other hand the verbal surface is already completely rich, as a perception is completely rich, full and eternal in a sense, because that moment of flash is going on all the time, in a timeless sort of nonpersonal world. So the real practical nuts-and-bolts problem is, how much can you get down if things are moving so swiftly?

RH: Well, the reason I brought this question up is that I encountered this problem myself in my own practice and teaching and studying. Your type of poetry, *sumi* poetry, is, I think, something that very gifted people can do easily, but not everybody can do this. You see a lot of bad spontaneous prose, right?

AG: It really is not spontaneous. It's really third, fourth, fifth . . . It's reflections about thoughts.

RH: Right. So there has to be a yoga to develop that technique.

AG: Well, hopefully, by this generation that poetic yoga is more developed as people get more and more into the idea of spontaneous recollection of mind. I think by now the gist of that idea has gotten across, at least from the evidence of that young kid I told you about that I saw the other night. So it's a question of practice that becomes traditional after a while, is embedded in the poetic culture.

So the poets develop their own wisdom, and once the idea of spontaneous poetry is introduced, like through Kerouac, and then reinforced by a growing awareness of Zen practice or tai chi practice or any kind of mindfulness exercise practice, and once it's kind of laid out again, officially, as whispered transmission in the Kagyu sect, like from Chögyam Trungpa, who's really concerned with aesthetics, there gets to be a growing body of really hip people in this area. So there's a cultural accumulation of granny wisdom or nuts-and-bolts practicality on the subject, where there's no more mysterious thing of, "Oh, my mind, I'm going to . . ."

RH: Kerouac really gave us all his shots, not just the ones that hit the bulls-eye, and I always respected him for that.

AG: I think he's a very great poet. He's not well-enough known as a poet yet, but poets like McClure and Snyder really respect his poetry, as a peer of theirs. *Mexico City Blues* is sort of a classic, a little book of scattered poems—there's a lot of beautiful things in that. And then there's another manuscript that's here in this office, called . . . let's see . . . *Pomes*, all except the blues, which Ferlinghetti has published.

RH: Well, also there's all the things he did in *Desolation Angels*.

AG: Little haikus. He was a perfect haiku man. Because it required two flashes with a gap in between, which the mind of the reader connects just as they were connected sequentially in the mind of the writer. So I think he broke the ground on that here in America. He taught me . . . I think that's where I got it. Specifically, because I had early writing that was very deliberate, the Michelangelo form, where a poem would take weeks, a poem would take half a year to complete, and I actually had no subject matter but an idea. I didn't have a direct perception of something, rather, an idea. An idea of eternity as subject matter rather than an actual . . . tire rolling across a piece of brown paper.

RH: Okay, but Gary Snyder still writes like that sometimes, doesn't he?

AG: Well, Gary does it a different way. He builds his poems out of little bits of flashes taken in over from different times, because he has an idea of the whole continuum of mind, and so he can pick one haiku or phrase or flash from one season and put it next to another and assemble a long scroll of poems like mountains and rivers without end. He can go in and out of time sequence, although he also composes directly.

RH: Well, what about your training at Columbia with Van Doren. Was that wasteful?

AG: Well, no it wasn't useful in this area because it was training the mind in a different way, in a sort of symbolic way rather than in a direct way, but on the other hand I got lots of experience in ear and in traditional rhythms and forms so that when I carried over I had lots of preparation. Discipline, so to speak, although I don't like the word. It's such a weak discipline, the traditional discipline, and the spontaneous mind is so strong . . . Practice. I think the word "practice" is better than the word "discipline." The traditional Buddhist word "practice," like sitting practice. In fact, I forget the word "sitting" and use "practice." So it's sort of like the practice school of poetry rather than the discipline school.

MG: Well, the traditional idea of a poet is somebody who sits down and works and works and works on a poem until he gets the thing perfect.

AG: Well, that wasn't Shakespeare's method. Shakespeare never blotted a line, as Kerouac always used to say. Ben Jonson said he never blotted a line. Yeah, Shakespeare was very much a spontaneous poet, writing fast for the theatre.

EW: He had to have it out tomorrow.

AG: Yeah, he was like a working practice poet. He was so good at it that he must have had some direct connection with his own head. "In the dread vast and middle of the night," phrases like that.

RH: I don't know, I'm such a traditionalist, I can't help but see in your poetry so much that has carried over from your formal practice. I have Van Doren's big book that he edited, and some of his other anthologies . . .

AG: There's a certain weakness in Van Doren's writing itself, his own . . .

RH: I don't mean as a writer, I don't like his own poetry very much, but . . .

AG: It's very wise, and very mellow, and very personal and clear, but there is a lack of spontaneous madness, like in Shakespeare, which he himself admired.

RH: Was it you or Gary Snyder who said somewhere that a poet has to decide whether he's going to go into scholarly work with old manuscripts or crazy blood skull?

AG: I don't think there is a contradiction between scholarship and spontaneity. There wasn't in Zen, or in the Tibetan tradition, or even in the Western tradition. Because Kerouac was quite a great scholar on Buddhism and on Shakespeare. Kerouac would, especially in his later years, read a lot of—well, he would read Kerouac and Shakespeare— he'd put his own books on the shelf because he'd want something to read—but he'd read Sir Thomas Brown, Richard Burton. He would read Rabelais a lot. He read all the great expansive prose writers that had a lot of improvisation in them. A lot of his writings are informed with that. I mean you can hear it underneath, like a lot of Thomas Wolfe, and I hear a lot of Louis-Ferdinand Céline, and maybe some Dostoyevsky, and you can hear some Sir Thomas Brown and Shakespeare under Kerouac. Whitman, lots of Bible.

That was something we didn't get to, more back to the nuts and bolts of transcription. Which is a problem, you see. I have small notebooks, so that determines, in a way, the line length. And then I have this great big notebook around which is for like at home or at night, and that allows for a much larger line and more expansion. Not only in prose but—I mean, looooonnng things . . . (Ginsberg leafs through his big notebook.) See: "Vibration breath of atom, life-breath's passing auras and waves, gross plane to past distant soul my star Jupiter distance." One line.

MG: Do you ever compose on the typewriter?

AG: Yeah. I wrote "Howl" on a typewriter. It doesn't really make very much difference. I could use a small notebook, a big notebook, a typewriter, a tape recorder.

MG: I find I write completely differently if I'm using longhand or a typewriter.

AG: Well, I like longhand a little better because you just cross out, you don't have to worry about the mechanics of a typewriter, an eraser, and things like that. Particularly if you're writing fast sometimes you hit a wrong note or write the wrong letter for the beginning of a word because you have another word in your mind too, and with a pen you can get them both and you can move it around and do things fast and go on ahead, whereas typewriter's a little heavier to work with. But it doesn't make too much difference one way or another. I wrote "Howl" on a typewriter and "Kaddish" out in longhand and "Wichita Vortex" on a tape recorder. The original text of "Wichita Vortex" is *really* close to the final text. The original text of all three are close.

MG: So there's very little revision.

AG: There's a question of tinkering. Once you have the practice down, see, you can trust yourself. Then the difficulty there is that your attention can lapse, just like in sitting. There might be a lapse of attention and your mind goes out of focus and you might be no longer on your first thought but on your third thought or your fifth thought or on the telephone or your foot or hard-on or whatever. Which may or may not be included in the scheme, in the unconscious scheme, in the spontaneous scheme. So there's lapses of attention, or maybe over-heavy solidification of the idea of spontaneity, and you get all hung up describing red white and blue flag fluttering off A-frame brown porch on the white-lined highway. You get too hung up on some mechanical way of flashing or describing flashes. Which is what happens when attention falters, and you're into really following the swiftness.

So where I find there's that lapse of attention, I try to eliminate a line or tinker a little and put the loose strands back together. When the thread of attention is broken, I try and tie it back together. Or, very often, just eliminate the line, put three dots at the end of the solid stuff and three dots at the start of where it begins again. Which I do in "Wichita Vortex" a number of times. Like in Pound, where you have some abrupt beginnings and endings, maybe a sentence unfinished, which I got from Williams. I showed him a poem once that had a good solid middle and a weak beginning and end, and it was self-conscious, so he said to eliminate everything that wasn't active. "Better to have one active line than whole pages of inactive, inert material." One sample of activity like real writing activity. You don't have to finish a poem, all you need is a little fragment that's genuine, solid, and active. His word was "active." So that actually it's not necessary to clean up too much. Blue-penciling and elimination I do quite a bit, sometimes tying threads together that are separated by inattention and inaction.

MG: Do you ever change words, like think of a better word after a while?

AG: Yeah. I did in a poem the other day. The poem is:

When I sit I see dust-motes in my eye
Ponderosa pine-needles trembling in the breeze shiny green
And blue sky
Wind-sound passes through treetops
Distant waves of windiness flutter black oak leaves and leave them
    still
Like my mind, which forgets why the crow
Across the woods-clearing squawks in mid-afternoon[135]

So I changed "crow" to "blue jay" because it actually was a blue jay and
when I was writing I couldn't remember the word "blue jay" and threw
in "crow" falsely. And I thought about taking out "squawks in mid-
afternoon" and put in "squawks mid-afternoon" because the "in" was
just a syntactical linking, a dull moment, so there'd be that cutting of
"in," the slowdown, and "blue jay" because I didn't have the right word
there. I didn't think of "blue jay" 'til two weeks later. I thought, "What
was the word, 'flicker?' No, it wasn't a flicker." I didn't think they had
crows around there, I just used the crow there so I wouldn't get
stopped. So, that kind of tinkering for literality, correction, amplifica-
tion of detail, but real detail.

MG: So the importance of getting the correct bird was for the idea of
the bird rather than the shape of the word or the sound of the word?

AG: "Blue jay" sounds better, I think. If you go back to the original
detail, it always sounds better, it turns out. Strangely, rightly, fortu-
nately. Which goes back to another aphorism like "First thought, best
thought." Ten years ago Kerouac and I cooked up, "If the mind is
shapely, the art will be shapely." If you actually get to the original
nature, the original detail—or in Kerouac's phrase, "Details are the life
of it," of prose, he's talking about—if you get back to the accurate detail,
it always seems to fit in right, whereas an inaccurate detail will cause a
blur. The crow there does actually cause a slight blur because . . . Well,
crow . . .

---

[135]These lines became a section of "Sad Dust Glories." See *Selected Poems 1947–1996*
(New York: HarperCollins, 1996), or *Mind Breaths* (San Francisco: City Lights, 1978).

RH: "Crow" makes a kind of caw. "Blue jay" squawks.

AG: Well, "blue jay" adds in a kind of funny flash of color, and it's accurate to the northern California landscape.

MG: It's also nice with the blue of the sky.

AG: Yeah, that's right, I had that. Forgot that. It's more likely a blue jay across the way that makes that sound. Blue jay squawking is funnier than crow squawking. And "why the blue jay across the woods clearing."

>*da-da DA-da d'da d'da da-da*

is better than:

>*da-da DAAAH d'da d'da da-da*

so that rhythmically it fit better.

EW: Sounds like bebop.

AG: Well, mind is bebop. Vowel. Vowel. Actual speech, *our* speech. Bebop came out of paying attention to the actual rhythms of breathing and speech. The way people talk, the way blacks talk in Harlem.

MG: Boppers especially.

AG: Well, the black language is always ending, "Gotta take that girl, the other one, *one*." There was always a correlation between bop rhythm and speech. Between the note movement, the note flow, which is what Kerouac picked up on and applied to speech. Very consciously, from (Charlie) Parker and some from (Thelonious) Monk. Well, varying rhythms which were like his own, a simulacrum of his own speech, which he had a really perfect ear for.

So the original detail always seems to provide greater music, solider music, and solider eyeflash, phanopoeia flash, phanopoeia being the casting of the image on the mind's eye. And also greater wit. Logopoeia has greater intelligence in it: A crow is almost a generalization, while a blue jay is really particular. So you have logopoeia, the dance of intelligence among words; phanopoeia, the casting of intelligence upon the mind's eye; and melopoeia, the original detail gener-

ally provides a better melody. And those are Pound's three classifica-
tions for the components of poetics.[136] But what's really miraculous,
interesting, is that if trusted, the natural mind, observant, is enough.
You don't need to be arty, you don't need to add on your own egotism
onto it, add your own opinionation and editorial, you can just work
with the natural mind and it's like working with something objective.
So in a sense the most subjective way of writing is also—because it
treats the mind as an object to be examined—is also one of the most
objective ways of writing.

A faking of that spontaneity, like getting into a ga-ga, aggressive,
assertive—"You fuck my mother, I fuck your bloody . . ."—it always
turns out to be the superficial thought rather than the actual
thought. So it's a question of grounding people back to their own
heart-felt mind.

RH: How about Pound?

AG: Well, Pound, I think, was a great practitioner of, or originator, of
the focusing of perception on the original images.

RH: I'm not too familiar with his poetry, but I know his translations
from the Chinese. They're incredibly accurate. You can take an Arthur
Waley translation of something—and he's probably the most respected
Chinese translator—and then the Pound translation, and read them,
and in Pound's translations the poetry is so much better, but he has also
done it at no expense to the literal translation of the poem. And that's
very difficult to do. It requires more than just the spontaneity yoga, it
requires that other yoga I was talking about that I think you have.

AG: Well, Pound had tremendous training, but Williams said Pound
had a mystical ear. Some of the translations, like "The stone drum's
tone." Talking about "stone drum's," you can hear the vowels. So his
basic thing was paying attention to the tone leading of the vowels. And
so he thought in terms of his body thought movement. His heart-
thought, his heart-*felt* thought, was in terms of vowels—"heart-felt
thought" is a nice phrase—and you write with that. So that comes by

---

[136]See "How to Read," included in *Literary Essays of Ezra Pound* (New York: New
Directions, 1968).

itself when you're trying to be accurate. So what we were doing, what I am doing, what everybody who's been involved in the poetic renaissance since the fifties has been doing, has to a great extent been based on Pound and Williams.

MG: And yet Pound, I think, was one of the people that I hold responsible for scaring a lot of people away from poetry.

AG: Well, when you get familiar with it, it's not so bad. I have a long thing of conversations with Pound in the new *City Lights Journal*. Maybe you can take a look at that. He was extremely interesting, but he was working—You gotta remember that he was the one who single-handedly had to drag literature out of sort of a metronomic, automatic desensitization it had fallen into because of the overuse of iambic, repetitive patterns in the nineteenth century. So the way he did it was to go through the treasuries of all the different kinds of poetry, like Provençal and Chinese, to see all the different ways people used to practice and then apply that to the American language. So he had to do all that historical work, get entangled in a lot of intellectual labor to free us from pedantry.

RH: I called you once when I was in high school after reading "Kaddish" with a friend who lives near you.

AG: So what happened when you called?

RH: We called you on the telephone . . .

AG: Was I there?

RH: You had just gone out, and Peter answered the phone and he said, "Yeah, he'll be right back, why don't you call back?" And we called back and you talked to us for 15 or 20 minutes, and Gregory Corso was there, and he talked to us, and it was like a 14-year-old girl meeting the Beatles or something. We didn't know what to do.

AG: I don't remember the conversation.

RH: Well, we talked about parents and family and real important things, about tradition . . .

AG: I dig that situation. Being available.

RH: Yeah, well finally I said how is it that a famous person like you with this huge sack of mail, why is your name and number in the phone book?

AG: It's not anymore.

RH: It's not?

AG: It got too much. I couldn't lie down without the phone ringing.

RH: Yeah, I remember what you said was that, "When I was your age, I would have really liked to speak to Blake on the telephone."

AG: Well, when I was your age I was able to talk to Williams, and it was really like talking to a Zen master who was able to point out a direction to ground me, and it really was helpful.

RH: Then you told us about a poetry reading you were having, and you read us "I Am a Victim of Telephone."

AG: Yeah, I was beginning to feel it then. It's a difficult thing. Dylan doesn't read all his mail. I don't think he can.

RH: That was one thing I was going to say. There are two people I respect for being available, and you're one of them and Pete Seeger is another one.

MG: There are people you do interviews with and there are people you talk to.

AG: Well, a long time ago I figured out that the interview and the media was a way of teaching. If you talk to people as if they were future Buddhas, or present Buddhas, that any bad karma coming out of it will be their problem rather than yours, so you can say anything you want, and you talk on about the highest level possible. And that's worked out. It's left me open to a lot of trouble, though, like with newspaper interviews as a sort of creepnik. But the other thing is that I don't get hung up on any of the interviews or anything like that. I never answered any because I don't go back. Let time take care of karmic tangles.

PETER BARRY CHOWKA
February 10, 1976,   On the Road Between Washington, DC and
    Baltimore
February 28, 1976,   New York City

"This Is Allen Ginsberg?"
New Age Journal, April 1976

Chowka first met Allen Ginsberg when he was a student at Georgetown University and the producer and host of a weekly radio program on WGTB, the radio station owned by Georgetown College. In 1972 Allen and a number of friends, including Peter Orlovsky and poet Nanao Sakaki, were Chowka's guests on his radio program. Allen and Chowka stayed in contact afterward, and when a major reading was planned at Washington's Corcoran Gallery of Art, Allen called Chowka and asked him to assist with publicizing the reading. Writing in 1999, Chowka explained that "the Corcoran represented a prestigious venue for Allen, and he wanted everything to be perfect, especially since Burroughs would be reading as well. Allen was very protective of William. . . . This appearance was part of a re-emergence of Burroughs, a legendary figure, in person in the US . . . after years of self-exile, so there was an additional aura of historical importance about the evening."

Chowka's introduction to his 1976 article explains how the interview was done: "Most of it was recorded during a drive from Washington, D.C. to Baltimore, the morning after a powerful reading by Ginsberg and William Burroughs at the Corcoran Gallery. . . . At the end of February I traveled to Manhattan . . . to complete the interview. . . . Spending the greater part of a day together, we were able to complete the interview in segments, while traveling crosstown in taxicabs, entering and riding the subway, walking down the street, taking advantage of any spare moment which would allow a question and considered answer."

Before the interview, Chowka had decided to focus on the background of Allen's recent work, part of which included Allen's participation in Bob Dylan's Rolling Thunder Revue. The Rolling Thunder Revue had been born in 1975 when Dylan strove to put together a musical tour of New England that would combine spontaneity with intimacy, allowing him and his musical and artistic friends and associates to play in small venues. As more and more artists became part of the entourage, a decision was made to make a film during the tour with members of the tour playing fictional characters in improvised skits. Although Sam Shepard was the official scriptwriter for the film (*Renaldo and Clara*), Allen also assisted in setting up a number of the film's scenes. While Allen had a limited role on stage during the tour, usually joining in the chorus for the show's closing number, "This Land Is Your Land," he was featured in a number of the film's key scenes. Dylan's respect for Allen was shown by his listing Allen in the film's credits as "The Father" and by telling him at the tour's beginning, "You're the king, but you haven't found your kingdom. . . . I'm presenting you. It's about time. This country has been asleep. It's time it woke up."

—DC

Peter Chowka: Allen, since we're in this automobile setting, I want to ask you: much of your poetry, especially in *The Fall of America*, was composed in cars on your various travels. I wonder if, lately, you're writing poetry while on the road in automobiles?

Allen Ginsberg: Not so much. Occasionally, I still write travel poems in airplanes, but not as often.

PC: It might be that the times have changed. In so many of the poems which came out of automobiles in the sixties you really captured the essence of the times, the Vietnam war reports on the radio, the lyrics of the rock music happening then.

AG: Also, we were doing a lot of cross country traveling in cars in the early and mid-sixties. More than now.

PC: A lot of your most recent poetry, especially some that you read last night (Corcoran Gallery, Washington, D.C.) contains very spiritual, and specifically Buddhist, imagery.

AG: Not so spiritual; it's more practical observations during the course of meditation or after.

PC: "Down-to-earth" spiritual, then. You don't like the word "spiritual?"

AG: Yeah, I'm not even sure if the word is helpful because it gets people all distracted with the idea of voices and ghosts and visions. I used to get distracted that way.

PC: How do you select which poems you're going to present at a reading? Do you consider what type of audience you feel will be there?

AG: Well, I read there years before with my father in a celebrated moment, for a Washington society lady who invited us. I met Richard Helms, the head of the CIA, at the last reading . . .

PC: Helms was at the last reading?

AG: Yes. And so this time I was all hyped up, 'cause Burroughs was coming along, too: Burroughs, who's the great destroyer of the CIA, with his prose.

PC: Were you able to sense any reaction from the audience last night to the kinds of things that were being read?

AG: I don't think there were any CIA people there this time, (laughs) I was a little disappointed: there were only secret agents—no big fish. I prepared poems that I hadn't read in Washington before, or poems that were extremely solid; I wanted a solid, good reading of high-quality poems rather than just sort of random poems, daily journal poems. So I picked "pieces" that were complete in themselves. For me the high point was a long, ranting, aggressive, wild poem ("Hadda Be Playing on the Jukebox") linking the CIA and the Mafia and the FBI and the NKVD and the KGB and the multinational cash registers.

PC: One line I especially liked was "Poetry useful if it leaves its own skeleton hanging in the air like Buddha, Shakespeare and Rimbaud." Would it be correct to say, from this line and from some of the other poetry you read, that your sadhana now is the spreading of the dharma through poetry?

AG: Well, I've been working in that direction with Chögyam Trungpa, especially influenced by staying all summer at Naropa Institute at the "Jack Kerouac School of Disembodied Poetics," which is ideationally modeled on Kerouac's practice of spontaneous utterance and Milarepa's similar, or original, practice.

PC: It was Kerouac who originally turned you on to Buddhism, wasn't it?

AG: Yeah, he was the first one I heard chanting the "Three Refuges"[137] in Sanskrit, with a voice like Frank Sinatra.

PC: And he wrote that as-yet unpublished volume *Some of the Dharma*,[138] which, I think, consisted of letters he wrote to you about Buddhism?

AG: Yeah, and he also, in the mid-fifties, wrote *Mexico City Blues*, which is a great exposition of Mind—according to Trungpa. I read aloud to Trungpa halfway through *Mexico City Blues* on a four-hour trip from Karme-Chöling,[139] Vermont, down to New York, and he laughed all the way. And I said, "What do you think of it?" And he answered, "It's a perfect exposition of Mind."

PC: Trungpa is a recognized poet in his own right. Do you think you've become so close to Trungpa because you're both poets?

AG: Oh, yeah, that's a big influence. He encouraged me originally to abandon dependence on a manuscript and to practice improvisational poetry. He said, "Why don't you do like the great poets do, like Milarepa: trust your own mind."

PC: Compose it and then forget it; not necessarily write it down?

AG: It's unforgettable in the sense that it gets on tape. The best thing I ever did was a long "Dharma/Chakra Blues" in Chicago last year, but the tape is completely incomprehensible and I can't transcribe it. That is an old tradition, like Li Po writing poems and leaving them on trees, or Milarepa singing to the wind with his right hand at his ear to listen to the sound, *shabd*.

---

[137]Part of a daily Buddhist practice. By chanting the "Three Refuges" a follower of the Buddhist path recognizes himself as such by taking refuge in the Buddha, the *dharma* (the teachings), and the *sangha* (the community of practitioners). It is the distinctive sign of entering into the Buddhist path: "Om Namo Bhudaya, Om Namo Dharmaya, Om Namo Sangaya."

[138]*Some of the Dharma* is now in print; ed., David Stanford (New York: Viking, 1997).

[139]Buddhist retreat center near Barnett, Vermont, established by Chögyam Trungpa, Rinpoche.

PC: How long have you known Trungpa now? He seems to have become a great influence in your life.

AG: An enormous influence. We first met on the street in 1971, in front of Town Hall (New York City). I stole his taxicab; my father was ill and I wanted to get my father off the street.

PC: It was purely an accidental encounter?

AG: Yeah. I said *"Om Ah Hum Vajra Guru Padma Siddhi Hum"* and gave him a *"Namaste"*[140] when he was introduced. I asked him years later what he thought of my pronouncing the Padma Sambhava mantra[141] to salute him, and he said, "Oh, I wondered if you knew what you were talking about." (*laughs*) He's been pushing me to improvise, to divest myself of ego eventually, kidding me about "Ginsberg resentment" as a national hippie characteristic, and warning me to prepare for death, as I registered in a poem called "What Would You Do If You Lost It?" published by the Lama Foundation.

PC: As far as the "resentment" aspect, has he influenced you in that direction? For example, many of the poems you read last night seemed more contemplative, meditative.

AG: He has provided a situation in which I do sit, like at the Naropa seminaries or at the intensive sitting meditations where Peter (Orlovsky) and I have gone and sat for a week at a time in retreat cabins in the Rocky Mountains, or where I've sat weeks alone, and he's suggested that I not write during those weeks when I'm in retreat—which has resulted in a lot of post-sitting, meditative, haiku-like writings. He's also made me more aware of the elements of resentment, aggression, and dead-end anger in my earlier poetry and behavior, which is useful to know and be mindful of. It doesn't necessarily curb it, but I'm able at least to handle it with more grace, maybe, as last night, where I read a whole series of meditative poems and then this outrageous attack on the CIA-Mafia-FBI connection. But it was put in a context where it was like the normal explosion of, maybe even, *vajra*-resentment, so that it doesn't become a dominant paranoia but is seen within the greater space—the flow of Mind Consciousness while sitting—of

---

[140]Sanskrit for "salutation" (from *Namo*, salutations).

[141]One of the two most famous mantras in Tibet. It is said that reciting this is equivalent to practicing the whole teaching of the Buddha.

continuing mindfulness over the years. Trungpa's basic attitude toward that kind of political outrage is that things like gay liberation, women's liberation, peace mobilization, have an element—a seed—of value in them; but it depends on the attitude of mind of the participant as to whether it's a negative feedback and a karmic drug or a clear healthy, wholesome action.

PC: Often those political movements can become so mutually exclusive that they serve to isolate one from a lot of the potential . . .

AG: Or so filled with resentment that they become dead-ends. More and more, by hindsight, I think all of our activity in the late sixties may have prolonged the Vietnam war. As Jerry Rubin remarked after '68, he was so gleeful he had torpedoed the Democrats. Yet it may have been the refusal of the Left to vote for Humphrey that gave us Nixon. Humphrey and Johnson were trying to end the war to win the election, while Nixon was sending emissaries (Mme. Claire Chennault) to Thieu saying, "Hang on until I get elected and we'll continue the war." Though I voted for Humphrey in '68 I think a lot of people refused to vote, and Nixon squeaked in by just a couple of hundred thousand votes.

PC: And now, eight years later, we might get Humphrey again anyway.

AG: So that might be the karma of the Left, because of their anger, their excessive hatred of their fathers and the liberals, their pride, their vanity . . . *our* vanity, *our* pride, *our* excessive hatred. It may be that we have on our karma the continuation of the Vietnam war in its worst form with more killing than before. We may have to endure Humphrey so that we can take the ennui or boredom of examining what we've wrought when we got "exciting" Nixon. In a way it all balanced out; maybe it was better that Nixon got in because then we had Watergate and the destruction of the mythology of authority of a hypocrite government.

PC: Since this is 1976, a year of inevitable increase in political discussion, I'd like to ask the following question. Your Buddhist practice seems not to have interfered with the acute political concern, for the CIA and other issues, which you continue to display in recent poems like "Hadda Be Playing on the Jukebox." How, if at all, has your work with Trungpa—your extensive meditation practice—changed your outlook on North American or world politics?

AG: It has changed it somewhat from a negative fix on the "fall of America" as a dead-end issue—the creation of my resentment—into an appreciation of the fatal karmic flaws in myself and the nation. Also with an attempt to make use of those flaws or work with them—be aware of them—without animosity or guilt; and find some basis for reconstruction of a humanly useful society, based mainly on a less attached, less apocalyptic view. In other words, I have to retract or swallow my apocalypse. *(laughs)*

PC: That's a lot to swallow. Do you have any specific thoughts on the American political scene in this presidential election year?

AG: Governor Jerry Brown.

PC: Is the condition of the Left refusing to support Humphrey in '68 the main thing that comes to mind in talking of the mistakes of the sixties, or are there other things that you've realized, as well?

AG: Well, that's sort of a basic mistake you can refer to that everybody can remember in context, I think, so it's a good, solid thing. What was the point of the Left? It was saying, "End the war." What was the action of the Left? It refused to support Humphrey because he wasn't "pure" enough, *(laughs)* so there was an apocalyptic purity desire which maybe was impractical, or "unskillful means."

PC: Which seems to go along with what I know about Trungpa and his teachings, in general: that it should be a very down-to-earth, practical sadhana, which doesn't include requirements of stringent vegetarianism or giving up cigarette smoking.

AG: And which is mindful of that quality of resentment which he characterizes as "Ginsberg resentment" or "America Ha-Ha." I was resentful, at first, when he came on with that kind of line. Actually, I voted for Humphrey, so I wasn't dominated that much by resentment, but it seems to be a stereotype, maybe 'cause Trungpa reads too much *Time* magazine. He's entitled to his opinion, and I'm surely profiting psychologically from him because there's enough insight in that to make me halt in my tracks and think twice, thrice.

PC: Do you see his movement in contemporary Buddhism as the most vital one in America at this point?

AG: Shakespeare has a very interesting line—"Comparisons are odious." So to say "the most vital," well, everybody's doing a different kind of work—some quiet, some more flashy. I seem to be able to relate to Trungpa best, although I must say that it may be that the looseness and heartiness and charm of his approach is not necessarily the deepest for my case. I notice I'm very slow in getting into my prostration: of 100,000 prostrations, I've done only 10,000 and I'm way behind, maybe the last in the class. But I guess he's gotten a lot of people more deeply into foundation practices, perhaps on more of a mass scale than any other Tibetan lama, if that's any good count. I suppose it's the quality of the student that counts. Trungpa's movement is a very rational and classical approach to Buddhism, in his real serious attention to sitting: "Go sit, weeks and weeks and weeks, ten hours a day."

PC: It's primarily silent meditation?

AG: His basic approach is to begin with *shamatha*,[142] a Sanskrit word meaning peaceful mindedness, creating tranquillity of mind. It consists of paying attention to the breath coming out of the nostrils and dissolving in space, the outbreath only, and is a variety of *vipassana* practice,[143] which begins with concentration on the breath passing in and out just at the tip of the nose, or Zen practice which involves following the breath to the bottom of the belly.

PC: What is it about the Tibetan style of Buddhism that first attracted you?

AG: Originally it was the iconography: the mandalas, the Wheel of Life,[144] and the Evans-Wentz books, some of which were recommended

---

[142]*Shamatha* or *samatha* is Sanskrit for tranquillity or peacefulness. According to Trungpa, one develops this through concentration and focus on simple acts.
Vipasyana takes this concentration and expands it into awareness of space. See *Cutting Through Spiritual Materialism*, Chögyam Trungpa (Boston and London: Shambhala, 1987); *Meditation in Action*, Chögyam Trungpa (Boston and London: Shambhala, 1992.)
[143]*Vipassana* (see prior note) in the Gelugpa school is insight brought about by a form of analytic examination.
[144]This mandala is very widespread in Tibetan Buddhist iconography, and depicts the cycle of existence. It is divided into six realms of experience, the upper half consisting of the realm of the gods, and the human realm, and the realm of the jealous

by Raymond Weaver, who was a professor of English at Columbia University in the '40s. Weaver gave Kerouac a list of books to read after he read an unpublished early novel of Kerouac's titled *The Sea Is My Brother*—a list which included the early gnostic writers, *The Egyptian Book of the Dead* and *The Tibetan Book of the Dead*. The *Herukas*—the many-armed, fierce guardian deities—reminded me of visions I'd had in 1948 relating to William Blake's poem, "The Sick Rose"; visions of terror—of the universe devouring me, being conquered and eaten by the universe. I used *The Tibetan Book of the Dead* while ingesting *ayahuasca* in New York City in 1960–61. Later on, some contact with Dudjom Rinpoche[145] in India reinforced this interest in Tibetan Buddhism.

PC: How has your study of Tibetan Buddhism, and your work with Trungpa Rinpoche, altered or expanded your own awareness?

AG: The *shamatha* meditation which I've practiced for a number of years under Trungpa Rinpoche's auspices leads first to a calming of the mind, to a quieting of the mechanical production of fantasy and thought forms; it leads to sharpened awareness of them and to taking an inventory of them. It also leads to an appreciation of the empty space around into which you breathe, which is associated with *dharmakaya*. In the tradition of the *vipassana* practice, this leads to insight into detail in the space around you, which is exemplified in William Carlos Williams' brief poems noting detail in the space around him. I'll paraphrase his poem "Thursday"—"I've had my dreams, like other men, but it has come to nothing. So that now I stand here feeling the weight of my coat on my shoulders, the weight of my body in my shoes, the breath pushing in and out at my nose—and resolve to dream no more." In terms of external manifestation rather than just subjective awareness, an observer could see in me some results of that "widening of the area of consciousness," which is a term that I used at the end of "Kaddish." For example, since 1971, I've come to improvise poetry or

---

gods; the lower half's three realms are the animal realm, that of the hungry ghosts, and a realm consisting of various hells. The wheel itself is in the claws of Yama, the god of the underworld, representing death. At the center are the three driving forces of existence: desire, ignorance, and aggression, which are represented by cock, pig, and snake, respectively, each chasing the other by the tail.

[145]Rinpoche: Tibetan Buddhist term literally meaning "greatly precious" and a title for particularly qualified lamas, a lama being the Tibetan equivalent, but not limited to, the Sanskrit word "guru."

song on the stage, trusting my own mind rather than a manuscript. Also, I do a lot of sitting, which is, in itself, a self-sufficient activity.

PC: Before you began to study with Trungpa, you'd never associated yourself with a spiritual master?

AG: I had worked with Swami Muktananda[146]—"Kundalini Swami," as Gary Snyder calls him, and sat for a year and a half with a mantra that he had given me.

PC: You knew Swami Bhaktivedanta (leader of the International Society of Krishna Consciousness) as well.

AG: Since '66 I had known Swami Bhaktivedanta and was somewhat guided by him, although not formally—spiritual friend. I practiced the Hare Krishna chant, practiced it with him, sometimes in mass auditoriums and parks in the Lower East Side of New York.

PC: You really did a lot to popularize that chant. Probably the first place I heard it was when I saw you read in '68.

AG: Actually, I'd been chanting it since '63, after coming back from India. I began chanting it, in Vancouver at a great poetry conference, for the first time in '63, with (Robert) Duncan and (Charles) Olson and everybody around, and then continued. When Bhaktivedanta arrived on the Lower East Side in '66 it was reinforcement for me, like "the reinforcements had arrived" from India.

PC: You mentioned your trip to India in the early sixties. Do you consider that to be very significant in your orientation afterwards toward your present spiritual goals?

AG: My trip wasn't very spiritual, as anybody can see if they read *Indian Journals*. Most of it was spent horsing around, sightseeing and trying the local drugs. But I did visit all of the holy men I could find and I did encounter some teachers who gave me little teachings then that were useful then and now. Some of the contacts were prophetic of what I

---

[146]Also known as Swami Muktananda Paramahamsa. In 1970 Ginsberg went to Dallas on Muktananda's invitation and practiced his particular style of meditation and the mantra *Guru Om*. Snyder's comment, "Kundalini Swami," refers to the meditation's focus on awakening the Kundalini, or life force, once initiated by the guru. See Muktananda's autobiography, *Play of Consciousness*, Chitshakti Vilas/Swami Muktananda, (San Francisco: Harper & Row, 1978).

arrived at later here in America, because I met the head of the Kagyü order, Gyalwa Karmapa there, and saw the black crown ceremony in Sikkim in '62 or '63. He subsequently visited the U.S. with Trungpa as host. I went to see Dudjom Rinpoche, the head of the Nyingma sect and got one very beautiful suggestion from him about the bum LSD trips I was having at the time, which I'll quote again: "If you see something horrible, don't cling to it; and if you see something beautiful, don't cling to it."

PC: Has LSD been less of a factor in your life lately?

AG: Less, though it was a strong influence and I think basically a good influence. I went through a lot of horror scenes with it. Finally, through poetic and meditation practice I found the key to see through the horror and come to a quiet place while tripping.

PC: Do you ever find it possible to do serious meditation while under the influence of drugs, or do you find the two exclusive?

AG: I haven't tried since I've been more deeply involved in meditation. The last time I took acid, I went up into the Teton Mountains, to the top of Rendezvous Mountain, and made a little sitting place on the rocks, near the snow. Just sat there all day, unmoved, unmoving, watching my breath, while white clouds pushed casting shadows on the stillness of the white snow. It was like sitting up in the corner of a great mandala of the god-worlds thinking of the hells—bombing Cambodia—going on down the other side of the mandala, the other side of the round earth; and then breathing, and the thought dissolving, and the physical presence of the place where I was resuming, sitting in a white snowy place in the middle of the whole "empty" vast full universe.

PC: The reason I asked is that most teachers I've heard of have counseled against using drugs or have said they're an impediment on the path, although occasionally my own experience has been to have felt really profound mystical meditative states while under the influence of certain drugs, by combining them with sitting or reading certain texts, for example. I feel they've opened me to a more expansive consciousness.

AG: I think that even those teachers who disapprove of the use of drugs by their students do credit the LSD wave with opening up people's awareness to the possibility of alternative modes of consciousness, or at least a search for some stable place, or perhaps leaving their imaginations open to understand some of the imagery, such as the wheels of life. Trungpa's position is that "psychedelics" are too trippy, whereas people need to be *grounded*; everything is uncertain enough as it is. The world, societies, mind are uncertain. What's needed is some non-apocalyptic, non-ambitious, non-spiritually materialistic, grounded sanity, for which he proposes *shamatha* meditation and discourages grass and acid, which is logically sensible. I think he may have some more ample ideas about that for specific situations.

PC: I want to talk a little about the concept of "egolessness," which is something a lot of us have trouble defining and practicing. Last night you mentioned the three marks of existence are "change, suffering and egolessness."

AG: Trungpa lectured on that at Naropa last year, very beautifully, and I turned it into a stanza:

> Born in this world
> you got to suffer
> everything changes
> you got no soul

representing suffering, change/transiency, and *anatma* or no permanent essential identity, meaning, in a sense, nontheism, or non-selfism. It's a description of the nature of things, by their very nature. It might knock out Krishna and Joya and God and some notions of Christ and some notions of Buddha. It may not necessarily knock out devotion or the quality of devotion, though.

PC: How long ago was your poem "Ego Confession" written? I'm curious, because the line in it that I picked up on was the first one: "I want to be known as the most brilliant man in America."

AG: Yeah, it's obviously a great burlesque, a take-off on myself, shameful, shocking. *(laughs)* I wrote it in October '74, listening to Cecil Taylor play in a nightclub in San Francisco, sitting next to Anne Waldman, who is the co-director of the Kerouac School of Poetics at Naropa. And I was so ashamed of what I wrote down that I wouldn't let her see it, I hid my notebook from her with my hand. Within a month I realized that the poem was funny.

PC: Do you have any new poems (in your notebook) that you'd care to read for us while we're on this trip to Baltimore?

AG: I think the text of the "Gospel of Noble Truths" hasn't been printed anywhere. It's a gospel style song, for blues chord changes one/four/one/five/ and next stanza return to one. There's another reflection of that theme in a poem I wrote along on the Rolling Thunder Review.

> Lay down Lay down yr Mountain Lay down God
> Lay down Lay down yr music Love Lay down
> Lay down Lay down yr hatred Lay yrself down
> Lay down Lay down yr Nation Lay yr foot on the Rock
> Lay down yr whole Creation Lay yr Mind down
> Lay down Lay down yr Magic Hey Alchemist Lay it down
>   Clear
> Lay down yr Practice precisely Lay down yr Wisdom dear
> Lay down Lay down yr Camera Lay down yr Image right
> Yea Lay down yr Image Lay down Light.
>   —Nov. 1, 1975

PC: Is Dylan the "Alchemist" in those lines?

AG: Yeah, the poem is directed to him, because we were considering the nature of the movie we were making, which will be a nice thing, a sort of "dharma movie," hopefully, depending on how it's edited. The movie made along the Rolling Thunder tour (120 hours of film which will have to be reduced to three) has many "dharma" scenes. It was like a Buddhist conspiracy on the part of some of the directors and film cameramen; the director Mel Howard was out at Naropa last year. In every scene that I played I used the Milarepa mantra "Ah" and kept trying to lay it on Dylan or the audience or the film men.

PC: Much of Dylan's music, even from the middle, electric period of his career, has impressed me as being very Zen-like in a lot of its imagery. Knowing him well as you do, do you think he has been influenced by Zen or Buddhism?

AG: I don't know him because I don't think there is any him, I don't think he's got a self!

PC: He's ever-changing.

AG: Yeah. He's said some very beautiful, Buddha-like things. One thing—very important—was I asked him whether he was having pleasure on the tour, and he said, "Pleasure, Pleasure, what's that? I never touch the stuff." And then he went on to explain that at one time he had had a lot of pain and sought a lot of pleasure, but found that there was a subtle relationship between pleasure and pain. His words were, "They're in the same framework." So now, as in the *Bhagavad Gita*, he does what it is necessary to do without consideration of "pleasure," not being a pleasure junkie, which is good advice for anyone coming from the top-most pleasure-possible man in the world. He also said he believed in God. That's why I wrote "Lay down yr Mountain Lay down God." Dylan said that where he was, "on top of the Mountain," he had a choice whether to stay or to come down. He said, God told him, "All right, you've been on the Mountain, I'm busy, go down, you're on your own. Check in later." *(laughs)* And then Dylan said, "Anybody that's busy making elephants and putting camels through needles' eyes is too busy to answer my questions, so I came down the Mountain."

PC: Several of his albums have shown his interest in God, especially *New Morning*.

AG: "Father of Night," yeah. I think that is, in a sense, a penultimate stage. It's not his final stage of awareness. I was kidding him on the tour, I said, "I used to believe in God." So he said, "Well, I used to believe in God, too." *(laughs)* And then he said, "You'd write better poetry if you believed in God."

PC: You've been fairly close to Dylan for a number of years now . . .

AG: No, I didn't see him for four years. He just called me up at 4 A.M. and said "What are you writing, sing it to me on the telephone." And then said, "O.K., let's go out on the road."

PC: He was encouraged by a letter you'd written him about your appreciation of his song "Idiot Wind?"

AG: Denise Mercedes, a guitarist whom Dylan admires, was talking to Dylan, and he mentioned to her that he was tickled. I had written a long letter to him demanding $200,000 for Naropa Institute, to sustain the whole Trungpa scene, just a big long kidding letter, hoping that he'd respond. He liked the letter, he just skipped over the part about money. (He doesn't read anything like that, I knew, anyway.) But then I also explained what was going on at Naropa with all the poets. I said also that I had dug the great line in the song "Idiot Wind," which I thought was one of Dylan's great great prophetic national songs, with one rhyme that took in the whole nation, I said it was a "national rhyme."

> Idiot wind
> Blowing like a circle around my skull
> From the Grand Coulee Dam to the Capitol

Dylan told Denise that nobody else had noticed it or mentioned it to him; that the line had knocked him out, too. He thought it was an interesting creation, however he had arrived at it. And I thought it was absolutely a height of Hart Crane-type poetics. I was talking earlier about resentment. "Idiot Wind" is like Dylan acknowledging the vast resentments, angers and ill-temper on the Left and the Right all through America during the sixties, calling it an "idiot wind" and saying "it's a wonder we can even breathe" or "it's a wonder we can even eat!"

PC: Right, and directing it at himself, as well.

AG: Yeah, talking about it within himself, but also declaring his independence from it. There's a great line in which he says, "I've been double-crossed now for the very last time, and now I'm finally free," recognizing and exorcising the monster "on the borderline between you and me."

PC: You've obviously been impressed by Dylan and his music during the last decade.

AG: He's a great poet.

PC: Is it possible for you to verbalize what kinds of influence he's had on your own style of poetry?

AG: I've done that at great length in the preface to a new book, *First Blues*, which has just been published in only 1,500 copies, so it's relatively rare. I wrote a long preface tracing all the musical influences I've had, including Dylan's, because I dedicated the book to him. He taught me three chords so I got down to blues. Right after Trungpa suggested I begin improvising, I began improvising and Dylan heard it, and encouraged it even more. We went into a studio in '71 and improvised a whole album.

PC: Which has never been released. Do you think it ever will be?

AG: Oh, on Folkways, or something.

PC: Back to the Rolling Thunder Tour. Perhaps you can place it in the context of the Beat movement of the fifties and the consciousness expansion of the sixties. Something you said while on the tour indicated that you saw it as being perhaps that important; you said that "the Rolling Thunder Revue will be one of the signal gestures characterizing the working cultural community that will make the seventies."

AG: Wishful thinking, probably, but at the same time wishful thinking is also prophesy. It seemed to me like the first bud of spring. I thought that the gesture toward communalism—almost like a traveling rock-family-commune that Dylan organized, with poets and musicians all traveling together, with the musicians all calling each other "poet"— "sing me a song, poet"—was a good sign. The fact that he brought his mother along—the "mysterious" Dylan had a chicken-soup, Yiddish Mama, who even got on stage at one point . . .

PC: Not to mention bringing his wife Sara and Joan Baez.

AG: Sara came, and his children came. And Sara met Joan Baez and they all acted in the movie together, and Joan Baez brought her mother and her children, and Ramblin' Jack Elliot had his daughter. So there was a lot of jumping family.

PC: Sounds like Dylan tying up a lot of loose karmic ends.

AG: Right. As he says in the jacket notes to the *Desire* album, "We've got a lot of karma to burn." To deal with or get rid of, I think he means.

PC: It was really a unique tour, bringing you primarily to small towns and colleges in New England . . .

AG: The great moment was arriving at Jack Kerouac's natal place, Lowell, Massachusetts, and going to Kerouac's grave.

PC: Was Dylan moved during that experience?

AG: He was very open and very tender, he gave a lot of himself there. We stood at Kerouac's grave and read a little section on the nature of self-selflessness—from *Mexico City Blues*. Then we sat down on the grave and Dylan took up my harmonium and made up a little tune. Then he picked up his guitar and started a slow blues, so I improvised into a sort of exalted style, images about Kerouac's empty skull looking down at us over the trees and clouds while we sat there, empty-mouthed, chanting the blues. Suddenly, Dylan interrupted the guitar while I continued singing the verses (making them up as I went along so it was like the triumph of the Milarepa style) and he picked up a Kerouac-ian October-brown autumn leaf from the grass above his grave and stuck it in his breast pocket and then picked up the guitar again and came down at the beat just as I did, too, and we continued for another couple of verses before ending. So it was very detached and surrendered; it didn't even make a difference if he played the guitar or not. It was like the old blues guitarists who sing a cappella for a couple of bars.

PC: Has Dylan ever acknowledged to you that Kerouac was an influence on him or that he's familiar with his work?

AG: Yes, oddly! I asked him if he had ever read any Kerouac. He answered, "Yeah, when I was young in Minneapolis." Someone had given him Kerouac's *Mexico City Blues*. He said, "I didn't understand the words then, I understand it better now, but it blew my mind." So apparently Kerouac was more of an influence on him than I had real- ized. I think it was a nice influence on him.

PC: Which poem was he reading from Kerouac's *Mexico City Blues*?

AG: It's one toward the end of the book, which he picked out at ran- dom. I had picked out something for him to read and, typical Dylan, he turned the page and read the other one on the opposite side of the page. *(laughs)*

PC: Which one did you pick out for him to read?

AG: "The wheel of the quivering meat conception turns in the void," the one that, I think, ends, "*Poor!* I wish I were out of this slaving meat wheel and safe in heaven, dead." There was another one I picked which lists all the sufferings of existence and ends, "like kissing my kitten in the belly, the softness of our reward."

PC: Was it your suggestion that Rolling Thunder include Lowell on the tour?

AG: No, Dylan had chosen it himself. We did a lot of beautiful filming in Lowell—one of the scenes described by Kerouac is a grotto near an orphanage in the center of red brick Catholic Lowell near the Merrimac River. So we went there and spent part of the afternoon. There's a giant statue of Christ described by Kerouac. Dylan got up near where the Christ statue was on top of an artificial hill-mound, and all of a sudden he got into this funny monologue, asking the man on the cross, "How does it feel to be up there?" There's a possibility . . . everyone sees Dylan as a Christ-figure, too, but *he* doesn't want to get crucified. He's too smart, in a way. Talking to "the star" who made it up and then got crucified Dylan was almost mocking, like a good Jew might be to someone who insisted on being the messiah, against the wisdom of the rabbis, and getting himself nailed up for it. He turned to me and said, "What can you do for somebody in that situation?" I think he quoted Christ, "suffer the little children," and I quoted "and always do for others and let others do for you," which is Dylan's hip, American-ese paraphrase of Christ's "Do unto others . . . ," in "Forever Young."

So there was this brilliant, funny situation of Dylan talking to Christ, addressing this life-size statue of Christ, and allowing himself to be photographed with Christ. It was like Dylan humorously playing with the dreadful potential of his own mythological imagery, unafraid and confronting it, trying to deal with it in a sensible way. That seemed to be the characteristic of the tour: that Dylan was willing to shoulder the burden of the myth laid on him, or that he himself created, or the composite creation of himself and the nation, and use it as a workable situation; as Trungpa would say, "alchemize" it.

We had another funny little scene—I don't know if these will ever be shown in the film, that's why I'm describing them—with Dylan playing the Alchemist and me playing the Emperor, filmed in a diner outside of Falmouth, Massachusetts. I enter the diner and say, "I'm the emperor, I just woke up this morning and found out I inherited an empire, and it's bankrupt. I hear from the apothecary across the street that you're an alchemist. I need some help to straighten out karmic problems with my empire . . . I just sent for a shipload of tears from Indo-China but it didn't seem to do any good. Can you help, do you have any magic formulae for alchemizing the situation?" Dylan kept denying that he was an alchemist. "I can't help, what're you asking me for? I don't know anything about it." I said, "You've got to, you've got to be a bodhisattva, you've got to take on the responsibility, you're the alchemist, you know the secrets." So he asked the counterman, who was a regular counterman at a regular diner, to bring him some graham crackers and some Ritz crackers, ketchup, salt, pepper, sugar, milk, coffee, yogurt, and apple pie. He dumped them all in a big aluminum pot. Earlier, I had come in and lay down my calling card, which was an autumn leaf, just like the one Dylan pocketed in the graveyard—the leaf which runs through many of the scenes in the movie, representing, like in Kerouac's work, transiency, poignancy, regret, acknowledge-ment of change, death. So I threw my calling card leaf in the pot and Dylan threw in a piece of cardboard, and then he fished out the leaf, all muddy, and slapped it down on the counter on top of my notebook, where I was taking down all the magical ingredients of his alchemical mixture. Then I said, "Oh, I see the secret of your alchemy: ordinary objects." "Yes," he said, "ordinary mind." So that was the point of that. Next I said, "Come on, look at my kingdom," and he said, "No, I don't want anything to do with it" and he rushed out of the diner. I followed him out, like in a Groucho Marx movie, and stopped: turned to the camera, lifted my finger, and said, "I'll find out the secret." Then we redid the scene and, coyote magician that he is, with no consistency, he suggested towards the end of the scene, "Well, why don't we go look at your kingdom?" So he led the way out and we went to see the "empire." He was completely unpredictable in the way he would improvise scenes. All the scenes were improvised.

PC: During the Rolling Thunder tour some of the participants expressed the hope that it might continue as some sort of functioning community. Are there any indications now, several months later, that that may come to pass, either through the film or another tour of the Midwest?

AG: I don't think it was intended to be a continuously functioning community in any formal way, like people living together. I don't think the energy would depend on that group of people continuing any more than, say, all the San Francisco poets living together. I think it might be necessary for those people to disperse and de-centralize, and also for Dylan to try something new—not do just one thing, but continue open-hearted experimenting.

PC: With (by now) ten years added perspective to your heralding a "new age" in *The Fall of America*, what are your present views on what the artist and the poet can do to hasten the advent of that "new age?"

AG: To paraphrase the poem: "make laughing Blessing." That particular quotation is probably the happiest and most optimistic, and at the same time the most egotistically righteous, lyric in *The Fall of America*. It invokes the spirit of both Hart Crane, who committed suicide, and Whitman, who didn't commit suicide, in building an American bridge to the future. I don't know, though. I don't have any simple answer to what the poet can do or should do.

PC: Theodore Roszak's chapter on your work, in *The Making of a Counter Culture*, quotes Wordsworth:

> We poets in our youth begin in gladness;
> But thereof in the end come despondency and madness.

As you approach your 50th birthday, your life outwardly seems to be the opposite of Wordsworth's dictum. Would you credit your Buddhist viewpoint and practice with having made the difference?

AG: My own common sense, and my experience of my mother's madness as a kind of preventive and toxin, as well as the ripening of my own awareness and peaceableness through *shamatha* meditation.

PC: Prior to your vision of Blake in 1948, had you ever gone through an agnostic, or questioning period concerning religion, spirituality, God . . . and, if so, did that vision bring you back to the realization of the imminent transcendence of God within everyday reality?

AG: I had absolutely no interest in religion, God or spirituality before the vision, although Kerouac and I had concocted a search for a "New Vision" back in 1944–45.

PC: An aspiritual "New Vision?"

AG: Yes. We didn't have any idea what we were looking for.

PC: Your experience seems to parallel what many young people underwent in the sixties and seventies. First, de-programming themselves from heavy religious conditioning they had undergone as children, and then coming back to a spiritual sensibility, either through drugs or . . .

AG: I never *had* any religious conditioning and I never *came back* to any.

PC: You're fortunate in that case.

AG: Yeah, thank God!

PAUL PORTUGÉS AND GUY AMIRTHANAYAGAM
Portugés: July 1, 1976,   Boulder, Colorado
Amirthanayagam: October 17, 1977,   Honolulu, Hawaii

The Portugés portion of this interview was published in his book, *The Visionary Poetics of Allen Ginsberg* (1978, Ross-Erikson) in a chapter entitled "On Tibetan Buddhism"; both parts were published as one chapter (titled "Buddhist Meditation and Poetic Spontaneity") of a book edited by Amirthanayagam, *Writers in East-West Encounter* (1982, Macmillan).

The first part of the following dual interview includes an excerpt from one of a number of interviews done by Paul Portugés over a period of several weeks in the summer of 1976 for a book Portugés was writing. The Portugés interviews were done in Allen's apartment in Boulder, Colorado, while Ginsberg was at the Naropa Institute. Portugés taped and transcribed the interviews. In 1999 Portugés remembered, "Interviewing Allen was very relaxed. Usually we would eat, drink, and just talk, quite casually. He was spontaneous, easygoing, and very articulate. I often saw Allen after that period, whenever he came to Santa Barbara. I wept when I saw a photo of him on his death bed."

The Amirthanayagam interview took place in Honolulu, where Amirthanayagam was director of the Literature and Culture Program at the East-West Center. The two men began talking outside Amirthanayagam's home, which has a view of Waikiki Beach, while Allen sucked on passion fruits from his host's garden. They then retreated to Amirthanayagam's office where the interview was video-taped. Amirthanayagam transcribed the tape.[147]

—DC

_____

[147]The date of the Amirthanayagam interview is established from a typed and hand-written list of interviews found in Ginsberg's former Union Square office.

Paul Portugés: What do you think about Trungpa's approach to meditation?

Allen Ginsberg: The Tibetan monks I've talked to all report that Trungpa's meditation is very good. His teaching of meditation is excellent; acute, practical, experienced—seems to know all the angles. From his own experience, he's gone to the center and is able to teach it well. He said some amazing things to me, like I was hung up on where does my breath begin and end. I went through it very early, and he gave me the image of the breath continuing, sort of, from one breath to another like an opening up of a telescope. Beautiful. I mean one breath leading to another, like the unfolding or opening up of a telescope. Very beautiful, precise image; and once I thought of it in those terms, it seemed to resolve a psychological, mental thing I had, or a self-consciousness I had in proceeding from one breath to another.

PP: But you've always been concerned with breath, much longer than you've been studying in the Tibetan tradition.

AG: Yes, that's true; it was implicit in the long-line poems, like "Howl."

PP: Has it changed, the poetics of breath, since you've been practicing shamatha, etc.?

AG: No, because poetry, poetic practice, is sort of like an independent carpentry that goes on by itself. I think, probably, the meditation experience just made me more and more aware of the humour of the fact that breath is the basis of poetry and song—it's so important in it as a measure. Song is carried out on the vehicle of the breath, words are carried out through the breath, which seems like a nice "poetic justice," (laughs)—that the breath should be so important in meditation as well as in poetics. I think that must be historically the reason for the fact that all meditation teachers are conscious of their spoken breath, as poets are. That's the tradition, the Kägu tradition, that the teachers should be poets. And that's the reason for the Naropa Kerouac School of Disembodied Poetics; originally, Trungpa asked me to take part in the school because he wanted his meditators to be inspired to poetry, because they can't teach unless they're poets—they can't communicate.

PP: In the tradition of Milarepa . . .

AG: Or any tradition, really: sharp, acute, flavorful communication.

PP: What is the influence on your recent poetry of your dedication to the *shamatha-vipasyana* practice? For example, in terms of the focus of attention, the ability to actually see, record, etc.; how the mind operates, as well as becoming conscious of your perceptions of the quotidian world . . .

AG: It's more in the area of observing the mechanical nature of certain passions, like anger and sex, so that they become more transparent. I get less entrapped in them, like political anger—it affects me poetically. One thing Trungpa said when I was yelling at him for smoking and drinking was that any trip I lay on him proceeds from my own anxiety, and creates more anxiety in him. And so does not resolve the problem, but increases the problem.

PP: So, his instructions had an effect on your political . . .

AG: Well, it had an effect on my entire political strategy. I had the same lesson with my father. The more I attacked my father, the more I drove him into the wall, so that he wouldn't oppose the war, until one day I heard him arguing against the war and taking my side with someone else. I realized that the argument I had with my father had nothing to do with the war. It was a wall of frustration between us. In order to get my father really to oppose the war, I had to soften down and talk about it reasonably, without attacking him, without animosity. I just had really to reconsider the whole thing, present my facts in an orderly way, in a way that he could understand and receive without it offending his ego. That's *upaya*, skillful means, I suppose; but that depends on curbing your anger and being able to communicate what you know. When you're shouting and angry, you don't really communicate the details of what you know. Very often you mistake opposition for evil, when it's misunderstanding.

PP: But you still are writing political poetry, as I noticed going through your new manuscript, the one you are preparing for City Lights—what's it going to be called?

AG: *Mind Breaths*, maybe. Yes, I'm still writing political poetry . . . Life is too dangerous to get angry. The anger only comes back on yourself; also the anger only comes back on me—in the form of sickness, finally. I think all the anger I accumulated was responsible for the illnesses I have had in the last few years. I had a really total object lesson in that, breaking my leg in 1971 I think it was.

PP: Didn't you slip on some ice or something?

AG: Yeah. I was up in Cherry Valley on the farm with Gregory (Corso). I was a little irritated at having to take care of him. Denise (Mercedes) and the other people had left a lot of dogs around, and I was responsible for taking care of the dogs too. I don't like dogs that much. I didn't want to take care of more than one or two, and there were four or five. I had to keep interrupting my "poetic beauty" to go outside every day and take care of the dogs—feed them, water them. One really cold day, I went out to the barn where they were, to bring them food. I was carrying their water and food, but I was really irritated and angry—stomping out angrily, and I stomped right out on the ice and slipped and fell. As I was walking I was thinking—resentfully—"Why did they leave those dogs with me, rah, rah, rah. . . ." So I wasn't watching where I was going. I wasn't being mindful of the fact that I had slipped on the ice because of these slick tennis shoes I was wearing. I should have been slower and more deliberate, and enjoyed what I was doing; or, at least, been aware of what I was doing, and put some good boots on to go out there on the slippery ice. Done it right; but I thought, "Awh, fuck it, I'll get out there. I'll do it and get back here. . . ." It was a direct object lesson that while the mind was clouded with resentment and anger, I could get hurt! I mean it was just totally direct. There was no way out. I just lay there on the ice, fallen down, having slipped on the ice while realizing something disastrous had happened. Also, I was seeing very clearly the chain of emotional events which led me to go and slip on the damn ice. It was no accident, in the sense that one of the conditioning factors of slipping was that I wasn't being careful where I was walking because I wasn't observing the ground, since my eyes were rolled in my head in anger. It's just direct cause and effect, not an ideological matter. There was no way out. Like Anne Waldman says, "No escape."[148] So that's totally related to mindfulness practice.

PP: So meditation helps create a general awareness of the outside world, the space around us, as well as the space inside—learning how to deal with emotions, becoming aware of the currents of feeling. Does that kind of awareness enter into your poetry practice?

[148]The poem "No Escape" is in Anne Waldman's *Fast Speaking Woman* (San Francisco: City Lights, 1994).

AG: Well, let's see. Yeah, I think so, because rather than settling for accusatory generalization, like "God damn finks!", I have to research my anger and find what was the original fact that I was thinking about, and present that and see if that looks as bad as my "God damn fink!" epithet . . .

PP: or "God damn that fucking dog . . ." *(laughs)*

AG: *(Laughs)* Yeah. I have to really go back to the ground of the situation, get back to the fact and say what it is I'm resenting, and write that down as an image. "Dogs barking in the barn calling for their food in the icy hay," or something like that. Is that something to get angry about? Otherwise, I might have said, "Having to go out and feed the dirty dogs of reality" or something rather than, "the dogs lonesome in the barn barking in their icy hay." When you have to go back to the "icy hay" you come up with poetry. I mean the "icy hay" is much better than "dirty dogs." That's why I said, in "Broken Bone Blues:" "Broke my body like a dog / Like a scared dog indeed / Broke my dumb body . . ."

PP: So instead of writing abstract poetry coming out of an unconscious, unmindful energy, you are more concerned with writing poetry that is mindful in the *vipasyana* sense . . .

AG: Well, my poetry was always pretty mindful anyway. I always had based it on elements of William Carlos Williams' elemental observations. Since I went through the same kind of crisis in having an abstracted, visionary experience [in 1948] and clinging to that in abstract forms and then having to take refuge in Williams' "No ideas/but in things" back in '53 . . . so I've gone through a lot of cycles of the same . . .

PP: Realization . . .

AG: Different aspects, and closer and closer to the known, closer and closer to my own life. Instead of having a generalization rising out of the anger, a sort of surrealist image, you get an even better surrealist image if you go back to the lonesome dogs barking in the "icy hay," and coming up with a phrase like "icy hay" which is really good. I just came up with it now, trying to research back. You also get a better balance in the poem—better humour, better balance of attitude in what you're talking about. Wakefulness leads to vipasyana, detailed awareness or awareness of detail around you, the space around you. So there seems to be that correlation to poetry, that practicing mindfulness in meditation provides haiku-like detail.

PP: You've been writing a lot of haikus lately. In fact, many of the ones you've shown me come directly out of the meditation practices during and after retreats, like the King Sooper haiku in your new book *Mind Breaths*.

AG: Something interesting here is that I went into a retreat in September 1975, with the suggestion by Trungpa that I not do any writing while on the retreat. At first, when I heard the notion, it really offended me, thinking it was a philistine notion. But, on the other hand, when I came around to it, it seemed great because I realized I was obsessed with writing, and transforming everything in writing, and that obsession was inhibiting the writing. Not conducive to good writing. It was only conducive to a lot of mental friction, and self-consciousness, and inattentiveness to detail outside. So I spent two weeks not writing. For the first time in thirty years, two weeks where, suddenly, I had the monkey of writing off my back! And it cleared a huge space in my head. Took a lot of anxiety away, because I realized I was writing like a guilty Puritan or a guilty Protestant—you know, I felt that I had to write a poem all the time. That affected my seeing things, because I was seeing things in terms of how you verbalize them.

PP: Everything was immediately translated into words for poetry, or everything was poetry?

AG: Well, no, not as good as that. It was looking at objects, always trying to verbalize them, like the trees waving in the late afternoon outside here: "trees waving in the late afternoon breeze under the telegraph wires." Actually, it was just sort of stereotyping everything constantly into mind, into words. I was missing some finer, organic detail.

PP: So what happened after two weeks of all day meditation, without resorting to writing?

AG: When I found that certain situations were so strong and pungent visually, I was able to remember them anyway—even if I didn't write it down; real perceptions, really acute imagist perceptions stayed with me without my having to work on it. The day I came out of the retreat, then I wrote down a whole series of situations, haiku situations. Very briefly, in ten minutes.

PP: Things you felt or that occurred during meditation?

AG: Haiku situations that occurred during the meditation which were so deep-rooted as insights that they stuck with me. I didn't have to worry about writing them down. I could always write them down later because they were permanent insights, like: "Sitting on a tree stump with half a cup of tea / sun down behind the Rockies nothing to do." Or another one: "Not a word! Not a word! / Flies do all my talking for me— and the wind says something else." Or: "Fly on my nose, / I'm not the Buddha, / There's no enlightenment here!" I was trying to get the fly to go away. *(laughs)*

PP: In your last book, the little pamphlet of work poems, *Sad Dust Glories*, there is that cutting through to the suchness, little epiphany-like poems like the "King Sooper" one . . .

AG: Maybe, maybe some of them. I don't know how much of that book I'll include in *Mind Breaths:* It's too fragmentary. I've written enough now that the new book will have just the best poems, not everything I've written. I'll relegate lesser notations to a journal, a lot of the rhapsodic stuff that doesn't make sense, that doesn't hang together.

PP: The focus of your new book seems to be dharma, exemplifying and talking about dharma. Is that kind of . . .

AG: I didn't *intend* that, but it seems to be the development—dharma preoccupation, the preoccupation I've had during the time of writing it. Trungpa's in it a lot, as a reference point.

PP: There's also a different kind of confessional element in the book, like the poem "Ego Confession," which starts out with: "I want to be known as the most brilliant man in America / Introduced to Gyalwa Karmapa[149] heir of the Whispered Transmission Crazy Wisdom Practice / as the secret young wise man who visited him and winked anonymously a decade ago in Gangtok / Prepared the way for Dharma in America without mentioning Dharma . . ."

AG: I like that poem. I dig that, but it's tricky.

PP: It's like a total, absolute honesty—without any accouterments, without trying to impress, just nakedness.

---

[149] 16th lama head of Milarepa lineage, Kagyu order of Tibetan Buddhism.—AG

AG: Except it's sort of sly; I mean it's finally so I can get away with all that—because they're all poems, ultimately.

PP: Trungpa, as you say, definitely does seem to loom throughout the book, like in the poem "What would you do if you lost it all?" That's the question Trungpa asked you when he saw your harmonium traveling-case full of *City Lights* books, and bells, and incense.

AG: I lost it already.

PP: You've given up your prophetic, messianic identity; I think that's how you characterized your role as a poet once.

AG: Yeah, that's lost, and I could give up my harmonium. I mean, it would be bad if the harmonium was just another prop, a crutch.

PP: So, what we're talking about now is surrendering, in Trungpa's sense of it.

AG: Yeah, that's the word. I haven't thought of it much in those terms. Somehow I don't like that word; it's too "icky," the associations are too weird.

PP: How about "giving up," letting go of ego, completely . . .

AG: No, that's too self-virtuous. I think the process is more discoverable, like including more embarrassing reality as in the "Ego Confession" poem. I think "giving up" is including more embarrassing reality. In other words, I'm including, in "Ego Confession," thoughts and formulations that pass through my mind that once seemed too extravagant, like the idea in the first line, of thinking I want to be known as the wisest or most brilliant man in America. Well, I did secretly have that image of myself.

PP: "Did," or still do?

AG: Probably do. But to say so, in a serious context, would expose me to the absurdity of that ambition, and the fact that it's simply not true. So, except by willing to say so, that does make me a little more brilliant! *(laughs)* By willing to be so playfully, without being scared of being stuck with it. It's just a little aspect; it's a thought that becomes transparent. You don't have to be scared of it being there in your head. It turns out that everybody has that feeling, or sixteen million people have that feeling. So, it's a poem with sixteen million readers, instead of sixteen million people turning and saying "Ahhh, Ginsberg, Ickkk!" In other words, it turns out to be a somewhat universal fantasy—wide enough, broad enough area, archetypal enough fantasy, that other people see the humour of finally saying it. It takes the sting out of it now.

PP: It's learning non-attachment to ego . . .

AG: Well, learning non-attachment to specific facets of the image of myself that I created through my poetry and through my own mind, and to my friends. Learning to break those stereotypes, allowing those stereotypes to fall apart naturally.

PP: You've been teaching the idea of writing spontaneously to your students here at Naropa. Has the Tibetan influence entered into that notion as well?

AG: I'm good at teaching it and formulating it, the idea of spontaneous composition.

PP: You're good at doing it too, not having to write a crafted (in the Western sense of craft) poem, just letting it happen, naturally, spontaneously.

AG: That was always a basic principle, to write a poem by not writing a poem. It's Williams' practice. So that was inculcated early in my poetry, especially in *Empty Mirror*. Things I didn't expect were important turned out to be the best poetry, because the spontaneous mind was more straightforward, full of strong detail.

PP: The association I was trying to make is Trungpa's definition of poetic practice as "First thought, best thought."

AG: I'm not sure whether he said that first or me. I think he appropriated it, but we probably worked it out together. I am not sure. It does involve accepting your thoughts, being able to work with your own thoughts as they are, without pre-selecting the tendency that you want to emphasize, on account of a specific kind of spiritual bias. It's more like having an open mind about your own thoughts, so that you don't formulate them into a romantic stream, just picking out the romantic ones, or the anti-war ones, or the self-humiliating ones, or the self-glorifying ones. For instance, "Not a word! Not a word / Flies do all my talking for me." I think I would have been ashamed of that attitude twenty years ago. Not realizing how sharp it was. I would have admired it as a haiku, but I think I would have been ashamed of settling for that.

PP: I remember at one point in your development you said that language itself might be a barrier to suchness, to further awareness.

AG: Well, that's a tradition. I was preoccupied with that notion in Burroughs' cut-ups in the 1960s.

PP: Did the Tibetan practice help you out of that dilemma?

AG: Trungpa said, "Don't write a poem, don't take any writing with you. Don't bother your head with language, with formulating language." "Don't bother your head with trying to solidify perceptions, prematurely . . ."—really is what he was saying. I think it finally means let the perceptions flow as they come, and write them down when you write—but don't be straining aggressively to solidify perceptions, just to have solidified perceptions on a piece of paper, so you can have a poem and call yourself a poet! Which is just common sense, also.

PP: Have you ever been hung up on trying to better your previous work? A lot of writers have one great masterpiece, and then live in fear of it, always trying to outdo themselves.

AG: I never did have very much of a preoccupation of having to equal early work—that's an obvious trap, a vulgar trap. And there are lots of examples, like Blake and Pound. Once they start an autobiographical curve, any point in it is interesting—because it's interesting. Even Whitman in his old age, just writing *Sands* at age seventy. He's writing great *Leaves of Grass*, but he's also writing the necessary "Sands At Seventy"! Or old leaves from a tree. There was no need for Whitman to feel guilty that he was still not working on a new conception while he was falling apart. In fact, it was necessary to register his falling apart! Rather than maintaining a fixedness with things still building toward some orgasmic, youthful expostulation.

PP: There's one last thing I want to get into. It seems that the parallels with the vipasyana technique and what seems to interest you in Williams and in a lot of other poetry is the ability to write down, exactly or as accurately as possible, what is happening right here and now. You said you admired Williams' dictum about concentrating on square, eye-on-the-thing reality—not ideas or abstractions, but on the things, existence itself, a poetry writing based on the here and now of quotidian reality . . .

AG: Yeah, located or defined as here and now, or quotidian, reality because that's what people need, mostly. But when Williams is dancing in his attic, waving a shirt around his head, naked, admiring his body, saying "I'm lonely . . ." that's also quotidian reality. It's not everybody's; but it's his. "Quotidian" in the sense that he's in his attic, and he's a family man; but everybody does that anyway. And it's also an extraordinary reality; so, quotidian reality really is extraordinary—a lot of it. It's actually discovering the goofiness of actuality, rather than the boring . . . or the goofiness of the boring actuality.

PP: The suchness?

AG: Well, yeah. The flavor of suchness is goofiness. That's what I mean by "goofiness." That's the way suchness appears—very often as "goofy." As Kerouackian, as Kerouacky.

PP: Does sitting help you incorporate these Buddhist ideas into your poetry?

AG: I don't correlate it too much, actually. I correlate *vipasyana* and *shamatha* practice to poetry in the class I teach here at Naropa, because it seems like a clear way of pointing out direction to a group of students involved in Buddhism, actually practicing these meditation techniques, as I am. But I don't necessarily think of the correlations—it's too logical to put it down like that.

PP: Too one to one? Too simplistic?

AG: Yeah, too one to one. Really, the best advice is mellow character—rather than vipasyana. *(laughs)* It means the same thing, accommodating humour, mellow character toward one's quirks—and others'. Like my ideal in poetry, or "Buddhist poetry," was: "the autumn moon shines kindly on the flower thief."

PP: Issa?

AG: Yeah, I'm misquoting it, but it goes something like that. It is like Whitman's "Not until the sun rejects you, do I reject you." Awareness continually shining, or mellow accommodation-mind continually shining. It's somewhat equivalent to Buddhist terminology, but I think it's more Americanese when you say "mellow"—more understandable. "Mellow character," the ideal of which is W. C. Fields—in a way. *(laughs)* Yeah, the Fieldsian aspects of accommodation . . .

PP: Playfulness . . .

AG: Yeah, noticing detail, and—at the same time—surprised, horrified and shocked; you are humoring one's extremist reactions. One of the things I do in class is try and make all these Buddhist correlations, trying to make it orderly and rational, but making it too packaged, too Buddhist-like. At the same time, some of the Buddhist parallels fit, like Williams' poem "Thursday," or the descriptions of samatha leading to *vipasyana*, or attention and wakefulness leading to minute observation of detail. The parallels are useful reference points, but, maybe, impractical in a sense.

PP: Has meditation helped you achieve a finer, more detailed perception of reality, opened you up to clear perception—in the Buddhist sense of not laying your interpretation on the actual reality?

AG: I don't think so; hell, I don't know. *(laughs)* I really can't tell or measure it, because I already had the idea of experiencing detail and chasing it around since the early 1950s. At this point, it has made me more relaxed about chasing details. *(laughs)* I was always too heavy-handed and too theoretical about this. You can't really chase details— you have to remember them, experience them and then remember them, rather than chase after the perfect detail. You have to learn to see things, but without always the self-consciousness of noticing this detail and that one. It's more like a real process of recollection rather than automatic attention—so it's attention to your recollections, maybe. Attention to recollection, in the sense of seeing something curious, like "the icy hay"—funny sound, interesting image, clear . . . goofy! "Icy hay in the barn, the dogs lying in the icy hay in the barn . . ." In trying to describe that situation, I wound up with "icy hay"; then, a second later, I realized "icy hay"—ah, that's interesting, that's really interesting, I really got it. But I didn't realize I got it when I got it! It was only after a moment of recollection. Most haikus and most poetry, most images, are recollections of an instant, the thought of an instant ago, or a minute ago, or maybe an hour ago, or a year ago—the picture comes up that you hadn't paid attention to.

PP: Does that conflict with the notion of spontaneity, recollection versus spontaneity.

AG: Well, recollecting is something you can't do on purpose; it just comes up spontaneously. Recollection is spontaneous by itself. Spontaneous means allowing the mind to . . . remembering the mind, remembering the mind activities. And, using that, the mind's recollections, as a subject; using that kind of attention as a subject rather than a more fixated thing like having to write down everything that's happening right on the street, right this minute. Or the attitude: "I gotta write it down for a sonnet. I gotta have an idea about life and write it down in the form of a sonnet." So, the poetry is natural that way, and becomes a natural product of awareness, rather than a crafty synthesis—crafty synthesis because you're synthesizing your recollections. Maybe I'm just kicking a dead horse.

PP: The use of spontaneity in writing songs, as Milarepa practiced, and as you do now in your collection of songs, would be the ultimate in that practice . . .

AG: There's a very nice song in *Loka II*[150] that I wrote spontaneously, spontaneously recollecting the events of a recent illness. It comes at the point in a conversation between William Burroughs, Trungpa, myself and other people when Trungpa asks me to make up a song, on the spot. I did, and it was typed up later. It's not bad, actually.

PP: How does it begin?

AG: "Started doing my prostrations sometime in February '75 / Began flying as if I were alive / In a long transmission consciousness felt good and true / But then I got into a sweat while thinkin' about you / Fell down with bronchitis, the first illness that came / Pneumonia in the hospital was what they said was the name . . ." It's actually pretty humorous for such a serious set of events. Writing spontaneously while recollecting—I had to accept any thought. The

---

[150]A journal from the Naropa Institute edited by Rick Fields (Garden City, NY: Anchor Press/Doubleday, 1976).

whole point of spontaneous improvisation in a song is that you have to accept whatever thought presents itself to your rhyme—on the wing, so to speak. Otherwise, you have to break the rhythm, stop the song, start thinking. Once you do that, you're lost. You have to keep the impulse going—accept doggerel as well as beauty, because you are improvising and relying on the moment to moment inspiration. It means relying on moment-to-moment ordinary mind, whatever rises. It's absolutely necessary to take whatever you can get, if you're singing; settle for what's there, at the instant—otherwise, you break the chain. I do think *samatha* practice does help there because you become more minutely aware of what's rising in the mind, thought forms rising and disappearing. And you learn to look on them with less prejudice than before—like this thought is good and that thought's bad. Any thought will do! When you get to that equality of temperament or judgment, all the thoughts turn out to have their place, to form a sort of recognizable pattern, or chain of workable thoughts—as Trungpa says, "workable."

PP: That must take a tremendous ability for self-acceptance.

AG: I think it's much easier than you think. It's fun! *(laughs)* Fun in the sense of being with good friends, drinking and making up songs. You let your tongue go loose! Everybody accommodates to that. Nobody's embarrassed by anything. There's less anxiety once you've realized that there's nothing you can do about it. Why fight it? You can't change your mind—your mind is its own. And there's nothing heroic about that acceptance. Nor do I think it's a transition to another state of consciousness. That's the whole point—it's ordinary mind! The question is "Do you accept ordinary mind or not?" But even that question is too much of an either/or proposition. It's more like "Do you recognize what an ordinary mind is?" That's where the problem is, not willing to recognize it, having to be turned on to it. Most people don't recognize their ordinary mind; they use it all the time, but they use it selectively, just for certain highlights to fix up a specific thought pattern, image, ambition. They think that ordinary mind is just certain highlights of ordinary mind, rather than the whole thing.

PP: But in Western art and poetry, the tradition demands one respect the idea of high thoughts, like "What was so often thought, but never so well expressed."

AG: Well, it's true! I agree with that—just express what is so often thought.

PP: But what to do with the doggerel, the superficial—it wouldn't be very good poetry with all the ordinary mind trivial flow.

AG: That's too academic and stylized and way off in the distance. Nobody really takes that attitude too seriously.

PP: I'm not so sure.

AG: You shouldn't isolate what you do every day all day. Anyway, there isn't much difference, it's just a question of learning a sharper, more experienced way of recognizing and appreciating what's already in your head. It doesn't require a big breakthrough or anything like that.

PP: Stop trying to have visions . . .

AG: One wants to have visions because one thinks that one's ordinary reality, ordinary consciousness, is not visionary enough. Which is a big stink everybody has about themselves—everybody—that their body is awful, their mind is awful. Being who you are is awful enough without it being *that* awful!

Guy Amirthanayagam: Allen, the question I would like to begin with is, what effect do you think your interest in Buddhism has had on your recent poetry?

AG: Well, the title of my most recent book is *Mind Breaths*; and that relates to an increased awareness of mind, *bodhi*, awakening mind, through meditative attention to breath, which is the basis of Zazen,[151] or sitting meditative practice. So the poetry then becomes conscious of mind and breath; poems as thought-forms rising in the mind, projected outward into the world on the breath. Breath is a basic notion in poetry. Buddhist interest also brought my attention to older models like Classical Chinese poetry—I'll read you an example of that rather than talk about it. "Returning to the Country For a Brief Visit" was

[151]Zen Buddhist form of sitting meditation practice. See *Zen Mind, Beginner's Mind* by Shunryu Suzuki; edited by Trudy Dixon, with a preface by Huston Smith and an introduction by Richard Baker (New York: Weatherhill, 1996).

written on the margin of a book of poems by Sung Dynasty poets. I was imitating their style:

> Old One the dog stretches stiff legged,
> soon he'll be underground. Spring's first fat bee
> buzzes yellow over the new grass and dead leaves.

> What's this little brown insect walking zigzag
> across the sunny white page of Su Tung-p'o's poem?
> Fly away, tiny mite, even your life is tender—
> I lift the book and blow you into the dazzling void.

GA: If, for purposes of convenience, we adopt the unreal distinction between content and form, do you think your interest in Buddhism and Eastern spirituality has affected the forms of your poetry?

AG: Well, in this case you may have noticed it was just an imitation of Chinese poetry in translation, with a conversational tone and a very simple form. So in this particular case, it affected it; yes, you could say that. I have been using part of a Buddhist approach—remembering flashes of insight into the emptiness of things (or into the relation between things), and so I've been writing little short poems, maybe one or two lines that capture one single flash, moment of *satori*, that's in haiku style. I had been in a meditation retreat for several weeks, and this was the first poem I wrote when I came out. It's called "Walking Into King Sooper. After Two-Week Retreat."

> A thin redfaced pimpled boy
>   stands alone minutes
> looking down into the ice cream bin.
>   16 September 1975 Boulder, Colorado

This is like—from a Buddhist point of view—instant karma. A thin, red-faced, pimpled boy—he eats too much sugar—is standing alone in the supermarket, looking downward many, many minutes, spaced out with the desire to eat more ice cream.

GA: It does remind me though of the Imagists.

AG: You're thinking of the American-British Imagist school of poetry, which actually took its inspiration, around the turn of the century, from Chinese poetry, and Japanese "Ho-ku," as they called it in those days. The sources were similar, the impulse to deal directly with perception as quickly as possible—with as little bullshit as possible—in vernacular speech, or as Pound said in 1910, "direct treatment of the object."

GA: Would you agree that the main influences, if they could be called influences, on your poetic style, have been Whitman and William Carlos Williams, or is that a simplification of your style of writing? I am thinking especially of your long ambulatory line.

AG: I would add that for a long ambulatory line, which is a nice phrase—I never thought of it before—the best I've heard—an ambulatory line, a line that takes a walk from one end of the page to the other and, maybe, crosses around the margin and continues on its path. Actually, my earliest influences in that line is the eighteenth-century English poet Christopher Smart, who wrote a long poem called *Jubilate Agno* ("Rejoice in the Lamb"); and specimens of this poem, especially the little section beginning "I Will Consider My Cat Jeffrey," can be found in anthologies. Smart was a friend of Dr. (Samuel) Johnson, and he wrote his long poem in Bedlam because he kept falling down on his knees in the middle of central London and praying to his Jesus, and so they took him away to the mad-house. Dr. Johnson said, "I'd as lief fall on my knees and pray with Kit Smart as with any man in London," when he heard that Smart was dragged off to the booby-hatch. So actually Christopher Smart's "Rejoice in the Lamb" was the first major influence. Then Whitman, whom I read extensively after starting my own long-line style; Jack Kerouac, for spontaneous speech or "composition on the tongue," rather than on the page; William Carlos Williams as a direct teacher in person in the early 1950s, and as a living "poetry guru." It was from him I understood modern poetry—that aspect which we were pointing out before—"direct treatment of the object" or "Objectivism" or Imagism.

GA: To change the subject somewhat, do you think your interest in Eastern religion has anything to do with the fact that Christianity, say in the West, and particularly the Puritan heritage, has gone stale and sour in America?

AG: Well, I think it can be said that both Hinduism and Buddhism have gone stale in Asia too. I think every "modern" culture suffers from that problem. However, there is one thing in Tibetan and Japanese Buddhism: a transmission from generation to generation among the *ryoshis* (teachers) in Japan, and the *lamas* (some *lamas*) in Tibet and India. There has been transmission from generation to generation of the active mind—of "mind transmission"—basic understanding and practice of meditation, or how to meditate. I don't, or I have not found that, in the Western tradition, the actual sitting practice or meditative practice, traditionally handed down from, let us say, Christ's day on. You do have a transmission through the pope, but nobody ever claimed it to be an unbroken transmission directly from Jesus; that went through the Mafia somewhere, so we don't know what happened. With Zen Buddhism and Tibetan Buddhism, there does seem to be a lineage that goes back to Buddha, Ananda, Sariputra; and there are practices of sitting that are somewhat reliable from that point of view—they have been checked and counter-checked generation after generation. So I feel that there is a reliability in the Buddhist teaching through the persons, through the teachers—a living tradition of meditation.

GA: Are you not troubled by the fact that, let me put it this way, I can see that you are not troubled at all by it—but how do you retain this serenity, or seeming serenity, if you really believe that ultimately there is only the void? Or is that a fair question?

AG: Well, it's a fair question. Void is a Western word, and when you say "void" in the West you see something black, a big black hole like a quasar or something, rather than *śunyata*, a Sanskrit word, which means something "in the direction of emptiness." Or to put it another way, there is no claustrophobia, no walls closing in, no prefixed definition, no solidification of mind into some object that you can bounce on the floor and throw out the window. Void or *sunyata* means absence of fixed projections, the absence of fixed identity, the absence of limitation; and it's like raking the roof off the skull so that you have all this space to be free in. Because actually the void is the space that we're sitting in—we're sitting in the middle of the void, and there's nothing frightening about that. Where else would we sit? In the middle of a ball of iron?

GA: Good. I'll take that. So, in fact, for you nirvana is this . . .

AG: Well, not nirvana—"*dharmakaya*" is this. It's unqualifiable, and in that sense, you couldn't determine it, you couldn't describe it, you couldn't encompass it with language—"all conceptions of existence as well as all conceptions of non-existence" would be equally arbitrary as per the Diamond Sutra.

GA: So really you are not strictly a Hinayana Buddhist?[152]

AG: This is the Vajrayana interpretation of Hinayana. Hinayana Buddhism has its own terminology. Mahayana has its compassionate terminology—and Vajrayana has its crazy wisdom terminology. My study is basically Vajrayana, my teacher is Chögyam Trungpa who is a Tibetan lama of the school of Kagyu Buddhism, middle-school, which has as its teachers Tilopa, Naropa, Marpa the translator, Milarepa the Yogi poet, S'Gampopa and Tusumkhempa the first Karmapa, and then a series of sixteen lamas up to the sixteenth-century Gyalwa Karmapa.

GA: So you don't consider Buddhism to be inimical to art and poetry, as the stricter, earlier precepts and practice of Buddhism seem to indicate?

AG: Well, actually, Buddhism did develop a great many art forms in many different countries, depending on the cultures, as we know them. There was an infusion of Greek art, culminating in the Gandhara style; fantastic painting schools in Tibet, wheel of life mandalas, portrait paintings of the different saints, yogis, Siddhas, Kalas, etc., and lamas. In Japan there was a great flowering of different artistic and aesthetic styles in gardening, martial arts, calligraphy, landscape painting and in haiku poetics. So, the novel, *The Dream of the Red Chamber*, is a genre influenced by Buddhist works.

GA: I was thinking mostly of the puritan strain in Buddhism in certain Asian countries.

AG: That's the Hinayana, which is maybe interpreted as puritan, though it is perhaps a purificatory stage; but as I say, I come from the Vajrayana school, which says work with energy—work with mistakes,

---

[152]*Hinayana, Mahayana, Vajrayana—Yana* means vehicle, so literally the lesser, greater, and thunderbolt vehicle, the three vehicles by which a Buddhist practitioner can advance along the path to enlightenment. From the *Vajrayana* standpoint they can all be practiced simultaneously. In Tibetan Buddhism, the vehicle practiced depends on the ability of the student and teacher.

work with passion, transform passions into light energy rather than heavy chains of bondage. We're faced with suffering, we're faced with "passion, aggression and ignorance," so we have to work with our pride, we have to work with our anger and our stupidity. Thus, if we don't acknowledge our aggression, we'll never get out of it. So, beginning to work with aggression is learning to see it unprejudiced, not projecting it outward, but acknowledging it, and then perhaps transforming it with a lighter hand to useful energy.

GA: I'm very happy for you—I was not trying to involve you in any controversial discussion of Buddhism. I'm quite happy that you can continue to be a bard and a Buddhist.

AG: Well, there is a member of the *sangha*[153]—in the old tradition—namely Milarepa, whose lineage I follow. He was a great yogi poet whose style was spontaneous utterance, or "composition on the tongue," and whose book in English is translated as *The Hundred Thousand Songs of Milarepa*. So, he was a productive bard who fused song and dharma. Actually, it's an image, the "hundred thousand songs"; but he wrote or sang many. There's a picture of him with his hand to his ear, listening to his own voice.

GA: There seems to be no danger in your case of considering the writing of poetry as one of the last fetters that you haven't yet abandoned in your search for salvation?

AG: Well, actually, no. There's no danger because if there was, so what? Who cares?

GA: But I care. I care for you and for poetry.

AG: If I cared, I'd be in bondage to poetry. On the other hand, the tradition is that teachers who wish to turn the wheel of dharma and explain passion, aggression and ignorance,[154] have to have a good tongue; they have to be "poets" so that they can be understood.

---

[153]The community of Buddhist monks and nuns. The term also can be used to include lay followers.
[154]See also the wheel of life above, note 152. These are represented at the center of the wheel of life mandala by the cock (desire), pig (ignorance), and snake (aggression or hate) as the causes of the life cycle of suffering.

GA: But to go back to a technical question again. Do you think you are moving towards a more formal poetry, more structured utterance—like songs?

AG: I would say, yes; I'm moving towards the more formal and classical modes, as befits a gentleman of my age. On the other hand, those particular formal modes that I'm moving towards are those that involve more spontaneous utterances, such as American blues, a Western improvised form; or haiku, and *Dharmachakra*, utterances expounding Dharma, which involve spontaneous explanation on the moment.

1980s

NANCY BUNGE
March 18, 1981,   Boulder, Colorado

From the book *Finding the Words* (1985, Swallow Press/Ohio
University Press)

Bunge wanted to interview Allen as part of a series of interviews she
was doing with poets and fiction writers who teach. The interview had
been set up by correspondence so that Bunge would arrive in Boulder
while Ginsberg was teaching there. Bunge's 1999 account of the inter-
view explains that after she carried a letter to Naropa saying she had
arrived, Allen knocked on her motel door early on a Saturday after-
noon and "asked if I wanted to get some breakfast. When I pointed out
that it was two in the afternoon, he suggested lunch. We talked during
lunch at a restaurant that was on the second floor of a corner building
on Pearl Street. We started talking at a table in the middle of the room,
but when a window opened up, Allen suggested that we move there.
The interview was audio-taped on a recorder I got on sale at K-Mart.
He told me, correctly, that I needed to get a mike. After the interview
he walked me around Pearl Street, pointing out the best bookstores."
Bunge transcribed the interview.

—DC

Nancy Bunge: I was confused by the tapes of your classes. Are they lit-
erature classes or writing classes?

Allen Ginsberg: I don't make a distinction. Although in writing classes
people presumably write poems and bring them in and have them criti-
cized and in literature classes people presumably study other people's
texts and don't write, the best teachings I got from Kerouac and Bur-
roughs was hearing them pointing out gems of language and rhythm and
perception in world literature as well as in my own which turned me on to
say, "I can do that" or "I did that" or "This is just like my brain." So what

I'm doing is presenting texts which give the students permission to be as intelligent as they secretly are. So it's a writing class and it's also a litera-ture class, but I don't think the teaching of writing necessarily involves the full-time examination of the students' texts or the teacher's texts. I think it's a byplay of intelligence between the students and the teacher on anything, whether it's Shakespeare or a brick wall. It's indicating to the student how to use perception, not necessarily in written form, in terms of body English: how to sit in a chair, how to be aware of breath, how to walk across the street. Knowing how to walk across a street is the same thing as knowing how to write a haiku; learning how to walk across the street is the same thing as learning how to write *The Brothers Karamazov*.

NB: "If the mind is shapely, the poem (art) will be shapely."

AG: Yes. The thing is to get under the students' skin and arouse enough enthusiasm that they get under their own skin. This means allowing yourself to be yourself in class. My own best teachers were William Carlos Williams, William Burroughs, Gary Snyder, Gregory Corso, and Jack Kerouac. I learned by hanging around with them, from watching their reaction to cars going down the street or a story in the newspaper or TV or a movie image or a sunset or moon eclipse; when you see the intelligence of somebody reacting to the phenomenal world, you learn by imitation. You see beauty and you want to share it.

My best learning was just being myself with them and they giving me permission to be myself and then discovering myself with them—how funny I was. So you've got to encourage the student to discover himself and how funny he is and the only way you can do that is by let-ting yourself be yourself in class which means not teaching, but being there with the students and goofing off with them. The best teaching is done inadvertently.

The oriental theory is such, called darshan. People will go across India to visit Gandhi or Ramakrishna or Great Lama, to take darshan, which means just to look at them, see how they move their arms, how they carry themselves. It's not a mystical matter; it's seeing and exam-ining someone whose intelligence is unobstructed, whose breath is unobstructed, whose body is unobstructed, whose psycho-physical make-up is unobstructed or full of character. That isn't a rationalistic or mystical or mysterious matter—it's common sense.

You finally have to get down to cases and take a look at what [the students] put down on paper because you could talk beautifully and they could talk back intelligently and sympathetically and full of erotic clarity and you think they understood, but then when they write down something, it might turn out to be the most dreary, third-rate imitation of Rimbaud-Dylan . . . which happened to me the other day. I ran into a student who struck me as radiant and then I read what he wrote—he had only two vivid words out of a page and a half of romantic drivel. So you have to look at the work and see if students are on the level you think they are.

Williams taught me by going over my poetry, *Empty Mirror*. I sent him eighty pages of stuff and he separated forty pages that he thought were really good and he told me what he thought was "inactive" among those forty pages, said I ought to cut it out. I said, "But it's part of the writing, isn't it?" He said, "One active phrase is better than a whole page of inert writing because nobody will ever read or reread it, whereas the active phrase, even if it's not a complete sentence, is more interesting." He put two pages in front of me. One had an active phrase that didn't come to anything as a whole poem, but was fine as a fragment, like a Greek fragment, a piece of Sappho that's still brilliant even though it's only part of a clause versus a whole page of something that's not active. So he said, "Cut down to what's active."

There's another useful principle Mark Van Doren pointed out. He used to write book reviews for the *New York Herald Tribune* and almost every one of the reviews was intelligent and sympathetic; he was always talking about something absolutely marvelous. I said, "What do you do with a book you don't like?" and he said, "Why should I waste my time writing about something I'm not interested in?" So, in looking at students' writing and teaching poetry, it's a waste of time to try to tell them what they're doing wrong. It is not a waste of time to point out examples of active language to them and give them an arrow in the direction you think they should go. You might briefly give some explanation of what is wrong with a phrase like "a dim land of peace" or "I am suffering the terrible illusion of being born into the mystery of nature." You might try to analyze it, but it's like trying to explain what isn't there; so the best thing you can do is point to what is there. Use every situation to enlighten in the direction of what is practically apprehensible and useful. Point out that part of their nature that is already successful and apparent and concretized and palpable. Attacking what's impalpable is like attacking the ocean. Same thing in terms of poetry samples. Some teachers used to take great delight in mocking bad poetry; it might be more interesting to constantly uplevel the whole discourse by working with material that's active.

NB: As I understand it, you think that in order to write poetry well, one must let go of pretensions and accept whatever comes up.

AG: Yeah.

NB: That's hard to do.

AG: Then maybe you don't understand what it means because it seems to me easy. What's so hard about it?

NB: Well, you've said that Kerouac had to keep telling you to let go.

AG: Yes.

NB: And that it took you a long time to be able to do it.

AG: Yeah, because I thought I was supposed to do something different than what came naturally. It wasn't that I couldn't get to my nature. It was that I thought my nature was unacceptable for high class poetry. I thought that high class poetry meant something besides just ordinary mind;[155] I thought it meant some other kind of mind than the one we've got. Trying to fake another kind of mind or another kind of language or another kind of perception constantly leads poets into these paradoxical situations where they fake something that might be imitatively interesting, but, on the other hand, ultimately is uninteresting to them. They dry up at the age of forty or they have writing blocks because they're not making sense on any level, but they think they're supposed to make sense on some level they can't get to. It's like a Marxist with an idea that his writing is supposed to be social for the people, except he can't think of anything social for the people (*laughter*) and so he stops writing because he never understood what "social for the people" means to begin with. It's such a generalization, it has no real, immediate, practical application. But if someone is dominated by the conditioned reflex in that phrase and criticizes every immediate reaction he has according to this phrase he doesn't understand any-

---

[155]Ordinary mind in Buddhism is mind as it happens, generally much more interesting than one could expect (as Ginsberg points out, this is also in opposition to the idea of creating some other sort of mind than one's own). By "spontaneous mind" Ginsberg seems to mean both "composed on the tongue" as well as the spontaneity and freeness of awakened mind.

way, which is supposed to be an idea. . . . Like, "My ambition as a writer is to be elegant." Who knows what the fuck "elegance" is? Everything you write, nothing is elegant; so you feel that your writing is terrible or you stop writing. So "letting go" meant letting go of an arbitrary idea that didn't make any sense.

NB: Where did you get it?

AG: From the teaching of poetry at Columbia and the community around Columbia.

NB: Well, most of your students have been to school and some people have suggested that schools teach people to lie.

AG: Well, Socrates' school didn't teach people to lie. I don't think Black Mountain (College) taught people to lie. Run-of-the-mill schools do.

Waitress: Coffee?

AG: No thanks.

NB: Yes, please.

AG: Better watch out. You'll get cancer of the pancreas! I don't think the nature of school is to make people lie. I think it's the nature of capitalist-supported or communist state-supported schools; schools tied up with major exploitative economic and bureaucratic systems naturally are supportive of bureaucracy.

NB: Your students are not five years old; they've been to a lot of schools. So when they hear, "Write down your perceptions as honestly as you can," they have probably learned that they ought to have certain kinds of perceptions.

AG: Oh, so how do you get them to locate their own perceptions rather than imitating mine? How do I get them to write in a style that's different from "Howl"? By demonstrating in the classroom how those perceptions are arrived at by me, by arriving at such perceptions during the course of the conversation in the classroom so that we all arrive at the same perception together or by pointing it out to them when inadvertently they've let loose with a perception that's native to them, by checking out the perceptions of other poets and texts and pointing them out as examples, particularly Charles Reznikoff and Williams because their basic method is "ordinary mind," then check out the student's own writing, then in personal conversation and contact, or by lovemaking with the students in bed when appropriate . . .

NB: Are you serious?

AG: I'm totally serious.

NB: Okay.

AG: I believe the best teaching is done in bed and I am informed that's the classical tradition, that the present prohibitive and unnatural separation between student and teacher may be some twentieth-century wowser, Moral Majority, un-American obsession. The great example of teaching was Socrates, and if you remember *The Symposium*, the teaching method there involved Eros. So Eros is the great condition for teaching. It's healthy and appropriate for the student and teacher to have a love relationship whenever possible. Obviously the teacher can't have a love relationship with everybody in the class and the student can't have a love relationship with every one of the teachers because this is strictly a human business where some people are attracted to others, but where there is that possibility, I think it should be institutionally encouraged. The immediate question that arises in our environment is the exploitation of the student by the teacher and vice versa, but that problem arises in any love relationship and does not rise any more in the student-teacher relationship than in any other relationship. So it's a fake issue. It's a worry about what people will think rather than a native worry about the actual situation. Of course such a situation depends on the tact and intelligence of the teacher. If you have a gorilla professor or a man-eating Amazon professoress or a completely neurotic person, then the relationship will be neurotic and most relationships are neurotic and that's the way life is. On the other hand, I wouldn't worry about it because that's part of the learning process.

When you read the *New Yorker*, or other accounts of the academic or poetic circuit, it's commonly accepted that it does happen, if not universally, at least more often than not. So what we're discussing is not something specialized to me; we're just opening the discussion of normal, average behavior. We're talking about what goes on rather than what ideally should go on or what a minority of adventurous spirits persist in. The majority of ordinary people behave this way. That such a teaching relationship as an ideal would be considered reprehensible and scandalous, although it's universally practiced in the academy, means there's some basic lie as to educational method that's been universally accepted for conscious talk although the behavior is different. The difference between actual behavior and conscious discussion, where you do one thing and say another, is double bind, and that leads to schizophrenia and alienation and confusion. So the reason most academic and institutional teaching is difficult is because it creates emotional schizophrenia where the impulses and the behavior are in one direction and the speech and instructions are in another. That's a basic stress on both students and teacher: the natural overflow of affections and intelligence and energy are checked.

If you don't acknowledge the actual conditions, you can't, in class, point to ordinary mind, you can't point to epiphany and you can't point to the recognition of one's own nature. So while there is a great blank or amnesia or evasion or suppression or avoidance of a major area of emotion and human relationships going on, it makes it more difficult for the poet who is depending on intelligence about human nature for his subject matter. It makes it difficult for the teacher to point to instances present in the classroom and in the community. So it makes teaching almost impossible because one of the bases of poetry is frankness. When there's an obstruction of intelligence in classroom communication because of law or convention, then it's not quite possible to communicate. If you can't communicate, you're not teaching poetry to poets. So that's a fundamental insight to which we should relate. It doesn't mean you have to make love to your students; it doesn't mean you have to talk about it all the time in class; it just means that the teacher has to consciously relate to that situation in a creative and open way. Even if the relationship is an avoidance of it, it's got to be conscious. If it's subliminal, then the entire nature of intelligence becomes fogged and if you have foggy intelligence, you have imprecise poetry. That's why it's important to discuss it consciously; otherwise, you get platitudes that have no grounding in nature. Shakespeare, Chaucer, Milton, Blake, Whitman—those poets considered greatest and who are the staple of what is taught would laugh at the manner of teaching and the treatment of Eros in the teaching of their work. So everything has been turned upside down in the regular teaching situation.

The reason I'm going on so is that you seemed surprised when I said the best teaching is done in bed.

NB: At most universities it's the only thing you can be fired for except gross incompetence.

AG: Yes, of course that's it. It's the one thing you can get fired for. If you look at the great teachers, just in the ordinary academic scene, you'll find all the gossip and scandal and humor of Coyote, the American Indian god, who is a trickster hero and who's beyond the law, but isn't beyond human nature. Is human nature beyond the law? It's not perfectly normal for students and teachers to be afraid of each other erotically. It's just a social convention and if at this point, someone can't tell the difference between a dopey social convention and universal human nature humor, the whole discussion is hopeless—particularly when a poet is teaching and a poet is not supposed to love the girl he teaches or the boy he teaches? Ridiculous. That's absolutely absurd. Throw the homosexual shot in the pot too to make it even more outrageous and you have something totally ancient and classical and at the same time, totally outrageous from the limited middle-class, wowser, boo-boo, bourgeoisie point of view.

So if you open up the notion of teaching to the philosophy of the boudoir, you have a different angle. My own experience is that a certain kind of genius among students is best brought out in bed: things having to do with tolerance, humor, grounding, humanization, recognition of the body, recognition of ordinary mind, recognition of impulse, recognition of diversity. Given some basic honesty, some vulnerability on the part of teacher and student, then trust can arise. Mutually acknowledged vulnerability leads to mutually acknowledged trust—erotic vulnerability, scandal vulnerability, social vulnerability, the fact that raw human nature is vulnerable anyway which is characteristic of great poets like Keats and Hart Crane, that raw open heart that's so useful in poetry. Given a conscious acknowledgement of vulnerability, you have a basis to begin teaching poetry.

NB: Most of the people I've spoken with think it's important to read work out loud, and you've explained why that's valuable.

AG: Well, probably I've talked enough about vocalization in other interviews. Just one sentence—poetry and language exist in the dimension of sound as well as ideas and letters, so in order to have unobstructed intelligence, you have to be apprehending and hearing sound.

NB: Some people have run into trouble with colleagues in the English Department.

AG: It's the same thing like you're not supposed to make love. The poets who don't think you're supposed to make love and don't think you're supposed to read verse out loud are also the ones who have a limited idea of what's classic and what's traditional and don't like open form and don't like blues, they only like the closed forms that were practiced one hundred years ago. They only like the dead, closed forms like the sonnet; they don't like the rhymed, triadic, five-foot line exhibited in the blues, which is also a classic form, but they never heard of it as a classic form although it has a name, a nomenclature, and a practice and created what may be the largest and most sophisticated body of literature produced in America:

> I'll give you sugar for sugar that you'll get salt for salt
> I'll give you sugar for sugar that you'll get salt for salt
> Baby, if you don't love me, it's your own damn fault.
> Sometimes I think that you're too sweet to die
> Sometimes I think that you're too sweet to die
> Other times I think you ought to be buried alive.

That's Richard "Rabbit" Brown's "James Alley Blues." So there's this tradition of texts which will be classic in a couple of hundred years but which would be avoided as literature by the same people who don't believe in lovemaking or reading aloud (*laughter*) or talking frankly.

NB: A number of people have said it's important for students to receive exposure to lots of cultures, but that blues is excluded from the academy, American Indian literature is excluded from the academy, Chicano literature is excluded from the academy, so there's no American literature being taught in the academy. It's all derivative European literature.

AG: Or American high literature like Pound, Williams, Marianne Moore—which is great, but the actual community literature which in ancient times was considered the important thing, the Homeric community literature, is ignored in contemporary letters.

There may be some argument about American Indian literature

because it's a minority literature; Chicano literature because it's a minority literature; but blues is a majority literature that every white and black person in America knows by heart, has feeling for, listens to and does. Everybody hears blues; everybody sings it; everybody that listens to Dylan, the Rolling Stones, the Beatles, all rock and roll, as well as the people who are smart and go to the black originals. So it's a universally practiced form which is ignored as literature. Right there is a case of neurotic amnesia—obliterating this whole area of literary practice as if it did not exist. The reasons for it are complicated. It has to do somewhat with racism, somewhat with rigidity of consciousness, so that spontaneous and oral forms are not brought to the notice of those few students who would pick up on the form—it would alter their perception of the phenomenal world—because the blues are franker about human relations than most literary poetry.

It might be just a class thing, that upper classes, the nonlaboring classes, the upper bureaucratic, exploiting classes who live on paperwork but don't do actual physical labor, and whose paperwork exploits the physical work of other people, don't want to know about the physical world. Sound is the physical world, sex is the physical world, Eros is the physical world, spontaneous blues—the existence of a large body of universally practiced poetic forms in the actual physical world—all of these are eliminated and anesthetized. So it may be the practice of an elite group of paper-shuffling bureaucrats who are trying to suppress evidence of the existence of a suffering physical world.

John Crowe Ransom, in one of his greatest poems, addressing the graduating class at Harvard, ended his poem:

> And if there's passion enough for half their flame,
> Your wisdom has done this, sages of Harvard.

It's a Phi Beta Kappa poem written by the eminent, classical, conservative poet John Crowe Ransom, the most respected New Critic in America.[156] Even he criticized the education at Harvard for having dampened the physical passion of his students. So this is no bohemian

---

[156]The term comes from the John Crowe Ransom book *The New Criticism* (New York: New Directions, 1941), concerned primarily with T. S. Elliot, I. A. Richards, William Empson, and Yvor Winters. It has been argued that the book characterizes a certain kind of literary interest rather than any actual school of criticism or any actual traits held in common.

insight that we're arriving at; this is old granny wisdom, even in the academy. It was a complaint that William James took up when he was there. That was the cause of the whole pragmatic philosophy—checking out what was actually happening instead of stories about it. Check out the physical body. It would even apply to religious experience, that you have to examine the specimens, the visionary experience, rather than make generalizations deductively.

NB: And you certainly know about all kinds of literature, not only the blues.

AG: The idea that recognition of the body, recognition of Eros, and recognition of sound would exclude rational intelligence is an error of judgment that only someone locked into rational intelligence and nothing else, neither imagination nor body nor feeling, would make. That's Blake's classical division; there are four Zoas,[157] four basic principles of human nature: there's reason, there's feeling, there's imagination, and there's the body. If reason dominates the body and imagination and the heart, it becomes a tyrant and winds up a bearded old man inside the cave of his own skull tangled up in the knots of his rationalizations. If the heart tries to take over and push too far, then it becomes a parody of sentimental gush. If the imagination tries to take over and exclude reason and balance and proportion and body, you get some nutty LSD head, jumping naked in front of a car saying, "Stop the machinery!" and getting run over. The body trying to take over, you get some muscle-bound jock. So you have to have them all in balance. To have them all in balance would mean that reason would have its part to play as "sweet science." There's a line in Blake's *Milton* about reason as "sweet science" rather than "horrific tyrant." Reason has become a "horrific tyrant" in Western civilization and created the nuclear bomb which can destroy body, feeling, and imagination. So there's no sacrifice of intellectual labor in acknowledging the existence of other modes of intelligence, as in body, feeling, and imagination.

The difficulty with Urizenic thought is that it bounds the horizon— "Urizen," Horizon. It makes a boundary so you don't see beyond the limits of conceptual thinking. So you don't hear the sound of, poetry, you don't feel the rhythm of poetry, perhaps you don't even imagine

---

[157]See note 5 above and *The Complete Poetry and Prose of William Blake*, ed. David Erdman (New York: Anchor Books, 1988).

the vast implications of the poem; but you get pedantically hung up on some rearrangement of mental forms in the poem. The poem finally presents manners rather than the entire gamut of human feeling and intelligence and rhythm and prophecy.

You remember Socrates was given the hemlock for corrupting the young and Socrates was certainly a great educator; he was the acme of the teacher, both of philosophy and poetry. Socrates slept with his students, corrupted the young, was kicked off the campus, was driven out of society and made to take hemlock. Christ, the greatest teacher of all, was taken by the forces of law and order and crucified. Buddha was relatively successful, lived to a ripe old age; however, because he was a very great teacher, he had to deal with ignorance and there are some forms of ignorance that are pretty implacable, like the ignorance of the guy who killed John Lennon, Mark Chapman—who was attracted by Lennon's light, but also confused by the light and so struck out murderously, aggressively with his innocent stupidity. So old Buddha was given bad pork to eat by a jealous cousin. Buddha did quite well, but there still is a certain recurrent unreasonable ignorance in mankind which might be resolved after a long, long time; but confusion has its crises and its aggressions and there's no insurance against it. Actual teaching in the most traditional forms has led to conflict with ignorance in which ignorance has taken violent action, so it should be kept in mind that though true teaching won't necessarily provoke a vicious counter-reaction, when somebody gets into trouble, it may be that they are teaching well. Every scandalous teaching should be examined with that in mind as the background: that the classical teachers were scandalous or were perceived to be scandalous by those who didn't understand their teaching; the ignorant thought it was something extravagant rather than something obvious.

As this interview was submitted in questionnaire form and the answers were written by Allen, no attempt has been made to render Ginsberg's personal prose style conventional.

—DC

Helen: Many of our readers are unfamiliar with your personal history—how you made a choice to quit the advertising "straight" world to embrace the creative underground of the 1950s. Do you have any advice for others who are not quite sure how far to go?

Allen Ginsberg: Look in heart; check out your visions with your friends; be bold and careful at the same time; Mind includes both sides of any argument; balance body, feelings, reason and imagination: ALL 4 working together make whole man; read William Blake & Dostoyevsky; listen to old Blues (Leadbelly Ma Rainey & Skip James); learn classical Buddhist-style meditation practice; try everything; "If you see something Horrible, don't cling to it," sez Tibetan Lama Dudjon Rinpoche. See Charlie Chaplin Marx Brothers & WC Fields. Read PLATO's *Symposium*. Tell your friends everything. Give away all your secrets. "Be wise as serpents and gentle as doves." Feed everybody. Remember life includes suffering complete change and no ultimate personal identity, neither permanent Self or permanent God. Cheerful! Help everyone!

MICHAEL SCHUMACHER
March 11, 1982,   Milwaukee
March 23, 1982,   Phone: Wisconsin-Pennsylvania
Oui, June 1982

Although Ginsberg had long been associated with musicians and had even recorded his own setting of some of Blake's *Songs of Innocence and Experience*, his involvement with music deepened throughout the 1970s and the first years of the following decade, finally resulting in a two-album set, "First Blues." Before and after the release of this album, several interviews focused on Allen's musical work.

The following interview was done by Michael Schumacher, who would later write a critical biography of Ginsberg, *Dharma Lion*.

The initial interview took place, according to a March 25, 1982 letter the interviewer wrote to Allen, "in car, restaurant, and [local] poet's house with numerous interruptions." Schumacher remembered in 1999 that the restaurant was near the University of Wisconsin-Milwaukee campus, for "Allen was giving a reading at the college that evening. I recall its being rather awkward: my pile of Allen's books, all ready to be used as reference (and, as a fan, to be autographed), as well as my spiral notebooks and tape recorder—all filling the small surface of the table. Lots of background noise." The second part of the interview was done by phone, while Allen was in Pennsylvania and ill, according to the same 1982 letter.

—DC

Michael Schumacher: Why don't we begin with a discussion of your record, *First Blues*—

Allen Ginsberg: Well, there are actually two separate records called *First Blues*. The first is a Folkways record, recorded on an old Wollensack in 1972–73 by Harry Smith, who is both a filmmaker and Folkways archivist. He has done several boxed albums of folk music and blues which had a big influence on Dylan and the folk development in the fifties. He's a big eccentric. Do you know of him?

MS: I think so. Large, bearded guy?

AG: No, no. Not too big. There's a Harry Smith who's got a little magazine, and then there's another Harry Smith who is perhaps more famous in avant-garde circles as a filmmaker who originated the whole concept of mixed media way back in Berkeley in the forties. He also collected a complete Kiowa peyote ceremony for Folkways.

He was living at the Hotel Chelsea. I came in for a month or so and he recorded everything I ever composed, everything Peter Orlovsky ever composed, and a huge amount of Gregory Corso, as part of his collection, *Materials for the Study of Religion and Culture of the Lower East Side, New York, Circa 1970.*

There's another album called *First Blues* which is a compilation of ten years' work, a double-album with extensive liner notes by me, put together by John Hammond, Sr. This album is the result of three separate situations.

The first was in 1970. I went into the Record Plant in New York with Dylan, Happy Traum, David Amram, and Perry Robinson. The idea was to improvise in the studio without any paper in front of us. Over several days' work, there were three cuts that were any good: "The Jimmy Berman Gay Lib Rag"; a thing about flying down to Puerto Rico called "The Vomit Express"; and a little song telling people to come to San Diego for the Republican Convention which never took place. I paid for all that, got it together, and produced it myself.

MS: When did you first meet Dylan?

AG: I met him in '63. I was in India for about a year and a half, and there was a welcome home party for us when we got back. Al Aronowitz, the journalist who is a friend of Dylan's and whom I knew from the late fifties when he did a story on the Beat Generation for the *Post*, came and brought Dylan. He was coming that night from his meeting with the Emergency Civil Liberties Committee's annual banquet, who had given him an award. Dylan had declared a sort of an independence of any specific political allegiance and that upset them a bit. So we talked about poetry and politics, how poetry was just a reflection of the mind, independent of politics. We made friends that night.

MS: And you saw each other off and on?

AG: He invited me to go out with him and see how he worked, so we went to Princeton for a concert he gave there. There was a photographer there who took photos that he used on the back of his album, *Bringing It All Back Home*. It has a picture of me, without a beard, wearing a funny top hat.

MS: And the *First Blues* sessions came as the result of his hearing you improvise at one of your readings?

AG: Yeah. He came to a reading at NYU with David Amram and was standing in the back sort of anonymously. I didn't know he was there. We were just having a good time, making words up and rhyming. I was on harmonium, Peter had a banjo or guitar, and we both got into a long improvisation on the subject of why write things down on paper, when to do that you have to cut down trees. Why not sing songs into the air and lose them, rather than write them down on paper and cut down trees?

While I'd been in San Francisco, recording the second album of Blake, I'd met Chögyam Trungpa and we'd had a very funny conversation about poetry and music. I had shown him some mantras that I'd learned and he interrupted me in the middle of it by putting his hand on the keyboard of the harmonium. It caused the music to stop for a second, and he said, "Remember, the silence is just as important as the sound." Then I compared my travel and reading schedule with Trungpa's and I said that I was getting fatigued. He said, "Oh, that's because you don't like your poetry." And I said, "What do you mean?" He said, "Why don't you make up your poems on the spot, right up onstage. Why depend on a piece of paper? Don't you trust your own mind?" And that reminded me of advice I'd gotten from Kerouac. The next day, I had a benefit reading for Tarhang Tulku, a Tibetan teacher of the Nyingma sect—or "Old Sect"—and I got up onstage with a choral group behind me and chanted a lot of mantras and then I launched off into a two-chord song of how sweet it is be born in America, followed by some eating McDonald's hamburgers, breathing gasoline, consuming 60 percent of the world's raw materials, and causing wars and destruction in every direction, and living in the heaven world. It was both a lamentation and celebration—sort of sweet and sour about living in the *deva loka*, the heaven world, one of the many possible worlds to be born into. So this is the background.

When I got back to New York, I started improvising onstage with Peter, and Dylan was there and heard it. He phoned me when I got home that night and asked, "Can you always improvise like that?" I said, "Sure, I used to do that with Kerouac a lot under the Brooklyn Bridge." So he asked if I was going to be home that night, and he came over with David Amram and another Indian kid who played guitar. We started jamming. One of Peter's girlfriends (Denise Mercedes) had a guitar, and Dylan started jamming with that. I had a harmonium, a big pump organ, and he showed me a few chord changes that he used. We just did some jamming with Amram, particularly some kind of Latin rhythm which evolved into the samba, "Going Down to Puerto Rico." So then Dylan said, "Why don't we go into the studio and do this?" I called the Record Plant and we went in and recorded for two days.

MS: Were those sessions anything like the sessions Jack Kerouac had when he recorded his haikus with Zoot Sims and Al Cohn?

AG: The first five poems in the book, *First Blues*, were made out of whole cloth in the studio, right on the spot. "Jimmy Berman Rag" is the most successful because the original tape is the original manuscript. So that was more improvised than Kerouac, because Kerouac had his haikus written down. Kerouac would read his haiku and Sims would play and, inspired by his sax, Kerouac would read his next haiku from that particular tone and they would take off from his tone. You could see a certain amount of interaction between them. In the case of Dylan, I was much more conventional in form, with rhymed lyric verses and regular blues or ballad forms, but on the other hand, the words were made up on the spot.

MS: When was the second session?

AG: In 1976. John Hammond, Sr. produced an album for me at Columbia with David Mansfield, Jon Sholle, Arthur Russell, and Stephen Taylor. David Mansfield was from the Rolling Thunder tour. Jon Sholle is an old musician friend, a studio musician of great accomplishment, basically my music teacher, who had worked with me on an album I put out in 1968, *The Blake Songs of Innocence and Experience*. Arthur Russell is a classical cellist, and Stephen Taylor was a kid I'd just met, a trained musician in classical and rock 'n' roll. We recorded a full 45 minutes of music: "New York Youth Call Annunciation," "CIA Dope Calypso," "Put Down Yr. Cigarette Rag," "Everybody Sing," "Broken Bones Blues," "Stay Away from the White House," "Hardon Blues," and "Guru Blues" from the *First Blues* book, plus "Gospel Noble Truths" from *Mind Breaths*.

MS: All those were recorded in 1976?

AG: Yes. Hammond was working at Columbia and he wanted them to put the record out, but Columbia made a corporate decision not to put it out. I was told so by the A&R man who dealt with Dylan. It was decided that it was too dirty or, as I heard from one of the merchandising men who heard the album, they thought it was great, they listened to it at home, but they were afraid to play it for William Paley, the head of CBS. This merchandising man, slightly drunk, told me, "You've got a good record in you, Ginsberg, but this isn't it. You're shaking your cock around too much." So they never put it out.

MS: There is a lot of sex in *First Blues*, which is a topic record companies like to avoid.

AG: There's a lot of sex in a lot of the stuff now. They use much more dirty words than I used.

MS: But those people are cute, like *Three's Company*:[158] it's all right to laugh at it, but you can't take it seriously.

AG: Some of it is quite serious. That's what surprised me. There's a lot of serious sex in the Rolling Stones and other music now. So I really didn't quite understand. It may have been that my album was mixed very clearly so you could hear all the words. The language was articulate, not buried in the music.

MS: Do you think the homosexuality projected in some of the songs greatly turned them off?

AG: It never turned off the merchandising people. They probably found it cute. I think they thought it might turn off the public—not the entire public, because there's a huge gay public—but the people listening to this would be more specialists anyway.

---

[158]Popular 1970s television situation comedy.

MS: Companies are famous for looking at it all in dollars and cents. They're more concerned with making back production costs and a nice profit than they are with producing anything artistic.

AG: It'll make its production costs. They hardly have any because I produced most of it myself, except for the portions that were done by John Hammond. There's nothing they can do but make money on it.

MS: So you're not setting records making records. You've got all this stuff recorded, but very little of it out—

AG: By this time, I'd had the experience of putting out one record of William Blake, as well as a second Blake album which has not ever been put out. I'd recorded the record with Dylan and was not able to put that out. I also prepared sixteen volumes of my poetry—sixteen albums, complete, collected, assembled poetry vocalized, 1948–1970—which I sold to Fantasy, but they're just sitting on it.[159]

MS: And you still wanted to invest your time and money in the business?

AG: I didn't do any more recording because I didn't have the money until January, 1981, when I got a free scholarship for four days in an eight-track studio upstate New York. So I gathered a lot of the same musicians—Taylor, Sholle, and Russell—and went up to Fort Edwards to ZBS Studios, which is an old radical commune which formed itself into a studio in the sixties. We got into a little electric for the first time in "Dope Fiend Blues" and a new song, "Capitol Air"—

MS: From the book *Plutonian Ode?*

AG: That was done in February-March, 1981. Over the years, I had always done acoustic music in readings wherever I could. I'd either gone with musicians or picked up musicians on the road—just ran into kids in the schools or picked up any good-looking kid who could play funky blues and improvise. But once I got a taste of electricity, I began checking out rock bands, singing with a lot of little rock bands.

MS: Which brings us to your involvement with New Wave music. Certain members of the Beat Generation—you and William Burroughs in

---

[159]Some of this material has been issued in the four-volume set, *Holy Soul Jelly Roll: Poems and Songs 1949–1993* by Rhino (1994), with extensive notes by Ginsberg and musicians.

particular—have been associated with the development of New Wave. What's the connection?

AG: Well, all through history, you have a bohemianism which has gone through many artistic phases and taken many forms. There's always been a continuity on the margin of people working by themselves secretly on sex, dope, art, strange ideas, or anarchism. They thought politics was shit, which every working man does also. They didn't believe in the authority of the state. They were groups of friends who hung around in interconnecting bohemian circles.

The Beat movement of the forties was simply another branch of the traditional marginal bohemian insight into the fact that we were all personal, that life was not square. Jack Kerouac and Burroughs got together in the early forties, '44 or '45, and I met them at that time. I met Neal Cassady in '46 and Gregory Corso in '50 and Peter Orlovsky in '53. We all met up with Gary Snyder, Philip Whalen, and Michael McClure on the West Coast around '55. We'd already met up with the East Coast poets, and we met up with Philip Lamantia, who was a member of Kenneth Rexroth's anarchist surrealist circle. So there were these interconnecting bohemian circles: anarchists, Catholic pacifists-worker groups.

In 1957 On the Road was published. In about 1960 we were in New York and met a great intricate group of underground filmmakers. Robert Frank, a great still photographer, made a movie with Kerouac called Pull My Daisy in which I acted with Peter Orlovsky, Gregory Corso, and Larry Rivers. Larry was a saxophonist who turned on to musicians. There was Charlie Parker and Thelonious Monk, and they knew some of these painters and poets and so on. One interesting filmmaker was Jack Smith, who made a film called Flaming Creatures, which was the first New York, U.S., exhibition of sort of transsexual, transvestite, outrageous, scared everybody funny stuff, with people in veils, turbans, shaven heads, and strange costumes. One of the people in Flaming Creatures went out to San Francisco and started the group, Angels of Light, who did three shows of extravagant dressed-up the-atre, where everybody was all dressed up in elephants' costumes or Greta Garbo costumes. That same kind of glitter rock influenced the New York Dolls and helped create the early-1970's punk scene. So actually, it started in New York, went to San Francisco, and then washed back to New York.

In terms of lineage, Burroughs had a lot to do with the futuristic, extraplanetarial, extraterrestrial, Wild Boys, Heavy Metal, Soft Machine, Steely Dan consciousness. A lot of the early outrage bands took perceptions, as well as names, from Burroughs. As time went by, a lot went back to see him—David Bowie, Iggie Pop, and Jim Carroll. They knew Burroughs. Some of the New Wave musicians, like Lou Reed and Jim Carroll, were poets who were once part of the St. Mark's Poetry Project in New York.

MS: Your involvement in New Wave music has gone as far as recording. What was it like working with the Clash?

AG: Last summer, in June, I met the Clash and sang with them at Bonds in New York.[160] A friend had brought me backstage to meet them, and Joe Strummer said, "Hey, Ginsberg, when are you gonna run for president?" He was kidding me, of course. So I said, "I'm not, because my guru said if I was president I'd wind up in diamond hell."

MS: Diamond hell?

AG: We used the phrase in Dylan's *Renaldo and Clara*. It's the unbreakable hell of egotism. It's like your karma projected out and extended with your own double dealings, where you find yourself in the insoluble contradictions of your own making. By diamond, they mean the hardest; it will cut through everything.

So Strummer said, "Do you have any poem you could read tonight? We've been sending somebody up onstage to give a little lecture on [El] Salvador, but people throw tomatoes." I said, "I've got a poem, but it's got chord changes, if you want to run through it and sing it." So we were backstage and we ran through it for ten minutes, got the changes, and then went out after their encore set and did it. We had the sound turned up so the vocal could be heard clearly. And the crowd dug it and so did Strummer. I had to go back to Boulder the next day, and he had to go back to England, so I wrote him and said that I'd like to work with him and make a single. When he got back, he called me up in Boulder from New York. He gave me his lyric sheet for a song

---

[160]Ginsberg performed his poem "Capitol Air" with the Clash for one of their seventeen shows at Bonds Casino, New York City, in December 1980, the recording of which is included in *Holy Soul Jelly Roll: Poems and Songs 1949–1993* (Rhino, 1994). This collaboration resulted in Ginsberg's participation in "Ghetto Defendant" on the Clash album *Combat Rock* (Epic, 1982).

called "Ghetto Defendant" and asked if I could improve the lyric. So I tried, made some changes in it—something like tinkering with the lyrics of my students—and then they put me on tape, in a reggae type number. They needed something to sound like the voice of God, and so I wrote a verse for that and also did a lot of mantra/Sanskrit chanting at the end of the song.

MS: After that, you recorded "Birdbrain" with a Denver New Wave group?

AG: A 33-single with the Gluons, a local Denver group. We were originally going to record "Capitol Air," but they heard me read "Birdbrain" and thought they could put that together as a record. "Birdbrain" was conceived by Mike Chapelle, who is the head of the Gluons, when he heard me read it as a text, so we recorded it last December and put it out.

MS: What was your involvement in the New Wave publication, *Search and Destroy?*

AG: It was edited and published by one of the people who works at City Lights. He took me around to the New Wave places in San Francisco and turned me on to that in '76 or '77. He wanted to start a newspaper, *Search and Destroy*, but he didn't have the money to get it going, so I gave him some money.

MS: Just to keep it going?

AG: Well, that was the first issue, so I guess I was the publisher of the first issue of *Search and Destroy*. Where did you hear about that?

MS: I've seen copies.

AG: How did you know I had anything to do with it?

MS: They mentioned it in advertisements. . . . Getting back to your poetry, your readings have received mixed responses, one of the major criticisms being that this great poet who's written classical stuff like "Howl" is now dabbling with rock 'n' roll, New Wave, and blues. What's his point?

AG: The point is a classical point: poetry and music have always been allied. I'm a little late in practicing that. I'm a bit of an amateur at it, so I'm trying to sharpen my practice and get back to home base. When I was young, I heard a lot of Leadbelly and Bessie Smith. I heard a lot of that when I was in grammar school and high school in Paterson, New Jersey. I went to black spiritual churches along River Street and heard preaching and singing, so I heard a lot of spiritual popular music when I was young. So it seems like a natural extension, as I'm recovering my own person and beginning to enjoy my own body and speech and mind, that my poetry has come to a point of refinement where I also include music.

STEVE FOEHR
December 5, 1983,   Boulder, Colorado

Steve Foehr had first met Allen in 1962 or 1963 when he was the editor of Denver's *Straight Creek Journal,* an underground antiwar newspaper. Ginsberg had passed through town and needed a place to stay, so Foehr put him up. In 1983 Foehr conceived of a "project in which artists in different disciplines . . . would participate in discussions of common themes. I would initiate the discussions and be the clearing house to put people in contact with each other, etc. Sort of a mail salon." As Foehr was struggling to make a living as a writer, the first subject he chose was "Art and Commerce." Allen agreed to be the first interview. Foehr sent Ginsberg's interview "to a few artist friends, but the project went no further and died on the vine."

They met in the afternoon in a classroom at the Naropa Institute after Allen had taught a class on Emily Dickinson. Foehr sat in an old-fashioned school desk with an extended, broad arm for writing and Ginsberg in a big overstuffed armchair, wearing a tie, blue shirt, and dark blue sport jacket. "I asked the question, 'What is the relationship between art and commerce?' and Allen started talking. I simply hung on and tried to get it all written down. At one point my black pen gave out and I switched to a red pen. Later I typed up the notes."

—DC

Steve Foehr: What is the relation between art and commerce?

Allen Ginsberg: Art is an activity completely independent of commerce. It has nothing to do with publication or public acceptance. That's always been true; art has nothing to do with fame, publication, or public acceptance.

The basic activity, the basic insight, the basic energy that goes into art has to do more with inquisitiveness, curiosity, and exploration and a kind of gaiety and glee in composing feelings and doing whatever you want to do. One thing a poet can do is say anything he wants to say . . . on paper—not in public, but on paper and to his friends. Art does have to do with acceptance by your friends; your friends digging what you're doing and you digging what your friends are doing. But that is the extent of the sociability of art. Everything else is an accident. Any other acceptance beyond that is gratuitous, accidental, helpful, charming, useful, might make you a living, might give you more time, but irrelevant to the original visionary glee, or visionary energy, which comes from a vision of nature or a truth of your own heart which is beyond social acceptance or rejection.

If financial rejection dampens or snuffs out that visionary energy then it's not real art. Only in certain cases of certain hypersensitive types, maybe Hart Crane, among others, had that kind of problem, although he was very successful as an artist. His family put him down and his father wouldn't support him—they were millionaires from Shaker Heights in Cleveland. He committed suicide.

I think that the idea that financial consideration would dampen an artist's art would be like saying it would dismay a Franciscan monk or something. The whole point is the vision, is the vision of unity and a humor that is well and beyond any financial consideration—well and beyond *life* or *death!* If financial consideration is the basis of art, can affect art, then it's not a very deep art. Finances could distract your time, but that just means you'd have to find an art form that you could do at night.

I had considered myself professionally rejected. I didn't get anything published until 1955 or so, and then I didn't make any money on it. I still don't make enough to live on from my writing. That is hardly the consideration. I just got $300 from City Lights, my publisher, for the year 1983. This is December. That on top of $7,800 that I got in the spring when my royalties were due. They didn't have enough money so they paid me in pieces. This is the last piece of $7,900. If I wanted to make a nice living, I would be a grammar school teacher and make more than twice what I am getting.

I can make money different ways. I can go out and give readings, but poetry would not give me a living. After I published my first book of poems, which was called "a world success," in the sense that it was translated to a lot of languages on different continents within five years, my income from it was something like $2,000–$3,000 from 1956 to 1965 or so, and then I got up to maybe $10,000 to $13,000 for a couple of years out of five or six books, which, given my standard of living, was enough. Then getting older, I needed more money—doctor bills, insurance—so now what I'm getting for writing doesn't *pay my phone bill*, or my rent, much less food.

So I have never conceived of art as commercial. The commercial thing is, if you look at it, irrelevant. A hype by artists looking for excuses. Edgar Allan Poe was the only man of letters in the nineteenth century, though one of the really great poets, who made a living writing. Emily Dickinson didn't even publish her poems. Walt Whitman put out his own poems, lost his job on account of it. Herman Melville got a job in the New York Custom House and then wrote his poems after the failure of *Moby Dick*. His doctor told him to stop writing prose or he'd have a heart attack. Too heavy, too expensive, too much hard work. *Moby Dick* was written in a year. Heart failure! His next book a failure commercially, but a great book (*Pierre*). Mrs. Melville wrote to Hawthorne, "Don't tell anybody but Herman has taken to writing poetry." Then he got to be a very great poet. Then he wrote *Billy Budd*. *Billy Budd* was found in a trunk in an attic by Professor Raymond Weaver in 1920.

Did Herman Melville write *Billy Budd* for money? No! He hid it! Did Emily Dickinson write for money? No, she wouldn't even publish. She had five poems published in her lifetime.

The idea that any really serious art would be stopped by social considerations and rejection is a big mistake. It is a humorous mistake. Art is much funnier than that and much vaster and grander than a job. It's a religion which thrives on persecution.

There is William Burroughs. To this day Burroughs didn't make a living writing. Last year he tried to borrow money from me. Kerouac, in the late 60's, when he was a world-famous novelist, asked permission to sell my letters so he could get a down payment on a house in Florida, rather than sign a contract for $7,000 or $8,000 forcing him to write a novel he didn't want to write. He didn't have enough money to do his scene.

Burroughs didn't find it depressing. Nothing depresses Burroughs. Kerouac found it embittering, I think. But he wrote more. It might be depressing slightly, but not if you've got the spirit of a great writer. An artist by very definition means penetrating to the heart of the universe, i.e., your own heart, going beyond depression or exuberance.

Emily Dickinson originally wanted to be public, but then she realized the conflict between staying with what she wanted to do compared to what her advisors thought proper for her to do as a writer. She chose not to take their advice, but did what she wanted to do. Walt Whitman was told by Emerson, "Don't publish the dirty, gay section in *Leaves of Grass*," because he would never get away with it. They took a two-hour walk around the Boston Commons and Emerson gave him every practical argument. Whitman said, "I couldn't answer his arguments. He was brilliant, emotional, beautiful, intellectual, but at the end I said that I knew what I wanted to do and I'm going to do it."

There is huge literature in Russia that is not published. Publishing is a middle-class American professional idea. If you don't like it you can go into insurance; if you don't like insurance, teach; if you don't teach, you can write for magazines.

Art as I would define it in terms of "yes, you need Emily Dickinson, Walt Whitman, Jack Kerouac" has nothing to do with acceptance and rejection. I was pissed off at my first professional rejection. All those guys (editors) were running literature, and they didn't know what they were doing. I knew what I was doing, and I said so. Kerouac said so if you look at his early work. You realize that he wrote *The Town and the City*, a novel, unpublished. Then he wrote *On the Road*. It was put down by his favorite editor, Robert Giroux, a famous editor. So he went and wrote an even more wild novel, *Visions of Cody*, which is even much more unpublishable and then he wrote ten novels, assuming "Rack my hand for labor of nada! All these books are published in heaven."

When I wrote "Howl," I dedicated "Howl" to Kerouac. I listed fourteen books that he had written that were rejected and not published. All those books are published in heaven. I knew the worth of them. He knew the worth of them. City Lights and everybody else rejected my first book. So I went off and wrote "Howl."

Commercial publishing is run by the bottom line rather than great editors. There is a plethora of noncommercial publishing like City Lights, Black Sparrow, that is more active than ever.

Art is not commercial. Art is art. Commerce is commerce. An artist is lucky and fortunate if his art has commercial value, but the commerce is not the criteria. The criteria is the exuberance of expression. If you're an artist you make art. If you're not an artist you go around peddling your pen. There's nothing wrong with peddling your pen. It's fine. It's like digging ditches. It's work. Definitely harder and more unpleasant than being an artist. A true artist won't get discouraged. I doubt if very many true artists die during the quest because of lack of commercial success. True art is based on some kind of vision, a personal experience that is beyond success or failure, even beyond life and death. It's an accident. It happens through suffering. If you suffer enough you finally give up and realize, "Oh, the universe isn't suffering. Is there no one left? No, there's no one left. So I'm not suffering either. Then what are all these people walking around suffering for? Oh, they think they are somebody." Art is an expression of "I'm nobody, who are you? Are you nobody, too? Don't tell."

I go on faith that my materialistic needs will be met. If not, I can go out and wash dishes. I originally got seaman papers and sailed around and made money. William Carlos Williams worked as a doctor. He didn't get any money for his poetry.

You can't do art all the time. Take Van Gogh. Van Gogh couldn't sell a painting. Nobody wanted them. If he were recognized as an artist, he might not have committed suicide because he might have been able to get girlfriends so he wouldn't have cut his ear off. You might say that Van Gogh might have made another thousand paintings. He already made more than anybody else. It's amazing how many paintings he made in that short lifetime. He worked like a madman. He was totally unsuccessful. Cézanne. Nobody liked his paintings until his very end. They thought he was an awkward amateur. And yet he is certainly considered *the* classic of the last two hundred years. Cézanne was this upper-middle-class guy who could afford to do it himself. But he was not recognized as a painter. And his ambition was to paint a painting that could be put in a museum.

It's not a dilemma of how to be an artist and how to make a living. It's more how to find a form that will manifest the particular vision. The artistic problem is the big problem, not the financial. The financial problem is a problem. But everyone can solve that problem. Or they can't.

The dwelling on the commercial scene is a problem that comes to artists who haven't had a primo vision and therefore haven't gotten beyond the personal, or the complaining stage.

There are maybe some artists who need approval, and it may be very legitimate. Kerouac, I think his heart was poisoned by the attacks on him. I think that was responsible for his death. He identified himself with America. He realized if he were put down that meant America was going to move through an immense amount of suffering and would inflict a lot of suffering on the world. He realized that his personal tragedy of being rejected by the official arbitors of culture meant that America was in for a bad, bad scene. The vision of America that he and Whitman had was going to be a flop and there was going to be great destruction. So that point of view, rejection, did affect him, but it was more spiritual rejection. Rejection of the sacred heart rather than that he couldn't make money. It was on that level of spiritual warfare, not on the level of making money.

The reason someone is an artist is that they get out of the commercial trap by realizing that it is only a trap, that you're not going to die. You can find a little corner with pencil and paper or crayons or something. It's only if you think that the *sign* of art is commercial that you might get confused.

Artists do have some sort of genuine grounding in contact with nature, with a vast nature, their own hearts and external nature that most people don't have. Or others have but get discouraged from or don't realize it's for real, that it's subject and they don't realize everything is subjective. Maybe artists are people who realize that the subjective is the only actual data that we know. And then there is the majority of people who think there is an objective world that they have never obtained because all they feel is the subjective. So they had better follow what other people say is objective. And anyone who has a loud enough mouth and says, "My vision of the world is objective, I'd like you to give me $242,000,000,000 for my military to fight the Russians" and that is real, that is politics, like Kissinger laying that down as objective without realizing that it is his own imaginal projection on the universe. People who think that somebody has the authority, and that may come from God—who doesn't even exist, anyway—will get inveigled into this hype.

Then there is the artist who says, "This is all a hype. There is nothing but our subjective world. Reagan has a subjective world and I got mine. He is making his loudmouth, I'm making mine. Check it out. Let's check it out." It requires total uncertainty to say that. "Oh, I know my subjective world. I don't know anything. I'm totally confused." Then you've got Kissinger who says, "I know everything. I'm not confused, I'm objective. I'll tell you what." You've got all these people asserting that there is some absolute and not realizing that it is all totally imaginary. An object of art is absolutely imaginal. The artist says, "The only thing I know is I'm an idiot. The only thing I know is what I see subjectively, which can change from moment to moment, from art piece to art piece," but actually objectively reflects his subjectivity.

And then there are all these other people who say, "I'm not subjective. I'm absolutely real, and you're not real. And my feelings are real, and my thoughts are valid, and I got the statistics to prove it. And if those don't prove it, I've got guns." It's this presumption of this hyper-rationalistic objective which is a kind of Orwellian thought control hallucination laid on people that overlays their neutral consciousness of being slobs like everybody else, or being human, and being dopey and spacing out and not knowing anything except what they do know, which is what they feel. The artist is the person who knows that his subjectivity is the only thing that he's got that's real.

Actual art is different from museums. Museums come after work. The word "art" comes after work. Art isn't art, actually. It isn't intended to be art. Art starts as sort of fucking around, experimenting with whatever you're interested in, like somebody who makes a bridge out of toothpicks. It's on that level. It has nothing to do with museums or art. Can you write a poem that says what your feelings are; can you paint a picture that will give the optical impression of space without using perspective lines?

I had no intention of making a living off art. I'm surprised that I can at all now. I have never been discouraged by my art. That's like asking, "Are you discouraged by the weather?" It could snow and rain and you could take it as discouraging or amazing: "Oh, look, snow and rain!"

I was never interested in writing poems to write poems. I wrote poetry to express certain feelings. If there are no feelings to express, then there is nothing to express and there is nothing to be discouraged about. It's as simple as that.

I think the difference is between viewing art as an external activity to make the artifact called art and viewing it as a natural expressive function of certain states of subjective consciousness that you feel are so grounded and basic, natural, that their expression is inevitable and natural. But it depends on the experience of, say, the particular physical, mental experience that you have of what space is immediately after you learn that your father is dead. That deep a subjective thing about what is a sensation, about the nature of the universe and specifically of space, when you hear that your father is dead, which happens only once in a lifetime, that little sensation. If your art is based on that kind of experience, that depth of really rooted experience, really specific, so if that is the area art really deals with, it isn't really art so much as it is a way of finding some way of communicating that to others or to yourself. Then, later on, you like to call a piece of writing, or a poem, or a painting of it art. During the composition there isn't the thought of art in mind, you are absorbed into the material without thought of what you're going to do with it. It would be a distraction to think about its commercial utility. You might think of it as a social utility of communicating to others. It goes beyond just making money, but what are you going to do with the rest of your life? Should you take care of your body, change your job, prepare for death? So the consideration is much deeper than "can you make money out of the incident?" It's more like "how can I live?" From that point of view, where you have real art, I don't think the acceptance or rejection of your realization is very important. Real art is beyond value, although it is invaluable subjectively. But it is beyond objective value. If the art is on that level, then that is where it really gets interesting.

SIMON ALBURY
June 3, 1986,   Orono, Maine

This interview was done in Orono, Maine, as research for a Granada Television documentary *Sergeant Pepper—It Was Twenty Years Ago Today* about the Beatles' masterpiece and the Summer of Love. The interview took place while Ginsberg, Anne Waldman, and Robert Creeley were teaching for poet and scholar Carroll Terrell at the University of Maine, Orono. The interviews were conducted in three sessions on June 2 and on Allen's sixtieth birthday, June 3. The tape from the June 2 interview has not survived.[161]

Simon Albury had known Allen since 1964, when he organized a reading that featured Allen and Peter Orlovsky at Brandeis University. Over the years they remained friends, and Simon helped to organize readings whenever Allen went to London. Albury was among the contacts Allen used to assist him in researching the CIA's involvement in heroin trafficking.

The interview was transcribed by the current editor.

—DC

---

[161]Although the tape of the June 2 interview is lost, Albury did transcribe from the interview some of Allen's quotes on *Sergeant Pepper*: "There was this element of adolescence opening up and discovering the universe. There was an individual life in the life of friendship or imagination that was cheerful and uplifted or humane with some rediscovery of humor.

"After the apocalypse of Hitler and the apocalypse of the Bomb, there was here (in *Sergeant Pepper*) an exclamation of joy, the rediscovery of joy and what it is to be alive. They showed an awareness that we make up our own fate, and they have decided to make a cheerful fate. They have decided to be generous to Lovely Rita, or to be generous to Sergeant Pepper himself, turn him from an authority figure to a figure of comic humor, a vaudeville turn.

"Remember, this was in the midst of the sixties, it was 1967 when some of the wilder and crazier radicals were saying 'Kill The Pigs.' They were saying the opposite about old Sergeant Pepper. In fact the Beatles themselves were dressing up in uniforms, but associating themselves with good old time vaudeville authority rather than sneaky CIA, KGB, MI5, or whatever. It was actually a cheerful look round the world. . . . *for the first time, I would say, on a mass scale*.

"They were giving an example around the world that guys can be friends. They had, and conveyed, a realization that the world and human consciousness had to change."

Simon Albury: Some people think that the Be-in was largely (Timothy) Leary's event.

Allen Ginsberg: No, not at all. He was our guest. It was an event that he attended, but the organization and the promotion was mostly Gary Snyder and the Haight Ashbury group. If you read Jane Kramer's book[162] there is considerable discussion about how much time we would allot to Leary, and whether he should be considered a politician, a spiritual leader, or a poet, and I urge that he be considered a word-man, a poet, so given the time accorded to the rest of the poets. There was a long discussion of that in Michael Bowen's house with Snyder and myself and Ron Thelin and the people who were planning it.

SA: Were you living there then?

AG: At the time I was visiting for about half a year. I had an apartment in the Haight Ashbury.

SA: So the Be-in was an event that was built around acid?

AG: No, not altogether. Part of it is acid, part of it is spirituality, part is a ten years or so after the Beat Generation San Francisco Poetic Renaissance, because it was the ground for all that to grow. Because we were sort of the fathers of that scene we were invited to be the masters of ceremony and the speakers to open it and close it. The poets opened it and closed it with a little ceremony, and then in the afternoon Snyder and I did the circumambulation and then it ended with me asking everybody to do kitchen yoga and clean up the park, and then chanting Om Sri Maitreya as the sun set. You'll find some of that in Kramer's book and the other in the book by . . . it was the *Rolling Stone* guy . . . Perry![163]

Now the Diggers were using Gregory and Gary and myself and others as sort of the theoreticians and prophets of the situation for the Diggers and Haight Ashbury scene. The *Oracle*, which was the newspaper, was looking to Alan Watts and Leary and myself and Snyder as the sort of wise elders to have a conference afterwards, and then they published this complete conversation that took place on Watts's houseboat. The whole notion of the Be-in was a Snyderesque medita-

[162]Jane Kramer, *Allen Ginsberg in America* (New York: Vintage, 1970).
[163]Charles Perry, *The Haight-Ashbury* (1984).

tion: be-in rather than a sit-in, to merely be there, like "be here now." I think it was a psychedelic thing, but with the input of the imagination of the poets and Zen meditation. Suzuki Roshi was the first great teaching Zen master in America who established a meditation hall, although there had been others since the world Parliament of Religions in 1893 in Chicago. D. T. Suzuki was also there as well as others. He was there when I first came in '58 and worked in rapport with Joanne Kyger and many of the people that Trungpa now teaches, and Snyder and Diane DiPrima were old members of the Zen Center.

SA: To what extent was the San Francisco thing really about acid more than anything else?

AG: It was a public surfacing of acid in the set and setting[164] of the San Francisco Renaissance and Beat Generation and literary movement and poetic imagination that emerged from 1955 on in San Francisco with City Lights and with the ecology movement. And that was spearheaded by Michael McClure and Snyder, the ecological notion, the notion of a be-in to begin with, the notion of an art newspaper featuring poetry and featuring the *On the Road* generation philosophers and poets. And the Be-in itself was planned and coordinated by artists and poets: Ron Thelin was the business man who didn't want to be a commercial businessman, the Diggers who emerged around that time, the Hells Angels brought in by (Ken) Kesey and his troupe,[165] who had been some of the promoters of acid, after all, but Kesey the novelist and his Merry Pranksters was again poetic imagination.[166]

The formulators of the actual people on the stage were a committee of Thelin, myself, Snyder, Michael Bowen, and Allen Cohen of the *Oracle* who is also a poet. We were the ones who decided who would have a place to speak and how long they would speak and who invited Suzuki Roshi to sit on the platform and join us. And he did, which

---

[164]"Set-and-Setting" was the title of a presentation given by Timothy Leary at a 1962 American Psychological Association conference.
[165]The San Francisco Mime Troupe—a radical play group that had been cited for performing in Golden Gate Park. Bill Graham organized a benefit to pay for their defense. See *Summer of Love: The Inside Story of LSD, Rock and Roll, Free Love, and High Times in the Wild*, by Joel Selvin (New York: Plume Books, 1995) and Peter Coyote's (once a member of the Troupe) *Sleeping Where I Fall: A Chronicle* (Washington, DC: Counterpoint, 1998).
[166]See Tom Wolfe's *The Electric Kool-Aid Acid Test* (New York: Farrar, Straus and Giroux, 1968).

Snyder thought was quite a historical event because it's very rare for the Zen people to come out into large crowds, for they generally were withdrawn into their meditation hall and taught in small groups. So the set and the setting was determined by the poets, the catalyst was acid, the ground was established by the San Francisco Renaissance and the whole Beat movement of '58 and *Beatitude* magazine and City Lights and all the poetry activity. The first activities by Bill Graham featured the poets in the mind-food benefit as among the major attractions that opened and closed and gave dignity to the whole benefit in a kind of spiritual direction. What I'm pointing out is that the imaginative and spiritual direction of '67 in San Francisco was a great deal the creation of the poets. And one of the great culture heroes was Antonin Artaud of the supremacy of the imagination. The levitation of the Pentagon was a further extension of Gary Snyder's conception of the Be-in and an outgrowth of our circumambulation of the Meadow.

SA: The Merry Pranksters and Kesey, what was positive about them?

AG: Well, first of all they didn't reject the American flag but instead washed it and took it back from the neoconservatives and right wingers and war hawks who were wrapping themselves in the flag, so Kesey painted the flag on his sneakers and had a little flag in his teeth filling. Actually his Merry Pranksters were Korean War veterans, Ken Babbs among them, who were sort of like big macho Americanists. And it was an Americanist movement to regain the old American tradition of Prankster, Voyager, Explorer, Davy Crockett, individual enterprise can-do, but can-do in a peaceful rather than in a warlike way. Family-oriented, Kesey coming from a big family and the gatherings being at La Honda at his family home, so it was an extended family notion. There was also the idea of humour and art being the basis for political gatherings and the Acid Tests.[167] Humour and daring and adventureness rather than paranoia and fear, cutting through paranoia, and we're going right into the paranoia and seeing its emptiness. On a political level in '64 Neal Cassady, who was the hero of the earlier Beat thing, drove Kesey's bus cross-country during the Goldwater-Johnson presidential campaign with a slogan painted on the bus, "A vote for

---

[167]See *Acid Dreams: The CIA, LSD, and the Sixties Rebellion* by Martin A. Lee and Bruce Shlain, (New York: Grove Press, 1985).

Goldwater is a vote for fun,"[168] and went through Texas and all through
the South with that slogan, exploding the whole serious hawk and war
issues. It was Kesey also who had turned the Hell's Angels onto acid
and warded off their attack on the 1965 peace march that Jerry Rubin,
Snyder, myself, and others participated in,[169] as distinct from the
heavy political people whose ground was that this is the last conflict
with the black-shirted fascists and we should attack them with chains
instead of having a flower power march as a demonstration as theater.
So he had a good idea of public theatre.

SA: Was flower power all about theatre?

AG: Oh, yeah. About '67, sticking the flower into the barrel of the Pen-
tagon gun during the levitation,[170] and flower power actually meant the
power of green, growth, and ecology. It was planet news, so to speak,
for an ecological statement. It didn't mean idiot sentimentality, it
meant the basic power of matriarchy, the feminine, Mother Earth, vul-
nerability, and the vulnerability of earth itself, and also the long-last-
ing strength. And a lot of the influence, brought in to some extent by
Snyder and the anthropological poets, was American Indian, the head-
bands and ponytails of the Indians. Then there was also the added
influence of oriental thought, both Japanese and Indian, Buddhist and
Hindu. The Hindu iconography was very powerful in the *Oracle*, like
the Be-in issue had a giant poster of a Saivite sadhu smoking grass so it
was a confluence of grass, psychedelics, and Eastern Indian and Amer-

[168]Barry Goldwater, an ultraconservative Republican and Arizona senator, was the
Republican candidate against incumbent President Johnson. He took a harsh stance
against the Soviet Union and opposed any arms-control negotiations. Many feared
his extreme anticommunist stance might cause a war with the Soviet Union.
[169]Berkeley activist Jerry Rubin, heading the Vietnam Day Committee, planned a
series of marches from Berkeley to Oakland protesting the draft and the Vietnam
War. Ginsberg and Gary Snyder participated in the first October march chanting
mantras to calm the protesters. The police stopped them at the Oakland city limits.
As future clashes with the police and Hell's Angels seemed inevitable, Ginsberg
offered advice in the form of a handbill titled "How to make a March/Spectacle" (see
Deliberate Prose, New York: HarperCollins, 2000).
[170]The Pentagon Levitation. On October 26, 1967, tens of thousands joined in a
march on the Pentagon. Ed Sanders and Tuli Kupferberg, of the avant-garde rock
group the Fugs, and many protesters performed the Pentagon exorcism for which
Ginsberg provided the text titled "No Taxation Without Representation." The event
is described in Norman Mailer's Pulitzer Prize–winning *Armies of the Night: History as
a Novel, The Novel as History* (New York: New American Library, 1968).

ican Indian peyote ceremony that was influential on the style of dress and demeanor and earth thinking.

SA: Where do the East Indian traditions fit in to what was happening in 1967?

AG: First of all, on a very basic level, you have the notion from Plato, "when the mode of music changes, the walls of the city shake":[171] if you remember around the middle of the 1960's, Charlie Mingus and Ornette Coleman began experimenting with monochordal music and began listening to Indian music. It was also about the same time that the Hare Krishna movement settled in the Lower East Side where I was working with them for a while and then settled with Prabhupad (A. C. Bhaktivedanta Swami) in San Francisco right after the Be-in in Haight Ashbury. Prabhupad was living there for some time, right off Haight Ashbury on Frederick Street, I think. Then there was the singing in the street of the Hare Krishna people, the part of the street action or the notion of street: street-fighting man or street smarts or street people. There had been that early in '64.

Nineteen sixty-two, '63, '64, there had been a trip that Gary, Peter, Joanne Kyger, and I took to India which got to be sort of a catalyst for a whole migration across. Especially when the media picked up on it and *Esquire* had a big cover story. There was a fake Allen Ginsberg with a beard. I don't know if you remember that? *Esquire* sent somebody to India to take pictures of us in the Ganges in 1962 or '63, and sometime around that time featured it on "*Esquire* Throws a Party." They had this cover where this guy in an Indian cloth looking like me is with a pretty girl model saying, "Oh, Mr. Ginsberg, how nice of you to come all the way from India to join our party." That was on the cover and apparently a lot of people read about India that way. So that was also the beginning of the transcontinental hegira to India where people were trying to get free or legal hash and grass. So it was partly a psychedelic and partly a spiritual search. And then a lot of that was in the poetry from the late fifties on, Snyder's and my own.

Also the notion of loose, light clothes that allowed sexual freedom to the genitals, of paisley and cashmere, like cashmere shawls. The organic flower-power paisley design had sperm symbolism in it, combining the erotic, spiritual, and ecological. A whole new style of

---

[171]*Annotated Howl* erroneously attributes this quotation to Pythagoras.

those long loose shirts and pajama pants came out of that. That was partly the work of Mrs. Pupul Jayakar who was the head of the Home Industries in India, with whom Snyder and I stayed, and we advised her to try and open up shops to sell the Gandhi Khadi cloth[172] in the West, which she did. So the Indian handcraft industries began seeing an opportunity to transfer both music and dress and styles and statues commercially to the West. That begins around 1964.

SA: But what particular quality did the Indian influence lend to the summer of '67 besides the dress?

AG: In regard to the use of kif, grass, ganja? The religious background and heritage: the sacred use of soma[173] and the divine herbs, and grass is one of them. It influenced the whole set. The approach towards psychedelics which Leary had already adapted in *The Tibetan Book of the Dead* style—we had *The Tibetan Book of the Dead* as a guide by 1960—changed. India had a big influence in music when Ravi Shankar and Ali Akbar Khan became notable in the West and began influencing Western jazz. In fact, there was like this fusion for a while so that it entered into the Beatles' music even. Of course in terms of poetics I brought up Dylan's influence, which is more powerful, because that was the text everybody was listening to from '65 on. And that's in a lineage from Kerouac, according to Dylan.

Other Indian influences were the notion of polytheism rather than the fixed Christian monotheism of the West, the entry of animism and polytheism and different American Indian as well as East Indian religious notions along with vegetarianism and the growth of Indian restaurants in the United States and the switch of diet from the heavy meat-eating Texas war diet of aggressive meat filled with adrenaline of frightened animals. You have a change of music, change of diet, change of clothing style, change of religion, a change of recreational drugs, all influenced by East India and American Indian: all that relating to a notion of the world is one consciousness: one world, fresh planet. And the phrase "fresh planet," introduced around that time, was Gary Snyder's. I think it evolved from "fresh planet" to

---

[172]Homespun coarse cotton cloth made in India.
[173]In Hindu mythology and ancient Indian cult worship, soma is a powerful elixir of the gods. See *Soma: Divine Mushroom of Immortality* by Robert Gordon Wasson, and *R. Gordon Wasson On Soma: And Daniel H. H. Ingalls' Response* (New Haven: American Oriental Society, 1971).

"planet news" to "planet waves." There was the practice of meditation
in many forms, whether the mantra chanting which related to the
music or mantra chanting which related to meditation, or actually sit-
ting meditation practice which was then beginning to get very power-
ful in San Francisco as indicated by Suzuki Roshi's presence and the
respect accorded him in his presence on the Be-in scene, and that's
something that had been cultivated for a long time from Kerouac on
through Snyder: *Dharma Bums*. In 1967 the notion of mantra and
meditation had arrived in London so that, around the time I left, the
Beatles had encountered Maharishi (Mahesh Yogi) in '67 and were
seriously interested in spiritual improvement. The whole notion of
rock—this is a little off the Indian thing but it's related—the whole
notion of rock and roll as a spiritual message, as a mental trip, as a
change of consciousness or as messages and signals, prophetic
singers, bards, and minstrels, which, translated in modern terms,
would be the minstrel show or the variety show or vaudeville or *Sergent
Pepper*: Making use of the vaudeville tradition for hermetic messages
or psychological messages. And just at that time or just before that
time the Birch Society and the neoconservatives in the West and the
old Stalinist Marxists in Eastern Europe began attacking rock and roll
and reacting against it as a youth movement which was antistate, anti-
authoritarian, anti-heavy metal science, and pro-ecology.

SA: How important were marijuana and acid to what happened in
1967? Really, are they *the* key to understanding what happened?

AG: I think they're *a* key in that they were a catalyst to a change of con-
sciousness, a deconditioning from the old authoritarian work con-
sciousness cultivated by the Vietnam militarization, a deconditioning
from the authority of the government and all apparent reality, catalyz-
ing a reconsideration of what was real and what was unreal, what was
illusory and what was hallucination and what was eternal and what was
perennial: nature versus hyperindustrialized civilization. The same
struggle that Blake had from the very beginning when he said "remove
those dark satanic mills" or when he referred to "dark satanic mills" as
"the dark satanic mills of the mind" as well as the industrial mills that
were blackening the English landscape. So Blake was a key reference
there, even in the use of drugs. But what's important is the imaginative
set and setting in the taking of the drugs, considering the drugs as a
deconditioning agent as Burroughs would call it, or as a glimpse of

Eden or a glimpse of the natural state of mind before the authority of Jehovah and the governments intervened, was the setting: the use of drugs for spiritual search. And the use of drugs for imaginative expansion and the use of drugs for reconsideration of the authority of the state. But those thoughts, those considerations, were first proposed by the poets, from Blake through Rimbaud and Artaud up through the San Francisco poets who introduced the notions of sacred use of drugs and sacred spiritual search and return to nature and deindustrialization and appropriate technology and modern city as Moloch. So, to the extent that the psychedelic trip depends on the set and setting, I would say that the drugs were the catalysts but the actual setting and ground was the poetic and spiritual imagination of the artists of the time.

And drugs, particularly marijuana, played a very important role in this way: that since the government with its authority announced that marijuana drives you mad and since by the early sixties a huge percent of the population had gotten high on marijuana, they saw that the government cover story was not real but was a hallucination publicly spread by the government, and that the actual experience was different from the government's story and therefore people began thinking, "Well, if the government's story on drugs is inaccurate and inflated, then what's their story on money?" So the notion of free money and the Diggers came up. "But what's their notion of the war in Vietnam?" That could be an equally large misconception of the world: "What's their story on the military? What's the government's story on almost everything in the government? What is government authority?" And that's what led to the levitation of the Pentagon notion: a demystification of the authority of the government, catalyzed by the direct sensory experience, not the *ideological* experience but the direct *sensory* experience of a state of mind that was slightly more sensitized and perceptive with marijuana and amplification of sensory input, let us say, compared to the government notion of hallucinatory distortion psychosis and frothing dogs mad in Algeria. So it played a large part in the direct experiential demystification of the authority of the state on some of the most sacred subjects: home, family, dope, dope fiends. So much so that a whole generation began calling themselves dope fiends, unafraid of the terminology and taking it with a sense of humour. In fact, they talked about themselves as freaks, like a *reversal* of values.

However, all that still depends on the whole generation reinterpreting history. As Charles Olson announced at the Berkeley Poetry Conference,[174] history is over: this is a post-historical period, a field open to reinterpretation, no longer closed down by the old dog-eat-dog military imperial policy, which is an echo of Einstein's original statement that the absolute nature of the bomb would have to make man reconsider his values and his consciousness and his whole frame of mind.

So it's the set and setting that determines the direction of the psychedelic trip. And in fact the hippie era is criticized now by hindsight for having an idealistic set and setting which set the whole generation on the wrong turn to an idealistic nature movement: instead of taking over the reigns of power, it renounced power: turning on, tuning in, and dropping out rather than turning on, tuning in, and dropping in. So the hippie movement is criticized for an inadequate historical grasp of the power of bureaucracy and military and state, an inadequate grasp of its rootedness, and an inadequate grasp of that rootedness in the family.

SA: Some people feel that the hippie movement lost its momentum. At the end of 1967 there was a ceremonial death of the hippie.

AG: That was because the media had taken over the hippie movement and distorted it into a Frankenstein as they had done ten years earlier with the Beat movement, turned the poets into the television image of Dobie Gillis[175] who was some kind of bumbling, likable fool who couldn't get his life straight and was like the family black sheep, and that was the image spread on television and in radio soap opera series. So Thelin and the Diggers, who had ideas of a more free society, said the only way to do that is to have the hippies bury the hippie movement so that the media could not exploit it any longer.

SA: How was the media exploiting it?

AG: By exaggerating the sensationalist aspect, pushing for illegalizing LSD, and taking fake stories like that Sergeant Jeffrey McDonald's family

---

[174]At this contemporary poetry conference in July 1965, many of the most important young poets of the time attended, including Creeley, Olson, Duncan, Wieners, and Snyder.
[175]*The Many Loves of Dobie Gillis* was a television sitcom featuring the cliché beat character, played by Bob Denver.

was murdered by a band of hippies marching around the room saying "kill the pigs" when McDonald himself had murdered his family; by playing up the horror stories and creating another set and setting which was one of hallucination, horror, and violence; by later using the Manson Family[176] as symbolic of the hippie psychedelic movement rather than the more grounded audio engineers who were creating a new music. Also simply sensationalizing the hippie movement, like there was a *Life* magazine cover that showed somebody in the throes of some awful swirling photo montage hallucination that had no relation to the microscopic clarity of possible trip but emphasized instead madness and murder.

SA: But there was some madness and murder which . . .

AG: Yes, there was, but very little compared to what was going on in Vietnam, for one thing. Very much a reflection of some of the social stress, like Manson himself was not exactly a left-wing hippie, he was anti-Negro. He thought there was going to be a massacre of Negroes if necessary. It was like an inversion of that. And the media grabbed onto that and Altamont[177] as a way of burying the hippie movement, I think. The government certainly did. But it was more like hyping the more garish and the least spiritual aspects and de-emphasizing the artistic and spiritual accomplishments.

SA: Looking at it today, what remains from 1967 and that spirit?

AG: There is a permanent change in civilized consciousness so that it includes the notion of one world, fresh planet, the awareness of the fragility of the planet as an ecological unity, the absorption of psychedelic styles in dress and music into the body politic, the sexual liberation movement, the black liberation movement, the women's liberation movement, all of those slight, affirmative, permanent alterations in all lifestyles. I think also a permanent change in awareness: remember, the notion of Armageddon apocalypse before the sixties

---

[176]See Ed Sanders's *The Family: The Story of Charles Manson's Dune Buggy Attack Battalion* (New York: Avon, 1972).

[177]Often called "the end of the sixties," the December 1969 free music festival held on the heels of Woodstock at Altamont Speedway outside San Francisco went bad when the Hell's Angels, hired by the Rolling Stones as concert security, turned on the crowd. In an altercation, one man who approached the stage with a gun was killed and many others were brutalized. The Rolling Stones segment of the concert was filmed and released as *Gimme Shelter*.

was considered eccentric, whereas now as a possibility it's a universal awareness. In Einstein's time in the forties it was only glimpsed by geniuses who knew the implications of the bomb, or geniuses who knew the implications of acid, or geniuses who knew the implication of the union of eastern and western consciousness. So, as Kerouac said back in 1960, walking on water wasn't built in a day.

I think there was a glimpse of possibility of survival of the planet, partly through spirit, partly through imagination and poetry, partly through psychedelic insight. Then to substantiate that glimpse and actually rework the material world would be a matter of decades and ages because the Industrial Revolution itself is 200 years old, and it would take another 200 years to readjust and refine and return to basic values, meaning compassion towards sentient beings, respect for earth as the American Indians had it, realization of the fact that we could pollute the earth. If not destroy it in one flash, we could pollute it slowly as Gregory Bateson pointed out at the 1967 Dialectics of Liberation Congress[178] when he pointed to the greenhouse effect being a slow poisoning of the atmosphere.

All that is left as an overwash, as well as an improvement in perception of lyric poetry so that for generations now people actually read lyrics and songs and examine them for what the symbolism is. In that sense there has been an improvement of literate culture, again co-opted or exploited so that by 1969 or 1970 MGM was taken over by right-wing politicians—Mike Curb among them—who banned all psychedelic music and anybody who had anything to do with drugs in an attempt to roll back or squash the intuitions of the youth movement and their statements.

SA: Some people might feel looking at America today and Reagan that things *have* been pretty well squashed.

AG: Yes, they have. I would say the military is going to get bigger. So, it's a kind of paradox that nobody really believes in the military anymore in the way they used to except maybe like right-wing, really neo-conservative fundamentalists who still want us to salute the flag and

---

[178]The Dialectics of Liberation Congress was a London conference organized by radical psychologist R. D. Laing. The summer 1967 conference was the occasion where Gregory Bateson first introduced the greenhouse effect. Among the many notable speakers were Paul Goodman and Stokeley Carmichael. Ginsberg spoke on the theme "Consciousness and Practical Action."

return to the old discredited and chauvinist nationalism and even imperialism of big stick, which by 1940 had become démodé. And even more miraculous, the old Baptist, money-grubbing preachers that had been satirized by Sinclair Lewis in *Elmer Gantry*—and one would have thought permanently wiped off the psychological map as an elemental force in the nation—have come back redoubled on television with network power. So that's kind of a paradox, that, for fear of the chaos and the unknown and the new scientism and the apocalypse, there is a Bible Belt fundamentalist religious revival, sort of the last straw to which people can cling to to stay in the old nineteenth-century order, rather than entering the space of the present and future.

(Examining the film treatment:) I think there's too much drugs and not enough culture in here.

SA: And you think the drugs need to be reduced?

AG: Well, I don't think that's the whole story. It's the change of culture that's important. The drugs were a catalyst, but they weren't the change of culture itself. The change of culture was the change of ideology and a change of basic awareness of the planet. To some extent flower power and grass power are identical but the other half of the equation is flower power as well as cannabis power. In fact, the reason that cannabis was so interesting was that it was a natural flower, and the reason why mushroom peyote was interesting was that it might be considered an extension of diet: rather than drug taking, it's simply eating a different food.

SA: So there should be more in the film about the set and the setting.

AG: Yes, if you're going to deal with it in terms of the trip, then you have to consider the set and the setting of the trip as determining the trip's nature. So far into this on page six, you wouldn't even know there was a Vietnam War going on in America and that there'd already been vast mobilizations and that the Be-in was basically a peace be-in, that's the meaning of Be-in: that the aggression of the sits-ins—or to the extent that the sit-ins represented a material force against a material force—this was spiritual force and breath against a material force: *Be*-in rather than *sit*-in.

SA: So the Be-in was in part directed or provoked by . . .

AG: Toward meditation and appreciation. It was a reaction to the speeding hypotropic militarization of the war and the aggression of the war. Actually the progression of the Beatles from "All You Need Is Love" to actually getting into some kind of direct spiritual involvement with Maharishi Mahesh Yogi is a paradigm of the whole era.

SA: But was it not only a minority that was involved in meditation?

AG: Yeah, it was a minority. I don't think you could ever get the mass of mankind involved, not for a couple of hundred years, but it was the active artistic minority, the antennae of the race: the poets who were the legislators of the world, the artists who determined the rhythms of dance and talk and singing and recreation, the painters who retrained the eye, the psychiatrists who retrained some of the mind, the scientists who realized nuclear winter.

After all, the reaction to the sixties is not really a reaction to taking drugs, it's also a reaction against women's liberation, it's a reaction in America against black power, the reaction on the part of the born-again against sexual revolution, a reaction against liberation of language and pictures in print, and a move towards censorship. So, in a sense, it's not merely the stamping out of drug excesses, it's also an attempt to stamp out social currents that were implicit before and after the drugs and would survive the drugs, including an attempt to override the peace movement with giant military budgets and overriding the SALT Treaty: break the SALT Treaty and you go back to nuclear competition. So it's not that the change was only drugs and that only drugs are being stamped out, but it's the whole culture they're trying to roll back. In Eastern Europe you'll find that the Stalinists and heavy-handed Marxists also disapproved of the rock 'n' roll and the youth movement and the free poetry movement and abstract art and saw the generational conflict as youth rebelling against Stalinism and the Stalinist bureaucracy. After all, 1968 was also the Carta[179] in Czechoslovakia and Hungary where Eastern European intellectuals asked for more freedom and protested against the arrests of writers, artists, and politicians who were libertarians. The year after 1967 was

---

[179]When I spoke with Josef Jařab to confirm the spelling of this word, he informed me that there is no such word in Czech. He opined that Allen was likely conflating the Charter movement begun by Vacláv Havel in 1977 with the 1968 Czech uprising. It does seem likely that Allen is referring here to the entire spirit of the 1968 Czech resistance to Russian domination.—DC

the year the Russian tanks rolled into Prague to stamp out a whole new wave of liberty. The tanks weren't rolling into Prague to stop drugs, although LSD had been a catalyst even in Prague with Doctor Rubichek's experiments on a very small scale. The counterrevolution that began in 1968 with the state trying to reimpose its authority on the Whitmanic, Thoreauvian individual didn't involve drugs in Eastern Europe, and it doesn't involve only drugs in the Western world. Margaret Thatcher didn't campaign against drugs, she campaigned against libertinism, welfare state, communal socialism, and old mining traditions. Reagan came in on trying to reimpose a Baptist God. The spiritual attack in the United States was on secular humanism, and the reaction was an attack on modern reevaluation of nationalism, having seen the effects of nationalism on Russia and Germany and on American nationalist imperialism in Latin America: there was the Alliance for Progress with Kennedy. There's been a rollback. In fact, the American government and Cold War theory has been that you have to have tyrannies to fight Communists, so that an approval of tyranny is part of the package deal of rollback.

If you're talking about what survives of the sixties then it's not merely a few scattered institutions, in which I would include things like Naropa, which is actually the first accredited Buddhist college in the Western world, or the great spread of meditation forms, or holistic health forms, even unto jogging, as a byproduct of that ethos. You'd also have to account for basic changes in attitude which have been adopted by the majority: not only the renunciation of the Vietnam War or the cutting off of money by congress at the end of it, and not merely the military but Watergate, the final demystification of the administration. Beyond these, the more prominent substrate of changes, like women's liberation, sexual liberation, black liberation, liberation from censorship, notions of cooperative control of industry, Solidarity—*Solidarnosc*—in Poland and whatever equivalents there are in the United States. So some permanent changes have now been built in so subtly that they are unnoticeable as byproducts of the expansion of mind and the liberation movement of the sixties. So it's the whole cultural change.

That still leaves unaccounted the scale and weight of the counter-reaction, which sees the sixties ethos as the enemy of Cold War, the enemy of nuclear power, the enemy of military solution, and sees gay liberation as Hitler did, as a weakening of the will to power of the nation: "a softening of the moral fiber of the nation," is their own words.

Since we're not going to have a nuclear war—I think common sense of humanity will go that far—that means in the long run there's going to have to be a slow adjustment to the insights of the sixties, of 1967, particularly the Be-in, of fresh planet, a vulnerable planet, alterations of human behavior patterns in order to survive biologically, notions of appropriate technology such as Shumacher and E. P. Thompson have proposed.[180] The limitation of nuclear power, even for peace, is necessary because it permanently poisons the planet as now exemplified in the announcement that they're building this tomb or sarcophagus around the Chernobyl nuclear plant, noting that it will have to be buried for hundreds and hundreds of years in this concrete egg case because whatever would hatch out of that egg is just too poisonous for the planet to handle. So the whole problem of nuclear waste which has been postponed for consideration all these years is now going to have to be taken up, and that'll have to be solved. Otherwise we leave a legacy of permanent poisoned eggs all over the planet as one by one the present nuke plants have to be shut down and encased. Right now Maine, where we are, is going through throes of joy that it's been eliminated from a list of places where the government is going to make their nuclear waste dumps and that the Texas Panhandle has been selected. And in Texas—I was just there—the Panhandle people are now beginning to rise up and revolt because they don't want it either. So actually nobody wants it. When it comes down to the bottom line, what do you do with the waste? Nobody will accept it.

SA: That's just what's happened in England. Conservative MPs one after another, when the waste comes to their area, get up and say, "No, we're not going to have it."

AG: That's an admission that the scientific equation has not been solved, and yet we're still like the sorcerer's apprentice, commanding

---

[180]Ernst Friedrich Schumacher (1911–1977) wrote *Small is Beautiful: Economics as if People Mattered* (New York: Harper, 1973), reprinted, 1989; Edward Palmer Thompson's (b. 1924) best known work is *Making of the English Working Class* (New York: Random House, 1966).

the broom to carry the water for us, not knowing how to stop the broom. Sooner or later some kind of biological saving grace or instinct for survival is going to determine an end to even peace nukes to reduce the high price of existence in the hypermaterialistic, hyper-scientific age. So I think the implication of the sixties was, as Blake pointed out, that hyperrationalism has created chaos so that the body's existence, the existence of feeling, and the existence of the poetic imagination, the fantasy and dream, the poetry, have to be acknowledged as equal partners in Albion, the whole man. That was Blake's formulation, and the Blake quote that was used as the motto for the original Albert Hall Poetry Incarnation[181] was "England, awake, awake, awake," so as to awaken Albion. I forget the rest of it, but it mentioned Albion, defined by Blake as the whole man, not only reason, but reason in balance with heart-feelings, the body, and the imagination.

So I think the implication in 1967 was a breakthrough of imagina-tion, a breakthrough of recognition of the health of the body of the individual and the planet as against the poisons of hyper-industrial-ization, and breakthrough of human feeling again: a return to human feeling after the homogenization and objectification or reification of human feelings due to automatic reproduction in the computerized machine age. So those are the stakes, not just drugs: drugs were the catalysts to recognize some of it.

---

[181]The Albert Hall Poetry Incarnation, also known as Wholly Communion or the International Poetry Reading, was held on June 11, 1965, at the Royal Albert Hall in London. Nineteen poets read, including Lawrence Ferlinghetti, Harry Fainlight, William Burroughs, Ernst Jandl, and Gregory Corso.

"The Puritan and the Profligate"
Chronicles, December 1989

The Lofton interview took place at the home of *Washington Times* columnist Suzanne Fields. After it was published in *Chronicles: A Magazine of American Culture*, *Harper's* published a shorter version of the interview. A former columnist with the *Washington Times*, Lofton wrote in 1999 that he wanted to interview Allen "to confront him with the Truth of God's Word." The interview was transcribed by Lofton and the present editor.

—DC

John Lofton: In the first section of your poem "Howl" you wrote: "I saw the best young minds of my generation destroyed by madness." Did this also apply to you?

Allen Ginsberg: That's not an accurate quotation. I said the "best minds," not "the best young minds." This is what is called hyperbole, an exaggerated statement, sort of a romantic statement. And I suppose it could apply to me too or anybody. It cuts both ways. People who survived and became prosperous in a basically aggressive, warlike society are, in a sense, destroyed by madness. Those who freaked out and couldn't make it, or were traumatized, or artists who starved, or what not, they couldn't make it, either. It kinda cuts both ways. There's an element of humor there.

JL: When you say you suppose this could have applied to you, does this mean you don't know if you are mad?

AG: Well, who does? I mean everybody is a little mad.

JL: But I'm asking you.

AG: Everybody is a little bit mad. You are perhaps taking this a little too literally. There are several kinds of madness: divine madness, and in the Western tradition there is what Plato called . . .

JL: But I'm talking about this in the sense you spoke of in your 1949 poem "Bop Lyrics," when you wrote: "I'm so lucky to be nutty."

AG: You're misinterpeting the way I'm using the words.

JL: No, I'm asking you a question. I'm not interpreting anything.

AG: I'm afraid that your linguistic presupposition is, as you define it, that "nutty" means insanity rather than inspiration. You are interpreting, though you say you aren't, by choosing one definition and excluding another. So I think you'll have to admit you are interpreting.

JL: Actually, I don't admit that.

AG: You don't want to admit *nuttin'!* But you want me to admit something. Come on. Come off it. Don't be a prig.

JL: No, I just want to ask you a question.

AG: No, you're not just asking me a question. You're *first* interpreting the language and wanting me to use the idea the way you use it. [But] it's my words. And I'm trying to explain to you what it meant.

JL: On the contrary, I was asking you what you meant by what you wrote.

AG: Oh, I see. It's a double use of the word "madness" or "crazy" or "nutty." But if you'll listen to this tape you'll find you asked to exclude one aspect and wanted "nutty" to mean "crazy" or "insane." And that's why I say you are interpreting and not wishing to use the language as I had originally set it out. And you weren't interested in my explanation. Are we communicating or just sparring?

JL: I think you can do both. It's not either/or.

AG: All right. All right. But you have to remember what we're saying. You can't amnesize what we were saying. I feel you're trying to avoid recognition of the fact that you were trying to exclude *both* meanings of the word "crazy."

JL: No, I'm just trying to understand what you meant by what you wrote. But this question of madness—

AG: There's also another background. In Zen Buddhism there is wild wisdom, or crazy wisdom, crazy in the sense of wild, unlimited, unbounded. Or as in jazz, when someone plays a beautiful riff or extemporizes a great chorus, they say "crazy, man."

JL: But I am interested in this question of your possible madness. It's not a gratuitous question. There is a history of madness in your family.

AG: Very much so.

JL: Your mom died in 1956 in a mental institution. So my question is not cute or facetious. And the *Current Biography Yearbook* for 1987 says that when a roommate of yours, in 1949, stole to support his drug habit and was arrested, you were implicated circumstantially and pleaded a psychological disability and spent eight months in the Columbia Psychiatric Institute. What was this psychiatric disability and why did you spend just eight months in this institute?

AG: Well, I had sort of a visionary experience in which I heard William Blake's voice. It was probably an auditory hallucination but was a very rich experience.

JL: This happened while you were masturbating, right?

AG: Yes, but *after*. Did you read this in the *Paris Review?* (*see Clark*)

JL: No, in the *Current Biography Yearbook* for 1987. But I want to ask you about your roommate, this drug arrest and this psychiatric disability you pleaded.

AG: No, no, no, no, no, no, no, no. Sir, first of all your tone is too aggressive. You have to soften your tone because there's an element of aggression here. There's an element almost like a police interrogation here.

JL: But that's not all bad. The police, in some instances, do a good job, particularly in dealing with criminals.

AG: Sir, in this case it's a little impolite. You're being a little harsh and unfriendly and making it very difficult to relate to you gently and talk unguardedly and candidly because there is an element of manipulativeness in the way you're asking questions. You don't feel that at all?

JL: No.

AG: No element of manipulativeness or coercion or aggression or a hard edge here that is supposed to put me at a disadvantage?

JL: There's no doubt that from what I've read about you, I don't like what you have stood for over the years. I don't like your politics, the kind of sex you engage in. So, if you mean there's a hostility here towards what you are, absolutely [there is].

AG: But you're talking to me as if I'm an object of some kind and not a person in front of you. There's an element of abusiveness in the way you're talking to me—your tone of voice and interrogative method in which I seem to have to answer yes or no.

JL: But an interview is an interrogation.

AG: Or it can be a conversation, a little warmer. It can have a little more sense of respect and less sense of hard-edged, police-like interrogation. I'm asking you, in a sense, to watch your manners.

JL: That's interesting because I'm not asking you to respond in any particular way. Why are you telling me how to ask questions? You can use any tone you like. . . . So, can we return to my question? What was this psychiatric disability that put you in an institute for eight months?

AG: Well, I'm not sure it really was a disability to begin with. So I can't answer the question the way you propose it.

JL: But I'm asking you if it's true, as the *Current Biography Yearbook* has reported, that you had this disability?

AG: It's neither true nor not true. Now, do you know about that?

JL: But it is true that you were in an institute?

AG: Yes, I was. I had a kind of visionary experience relating to a text by William Blake, "The Sick Rose."

JL: "The Sick Rose"?

AG: Yes, it went: "O rose, thou art sick! / The invisible worm / That flies in the night, / In the howling storm, / Has found out thy bed / Of crimson joy, / And his dark secret love / Does thy life destroy." So, it's a very mysterious, interesting poem that keyed off a kind of religious experi-ence, a visionary experience, a hallucinatory experience—whichever way you want to interpret it. All three descriptions are applicable and possible. Reality has many aspects.

JL: Were you using drugs while you masturbated?

AG: Not at all. I had been living very quietly, eating vegetarian diets, seeing very few people and reading a great many religious texts: Saint John of the Cross, the Bible, Plato's *Phaedrus*, Saint Teresa of Avila, and Blake. So I was in a kind of solitary, contemplative mood.

JL: Did you put yourself into this institute?

AG: More or less. Because I questioned my own sense of reality and I couldn't figure out what the significance of the illuminative experience was, whether it was a kind of traditional religious experience that you might find within (William) James's book *The Varieties of Religious Experience*, where there is a sudden sense of vastness and ancientness and respect and devotional awareness or sacredness to the whole uni-verse, or whether this was a byproduct of some lack-love longing and a projection of the world of my own feelings, or some nutty breakthrough.

JL: What made you think anyone at the Columbia Psychiatric Institute would have any light to shed on what you thought your problem was?

AG: Well, I was very young.

JL: Did they help you find reality?

AG: I don't think there's any single reality. They helped me relate to my own desires.

JL: You think you were better when you got out of there?

AG: I think they said I wasn't ever really psychotic or crazy, just an average neurotic.

JL: Did they think it was normal to have the experience you had after masturbating?

AG: No, but you could say perhaps normal or average, yes. People have very many extraordinary experiences in their lives, whether or not con-nected with having masturbated, although most people do masturbate.

JL: But, clearly, you thought something was wrong with you or you wouldn't have gone to this institute, right?

AG: Well, I wasn't thinking it was exactly wrong—

JL: But something was out of whack? You went there for some reason.

AG: You're putting words in my mouth.

JL: No, I'm asking you a question.

AG: No, you're asserting something—that I thought something was wrong.

JL: But someone doesn't go to a psychiatric institute because he thinks everything is OK, does he?

AG: You might very well do that, depending on what you mean by "out of whack" or "OK." Actually, I was more inquisitive as to what the nature of my experience was, and I thought I would be able to find out that way.

JL: Did you go to anywhere else besides this institute?

AG: Oh, later—I'm going to a psychiatrist now.

JL: How long have you been in therapy?

AG: About three years. I went through about five years when I was in my twenties, and about three years now.

JL: Why are you in therapy?

AG: To get down to the bedrock root and see what has conditioned me to be the way I am. To examine the texture of my feelings, whether my experience with my family has at all been digested and absorbed and whether I'm handling it properly and whether I know my own emotions, and sort of exploring the depths of my own feelings.

JL: What makes you think the people you are going to now know anything about this?

AG: Some sense of warmth, trust, and intelligence that goes back and forth between myself and my psychiatrist. It's an experiential matter that gives me confidence to have a trusting relationship and to discuss my problems with another person.

JL: I assume you're going to a secular humanist-type psychiatrist.

AG: I never inquired about her religious beliefs.

JL: Really? So you're going to someone whose religious beliefs, whose presuppositions, you know nothing about?

AG: Some, by body language and the response to the immediate situation in front of me, which is what I am really interested in rather than, say, in this conversation. I'm dealing with you in terms of how you display yourself here, not the history of your thoughts. I'm trying to deal with the evidence or manifestation of how you present yourself here—your harshness, aggression, and insistency and—

JL: Why not call it my perseverance? Isn't that a nicer word? Or guts? Or tenacity?

AG: I would say there is a little element of S&M in your approach. Power.

JL: No, I would say this is more like the kind of sex you like.

AG: And I would say this is the kind of power relationship you like judging from your behavior.

JL: Well, that's certainly what S&M is all about, power.

AG: And you seem to like that, don't you? Have your sexual fantasies ever involved that kind of power relationship?

JL: No, not to my knowledge. I'm a Christian. So I don't fantasize.

AG: Do you ever have any sexual fantasies?

JL: No.

AG: None at all?

JL: No. I said I am a Christian.

AG: You've never had *any* sexual fantasies?

JL: Before I was a Christian, I had them, absolutely.

AG: And since you're a Christian you don't?

JL: No.

AG: You have no erotic dreams, at all, that you remember?

JL: None that don't feature my wife, no. It's an amazing thing what Jesus can do for a person.

AG: Uh-huh.

JL: The power of the Holy Spirit. . . . Do you feel that after these years of therapy that you're any closer to knowing more—

AG: Yes.

JL: Like what?

AG: I think some of that is rather private.

JL: Fair enough.

AG: But it would relate to some of what I have written about in my poems "Kaddish" and "White Shroud." . . .

JL: Let's talk about some of your feelings over the years and see if they should be respected. An April 21, 1978 *Boston Globe* story says that when you were on a local TV show you shared your sexual preference for "young boys" and this caused an instant irate reaction from mothers watching—

AG: A few. Not that many.

JL: A few mothers who had children home on vacation from school. Is this an accurate report? That you do have a sexual preference for "young boys"?

AG: No, no, no. It's not accurate in the context of the broadcast.

JL: Did you say you had a sexual preference for "young boys"?

AG: We're not on trial here. I'm trying to explain the context—

JL: But, in a way, we're all on trial.

AG: Well then, you must excuse me if I don't adopt the submissive attitude that you wish.

JL: No, I would, I appreciate that—

AG: However, I actually would like to now explain: there was at that time an old district attorney, Byrne,[182] in Boston, who every time he ran for office would then have a crackdown on gays in Boston Common—he himself not being married and being about 80. He used it as a kind of political ploy to make sex scandals, and he busted a group of—I think it was a house in Revere where there were some gay people who had younger friends. The police sort of abused the kids in order to try to coerce them to testify against the older people, and the cases were mostly thrown out of court finally. So it was a sort of scandal of, again, this kind of sadistic manipulation of other people on the part of the district attorney as a little political ploy. I got on the air and said that when I was young, I had been approached by an older man in Revere and I didn't think it did me any harm, that I like younger boys, and I think that probably almost everybody has an inclination that is erotic toward younger people, including younger boys.

JL: How young were the boys?

AG: In my case, I'd say 14, 15, 16, 17, 18.

JL: That you had sex with?

AG: No, unfortunately I haven't had the chance. *(laughs)* No, I'm talking about my desires. I'm being frank and candid. And I'm also saying that if anyone was frank and candid, you'd probably find that in anybody's breast. The—

JL: But why? Why do you persist in imputing your own rottenness to other people?

AG: One moment. Your question is: why do I persist in imputing my own rottenness to other people?

JL: That's right.

AG: Is this a newspaper interview or is it—

JL: I'm a columnist who writes commentary and opinion.

---

[182]Ginsberg is referring to the late Garrett H. Byrne, longtime district attorney of Suffolk County, Massachusetts. He won quite a few elections. See *The Boston Sex Scandal* (Boston: Glad Day, 1980).

AG: You realize that you're using language which could be considered insulting.

JL: I hope so. I think it's a rotten preference to want to have sex with young boys.

AG: You feel that it's part of your role to sit here and insult me?

JL: Do you feel you should be insult-proof?

AG: No, but in this circumstance you might have better manners.

JL: You're a little hung up on manners, I think.

AG: You might couch your conversation with me a little more politely since—

JL: Why? Why should I be polite? I think to have sexual preferences for young boys is rotten. I can't say that?

AG: Yeah. But you want me to say it, too. . . . It seems you're being so abusive that it takes a great deal of patience to be your host and to allow you to try to manipulate the situation.

JL: But I don't think it's true that most people want to have sex with younger people.

AG: I didn't say that. You're taking words out of my mouth.

JL: That is what you said.

AG: No, I said that—we have this on tape, you know.

JL: That we do, and I won't burn this tape.

AG: What I said was that most people have erotic desires for young people. And that doesn't mean they would actually want to—

JL: What do you mean "most people"? Where do you get your data?

AG: Well, I'm just speaking as a human being poet who's been around 61 years, a little older than you.

JL: But that doesn't mean, necessarily, that you're smarter.

AG: No, but that's my experience. If you go back and talk to old folks they don't see it as horrible. They see it as part of the charm of—

JL: Yeah. Well, maybe you've been running around with the wrong people.

AG: Maybe you're not wanting to listen to the whole sense of eros here.

JL: I don't doubt that most of the people you've run with have wanted to have sex with young people. I'm just suggesting that their views don't reflect the views of all Americans.

AG: I didn't say have sex. I said most people have a kind of erotic attraction to younger people.

JL: You mean a fantasy they don't want to act on?

AG: Most don't act on it, of course not. And most wouldn't think of acting on it. But most people have in their breasts an erotic pleasure in younger people.

JL: You've repeated that several times. And I'm denying it. But I'm not denying the people you ran with felt this way.

AG: It's a kind of granny wisdom. And you're overreacting to it.

JL: I don't think your grandmother would have told you this.

AG: You're interpreting it in sort of a negative way. This is a part of the general spectrum of human charm and emotion rather than sin or rottenness as you say.

JL: I think most people reading this interview will agree that the desire to have sex with young boys is *rotten*.

AG: It depends on where you publish it.

JL: That's right. If you publish it in the pedophile community, you'll get a large amount of support.

AG: If you publish it in the *Times* or the *Washington Post*, people will say well, Ginsberg is just talking normal, humanistic, obvious old granny wisdom like you can find in Shakespeare.

JL: Okay. We obviously had different kinds of grandmothers. . . . Do you now have a desire to have sex with young boys?

AG: I have a sexual desire for them, I must say, yes.

JL: Still?

AG: Oh, the older I get, the more.

JL: And after six years of therapy, too. This therapy must really be doing a good job.

AG: You know what the therapy does?

JL: It probably tells you it's fine, just get comfortable with it, right?

AG: No, not quite. . . . Usually it's a discussion of where this comes from and trying to find the origin of it. And find what conditioning affected me that I arrived at this particular orientation. That's all.

JL: How about sin? Is it a possibility that you are a sinner?

AG: That doesn't come into play. What comes into play is an attempt to cultivate an awareness of the situation so that if there are any harmful consequences to myself or others, they can be seen through and avoided and altered. The attempt is to understand the situation, not categorize it with knee-jerk words like "sin," but to understand the roots of it historically and what the personal experience roots were.

JL: Well, sin goes back pretty far.

AG: And then relate to the direct personal experience in a way that will take the sting out of it in a sense of doing harm to others or yourself.

JL: But this sexual preference for young boys doesn't seem to be something you want to be delivered from. You smile when you talk about it. You don't want to be cured of this, do you?

AG: I should say my sexual preference is not exclusively for young boys, but also for middle-aged men, straight men, and women. I've occasionally had fantasies about making out with trucks as well as beasts. And maybe I'll be making out with you before it's all over. *(laughs)*

JL: Well . . . maybe I'd like to drive the truck while you made out with it, if you don't mind, an 18-wheeler, with the pedal-to-the-metal.

AG: Well, now, here we are with . . . there's your fantasy! *(laughs)*

JL: Excuse me, you raised the idea of having sex with a truck.

AG: You extended it.

JL: I'm just trying to accommodate you. And you're attacking me again.

AG: No, I'm not attacking you at all.

JL: I even offered to drive the truck.

AG: You sure did.

JL: But to hell with you. I *won't* drive the truck. Get your own truck.

AG: Oh, you can't get out of it that easily. You've already driven the truck in your mind. Gosh, you're funny. But you've got this sort of contentious obsession—God knows what's underneath all that.

JL: Well, yes, He does know. . . .

AG: You've got to remember that I'm talking on the basis of the experience of remembering my unconscious. And maybe you're not as aware of what's going on in your mind as I am. And therefore when you condemn impulses or fantasies that I'm willing to be candid about, you may not be so familiar with your own mind as to know that you do contain—

JL: Well, let's clear this up right now. If I had the fantasies you have, I would also consider them rotten. They are rotten not because you have them but because they are rotten desires, preferences.

AG: But what if you find human nature does contain such a great spectrum of fantasy, a different aspect of reality, as a kind of way of checking out all the possibilities—

JL: Mr. Ginsberg, the Book of Jeremiah says that the human heart is *desperately* wicked. You don't have to tell me—a born-again Christian, Calvinist, Reformed, Puritan—about the variety of evil fantasies human beings have. I read the Bible, sir, and it tells us all about this. And it tells what is to be done about this.

AG: But you don't read your unconscious, the contents of your mind, very carefully. You don't remember your dreams, your daydreams, subliminal thinking.

JL: You know why?

AG: Why?

JL: Because I'm not like you. The scripture, talking about unbelievers, asks the question, why do the heathen rage, and the people imagine a vain thing? The scripture tells us to trust the Lord with all our heart and lean not to our own understanding. You are an unbeliever. You lean to your own understanding. You're a heathen who imagines vain things. You have an overactive imagination, a kind of mental cancer.

AG: No, it's not quite like that. Do you know anything about meditation practice?

(At this point Ginsberg offers to show Lofton how to meditate. Lofton asks him if he plans to take his clothes off, to which Ginsberg answers no. They meditate.)

AG: *(Explaining at the end of the meditation the idea of concentrating on one's breath and letting go of thoughts:)* You may find that because you're thinking a lot that you forget you're breathing, and you space out or your mind wanders into speculation or subconscious gossip or planning or thinking. So you just rest again and acknowledge your thoughts. You don't push them away, you don't invite them in, but just turn your attention back to your breath again. In that way you're just coming back to yourself, so to speak, and to the room and to the space around you. At the same time you're aware of your thoughts and you just observe them: acknowledging them, taking a friendly attitude toward them, not participating, just letting them go by. That tends to lead to a kind of equanimity or peacefulness and, at the same time, some sense of observation of the situation around you in a kind of nonjudgmental peacefulness.

JL: What's wrong with being judgmental?

AG: Sometimes when you're being quick to judge, you don't notice the delicacy or softness or transient gracefulness of what is around you. In the overexcited mental turmoil of constantly having to judge every thought and perception, you might miss some of the luminousness of the phenomenal world or even not notice some of your own thoughts or feelings. The haste to rush to judgment over every perception might muffle or distort or even blank out awareness of what's around you.

JL: Well, see, that's your religion, that's not my religion. You're free to practice your religion, I guess. I might object to your attempting to impose it on me.

AG: Well, you were very cooperative in the sitting, that was nice. How did you feel?

JL: Oh, fine.

AG: But that didn't seem to be threatening, was it, or irreligious or anything? It was just contemplative.

JL: I'm all for that.

AG: It's just contemplating what is in your mind and what is around you and your own breath, as simple as that. Okay, now maybe we have some reference that we share that's a little bit more neutral than all this contentiousness.

JL: The 1970 *Current Biography* says that you aren't a proselytizer for homosexuality. What does that mean?

AG: I'm observing my own mind and consciousness and reporting on that and trying to be candid. Walt Whitman, who was a very great poet and, incidentally, gay, said he thought that for poets and orators of the future the great quality would be candor, frankness, truthfulness— like first-thought-best-thought, the notion of nonmanipulative communication rather than trying to dress it up and look good for the outside.

JL: Well, Walt Whitman suffered from, if I may say so, what might be called terminal candor—not unlike yourself. I mean, who cares? You don't have to tell us everything, instantly reporting on everything you feel. Who cares?

AG: Nobody could report instantly on everything.

JL: But some of you have tried hard. You have said you write down every dream you have. You report on every little impulse from the brain.

AG: But it's like a musician who has to practice and limber up for six or seven hours before he gives a concert. And so you get certain selected dreams and impressions you finally report on in a little book of collected poems. You have to do a lot of writing before you get sort of into shape and get practice enough to get an old dog attitude and are able to do it accurately.

JL: Well, some people do this and some people don't.

AG: Yeah. But the ability to transcribe comes from experience. And you build on it, learning your mistakes and blessings. So I don't think

Whitman wrote too much. He's considered a great classic around the world.

JL: I don't consider him as such.

AG: You don't like Whitman?

JL: No.

AG: Have you read Whitman?

JL: Some.

AG: What have you read?

JL: It really doesn't matter.

AG: Do you remember the name of the poem you read?

JL: Yes, one that says something like: "So, I make mistakes. I contradict myself. So what? I contain all things." This is absurd. Talk about arrogance.

AG: Dig this.

JL: I'm diggin' it.

AG: He says: "Do I contradict myself? / Very well. I contradict myself, / (I am large. I contain multitudes.)" Do you know what he meant by that?

JL: Probably nothing good. And I doubt if he knew what he meant.

AG: Yeah, he did. I know what he meant.

JL: How do you know what he meant?

AG: (laughs) Because I am large. I contain multitudes.

JL: But you might contradict yourself.

AG: Yes. And I certainly will contradict myself.

JL: This will be one of your multitudes—the ability to contradict yourself.

AG: That's what Whitman is saying.

JL: It's gibberish.

AG: That our own minds are so vast that we can wind up contradicting ourselves without having to freak out about it. It's very similar to what the poet John Keats said about "negative capability."[183] He said the quality of a very great poet like Shakespeare was his ability to contain opposite ideas in the mind "without an irritable reaching out after fact and reason." Meaning that that portion of the mind which judges and irritably insists on either black or white is only a small part of the mind. The larger mind observes the contradiction and contains those contradictions. The mind that notices that it contradicts itself is bigger than the smaller mind that is taking one side or the other.

JL: You speak very confidently about this. Where do you get your ideas about what the mind is?

AG: By the direct observation through meditation practice.

JL: But at most this would tell you only about your mind, wouldn't it? You were making statements about *the* mind.

AG: Well, then there is the experience of many people who have meditated and written about it over centuries also.

JL: But this would still be only the experience of these individual minds, wouldn't it?

AG: I should say I noticed this about my mind and John Keats noticed it about his mind. . . . Sure, you might want to check out which side is right, which judgment is right, but when you get irritable about it and insist on one or the other, black or white, it's likely you'll eliminate some information from both sides. Like, for instance, I'm Jewish and I've been to Israel—

JL: You're Jewish? You practice the Jewish religion?

AG: No. I consider myself culturally Jewish.

JL: I thought being Jewish was a religion.

AG: Einstein was not a practitioner, but he was Jewish.

---

[183]The source is John Keats's letter to his brothers George and Thomas Keats, December 21, 1817: ". . . at once it struck me what quality went to form a Man of Achievement, especially in Literature and which Shakespeare possessed so enormously—I mean *Negative Capability*, that is, when a man is capable of being in uncertainties, mysteries, doubts, without any irritable reaching after fact and reason." Thanks to Simon Pettet for providing this citation.

JL: What made him Jewish?

AG: He thought he was Jewish.

JL: Is that all you have to do, just think you're Jewish?

AG: Pretty much so. Or to be born of a family.

JL: Do you believe in the God of Abraham, Isaac, and Jacob?

AG: I don't think—my position is more nontheistic than yours, I think. You feel Einstein was not Jewish?

JL: As I say, my understanding is that being Jewish is a religion.

AG: No, it's also cultural. That's my understanding as a Jew.

JL: But this certainly doesn't comport with what Abraham, Isaac, or Jacob understood a Jew to be.

AG: But on the other hand, there is an old tradition of the European, rootless, cosmopolitan, delicatessen intellectuals like Heinrich Heine, Thomas Mann, Einstein, and Freud, who were always considered Jewish and heroes of nineteenth- and twentieth-century Jewish culture. They weren't necessarily orthodox, but socially Jewish.

JL: You know what? I'm more Jewish than you.

AG: It may be, because you have more adherence to what they call the Judeo-Christian tradition to the extent that it is defined as monotheism.

JL: No, no, no. As a Christian I am more Jewish than you because I believe in the Old Testament.

AG: I see. Well, I think the Old Testament is great. But—

JL: It's great? What does that mean?

AG: Some of the greatest writing done by human hands.

JL: Oh, don't patronize it. Is it *true?* Is it divinely inspired?

AG: The one thing I can say is that it was written by human hands.

JL: But how do you know this? How do you know chimpanzees didn't pound it out on a series of countless typewriters?

AG: I think they have some old Aramaic manuscripts, don't they? . . .

JL: Do you believe the human hands you say you believe wrote those manuscripts were guided by God's spirit, as these manuscripts say?

AG: *(shrugs)*

JL: Why do you shrug, as if this is a matter of no importance?

AG: You know what that's like? If you've got a full grown goose inside a bottle with a narrow neck, how do you get the goose out without breaking the bottle or injuring the goose?

JL: And how is the question of the divine inspiration of the Old Testament like your goose in this bottle?

AG: If you can get the goose out of the bottle, I'll answer your question. Can you get the goose out? . . .

JL: It's nonsense.

AG: No, it isn't. You want me to get the goose out of the bottle?

JL: No.

AG: You want the goose to stay in the bottle?

JL: I'd like to know how you got this goose in the bottle.

AG: The way I'm going to take it out. *(He claps his hands.)* It's out! It was put in with words. And it's taken out with words. . . . The point is not to confuse a verbal paradigm or conceptual statement with an event. . . .

JL: Did God give Moses anything?

AG: I don't know. I wasn't there. Somebody wrote down that He did.

JL: But you either believe He did or He didn't.

AG: No.

JL: No? What's the third possibility?

AG: Both. As Gregory Corso says, if you have a choice between two things, take both.

JL: Even if they contradict each other?

AG: Do I contradict myself? Very well, I contradict myself. I am large. I contain multitudes. That's the whole point of that.

JL: You do not believe in a law of mutual contradiction? You believe something can simultaneously be and not be?

AG: Let's put it this way. All conceptions as to existence of the self, as well as all conceptions as to the nonexistence of the self, as well as all conceptions as to the existence of the supreme Self, as well as all conceptions as to the nonexistence of the supreme Self—

JL: Yeah.

AG:—are equally arbitrary, being only conceptions.

JL: Says who? You?

AG: Well, you just have to listen to the—

JL: Can you prove that what we're doing now is not a dream but is real?

AG: I think it is part dream. There's an element of dreamlikeness in this—

JL: Please, don't go metaphorical on me. Are we here doing this interview, or is this a dream?

AG: We are both really dreaming it and doing it at once.

JL: You're being metaphorical.

AG: No.

JL: No? Okay.

AG: The quality this situation shares with dreams is the quality of transitoriness. Like a dream, this situation will have disappeared in a day and will be forgotten in a hundred years. So, it gives a certain emptiness to the situation as well as, simultaneously, the situation is quite real.

JL: Well, that certainly clears it up. I appreciate that.

AG: Both. *Both*. Are you following me?

JL: Yes. Yes. You said that earlier. I'm very quick. I have a very quick mind.

AG: But did you understand it? Or did you just say you heard it?

JL: I didn't say I understood it.

AG: If you'll slow down a little, you might actually find it's not so repulsive or repugnant, what I just said.

JL: My problem, of course, is that I would, first, have to find what you said comprehensible in order to judge it. That's my first challenge.

AG: Like it says in the Bible: all flesh is grass. . . . I'm saying that this interview shares the quality of a dream in the sense of vaporousness, emptiness, transitoriness. You know, there's a good deal of parallel between the rhetoric and the intention of that poem ["Howl"] and what you like so much in Jeremiah—telling the cities to repent. The Moloch section of "Howl"—have you read that?

JL: Long ago. Look, you attacked a lot in that poem that should have been attacked, all this Moloch worship, but the problem is that you have no answers. You're a darkness curser only who lights no candles.

AG: But the answer I was giving to the impersonality of Moloch was human sympathy. I thought that was an appropriate medicine and a positive suggestion for the culture. . . . It seemed to me that sympathetic attentiveness [to Carl Solomon in "Howl"] was the basic answer to the quality of dehumanization.

JL: No. There is nothing more hypocritical than an unbeliever like you attacking somebody else's idolatry. That is rank hypocrisy.

AG: Well, I think all of us agree that our high-tech, materialistic civilization has some deficiencies in human—

JL: You're not listening to me. I say I applaud some of your attacks on Moloch worship. But what *ought* to be worshipped? If not Moloch, who?

AG: The alternative I was suggesting was opening up to some kind of softness and sensitivity and human sympathy.

JL: No, no, no! This won't do it. This is only sentiment. Every human being will worship something, seeking power from either above or below.

AG: And I'm suggesting that we have the natural power of compassion within us without having to go much further.

JL: But this doesn't solve the problem of *sin* and why people worship Moloch in the first place! Sin is a serious problem.

AG: When people learn to recognize their own vulnerability and compassionate nature, and their own suffering, they will extend that sympathy to others.

JL: But I want to deal with what you so articulately attacked in "Howl"— Moloch-worship. I want to shake your hand and agree that Moloch-worship is bad. But when you tell a man to get off his face and stop worshipping his false god, you have to put something there in its place.

AG: Human sympathy. Basic human good nature and sympathy. That's what I was saying in the third part of "Howl."

JL: That's it? And since then, over thirty years ago, you have no more ideas about what man needs to do to stop committing idolatry?

AG: I've come a little further. Because in those days I was of a more monotheist persuasion than I am now, which is more nontheistic. By 1965, having been to India, and experiencing other religions and meditation practices, I felt the monotheistic attitude was perhaps an extension of egotism, an attempt to preserve one's ego forever by putting it in heaven, by projecting one's own notion of self into the sky and worshipping that, making an anthropomorphic notion of the self and saying, "Well, if I'm not permanent, at least that's permanent and I can worship that."

JL: But if ever there was a place where man created gods in his own image, it is in India. And also in the ancient world, Greece and Rome. Whereas the God of the Old Testament is a God no human being could make up. He is beyond the mind of man to invent. In other words, the God of the Old Testament, the God of the Hebrews, is the *opposite* of a god made in man's image.

AG: The Buddhist approach is interesting because it attempts to cultivate awareness by examining all your thoughts and your mind—to practice long periods of sitting, meditation, observing your thoughts and feelings and their texture, and actually observing the stuff of the mind. It's really interesting if you do it in a prolonged way.

JL: But this is bad because it is an egocentric religion. Whereas the Christian religion says that whoever exalts himself shall be abased, but whoever humbleth himself shall be exalted. The Christian religion is the opposite of self-exaltation. The idea there is to bring every thought *captive* to the mind of Christ. The idea is to empty self of self and put on Christ, the full armor of God. Yes, contemplate, but you don't think about your *self*: Self-centeredness is death. It is carnal.

AG: But when you examine the texture of your mind, you find after a while that the notion of the self dissolves. That there never was a self there.

JL: Absolutely. Because God made you that way. Saint Augustine said that God made us for himself and our hearts do not rest until they rest in *Him*. You don't know anything about your mind until you first know God. . . .

AG: But I'm suggesting that if you sat and examined your conceptions and their texture, you'd find that notions, conceptions of God also dissolve with the dissolution of the notion of a permanent identity of the self. It's more like in Heraclitus where everything is flux. Or more like in Heisenberg where when you look at a wave, it finally dissolves.

JL: And this is precisely the way the Bible says it will be for people like you who have no faith. Saint Paul says in the book of Romans that all men know God, his eternal power and Godhead. But the problem of the unbeliever is that he knows God but suppresses this knowledge, the truth of God. The unbeliever holds down this truth in unrighteousness. And instead of worshipping the Creator, he worships the *creature*.

AG: But I'm not recommending worshipping anything.

JL: But what you believe ends up as self-worship.

AG: You just examine the self and you find that the self disappears, that it isn't there.

JL: But the self makes sense only in terms of God Almighty. You start with God. Because if the starting point is you, you are dead in the water.

AG: It sounds like you are trying to preserve the self here.

JL: No. I'm saying the true road to inner peace and happiness lies in divesting the self of self.

AG: That could be. But I don't think you divest self of self by beating up on the self. That's more self. You just have to sit there and observe the self dissolve.

JL: But when you study the religions of ancient Greece and Rome, and contrast them with Christianity, you realize that the modern idea of the self, the person, originated in the Old and New Testaments. The idea that there is such a thing as the individual, with inalienable rights that the state cannot deny, that there is intrinsic worth and value in persons, is a Christian idea. . . . You seem to want to believe that humans are made in God's image, but this is irrational for you since you reject this faith.

AG: All this makes it sound like you're still trying to defend the notion of a permanent self.

JL: I believe the Bible, that we're all created in the image of God.

AG: Well, you're going to be stuck with yourself for a long time, if you want to be. (laughing) . . . It's much simpler than you're making it. You're having to drag in a supreme self to do the job for you. . . . Do you believe in a permanent hell?

JL: Absolutely. And I'll tell you, friend, you don't have to believe in hell to go there. This is not a condition of entry.

AG: Well, you know there's an old saying that all the constituents of being are transitory. A wise old Buddhist once told me that if you see something horrible, don't cling to it. And if you see something beautiful, don't cling to it. I think your problem is clinging and attachment, grasping, a kind of greed for heaven, an element of greed for selfhood in heaven.

JL: The difference between us is that you don't want to proselytize, really, for anything you believe in. In fact, if you were totally honest with me, you'd admit that you don't even want to impose what you believe in on yourself.

AG: I certainly would not like to impose any conception on myself.

JL: But I've read that you don't want to proselytize for homosexuality even though you're a homosexual. And I've read that you would like to have children. Would you want your sons to be homosexuals?

AG: If they would have as rich a life as I have had I wouldn't object.

JL: Rich? Rich by what standard?

AG: Appreciative of the phenomenal world and—

JL: Who do you thank for that?

AG: creative in the world of music and poetry and joining speech—

JL: Who do you thank for the phenomenal world?

AG: together with body and mind.

JL: Who do you thank for the phenomenal world?

AG: I'm just trying to answer—

JL: I know, I know—

AG: your first question while you're babbling away in the middle of my answer.

JL: Oh, look who's talking, Allen Ginsberg!

AG: I think we're both babbling, but I was—

JL: You've been at it—

AG: babbling back an answer.

JL: longer than I have.

AG: Well, I'm 61 and I can keep my temper probably a little bit better than you at this point.

JL: I think you major in minors, Mr. Ginsberg: keeping tempers, being mannerly, that's not what it's ultimately really all about, see?

AG: Equanimity is what it's about and a certain sense of compassion.

JL: Really? How about a righteous anger? Is there any such thing as a righteous anger? Some things people ought to really get angry about and pound the table?

AG: "Howl" is filled with righteous anger but let me tell you my experience of righteous anger is that whenever I get—

JL: You haven't experienced enough of it.

AG: into that state of righteous anger I find that I trample on other people's feelings, and I insult them and I really am just spreading my ego a little thick.

JL: Well, some people ought to—

AG: Let me finish!

JL: some people ought to be trampled on.

AG: Let me give you the experience of my age with righteous anger, that when I get righteously angry and begin insulting or denouncing or putting people down out of it I find usually it's an extension of my own vanity and I have to pay for it later—

JL: Well, fine, speak for yourself. I don't feel that way. Some people ought to be put down. Some people ought to have their feelings stomped on.

AG: *May*-be, but if you want—

JL: What do you mean "maybe"? Why do you say "maybe"?

AG: If you really want to change them, or enlighten them, or to communicate with them, you have to give them space to come out of themselves.

JL: No. Space. You've had too much space.

AG: You have to treat them kind of gently so that it isn't just a contest of your will or ego against theirs. In a sense what you do is present your feisty, cock-a-doodle-do ego—

JL: No, no. Nice try. Another ad hominem attack. And this is another sad thing about unbelievers such as yourself. Since you're egocentric, you assume every other person is acting like yourself.

AG: I'm trying to be candid with you.

JL: But I haven't, in this conversation, implied in any way that you are saying anything you don't really believe. My problem is with what you really believe.

AG: Why aren't you thankful that I'm at least being candid?

JL: No. Candid, schmandid. I'm concerned about your soul. Do you believe you have a soul?

AG: I'm not sure there is a permanent soul to worry about. That's one of those arbitrary conceptions so characteristic of monotheism. The Buddhists say you don't have to worry about that. The question is seeing the transparency of your thoughts and not getting attached to conceptions of the soul but leaving the whole space open so that what you think is soul can dissolve and open up to the great space that exists. Appreciate the actual vastness of the place where we are now.

JL: What does this mean?

AG: Out the window is all of Washington, the sky—

JL: Mr. Ginsberg, the earth is the Lord's and the fullness thereof. I don't need to be told by an unbeliever about the great marvels of the world and the universe.

AG: I think we always need to be reminded of the vastness of the place we are in. And we forget that when we get into these little snits and arguments about—

JL: About God? I think more important is Who created all of this, not its vastness.

AG: You know people could even believe in God but not really appreciate how vast He or the universe is.

JL: I agree with that.

AG: People can get hung up on the words, on the possession of the ideas and take pride and vanity in their belief.

JL: Sure.

AG: So you have that little problem to deal with.

JL: You think pride and vanity are bad?

AG: It depends whether you are aware of it or not. If you are aware of your pride, it probably does less damage than if you are not. Then, at least, you see through it and use it in a way that is conscious. . . .

JL: Let me close with a political question. In your 1967 poem "Returning North of the Vortex," you said, regarding the Vietnam War: "I hope we lose this war / Let the Vietcong win over the American Army!" And you went on to say that if you had your wish, we would lose our will, we would be broken, and our armies would be scattered. Do you still believe this and think it was a good thing we lost the Vietnam war?

AG: I think I lost my temper there. I was getting into a sort of righteous wrath like you're always doing, saying that we're sinners and should be punished for our sins. This was sort of like a poem, a double statement—

JL: But you were for the Vietcong. You were pro-Vietcong.

AG: Not actually.

JL: But you said "Let the Vietcong win over the American Army!"

AG: I was making a poetic curse, and it was probably bad judgment.

JL: Then you think it was bad we lost the war?

AG: No, I don't think it was that bad that we lost the war. I don't think it would have been any better had we won it.

JL: Really? But it certainly was bad that we lost for millions of Southeast Asians, wasn't it?

AG: Yeah, but it probably would still be bad for them—they would still be fighting over there if we had won it, probably.

JL: But there were more people killed there after we left, after you guys got your way, after your "peace" was given a chance! We got out but there was no peace. Millions of Cambodians were slaughtered. They're still fighting. What happened? You were wrong.

AG: We destabilized the government of Cambodia and precipitated that whole scene.

JL: So *we* were the bad guys and not the Khmer Rouge, who actually conducted the genocide?

AG: But we opened the way for them.

JL: How?

AG: By destabilizing the government.

JL: But why would this make the Khmer Rouge Communists commit genocide against their own people? Why do you blame us?

AG: I'm just pointing out that we had a situation where the lid was being kept on the Khmer Rouge and the CIA had a long period of time to destabilize this. The bombing made this situation worse.

JL: So the Communists weren't genocidal until we bombed them?

AG: That finally our people working with our intelligence agencies got rid of Sihanouk and installed really unpopular puppets. And that created an opening for the Khmer Rouge—

JL: So the Cambodians got what they deserved? Is that your point? What is your point? If everything is as you state it, so what?

AG: My point is that the US government and intelligence agencies destabilized the situation in Cambodia and created such chaos that the Khmer Rouge got in and committed genocide.

JL: Then you are blaming us for the Khmer Rouge genocide. Do you denounce what the Khmer Rouge did?

AG: Sure.

JL: Then why don't you say a couple of words about what you think of the Cambodian genocide committed by the Communist Khmer Rouge? Why don't we close on this with, perhaps, a little poem?

AG: Yes. Blood running inside the head of John Foster Dulles flowed through the bodies of Cambodians onto the streets of Phnom Penh.

JL: This was supposed to be a poem denouncing the Khmer Rouge, not John Foster Dulles!

AG: The whole problem was that Foster Dulles didn't want to recognize the Geneva Convention and refused to sign up for an election in Vietnam because Eisenhower thought that—

JL: But you're still excusing the Communists. After *all* these years you can't bring yourself to denounce the Communists. Why not?

AG: No. I've denounced the Communists—

JL: Baloney! I asked you to end with a little poem denouncing the Khmer Rouge and you attack John Foster Dulles!

AG: You asked me to denounce the Communists on your terms. But I'd rather denounce them on my terms.

JL: Yeah. By denouncing John Foster Dulles.

AG: I'd rather denounce them on *my* terms. But I'd rather not just denounce them but try to point out that everybody is complicit in this situation.

JL: Speak for yourself. I didn't have anything to do with Cambodian genocide. Did you? Do you take some blame for that?

AG: I would be willing to take some blame. And I think we should all take blame.

JL: What role did you play in causing this genocide? And have you apologized to any Cambodians?

AG: Yes.

JL: Who? Who did you apologize to?

AG: You're setting up something here which is not . . . it's like, again, you're going back to this black and white, either/or interrogation.

JL: But some things are black and white.

AG: And this comes from your monotheistic insistence.

JL: Is *nothing* black and white?

AG: Nothing is completely black and white. Nothing.

Allen Ginsberg's name has resonated with Josef Jařab since 1965. Writing in 1999, Jařab explained that "after [Allen's] expulsion from Czechoslovakia by the Communist regime . . . he became a vital symbol of freedom." Later Jařab was able to meet Allen a number of times outside Jařab's native land and once even smuggled copies of Ginsberg's books into Czechoslovakia for Allen's Czech translator, Jan Zábrana. (Because of Communist censorship, however, Ginsberg's first works in Czech were not published until 1990, after the translator's death.) Still, Jařab did not get a chance to interview Ginsberg until 1989, when Jařab was a visiting scholar at the Du Bois Institute for African American Research at Harvard. After Jařab told Allen that his research was in black literature, Allen invited him to visit his class at Brooklyn College on a day when poets Gwendolyn Brooks and Michael Harper were reading and discussing their work. After that class they arranged the interview, which would be held in Allen's East Village apartment.

Jařab recalled: "Allen took me to a Polish restaurant round the corner. . . . Then we shortly, at the door, greeted Peter Orlovsky and started talking about the Prague episode. I was always surprised how much it meant not only for us in Czechoslovakia but also for Allen. We spoke a lot about black poetry, the blues. Allen kept copying for me pages and pages of material on his small photocopy machine in the kitchen, and the pile of books I was to take home with me was also growing by the hour. . . . It was close to midnight when I realized there may never be an interview! But then we sat down and started recording. . . . The interviewing was done by, I think, two A.M. Outside there was a thunderstorm, and it was pouring rain. I packed all my belongings, and all the precious gifts of the host and, accompanied by him, I went to the street where we, luckily, got a taxi. . . . The young Pakistani driver asked me whether the man who accompanied me to the car was my relative, and after I explained that it was Allen

Ginsberg, a great American poet, he admitted his ignorance of the name, but wanted to prove his knowledge of poetry, and, while crossing the Brooklyn Bridge, recited by heart the whole of Shelley's 'Ode to the Western Wind,' which he remembered from his school years. An unbelievable ending to a wonderful day."

The tape was transcribed by the research assistant at the Du Bois Institute, Martha Moore.[184]

—DC

Joseph Jaŕab: How do you feel about people saying that the whole Beat movement started with you meeting Kerouac and Burroughs somewhere in the '40s and ended with Cassady's and Kerouac's deaths in the '60s? Can we really limit it this way?

Allen Ginsberg: Well, you know, the notion of the "Beat Generation" as a literary movement has a real basis, but it's also a journalistic stereotype. It's originally just a group of friends who *later* were called the Beat generation, but we did have some views in common. It was an extended circle, or an extended family. I would say that Kerouac and Burroughs were at the heart of it; myself learning a great deal from them, and other integral members were Gregory Corso, from the '50s on, and Peter Orlovsky. That would be an inner circle of friends. But from '48 on there were a number of other poets with whom we associated who were traditionally part of the Beat generation—the anthologies, and a friendship circle. You'd have to include Gary Snyder, Philip Whalen, and Lew Welch. Gary Snyder and Philip Whalen have gone on to become accomplished Zen practitioners—Zen *sensei*, which means teachers, and they have been given transmission so that if they continue teaching, they would be Zen masters, the first American poet Zen masters, who were, incidentally, integral to the Beat generation. So there's been a great deal of development out of what was in 1960 considered sloppy homemade Buddhism but which actually was most accomplished and sophisticated. *The Dharma Bums*, Kerouac's novel about Snyder in 1957, begins a progression from Snyder going to Japan to study, and now he winds up with transmission to teach. As for

[184]This interview also appears in *Race and the Modernist Artist* by Jeffrey Melnick and Josef Jaŕab, forthcoming from Oxford University Press.

myself, you know, I participated in founding the Jack Kerouac School of Disembodied Poetics at Naropa Institute which is the first Buddhist college in the Western world accredited along with other schools and universities. That's quite an accomplishment and a development, a ripening of early interests, or as Robert Graves wrote in *To Juan at the Winter Solstice*, "Nothing promised that was not performed."

Then if you survey the galaxy of writers associated with the Beat generation you find that only Kerouac died, Neal Cassady died, Lew Welch died early. Living and growing and more and more powerful artistically, and more influential finally are Burroughs, myself, Gregory Corso, Gary Snyder, Philip Whalen, Michael McClure, Robert Creeley, who's always been a friendly associate, Peter Orlovsky, who published a book very late in 1978, his first book, also LeRoi Jones/Amiri Baraka, and Lawrence Ferlinghetti in San Francisco. And I'm going to try to name the geniuses in a second generation: Among the best poets who were students and friends of Kerouac and shared influence from the Beat generation and the New York school: Anne Waldman, Ted Berrigan, Ron Padgett, Ed Sanders, Bob Dylan among others. And in a third generation beyond that: David Cope, Antler, Andy Clausen, Patti Smith, Jim Carroll, John Giorno.

JJ: Somehow then, there's quite an extension into the present.

AG: Yeah, and you'll find we have a better survival rate than most insurance men or professors. Certainly a lot better than the academic poets of our era who have wrecked themselves drinking, like John Berryman, Delmore Schwartz, Robert Lowell and Randall Jarrell who committed suicide.

JJ: Sylvia Plath.

AG: Yeah. So that as distinct from the academic poets perhaps we were saved by our interest in marijuana rather than alcohol. Those of us who died early were those who had problems with alcohol. The old American . . .

JJ: Monster.

AG: Yes, the old American monster. So that I would say that the major people of the Beat generation have gone on to the last forty years, the '40s to '80s, growing in authenticity and power. Burroughs's most recent work is his most powerful, a trilogy: *Cities of the Red Night* (1981),

The Place of Dead Roads (1984), and *The Western Lands* (1987). My later work will take time for people to catch up, but the late '70s work includes "Father Death Blues," "Plutonian Ode," and "White Shroud." Gregory Corso's 1981 book *Herald of the Autochthonic Spirit* is almost an international classic and will be when it's translated. It has been translated already into Russian, Chinese and Hungarian.

JJ: Well, we'll hurry up and have our own Czech translation soon.

AG: Now for the ideas involved—recollect that Gary Snyder and Michael McClure back in the 1950s poetry readings introduced early notions of ecology. It's taken the mainstream of American society 35 years to catch up to those insights introduced in original Beat poetry. In the tradition of Thoreau, Whitman, but formulating it as regard for nature, regard for the industrial wasting of the planet or up to the point where now we might even say that the planet itself has AIDS. The immune system of the planet is no longer able to restore the damage done by the human virus.

JJ: That's a powerful image.

AG: Yes, I mean, the immune system of the planet may not be able to cope with environmental degradation through hypertechnology and human overpopulation.

JJ: Terrifying.

AG: Well, on the other hand, there's a new theory among gay AIDS people of not "people dying of AIDS" but of "people living with AIDS"—as an attitude: not taking a negative attitude.

There's also—introduced somewhat by Kerouac—a main theme, the appreciation of Negro black culture and jazz. Another continuing theme or "motif" in ongoing "Beat" literature is the reconsideration of the nature of consciousness: the introduction in literary terms and in practical, social terms of "East meets West," i.e., the meeting of the eastern and western minds; the introduction of meditation practice, now familiar to many poets, more and more influential in the '70s and '80s in the American culture; even the founding of an institution to train young poets in that area of contemporary poetics, at Naropa (Institute).

So that what we've done is try to find permanent forms for these "open mind" insights. And it so happens that the particular "bohemian" insights that Kerouac had into spontaneous mind are also classical notions in Japanese, Chinese, and Tibetan poetry. It was for lack of sophistication and a sad provincialism among American and English literary scholars—even European scholars, but more the American and English scholars of the '50s and '60s—that they didn't recognize Kerouac's aesthetics of "First thought, best thought," as relating to calligraphy, haiku, Tao, Tibetan mind forms, the whole teaching practice of Zen and Tibetan Buddhism, in which seeing the mind is the guideline for wisdom. The discipline is in the *mind* first, not on the page revised over and over.

JJ: Yes, the apparent simplicity here is, in fact, complex.

AG: Yes, like improvised jazz. The mistaken notion of an "undisciplined" Kerouac, an "ignorant" Kerouac, by the academy, misled people who did not realize that Kerouac was extremely sophisticated, intelligent, and extremely learned and that he had a good grasp of ideas on wisdom practices.[185] And technically, writing insights. So you could say on several levels in the psychedelic, ecological, the contemplative aesthetics, and the advanced, post-modern aesthetics, the Beat generation early practices have developed and rightfully become much more understandable now. Even Burroughs cut-ups are understandable in terms of music television—MTV. And, of course, regarding Burroughs, Kerouac, to some extent, myself, the influence of these poets is very strong not only in Eastern Europe and China or Russia, but also in the younger generations in America. Every decade has a revival of Beat generation '40s, '50s, '60s texts; a reappreciation. It seems to be the main literature that survives of us in Europe, as the academic literature

---

[185]In many of his interviews Allen had to fight the impression that literature based on spontaneous mind meant that any person could become an accomplished writer by simply not censoring his thoughts. Ginsberg expressed his frustration with this common misinterpretation in an excellent interview with Christy Sheffield Sanford and Enid Shomer when he said, " 'First thought, best thought' is, after all, pronounced by a Tibetan lama to a Zen-influenced poet, and any academic who would interpret that as meaning 'any bullshit you utter without mental training is best' would have to ignore the entire history of half the world, as well as Gertrude Stein, Abstract Expressionism, modern art of the twentieth century, and Einstein and Heisenberg. They'd have to ignore Blues; and anybody on that level is not an academic but some kind of butcher who wandered off into the wrong profession." For the full interview, see "An Interview With Allen Ginsberg: The Mucous Membrane Barrier, Woman Writers, Rabbinical Rhythms and Hanging in the Void," in *Connecticut Poetry Review*, 1988.

begins to fade for lack of energy, vividness, and adventure—and appre-
ciation of the planetary new consciousness.

JJ: The vividness seems to be the key word with us, for instance. And it
has functioned this way for decades.

AG: Another element is that a lot of the literature of the Beat genera-
tion comes directly out of a classical lineage in American writing; it
didn't get itself born, it didn't hatch out of an egg from nowhere. It was
the line of . . .

JJ: Whitman?

AG: Whitman, of course, and Thoreau the ecologist, and Emerson the
individualist. But also, Robert Creeley was engaged in correspondence
about how to run a magazine, *Black Mountain Review*, with Ezra Pound.
And William Carlos Williams wrote prefaces to two of my books and
received me and Jack and Peter and Gregory in his living room and
warned us, as he pointed out the window, "There's a lot of bastards out
there." I knew Caresse Crosby and met Marianne Moore and spent
time with Pound. Burroughs and I visited Louis-Ferdinand Céline in
the last year of his life. And there was in America—in Chicago and New
Haven and in San Francisco—Jean Genet; and we met in Paris, not only
Céline, but also Tristan Tzara, Benjamin Peret, Marcel Duchamp and
Man Ray, many in the cafés in the late '50s. Actually we come from a
very sophisticated lineage, both European and American.

JJ: Apollinaire meant a lot for you, didn't he?

AG: Yes, when I lived in Paris I wrote an homage at his tomb. And it was
supposed to be a signal to the American academics where I was coming
from, but I don't think they picked up on it. So we were continuing that
open form "filiation" or "lineage" Whitman through Pound and
Williams; through (Charles) Olson, (Robert) Creeley and Kerouac,
which I think is a mainstream in American letters historically; we
joined that with music interest in black culture. The synthesis of all
these themes, after us, was the very remarkable Bob Dylan who inci-
dentally says that it was *Mexico City Blues* that "blew his mind" and
turned him on to poetry. Someone handed him a copy in 1958 or '59 in
St. Paul. Over Kerouac's grave, Bob Dylan told me that "It blew my

mind." And I said, "Why?" and he said, "It's the first poetry that talked American language to me." So you get a line in Dylan like "the motor-cycle black Madonna two-wheeled gypsy queen and her silver studded phantom lover"[186] which comes straight out of either "Howl" or Ker-ouac's *Mexico City Blues* in terms of the "chain of flashing images." Ker-ouac's spontaneous pile-up of words. And that's the way Dylan writes his lyrics. So poetry's extended itself in its own lineage afterward into John Lennon, the Beatles, named after Beats, and Dylan, so that it's gone around the world. And I think after the wave of Whitman and then maybe another wave of Pound, it's probably the strongest wave of American influence on world literature—the combination.

JJ: Definitely in our part of the world from the '60s onwards.

AG: Well, it's the same in China. And India and Japan somewhat. Eng-land is very resistant. This culture is widespread in France and Nor-way, Sweden.

JJ: I understand there was a wonderful French anthology put together on the Beat movement. Have you seen that?

AG: Yes. A number of them. I have seen them. And also, really influen-tial, most influential in Italy. A whole generation there. Not only the poets and writers but also painters and intellectuals. So I would say probably the presence—not the presence, but the influence of the Beat generation is probably stronger now than it ever was, because in the late '50s and early '60s it was "notorious"—What they were getting was a media-packaged Frankenstein version of people with berets and cockroaches and eyeglasses and bongo drums. Out of *Time* magazine and the CIA version. Rather than the original literature.

JJ: And the cultural movement that it became in a way. Would you call it a movement?

AG: Well, a movement in the sense that it had elements that lead to the Greens.[187]

---

[186]"Gates of Eden" from Bob Dylan's *Bringin' It All Back Home* (Columbia Records, 1964).
[187]Any number of environmentalist political parties, primarily active in Europe, but currently also in the United States (http://www.greens.org). The first official one (Die Grünen) was formed in Germany in 1979.

JJ: Yes.

AG: To ecological, psychedelic politics.

JJ: The name of John Clellon Holmes has not come up in our talk so far.

AG: Holmes was a very nice guy.

JJ: Was he a sort of Boswell of the movement?

AG: No, no. He was a good, close friend of Kerouac and a very good, close friend of mine. Back in the '40s. And early '50s. I don't think he was as great a genius in prose as either Kerouac or Burroughs, or even Hubert Selby, Jr.,[188] who I like. There's still a little, a slightly middle-class American novelist prose style. It isn't an invention. Holmes was very advanced in cultural appreciation, but I think the most important thing is the actual texture of prose; basically the Beat generation is a movement of people who began working together in the Ivory Tower, purely for art's sake. The reason it had a social fallout is that it's close to an Einsteinian "art for art's sake" wherein the subject is *the nature of the mind itself*. Candor! Whitman's word, for what he asked from future poets, was "candor." So in examining the texture of consciousness, the psychedelic aspect of the contemplative, a Buddhist aspect, naturally everything else rises—all the contents of the mind. Nothing human is alien: Erotic, ecological, political, and whatnot.

One thing I forgot was that with the trials of *Howl, Naked Lunch*[189] and the Grove Press's efforts to legalize *Lady Chatterley's Lover* and Henry Miller, we were part of the Liberation of the Word. I would say what happened in the '40s and early '50s was a spiritual lib, late '50s liberation of the word, legally, in America; breakdown of censorship. The last big censorship trial was *Naked Lunch* in 1966. After that the floodgates were open to anything! From that you have many different kinds of liberation; like Women's Lib, Gay Lib, even to some extent, Black Lib. You might know Abbie Hoffman and Tom Hayden went south to Birmingham to help Blacks gain the vote, and begin desegregation of races, in the early '60s with copies of *On the Road* in their

---

[188]*Last Exit to Brooklyn* (1964), *The Room* (1971), etc.
[189]See introduction to *Naked Lunch* (New York: Grove, 1959) for excerpts from this trial.

pockets. There was some "beat" influence; it wasn't the main one; but it was a substantial breakthrough that made people look around at America and begin to examine the texture of American life and society in stasis at the end of the '50s. So from "Spiritual Lib," I would say, to progression of Language Lib, it would have, as a fallout, other social liberation movements: the Gay Lib, sexual revolution, Sex Lib, to Women's Lib to some Black Lib and Old Folks Lib, Gray Panther Lib, and then Minority Lib and then appreciation of American Indian Liberation and appreciation of the quality of indigenousness; and individuality rather than homogeneity.

JJ: Did the black poets like Bob Kaufman, LeRoi Jones or Ted Joans spontaneously join the whole movement in its spirit, or would they bring something that was new, additional?

AG: Well, we were social companions. And literary companions by the late '50s. See, LeRoi Jones wrote me in Paris when I was living with Burroughs asking for material for *Yugen* magazine which he'd just started. His *Yugen* magazine (1958–1960) was one of the great magazines of that time. I would say three or four were: *Black Mountain Review*, which in its last issue carried the signal pages of Burroughs's *Naked Lunch*, Philip Whalen poems, Kerouac's "Brakeman on the Railroad" and his "Essentials of Modern Prose," my own poem, "America," Herbert Huncke's story "Elsie." All those were in the last issue of *Black Mountain Review*.

JJ: And *Floating Bear*?

AG: *Floating Bear* was edited by LeRoi Jones and Diane di Prima. Also *Evergreen Review* and *Chicago Review* to *Big Table*; and maybe one or two other mimeograph magazines, like *Combustion*, edited by Ray Souster, a disciple of Williams from Toronto. Also, LeRoi Jones had the grand "salon." Literary salon in the late '50s at which you could find all the contributors to *Yugen*. Three blocks away from here on 14th Street, I saw at one party, in one room, at one time Langston Hughes, Don Cherry, Ornette Coleman, Cecil Taylor, Franz Kline, Kerouac, myself, Orlovsky, Corso, A. B. Spellman, other blacks that I didn't know at that time; Frank O'Hara, maybe Frank's friends, Larry Rivers and Arnold Weinstein; maybe intersecting with Kenneth Koch, John Ashbery and others; Robert Creeley, Charles Olson; Olson wasn't there until Creeley was around New York.

JJ: This was an amazing combination.

AG: It was a real mad combination—"All American." The later jazz all based on spontaneous wisdom. The abstract expressionism, free jazz, open form poetry; or spontaneous mind poetry. Jones even went to visit a Buddhist teacher as well. And *Yugen* also had material by Snyder and Whalen. So there was an era of good feeling from the late '50s—and there was Peter Orlovsky; the poets Paul Blackburn, Joel Oppenheimer and Ray Bremser would all be there. Part of LeRoi's *Yugen* salon. There were some parties where we were all together, some beautiful moments. That was the cultural cresting of the Beat generation. It was also a joining culturally of black and white. Eldridge Cleaver and Jones appreciated Kerouac's appreciation of black culture. For that Kerouac was really put down by neoconservative whites—these ideologues called him a white chauvinist—actually Kerouac noticed that black culture was "mis-noticed" in America. That was his phrase—"mis-noticed"—very delicate phrase—by the white culture. You know the passage—in *On The Road*—where he passes a porch, and he sees wild, happy, funny blacks enjoying the jazz and the last sunshine. It's dusk on the porch. Eldridge Cleaver read that paragraph in jail and admired and thought it was great. But Norman Podhoretz, a neoconservative critic blind to Kerouac's prose beauty, cites it as an example of Kerouac's ignorance and white chauvinism, but blacks liked his élan in those days as being signs of sympathetic companionship and liked that particular passage in *On The Road*.

JJ: Yes, well that explains what one always feels, even from a distance. I think there was more of black and white being seen apart from within America than there is from the distance of, let's say, Europe.

AG: Maybe Kerouac as French-Canadian had a more European vision. Kerouac had a lot of experience in Harlem in the early '40s. He witnessed the development of bebop with Esoteric Records, with recording engineer Jerry Newman and Seymour Wyse, Kerouac's Horace Mann High School friend; when he was a student he went out and heard jazz at Minton's and saw in person Charlie Christian, Bird Parker, Lester Young, Roy Eldridge, Gillespie, Coleman Hawkins, Illinois Jaquete, others, Bud Powell; so Kerouac had a great immersion in black bop culture. He would listen to Bebop all the way, all night. Sometimes we listened in 1944–48 to Symphony Sid's radio broadcasts; Sid was a very famous guy. He was really a disc jockey and was broadcasting the new Bebop classics then.

So there was some mixture of cultures all along. I would say that it was like a confluence. A lot of black poets then and now by hindsight attribute to me or Kerouac a kind of breakthrough in poetry which empowered them to write in their own cultural idiom again, or reinforced their own values.

JJ: Right. It must be gratifying.

AG: At least—that's what I heard recently from Audre Lorde, June Jordan, and many of the African-American poets here who appreciated what we had done as a kind of liberation of language and a feeling, so that they could manifest their private world and not to imitate the academic white.

JJ: I think it's only fair and justified, I'm sure. That's the way we've seen it from the distance . . .

AG: And that's the way I see it, as a confluence.

JJ: Have we explained the name of Lawrence Ferlinghetti in our talk?

AG: Now, that's right. Ferlinghetti played a very strong role as publisher, appreciator, old Bohemian. There's a certain melancholy old world quality in his poetry that's inimitable and very valuable.

JJ: He's also the French connection, is he not?

AG: Yes, for (Jacques) Prévert and his Sorbonne education. Because of his Surrealist interests, Philip Lamantia's also a larger French connection, with (André) Breton. Yes, he definitely had that French café sense; Ferlinghetti is a champ. His poetry, though, seems a little too referential, dependent on puns, on atmosphere and French mood rather than on the kind of precision that Ezra Pound or Williams would, uh . . .

JJ: Ask for.

AG: Ask for. Yeah.

JJ: And Philip's? Philip Lamantia's?

AG: He's very valuable, particularly in the late '50s, his *Selected Poems* from City Lights. I understand he's had several books out since, all good. He was very ill for a while. He kept switching back and forth between Surrealism and Catholicism and junkyism and whatnot, now finally bird ecology. But he's always had an interesting career. When I went to China to teach I brought his texts as exemplary post-war American poetry with genuine surrealist flavor.

JJ: When was that?

AG: 1984 or 5—I was there two and a half months and I taught in the Foreign Language Institutes in Beijing, and in various colleges in Tunming, Suchow, Hangchou, Baoding and Shanghai's Fudan University.

JJ: Had there been any knowledge of your work and the Beats in general?

AG: Yes, I found that in the Chinese anthology of twentieth-century prose, volume two, used by the elite foreign language students, who study English and come to MIT, that the largest single selection of prose was Kerouac. Printed in China! With a very appreciative and very intelligent understanding essay on it. Better than I've ever seen in America.

JJ: Well, that's very surprising.

AG: Very funny. Also (Gary) Snyder and I had been influences on recent "misty" or "obscure" school of post-Cultural Revolution Chinese poetry.

JJ: So what did you teach in China? What was the message you wanted to convey?

AG: I had to decide what was the essence of American poetry that I would want them to experience. So what I brought was a bit of Pound, several poems by Charles Reznikoff, a lot of W. C. Williams: "the pure products of America go crazy"[190]—that's always in his poems; then I jumped to contemporaries. And I presented some brief prose by Burroughs, some of Kerouac's "Mexico City Blues," and a few prose sketches from *Visions of Cody*. Philip Lamantia—a few poems of his; Gregory Corso's "Bomb" and "Hello" and "Gasoline"; a long poem by Kerouac that had not been published, on Mao Tse-tung. A few poems

---

[190]From William Carlos Williams's poem "To Elsie." *See Selected Poems*, William Carlos Williams, edited with an introduction by Charles Tomlinson (Harmondsworth, New York: Penguin, 1976).

by Orlovsky, particularly one on recycling of human manure in Chinese rice fields. John Wieners—two or three poems by Creeley; so of the contemporaries it was: Wieners, Creeley, a little Olson, Gary Snyder, Philip Whalen, Lew Welch, Lamantia, Peter Orlovsky, David Cope, Michael McClure and one or two that I've forgotten. That was what I thought was the essence. With a copy of the Don Allen anthology *New American Poetry* 1945–1960.

JJ: Did you try Nanao (Sakaki)'s poetry on them?

AG: Yes, I brought one or two poems of Nanao. As part of the Snyder selection—a dozen poems. Those are not the standard American anthology, but the Chinese students really appreciated it.

JJ: Did you teach poetry in this country before you taught in China or is the current appointment at Brooklyn College the first one in the academic world?

AG: I had been teaching at Naropa Institute since 1974 and what I taught was, for four terms, line by line, the complete works of William Blake, from beginning to end. Up to the seventh book of the Vala, the seventh night.[191] By then all the students had gone on and I could start all over again. I also taught a course on William Carlos Williams.

JJ: There?

AG: Yes. Some of it's been published in *Composed on the Tongue*, you'll find several lectures on Williams there. And I taught a course on the English lyric from Pound's [translation of the Anglo-Saxon] "Seafarer," up through Corso. Lyric. You know, Thomas Wyatt and (T. S.) Eliot, (Alexander) Pope, (John) Dryden, Christopher Smart, (Thomas) Campion, all the way up. I taught a course in Sapphic poetry. And classical meters; quantitative meter as distinct from stress meter; then I taught a course in nineteenth-century American poetry geniuses—Poe, Dickinson, Melville's poetry, and his prose from *Pierre*, and Whitman. Then I taught a series of courses called Literary History of the Beat Generation. Those are being transcribed to be edited as a book.

---

[191]In William Blake's prophetic book, *The Four Zoas*, his system is the Vala, broken into nine (Vala) nights or books as Ginsberg refers to them here. See *The Complete Poetry and Prose of William Blake*, ed. David Erdman (New York: Anchor Books, 1988).

JJ: As for the teaching, however, do I understand correctly that the Naropa students would be people already motivated and interested primarily in studying poetry?

AG: Right.

JJ: Well, that would make them a different lot from the students at Brooklyn College, wouldn't it?

AG: No. I'm teaching the MFA Poetics program there, primarily.

JJ: Yes. And what about the current course in Afro-American Poetics?

AG: "Black American Literary Genius" is a volunteer course I invented for my own pleasure and to make a bridge to the students there and stir things up culturally, make a bridge over the alienation between the different cultural-social groups.

At Naropa the students attend (1) the Jack Kerouac School of Poetics; (2) the Naropa Institute Buddhist Contemplative College; and (3) summer programs with myself, Gregory Corso, William Burroughs, Anselm Hollo, Kenneth Koch, Robert Creeley, Diane di Prima, and Anne Waldman: those are the people that come and teach. What we've done is codify the same gang as in *Yugen* magazine and the *Black Mountain Review*. We've codified that as an institution for transmission from generation to generation. Which is something I don't think other generations of poets have done so well. The American "Fugitives" did it with Vanderbilt (College) in the South and Kenyon College. There was some of that with Eliot, but it didn't come to much poetry really, more for academic teaching and criticism. Whereas we combine poetics with the discipline and classicism of Tibetan Buddhism and Zen; we literally have Tibetan lamas and Zen masters coming there to teach. Connected with Burroughs, myself, and joining each other with a great deal of respect. So these are historically very interesting complementarities. Like the American Beatnik and black and now the American Beatnik black post-Beat connection with the Tibetan—with the East.

You see, Eliot studied Sanskrit and had some Himalayan wisdom, but I don't think he knew Zen masters and Tibetan lamas directly. Whereas we're actually working in conjunction, experiencing the actual texture of the mind with the originators of the Himalayan tradition. (William Butler) Yeats had a deal with third-hand Theosophy derived from Himalayan transmission gone through the Golden Dawn Society, Madame Blavatsky, and what mish-mash! Eliot had a "mélange adultère de tous." We have a more direct transmission. Also many of us actually went to India—the Whitmanic "Passage to India" was made literal in our generation.

JJ: Whereas for Whitman it was a daydream?

AG: Yes, a very intuitive and brilliant daydream. We had the chance to actually embody, manifest that.

JJ: I was going to drop this question, but now that you mention the Buddhists, the Zen and all that, someone, I think it was Carolyn Cassady, said that the Beat movement was much more than is generally believed a religious movement. Would you agree with that?

AG: Yeah, in the sense that it was mystically religious, yet practical and artistic. Kerouac spoke of it as "the Second Religiousness," a phrase he took from Spengler who points out that in a time of declining empires, a second religiousness arises. Kerouac's adaptation of the phrase. You should take a look at his great essay, "The Origins of the Beat Generation."[192] Ever seen that?

JJ: Yeah.

AG: It's a brilliant thing, very definitive.

JJ: One footnote question: the essay by Norman Mailer "The White Negro"—was it of particular importance for the movement?

AG: I don't think Kerouac dug it, because I think he thought it was relatively "ideological," it's clear. It probably was very intelligent, though, because he, Mailer, had a good grasp or glimpse of the great apocalyptic goof that middle-class white culture was making. You know, he had some sense of an apocalypse or transcendence, beyond ideology, some transcendent change of consciousness. So I think he's tuned in properly there. Except that Mailer, unfortunately, has, to some extent, "out-Hemingwayed" that macho business of thinking that "the cool psychopath" was more macho and more honest than a delicate artistic fairy like me, you know. Mailer has some sort of boxing-Hemingway-

---

[192]See "About the Beat Generation," included in *The Portable Jack Kerouac*, ed. Anne Charters (New York: Penguin, 1995).

macho-element that Kerouac was much too sophisticated to settle for. Kerouac, as a football player, didn't have to worry about that, you know, about physical boxing (or mental boxing). So that I think Kerouac disliked Mailer's and (John Clellon) Holmes's interpretation of Beat as "criminal" and "psychopathic" flavored, because Kerouac saw Beat as Christ-like; the Lamb, the emergence of the Lamb, not the emergence of the grand criminal savants. So he thought Mailer had it inside out.

JJ: On the terminology now, "Beat" and "hip"—what is the connection and what is the development of, let's say, the legacy of one into the other?

AG: Well, both words, etymologically, were introduced literarily by Herbert Huncke to us about 1944–45.

JJ: They were? So what is true about the word that Bob Kaufman introduced "Beat?"

AG: Kaufman came on the scene—well, I met him only in the late 50's, 1959, in San Francisco, though he was around before, but you know . . .

JJ: Barbara Christian, a black female critic, maintains it was Kaufman who invented that term and cited that as another example of blacks not being recognized for their contributions in America.

AG: No. John Clellon Holmes in 1952 had this article, "This is the Beat Generation," in the *New York Times*. Kaufman came to prominence later literarily, you know, San Francisco in the late '50s. And lived upstairs from me on 170 East 2nd Street in the Lower East Side and took psilocybin then in my apartment, with Kerouac, when Timothy Leary visited, 1960.

JJ: So it *does* come from Huncke?

AG: Oh, yeah. Our introduction is 1944-5-6 from Huncke. Ten years—really—fifteen years earlier. 1945 Huncke around Times Square.

JJ: And "hip?"

AG: He used that word also.

JJ: He did. Could you specify the differences between the terms?

AG: Now, the Beat Generation was, so to speak, a literary group. It was only later, by hindsight, called the Beat Generation in the *Times*. On the basis of a conversation with Kerouac. Kerouac was *un*naming generations, saying it's not a generation; it's no "Lost Generation;" it's just a beat generation. Everybody's too beat to be a generation. So Holmes then thought this was an interesting phrase. And then, in the *New York Times* magazine, wrote, "This is the Beat Generation," 1952. But again defining it with a side overtone in terms of violence and juvenile delinquency, i.e., mindless protest. So that was pasted on Kerouac by Norman Podhoretz later, that Beat folk were all ignorant criminals. But Kerouac was religious and talking about the Lamb of Jesus and spontaneous Buddhist mind.

JJ: Right.

AG: And I was having visions of Blake in 1948; we weren't illiterate "Barbarian" psychopaths.

JJ: And you were also cleansing of all the middle-class stuff, is that right?

AG: I don't think we were concerned with the middle-class values.

JJ: No?

AG: That's a middle-class paranoia. The middle-class writer people in *Time* magazine thought we were rebelling.

JJ: So the Beatniks were not rebels?

AG: Caught up by William Blake, you don't have to worry about cleansing yourself of "middle-class values."

JJ: All right, I am trying to understand.

AG: It's a minor matter, the "middle-class values" in this context. That's like a relatively primitive Marxist notion, you know, "bourgeois" and "middle class." That's some hangover from class war. Kerouac's whole point was that "beat" went beyond the old Marxist ideological battle of class warfare and into some *practical* attitude of transcendence. Practical had to do with, I mean, like dropping LSD or learning meditation techniques. It's like the bomb, you know. It's not cleansing yourself of the middle class, it's *cleansing the doors of perception themselves*; in which case middle-class notions and ego notions and everything else gets cleansed; personal identity as well as national or class or race chauvinist identity as well.

JJ: And the materialistic orientation?

AG: Yes, as in the Zen view or way. REAL is real. As well as simulta-
neously dreamlike. MIND goes beyond ideology. To me, it always did,
and to Kerouac. It was in the secondary explainers, such as Lawrence
Lipton, who put a Marxist trip or spin on the Beat ethos. Or Todd
Gitlin, an academic historian. Many ex-communist, Stalinist turn-
coats who became CIA agents; or political-hangup-people like Nor-
man Podhoretz—who was once a liberal and then became a right
winger—wrote all this babble gobbledygook, about "middle-class co-
option" and all that. That's their trip. It's an intellectual, ideological
trip, rather than a spiritual one involving an alteration of perception, a
basic turning about at the root of consciousness.

JJ: I dig your point, I think.

AG: Alteration of the source, the ground of perception. An examina-
tion, not of the ideas in the mind, but the texture of thought itself.
Follow?

JJ: Follow. The working of consciousness?

AG: Yes, the texture of consciousness. So that back in the 40s—'44, '45
and '46—Kerouac and I were talking about the "new vision" or "new
consciousness." By '48 both Snyder and myself and others had had
some kind of actual change of awareness. You couldn't quite call it
visionary or mystical experience, but it was some experience a little
more profound than the later psychedelic experience with acid. It's a
natural experience: the deconditioning from hyper-rationalistic,
hypertechnologic monotheistic heavy-metal bureaucratic homoge-
nized hierarchical aggression in thought processes—unnatural to
begin with—that create planetary ecological chaos, totalitarian monop-
oly of power, over-rigid centralized authority, and police state condi-
tions in response to the degeneration of the natural environment.

JJ: And back to the term "hip" . . .

AG: Well, then, by the '60s there was the Frankenstein image of the
Beatniks. "Bongo drums! Man, that's cool! That's hip!" Then by the
late '60s it became politicized with the SDS and a somewhat ideological
Marxist politicized mental thing laid down, rather than a change of
consciousness. Change of ideas, rather than a change of soul.

JJ: So those were the lifestyles of the hippies?

AG: Yes, "lifestyle" and all that. It had some elements. The '60s continued in a sort of lineage inheriting a lot of the Beat material—like Abbie Hoffman and Tom Hayden going south with copies of *On the Road*.

JJ: And the projection of the new style into the appearance of those people?

AG: Some of it did. So that was what they called the Hippie movement, I guess. And I see it as a historical lineage thereafter.

JJ: A projection, of a sort?

AG: I think it was a limitation of the original vision. The hippies still believed in "progress." Whereas, we had already read not only Spengler, but also earlier Rimbaud, who did not believe in progress when he said, *"La science, le progrès, la nouvelle nobilité, le monde marche; pourquoi ne tourne-t'il pas?"* Remember that? From *Season in Hell?* He says, "Science, progress, the new nobility; the world marches on, why the fuck doesn't it turn around? Why the hell doesn't it go backward?"

So, we already had that. We no longer believed in that sort of liberal, progressive notion of progress.

JJ: Not even in scientific terms?

AG: Oh, science has brought us to the end of the planet.

JJ: That's right, almost.

AG: The bomb, or planetary AIDS.

JJ: Did the active Vietnam protest movement unify various groups?

AG: Yes, everybody agreed there. Yeah, that's right.

JJ: And the civil rights movement?

AG: Yes, everybody agreed with that. There were noble pacifist Gandhian elements in the civil rights movement—I think the differences were that the progressive groups used subterfuge, manipulativeness and a stereotype aggression to get its political effects and that was kind of counterproductive. That's where Kerouac and I diverged mainly. When I was in Chicago I was in charge of mantra chanting, whereas Jerry Rubin was more in charge of rabble-rousing, so there's a difference.

JJ: In Walter Lowenfels's book, *Where Is Vietnam? American Poets Respond*, however, many streams seem to have joined in the protest against the very existence of the Vietnam conflict, wouldn't you say so?

AG: Oh, sure. And, you know, Lowenfels was around at LeRoi Jones's. Lowenfels was at those parties, also. Yeah. And Langston Hughes.

JJ: This is a very surprising fact for a few reasons.

AG: I met Langston Hughes at LeRoi Jones's party one night when Ornette Coleman was playing music and everyone was dancing. That's the only time I met Langston Hughes. In '59 or '60. A great touching moment in history. When Black Mountain poets and painters, Beatniks, the Abstract Expressionists, the free-form jazz, the Harlem Renaissance, all met together in one room.

JJ: That's great.

AG: Isn't that marvelous?

JJ: I find it particularly marvelous, because I am personally interested in Jean Toomer's and Langston Hughes's work, you know. Have you read Langston Hughes since? Have you read more of him recently?

AG: No, not since that moment. I'm just beginning now. I've started reading a lot more now.

JJ: So you wouldn't have an assessment of him and the Harlem Renaissance writers?

AG: I knew him, Toomer, Claude McKay, James Weldon Johnson, from the Untermeyer anthologies,[193] mainly, and I didn't think much of them, because I didn't understand what they were saying. I never knew the great poem, that great poem—that National Anthem called "The Black National Anthem" was written by James Weldon Johnson.

JJ: I am sure many more Americans did not know that.

AG: And I didn't know that Johnson was the head of NAACP as field

---

[193]Louis Untermeyer, anthologist and literary critic, edited many books of poetry, such as *Treasury of Great Poems: An Inspiring Collection of the Best-Loved, Most Moving Verse in the English Language*.

director. Nor did I know that he worked with (W.E.B.) Du Bois. Nor did I know that he was an American consulate in Nicaragua when the marines arrived in 1912. Nor did I know that he wrote, "If you like-a me, like I like-a you, da da-da da-da da-da," etc.

JJ: With his brother, Rosamond.

AG: "One live as two, two live as one, under the bamboo tree." They didn't teach us that T. S. Eliot got his "Fragment of an Agon's"[194] "Under the bam, under the boo, under the bamboo tree" from the darkies.

JJ: Modernism was never associated with ethnicity.

AG: That was rarely taught in white scholarship about Eliot.

JJ: James Weldon Johnson was quite a versatile man.

AG: Great man!

JJ: Yes. Though he pursued many exciting modern ideas one feels that in his ways of practical writing he still was a man of the previous century.

AG: Well, you know, he was updated to the twentieth century in the work he did with his brother, Rosamond. The shows.

JJ: Of course, of course. But also collecting all those spirituals and then collecting all the black music of the past.

AG: Yeah. Didn't he write, didn't he put together the first anthology of black poetry?

JJ: Yes, that's right.

AG: Also, he wrote a very important book which I haven't read yet . . .

JJ: Which one is that?

AG: *Black Manhattan*.

JJ: Right, he did, later in his life. It was published around 1930, I believe.

AG: And in the history of blacks in the United States—one fact is that it was the blacks who settled Greenwich Village first. Amazingly. So

---

[194]From T. S. Eliot's *Sweeney Agonistes*, an unfinished verse play.

there's a very funny form from ancient days to now, there's a funny parallelism going on. It was basically black culture, African-American Bebop, hipness, marijuana, jazz and blues improvisation that "turned on" the Beat writers. Kerouac's *Mexico City Blues* poems and my "Howl" are white adaptations of jazz-blues saxophone's ecstatic improvised choruses. And Bob Dylan's stylistic guru for his songs is the great mysterious blues poet minstrel, the African-American lyricist-guitarist singer Robert Johnson—died 1938 age 26!

1990s

THOMAS GLADYSZ
June or July 1991,   Phone: San Francisco?—Boulder, Colorado?
Photo Metro, August 1991

Although Allen had been taking photographs since the late 1940s, it was not until the 1980s that his photographic work began to receive serious attention. In early 1991 Twelvetrees Press published a book of Allen's photographs taken over the years, simply titled *Photographs*. However, before the publication of that volume, Allen's photographs had been shown internationally. It was in preparation for an exhibit at the Robert Koch Gallery in San Francisco that art critic Gladysz interviewed Allen. The interview was transcribed by Gladysz.

—DC

Thomas Gladysz: In the postscript to your new book, *Allen Ginsberg: Photographs*, you talk about the sacramental nature of life, and offer this concept as an aesthetic for your photography. Could you elaborate?

Allen Ginsberg: I think the notion is a Native American art aesthetic and life aesthetic, but my formulation of it is reinforced by a lot of Buddhist training. The notion is basically that the first noble truth most all of us acknowledge, especially senior citizens, is that existence is transitory—life is transitory. We are born and we die. And so this is it! It gives life both a melancholy and a sweet and joyful flavor. And from that point of view, you could say this experience is sacred.

TG: Would you go as far as to say that photography is sacred?

AG: Well, I think any gesture we make consciously, be it artwork, a love affair, any food we cook, can be done with a kind of awareness of eternality, truthfulness. The poet Louis Zukofsky said: "Nothing is better for being eternal nor so white as the white that dies of a day." One appreciates the poignance of the contemporary. In portraiture, you have the fleeting moment to capture the image as it passes and before it dissolves. And in a way, that's special for photography. It captures the shadow of a moment, so to speak.

TG: You said that "the poignancy of a photograph comes from looking back to a fleeting moment in a floating world."

AG: I was putting it more succinctly. The notion of the idea of a floating world is Japanese. A world floating on clouds—the world as a cloud, in a sense, floating through time and then dissolving.

TG: How does the sacramental relate to another idea you talk about, the quality of "ordinary mind?"

AG: This life is unique, and every aspect is unique and never will be repeated. There's a kind of charm and magic to that, we might call it ordinary magic, as Tibetan Buddhists do. There's a realization of death and poignancy and transitoriness—and the idea that the highest consciousness is ordinary consciousness.

TG: Do you feel ordinary mind is what Robert Frank achieved in *The Americans?*

AG: I think so. What he was noticing out of the corner of his eye were things that people see every day but didn't want to notice, or didn't notice, or didn't think were glamorous. They become totemic moments: a politician on a stand pressing his lips to kiss a baby; a black man all dressed up holding his chin at a funeral by the Mississippi. An unnoticed corner of the world suddenly becomes noticed, and when you notice something clearly and see it vividly, it then becomes sacred.

TG: In your postscript you also said, "I notice many things and notice that I notice, and instantly or eventually I might make a picture of it." I wonder how forethought comes into your making pictures.

AG: I never know what my thought will be next. The only forethought I have is the awareness that any thought might be interesting and be a surprise. I take forethought to be alertness.

TG: Is it a matter of practice?

AG: Yes, of cultivating an awareness of the fact that we're seeing things all the time as we walk down the street. And then, at another level of awareness, we notice what we notice.

TG: In terms of composition, do you have an idea of what you're looking for?

AG: Robert Frank gave me some very good suggestions, and so did Berenice Abbott. Robert said, "If you take someone's photo, more or less close-up, always include the hands." I asked why. He said, "The face is naked and the hands are naked. It gives a more complete picture of the action or the whole attitude of the body. If you just take a picture from the chest up—no hands—you don't quite get the whole gesture"

I was in New York at an art gallery where Berenice Abbott was showing some of her older photos. I approached her, pointed my camera at her, and she said, "Oh, don't be a shutterbug!" Then she said, "Forgive me. If you're going to take my picture, back away a little. You don't want to get too close, otherwise my forehead will bulge or the cheek will bulge and it'll be all out of proportion. Give a little space around the subject, so you see where it is and what the context is."

TG: In your book Berenice Abbott is credited for her "off-hand direction."

AG: Occasionally I used to visit her up in Maine where she lives. A mutual friend—Hank O'Neil, who edited her last book—took me to her, and I learned something. I told Robert Frank and he said with a wry smile, "maybe I could show you something too."

She was quite old, but quite sharp and alert. She knew people I admired, like Hart Crane, the poet, and Marsden Hartley, the painter/poet. And she knew William Carlos Williams in the 20's. From her I heard a lot of gossip about people I had read about. It was a pleasure to connect with that lineage and have that sense of old bohemia. It was good to see someone who had survived as an individual with her particularity of gender. Berenice herself liked ladies and, since I'm gay, it was nice to reinforce the fact that you could live a full life.

Also, what was quite interesting was her devotion to her elders—her lineage. She had rescued the glass plates of (Eugène) Atget and kept them for many years—all those delicate glass plates she brought from Paris. And of her admiration for younger generations she said, "I love that Robert Frank. I've never met him, but of all the younger photographers, he's just marvelous!" She had enthusiasm for a younger person and respect for an older person. That seems like the real missing link in American society.

TG: How did you come to meet Robert Frank?

AG: Through Kerouac, by accident back in the 50's. I think Robert knew Kerouac's work, or they knew each other and became friends. Kerouac trusted Robert, plain, glum, Swiss—this fellow with no heavy pretensions, very ordinary in a sense, not vain.

Kerouac was asked to write a preface to the American edition of Robert Frank's book. I think Walker Evans had written the preface to the French edition. Robert had the choice between a younger American writer and an older, established photographer and I think he wanted to gamble on the great reality of the present. He asked if Kerouac would please do it—and Kerouac did. They took a trip together. Robert drove Kerouac and his mother to Florida; that was a specialty of Robert's—on the road in America—as it was Kerouac's. Robert took a lot of photographs, many of which have never been seen. Their affection cemented because Robert was nice to Kerouac's mother—she was a very judgmental woman who didn't like many of Kerouac's friends.

TG: You've called Robert Frank your "kindly teacher."

AG: Well, I didn't really appreciate him as a photographer. I thought, he's just a photographer, I'm a poet. I had been taking pictures all along, but I hadn't thought of myself as a photographer. I hadn't been very curious about it. As the years went by, I began to see the value of his attitude. There's a kind of glum affection in him that's very nice. He's always very steady. He always said that he thought I was doing a good thing and admired the way I was consistently being a poet, and going about my world and not getting knocked out by alcohol or drugs or hysteria—just sort of plodding along doing my poetry. He liked that. He was plodding along doing his photography, encountering all sorts of personal tragedies, but continuing. We got to be quite affectionately friendly, family friends.

Maybe around 1984, I realized that I had accumulated a lot of photographs over the years. All I had were drugstore prints, so I asked Robert how to go about getting better and bigger prints. He introduced me to his printmakers, Brian Graham and Sid Kaplan, who made what you might call in high falutin' terms "museum quality prints." As it developed, I began taking pictures to Robert, asking, "Do you think it's any good, should I crop this?" He gave me a little advice. I began depending on him more and began to worship him as a photographer when I realized how much he knew. As it turns out he knows a lot. When he was 16, he was apprenticed to an industrial photographer in Switzerland: he had to learn about chemicals, how to light up a huge industrial factory two blocks long. He knows a lot, and so does his friend, the printmaker and photographer, Sid Kaplan. He's quite an expert. Sid works at the School for Visual Arts in New York and lives across the street from me. I go into his darkroom and ask him about things, like burning a detail in.

Another thing—Robert said that photography was an art for lazy people. If you're famous, you can get away with anything! William Burroughs spent the last ten years painting, and makes a lot more money out of his painting than he does out of his previous writing. If you establish yourself in one field, it's possible that people then take you seriously in another. Maybe too seriously. I know lots of great photographers who are a lot better than me, who don't have a big, pretty coffee table book like I have. I'm lucky.

TG: Has Elsa Dorfman been an influence?

AG: Very much so. I've known her since 1960. She used to work as an editor at Grove Press. She was into poetry and helped organize readings. Then she took up the camera. She got more and more into photography and was taking our pictures whenever we would come to visit. It was like family snapshots. An object lesson in ordinary mind! Sitting at the kitchen table, talking, reading the *Times* in the morning, eating a bowl of soup or sitting around on her couch, she'd just snap casually. And I liked that, it was what I was doing all along. She did a very interesting book called *Elsa's Housebook*.

TG: You and Gregory Corso once met Edward Weston. How did that come about?

AG: It was by accident. I knew his work, but not very well. When I was younger I had seen it in artbooks and at the Museum of Modern Art and at the Metropolitan Museum of Art. Gregory and I started going down to Los Angeles from San Francisco around 1956, at the height of the San Francisco Poetry Renaissance. In those days, Henry Miller was living in Big Sur. We decided that on our way to Los Angeles, we'd hitchhike to Big Sur and see if we could find Henry Miller. But we couldn't get a ride out of Carmel and were stuck on the coast highway.

We were walking and walking, and then we passed a big wooden sign saying "Edward Weston."

We decided that since we couldn't get a ride to Henry Miller, we'd go see Edward Weston. We knocked on the door of his beautifully constructed log house and this little old man came out. He had the beginnings of Parkinson's disease, so he was trembling. He was in his bathrobe, well groomed, small and bent over like a magical gnome or a dwarf in a fairy tale. He invited us to come in and gave us some tea. We asked, "Do you have any photographs we could look at?" He was very accommodating. He had this beautiful set of cabinets along the wall which were drawers, and it was there he had these favored prints. He pulled a few out. There were some photos from Point Lobos, the dead bird on the rock, and some shapes of tidepools. He was old and obviously fragile—that dead bird was almost an ironic comment. It was very dramatic in a sense. We spent about an hour and a half, kept him company, his son came in and said hello. Then he showed us to the door when it was time to go. We went down the steps. He waved to us first and then said, "Don't forget, I was once a young bohemian like you, too."

TG: When did you start taking photographs?

AG: I think I had a box camera when I was a kid. I took photos of my mother in the mental hospital at Greystone when I was 15 or 16. I had a whole series of photos I began taking at Columbia University in 1946 or 1947 of William Burroughs, Kerouac and some friends. I still have the drugstore prints of those, and some are quite well known. Some were in *Scenes Along the Road*.

TG: What kind of camera have you used?

AG: I bought a second-hand Kodak Retina in a Third Avenue pawnshop before I left for Europe, around 1953. I kept using that until the 60's. Then, as I left India, I bought a funny camera which has two pictures to one frame—a Ricoh, I think. I have some good pictures of Neal Cassady and Timothy Leary. And then I lost it, or left it at somebody's house. In the early 70's I bought an Olympus XA, a small camera you can fit in your breast pocket. I'm still using that quite a bit. It's an obsolete model—though well made, metallic I think.

In Poland, about five years ago, I bought an old C3 Leica. When I brought it home, Robert Frank said it was the same kind of Leica he used for *The Americans*. I look a lot of photos with that until I lost it in a cab. And then after visiting Berenice Abbott, I realized she had these panoramic views with minute details, like Brueghel. To get that clarity of detail and panoramic awareness of space you have to have a larger format camera. She had a camera called the Century Universal View Camera—which is a very romantic name for a camera. I decided I should get a bigger format, so I got a second hand Rolleiflex.

TG: Do you always carry your camera with you?

AG: I always carry the Olympus, like I always carry a notebook. I carry a notebook for writing little things, haikus, descriptions. And generally I carry my Olympus and one extra role of film. I don't tote around the Rolleiflex unless I'm really intending to take a picture.

TG: What is the state of your photo archives?

AG: It's a full-time job for somebody. My salary as a Distinguished Professor of English at Brooklyn College goes into maintaining an office and secretaries. The secretarial and print-making costs, and the archiving, updating, shipping and mailing costs, take up as much money as I get in from the prints. I was running pretty much in the red for many years but I may just be out of it this year because of the book. It's actually a very expensive hobby. But that's what's nice about it, because I'm not making any money, it is strictly an amateur hobby. It's just fun. Especially with the older photos, it's like having a telescope into time to see moments or instances that now seem sacred and glamorous, more tragic, poignant certainly, because the moment has passed. They are really very precious.

TG: What led you to write on your photographs?

AG: Well, for one thing, they all had a story, especially the old ones. A lot of them were taken before anybody was famous. They're sort of like funny, family photos. But what really turned me on to writing on the photos were the comments Hank O'Neil recorded by Berenice Abbott for the book they put together around 1984 or 1985. He went over the photos with her. Her comments give you information about the circumstance of taking the photo, or the character of the person photographed, or the situation. I always liked that. It seemed her little one-paragraph comments provided a model for me. I was already writing a little bit—but she had a good solid paragraph that was tape recorded. I'm a writer, so I could write it. Then I decided I could write on the margin at the bottom of the photograph instead of just signing it. I just improvised as I went along.

TG: You write different text on the same photo?

AG: I start with one line. Every time I write a new caption, I would write it more extended and with more information. Occasionally, I just repeat the last line. I have them all on a word processor.

TG: Lastly, do you feel photography has influenced your poetry?

AG: Well, to go back in time, there were the Imagist and Objectivist movements. They specialized in brief poems, flashes of the moment, visual poems like "So much depends / upon / a red wheel / barrow / glazed with rain / water / beside the white / chickens." They're little descriptive photographic worlds, almost cinematic in nature. William Carlos Williams was a friend of the photographer/painter Charles Sheeler and with (Alfred) Stieglitz. So there has always been some correlation between photographers and poets since the 1920's.

It's evolved to the point where Robert Frank and I taught a course together in Israel,[195] at the Camera Obscura School of Photography, called "Photographic Poetics, or the Poetics of Photography." It's basically the notion of the sacred moment or sacred thought or sacred idea or sacred perception. I don't know if Robert would use the word "sacred," but certainly that's what he did.

There is a great element of chance in Robert's photography, shooting from the hip. At one time, I think he even experimented with throwing the camera up in the air with a delaying click to see what would come out. Not setting things up, but accepting what passed before his eyes—that feeds into the notion of the spontaneous writing of Kerouac. So there are a lot of parallels. First thought/best thought or first glimpse/best glimpse, unpremeditated

[195]Ginsberg went to Israel in January 1988. While there he did readings with Natan Zach at Tel Aviv and Haifa universities and Jerusalem Cinematheque; met with Palestinian moderates Mubarak Awad and Hanna Senoria; addressed the Peace Now rally, attended by sixty thousand; taught "Photographic Poetics" at the Camera Obscura School with Robert Frank; and organized the PEN American Center protest of Israeli censorship of minority Palestinian literature.

awareness. It's like taking little flashes in your notebook, little flashes of thought. That aspect of chance Robert introduced into photography to some extent. The idea of ordinary chance or ordinary magic is the same as bohemian, Beat, Buddhist poetics. There are many parallels.

Frakes had first met Allen while a student at Boulder around 1987. Later the two became friends and would sometimes drive around Boulder together. The interview took place on the Naropa Institute campus in an adjunct building, a renovated old house on Arapahoe Avenue, in a second-floor room used for meditation instruction. In 1999, Frakes remembered that "it was an incredibly bright, sunny Boulder day. We meditated first for several minutes, sitting in chairs. Allen was quite animated and vibrant that afternoon."

—DC

Clint Frakes: You returned to Prague (in 1990) after 25 years for a celebration of the anniversary of your King of May event. You mentioned that you spoke with some of the students there and that they'd declared a strike. What were your perceptions of the changes in Czechoslovakia after a quarter century?

Allen Ginsberg: They never had another election for May King after that year, 1965. The Mayor of Prague, Jaroslav Kozan, was the anthologist of a book of American poetry and a translator of Gary Snyder and Gregory Corso. I'd met him in New York a while earlier and told him I'd like to go, and he said by all means to look him up. I went with Anne Waldman, Nanao Sakaki came from Japan with some musicians, and Andy Clausen came from the Himalayas through Hungary and Rumania, and we all converged. I had been in touch with the rock band, the Plastic People of the Universe.

In *Musician* magazine, October 1990, there's an interview with President Václav Havel and Lou Reed—really interesting. Havel traces the Czech revolution to rock and roll and delineates the influence of American counterculture. He felt it was really important to tell the Americans that in Czechoslovakia Communism was overthrown not by military but by cultural revolution. In this case the cultural revolution was, specifically, Kerouac, myself, the Velvet Underground, Warhol, Ed Sanders and the Fugs, psychedelic posters and student movements of '68 when he was here, and Frank Zappa and the Mothers of Invention, Dylan, and the Beatles. He traces it to this rock band, the Plastic People, in Prague and their arrest for public and private performances; their trial; his attendance at the trial; his protest at the unfairness of the trial and the intellectual trickery, double-crossing, and intimidation; his circulation of a petition called Charter 77; his persuading all the Czech intellectuals to support a rock and roll band—which was a hard job—their support; the arrest of the people who signed Charter 77; a formation of an underground in the Magic Theater and the creation of the student movement; the shooting of a student; mass movement against the government; the overthrow of the government; and his election to the presidency—all this in a straight line, from rock and roll to closing the offices of the secret police, in one interview.[196]

Havel told me—I didn't remember—but he'd hung out with me as a student in '65, so he asked me to get up and make a speech on May Day 1990 on national television on the main square, Moustique Place. Then I went over to the other, old square, Stare Maesto, where the students were having the May King and May Queen festival and Kozan introduced me as "the longest reigning King of May in history" and gave me my crown, a little paper crown. I put it on, read a poem with a little Buddhist content called "The Return of the King of May,"[197] and passed the crown on to the student who'd been elected this year to take my place. The whole thing ended like a big, nice circle. It began and ended in poetry.

CF: It's encouraging that though we often feel our statements fall on deaf ears in terms of structuring our government and policy, our relative freedom of expression trickles into other countries which are ripe for it.

AG: They don't fall on deaf ears—it's deaf media. The media blockade it. Around 1972 I stopped being invited on national television. I was, all

---

[196]Most of the information that Ginsberg relates here is not found in the Havel-Reed interview in *Musician* magazine. As Allen had only recently returned from visiting President Vacláv Havel in the Czech Republic when he gave the interview to Frakes, he presumably was conflating what he had heard while in the Czech Republic with what he had read in *Musician*.

[197]See "Return of Kral Majales," *Cosmopolitan Greetings*.

through the sixties, in and out. In '72, after (Vice President Spiro) Agnew denounced the media as "nattering nabobs of negativism"—I think that phrase was written by Pat Buchanan or William Safire—Agnew threatened to take over the media somehow or another. The media withdrew and got timid and no longer questioned the government, but became an ally of the government. There was a counterreaction to that in Watergate. That was so violent, the elimination of the president, that the media were afraid to do it again, like this new revelation that Bush and his people were involved with holding back the hostage release—"October Surprise"[198]—or even the latest, that the whole CIA and even Bush must have known about, Iran-scam. The media are letting things go by now that at one time would have been considered scandalous.

CF: Do you think there was an official declaration not to let you on?

AG: No—there is probably a blacklist, though unofficial, or a verbal agreement as to what kind of people should be heard from. There's a group called FAIR: Fairness and Accuracy In Reporting, and I am on their advisory board. It's a monthly magazine which examines the media—points out that, say in Ted Koppel's program (*Nightline*), the spread of experts is something like 98% male, 85% right wing from *New Republic* up to Pat Buchanan, 1% as far left as Buchanan is right. The people called in most often are Kissinger and others that represent the government. When these statistics were presented to Koppel and the producers, the reply was, "Well, we're mainstream and we're supposed to represent the 'newsmakers'—so we can't really represent dissent." So they have a real skewed nonrepresentative shot going there. So people like Noam Chomsky or representatives that would speak for the PLO (Palestinian Liberation Organization) or legalization of drugs or about government involvement in heroin trade and cocaine traffic, and CIA corruption are not considered legitimate for public discourse.

CF: That original May King incident, wasn't it viewed in Czechoslovakia as a small disaster?

---

[198]Allegations were made that Ronald Reagan's cohorts had arranged with the Iranian government to hold back the release of hostages taken in the U.S. embassy until after Reagan's election to the presidency, lest there be an October hostage release prior to the presidential election, possibly tipping the scales in incumbent Jimmy Carter's favor.

AG: No. That was only the view of Barry Miles, my biographer—he was Marxist oriented. He thought it was a disaster. And Gregory (Corso) had misunderstood and thought I had done something funny. Not only that—there was hardly any notice of it in the American press, so the story was never really told except there was a relatively intelligent shine in the *New York Times* magazine several months later by Richard Kostelanetz. There was finally some clarity. But Miles thought I was misbehaving even in Cuba by challenging the Communist authority.

First of all, I couldn't get a visa to go to Cuba. I had to sue the state department for it, and I won. Because of that, they put me on a Dangerous Security List on April 26, 1965. I'd gone to Cuba in February of that year, stayed a month, and was ready to leave a few days later with Nicanor Parra and people who were, with me, judges of this Casa de las Américas Inter-American Poetry Contest. While I was there I criticized Castro's gay policy in private and the monolithic press policy, and his antigrass policy, talking freely as if in America or anywhere else. This was considered bad, or something—I don't know what. Also while I was there, Nicanor Parra and I were in touch with a bunch of younger poets who liked the Beatles and who were disapproved of and were constantly being taken into jail and arrested by the police. And once when Parra and I and the rest of the judges went to a concert, these young kids came out to meet us by appointment after the concert and were intercepted by the police. My best friend there, Manuel Ballagas, was the son of a poet famous earlier in the century. After I left Cuba, we had some correspondence. He was raided by the police in '72. The writings in his desk were seized, and he was put on trial and one of the accusations was "passing information to the American provocateur, Allen Ginsberg"—an official accusation.

I'd complained about the gay thing because Castro had given an anti-gay speech in public at a cinema or theater school, which was a center of gay life, broke up the school, and sent them all to work camps in '65 while I was there. The Marxist-oriented people said "Oh, you shouldn't be complaining—look at the advances the revolution has made." This was true and I said, yes there have been certain advances here, and I'm on your side and that's why I'm complaining—don't fuck up your revolution. I even said that I have access to some communication with the Beatles; and what you really should do is have the Beatles come to Cuba and have a big concert. I was disdainfully turned down

by the minister of culture who said, "The Beatles have no ideology."
So it was a classic Stalinist shot. I was complaining in private, not in
public and, naturally, the police hear everything. Finally I was
arrested and held incommunicado and kicked out—and Miles thinks
it was my fault: "why didn't I just shut up?"

Same thing in Prague. It was even worse in Prague. The only
reason they got upset with me—I was in Prague for a month, went to
Moscow for a month, trained then to Poland for a month, and went
to Prague to leave for New York. I got back to Prague on April 26—
the same day I was put on the FBI Dangerous Security List—was
elected King of May on May first, was followed around Prague until
May 7, arrested, kept incommunicado, and put on the next plane to
London because the minister of culture and the minister of infor-
mation disapproved of an American gay beatnik, pot-smoking,
mantra-chanting Buddhist (or something) being a model for
Czechoslovakian youths. Because they wanted an ideology which
they wound up enforcing by 1970 with total Stalinist methods, so
that even Sartre, who was Marxist, said that Czechoslovakia had
become a "spiritual Biafra." I wasn't out to make trouble, I was just
acting normally.

CF: Didn't your recent trip to Korea create similar troubles?[199]

AG: They invited me to an international poetry conference. I checked
it out and found that it was official, a government conference, so I
faxed them that I'd be happy to come as long as they understood that I
was coming as a representative of the American PEN (Poets, Essayists,
and Novelists) Chapter Freedom To Write Committee which has criti-
cized the fact that Korean writers were in prison and that I was coming
with a list of Korean writers in prison and would feel free to talk about
it. If they had any objection to this, I should not come and I wouldn't
come. But as long as I was able to be free about what I was talking about,
I had no objection to working with the government. As soon as I got
there they asked me to shut up. They had wired me before that, saying I
would have total freedom of speech. The first half hour I was there,
when I was getting my hotel room, passes, etcetera, they said, "We
hope that while you're giving your poetry reading, you won't do any
criticizing." I said, "Well, I think I wired you that I would." They said,

"Well, we know, but we thought you would have better manners." I just exploded and said, "Listen, fuck you! I'm here, and I'm on my own, and I warned you."

The South Korean constitution says that anybody who has dialogue with North Korea has committed a crime that can be punished by life imprisonment because the South Korean government wants to do all the negotiating with the North Koreans—they don't want any individuals going there. There are a number of writers including one Reverend Moon Ik Qwan, a poet, who went to North Korea and shook hands with Kim Il Sung and came back and said North Korea is a police state but we have to figure out some way of communicating. So he was put in jail. And some leader of the student movement went up there, and she was put in jail. And there are a number of writers in and out of prison constantly—that's why they're having all the student riots—beatings of students, killing of students. And the students there are naïve—they think maybe there is a paradise up North, but it's the worst police state of all. I've met a lot of North Koreans in China who were totally paranoid, wouldn't talk to anybody, sat by themselves, had parties by themselves singing patriotic songs—they were completely brainwashed and scared. Even the Chinese thought they were totally "out of it" and said, "We went through this in the Cultural Revolution under Mao Tse-tung, and they're going through it again, but they can't see the difference. They're nuts."

So I began talking at press conferences about the specific list of prisoners in South Korean jails, and they got upset and they called me in again and said "please don't do it at your reading—your reading is supposed to be poetry, not politics." I said, "Well, if I do it in poetry do you mind?" They said, "If you do it in poetry, then ok." So I got up and began improvising like in my "CIA Dope Calypso"—a blues naming names and the name of the article of the law and things like that which got them upset again, but it was all poetry.

Then I went off and had a poetry reading with a dissident group who were the real poets. One of them had just got out of jail that day. It was organized by Reverend Moon Ik Qwan's son, who was an opera director in Europe for many years. He was a very sophisticated European kid. So there was this little poetry reading and the government hosts said, "We brought you here. You're not supposed to do that." I said, "Yes, but we're finished with our conference and I'm staying on another two weeks, and I'm free to do what I want now." It turned out that my translator was a member of the Korean CIA. They didn't bust me, they couldn't: I was announced in advance and was official. So I was behaving in South Korea more or less as I behaved in Prague or Cuba. I was a little more circumspect in Prague. I realized that a police state will allow just so much before you really get out of hand.

Then, when I got back to the U.S. in 1965, I was strip-searched at customs. I saw this little printout on the desk that said, "Ginsberg, Irwin, Allen—Orlovsky, Peter A.—these persons are suspected of international narcotics smuggling." This was a result of my winning the case to get a Cuba visa.

Later through the Freedom of Information Act I got all my papers. The funniest part was that the FBI translated a denunciation of me in *Mlada Fronta*, a Prague youth newspaper, for the Treasury Narcotics Bureau. On May 11th I'd gone to my local New Jersey congressman, Jo Elson, to complain about the Narcs trying to set me up illegally for a bust. In the answer to my congressman, they translated this denunciation of me in Prague as a drug addict and an alcoholic and not trustworthy, so that they wrote my congressman saying, "This person has this terrible reputation and you should not cooperate with him, he'll probably use anything you say against you publicly, you might as well ignore him." I realized that the Communist police and the FBI and Narc Bureau were just one big police bureaucracy that needed each other.

Meanwhile, Miles had a critical attitude, saying that I may have contributed to the conflict by a confrontational vanity or pride. On the other hand, I'm really happy I did it by hindsight. When I went back I found that a lot of the students thought I'd done the right thing and said that both my poetry and my actions there had been an inspiration—a good thing I opened my mouth. It became legendary and mythic that the King of May had been kicked out after being elected popularly and had said avant-garde things. Also everybody had my poems available, published in magazines, as well as a lot of interesting literary newspaper interviews I did while I was there on drugs, sex, change of consciousness, bureaucracy, what's wrong with capitalism—stuff like that.

CF: You've maintained a policy in these travels to try and speak "officially" with the governments . . .

AG: Not officially, but frankly with anybody I meet. If I do newspaper interviews or do conferences, best to say what's on my mind and be clear about it and find some way of getting under the skin of any problem I seem to have good intuition about. Like in 1985, I went with Arthur Miller and a group of American writers to a conference in Lithuania, representing the American Academy of Arts and Letters meeting the Soviet Writers Union. The official trickery was to hold it in Vilnius, Lithuania, but not to tell any of the local poets in Vilnius that we were there; it was a closed conference. The head of it was the ex-ambassador to China, Federenko—ex-ambassador to the UN, also—a big cultivated man, but a party-liner. We were each asked to give a twenty-minute lecture on where we were at intellectually—"Sources of Inspiration."

I began with Whitman, which they all applauded, then led on to Whitman as an individualist, which they sort of liked, as against materialist conformist capitalism, and Whitman as Gay—which they froze up on—then liberation movements of the sixties—they're still clapping—Spiritual Liberation—still clapping—Black Liberation and Women's Liberation—still clapping—then Gay Liberation—silence—and Psychedelic Liberation—nothing—then all of these as part of the anti-Vietnam war movement—they clap again. I then went to "Freedom of Speech" and on to article 92 of the Soviet Constitution which said that (1) any discussion of the territorial boundaries of the Soviet Union was considered against the law in a treasonable way, and (2) any discussion of socialist basis of the state was also considered out of bounds. This was a hot subject in Lithuania, but this was a closed conference and was not being broadcast out. They wouldn't even tell my Lithuanian translator what hotel I was in or that I was in town. He turned out to be a Mongolian Vajrayana student and showed me his secret shrine room with puppets and dolls and mandalas and everything classic. So, the points I picked on were Article 92, i.e., socialist basis of the state (which is nowadays a wide open discussion), and the territorial boundaries of the union which is also wide open for discussion now. I was pointing out that those two points made it impossible to have any reasonable public discourse and restrained any sense of liberty of expression and perhaps they should change their constitution. That brought down a storm of criticism from the Russian officials like Federenko who said, "Mr. Ginsberg has taken advantage of us and knows very well how to divert the discussion from literary matters." The guy from Kazakhstan said, "I represent the 200,000 minority peoples of Kazakhstan and they sent me here as a representative of the Soviet Union, and if they knew there was a discussion of homosexuality, they would not want to pay me, and they would laugh at me." I said, "I represent the 250,000 minority peoples of San Francisco who are gay and we want a nonaggression pact with your 200,000 people of Kazakhstan"—this was on the bus, nonofficial, so there was no problem.

Miller and I brought with us several case studies of writers in prison, including one Irina Ratushenskaya, and we presented that publicly. They didn't want to hear about it, but privately, Federenko asked us to slip the information under his door. She got out about a month after we left, so we actually delivered someone. (Yevtuchenko also'd intervened.)

CF: Since your mother and grandparents are from Russia, there is a special interest in participating in Soviet affairs . . .

AG: Yes. I kept telling them I was a Russian poet writing in a different language.

So it's Miles that gave you that odd idea. Miles makes several points with which I differ. He didn't like Trungpa. He doesn't really approve of meditation or Tibetan Buddhism because he sees it as Buddhist fascism or hierarchy. He doesn't seem to like Kerouac in this context, and you'll notice Kerouac portrayed as sort of a nuisance throughout the book. Same with Gregory Corso. He does approve of Burroughs and he does have affection for Peter (Orlovsky). But you know, most of my artistic practice is Kerouac-derived. It's not as good as Kerouac, even, but it's certainly inspired by Kerouac, and Miles doesn't give allowance for that affection.

CF: You felt it important to give Miles free reign on the biography, though.

AG: Yes, I don't want to be like a Soviet censor. He's an old friend, and I knew he had his prejudices. I was a little apprehensive, but I didn't want to interfere because my agent had asked for 10% of his advance to apply to somebody to put my archives in order, to index the archives for him to use, which Miles didn't feel was necessary. He felt he could do that himself.

I had quite a bit of preliminary argument correspondence with him that strained our relationship, particularly over Trungpa. Miles said that when I came to Naropa, my poetic production decreased and became enfeebled and that I wrote nothing better until I left Naropa and I regained my powers. I said, "Well, wait a minute. What do you think of 'Father Death Blues?'" He said, "That's one of your great poems." I said, "What do you think of 'Plutonian Ode?'" He said, "That's a major poem too." I said, "What do you think of 'White Shroud?'" He said, "That's also a major poem." I told him all of those were written during my time of most intense or year-round retreat at Naropa: 1975–83. He said, "Ok, I'll take that sentence out." I noticed that his cast of mind was thus biased, but there was nothing I could do about it except correct him as much as I could and try to persuade him.

The big argument was over the assertion that a million Tibetans had been killed during the Cultural Revolution. He said there's only 1,200,000 Tibetans altogether, so how could you do that? I had understood that there were 6 or 7 million Tibetans. There was this real confusion which I only resolved later on. There are only about a million Tibetans in Inner Tibet, which is considered somewhat autonomous, or should be. But there are 6 million Tibetans scattered through Mongolia, China, and Outer Tibet, including Trungpa's region near Szechwan. The Tibetan political office, which is chauvinistic, claims that whole area as Tibet. The Chinese resent that and might be willing to compromise on Inner Tibet. There are still another 5 million Tibetan speakers scattered around and a million Tibetans *were* killed. The confusion as to what the population of Tibet is was the source of argument, and I ignorantly insisted it was 6 million Tibetans without understanding the geography, and he ignorantly insisted there was only a million Tibetans without thinking of the geography.

CF: He seemed to focus a lot on Kerouac's drunken times, maybe after 1960, but a lot of people these days tend to focus on that side of him to discredit him.

AG: It's a big mistake, and it's part of the Reagan-Bush era. You've got to remember, after his alcoholism, he wrote the classic *Big Sur* on the subject of alcoholism; the classic *Desolation Angels* on the whole San Francisco era; his perhaps most self-critical book, *Vanity of Duluoz*; an enormous amount of beautiful journalism; half the bulk of his published but uncollected work I have in the library here; *Pic*; and innumerable poems and letters that were quite clear. So I would say his production wasn't as great as in the period of '51–'55, but it was as great as any writer writing in America at the time. How many people wrote three or four major books in a period of five years?

CF: It seems, hearing from people that were a little more objective toward him at the time, that he was, though seemingly a right-wing drunk, really as tender and lucid as ever.

AG: Kerouac got his reputation mostly for being a right-winger by insulting the left in two different ways: first, at Harvard, saying, "I'm not afraid of Mao Tse-tung." That was when the left intellectuals at Harvard were making a hero out of Mao Tse-tung when Mao was universally feared or hated in China. Later, Mao's totemic, mythic character was renounced by everybody—left and right (except the Revolutionary Workers Party splinter group). He also said that psychedelics were creating a race of cretins who couldn't sign checks or add up their own checkbooks, (laughs) which was a witty remark and made sense, which, now, everybody also agrees with. So he was just being witty, actually. But he was desecrating some false idols prematurely. He also said in 1968 that Jews on the left like Ginsberg, Abbie Hoffman, and Jerry Rubin were trying to "find new reasons for spitefulness." He was one hundred percent right about Jerry, about ten percent right about me, and thirty percent right about Abbie. But what "new reasons for spitefulness?" Think of "Kill your parents." He was really on the nose in his critique of the left. I think the reason the left failed in the sixties was because of its aggression and ignorance and the theory that anger was necessary for social revolution, and that extremism and radicalism like the Weathermen 1968–69 who would ignite a prairie fire of physical revolution. They were lunatics, yet they were being taken seriously by a lot of the left, and it was Kerouac who was saying, "These people are looking for new reasons for spitefulness." So, he was prophetically accurate, drunk as he was. But he was resented by the left and liberals, unreasonably, who were still intoxicated by the rhetoric of anger.

I think the reason that the left blew it was because of such slogans as "Kill your parents," "Don't trust anyone over thirty," "Prairie Fire," "Raising the banner of Mao Tse-tung," "Raise the banner of Castro," when all those ideas and people were already loathed in their own countries, and the CIA and anybody really experienced knew it and realized the left in America was a paper tiger that didn't know where it was at and trusting them in power would have been as bad as trusting Stalin in power, perhaps. So Kerouac was putting his foot in his mouth according to some leftish liberals, but actually, he was putting his finger exactly on what was wrong, and anybody that had read Dostoyevsky's *A Raw Youth* or *The Possessed* would know about that. Kerouac knew about it, I knew about it. Miles did not, at the time, and carried this mythic idea of Kerouac as a reactionary.

CF: Kerouac documented a historic split in *Desolation Angels* where you, Corso, and he were at a dinner table with other people and you were talking of "raising the lamb" and taking this Beat Generation thing out and changing America, and Kerouac reflects on that point as a critical departure where he goes back to Mom, and you guys take it to Moscow or wherever. With all these years of retrospect now, and given his untimely, drunken, yet accurate view of things, how do you perceive that point in your respective lives?

AG: Actually it's not that one critical point, because you can take it back all the way to the day I first met him, or two days later. It's always been the same ever since we met. When I said I wanted to be a labor lawyer, he said, "You never worked in a factory. What do you know about labor or law? They're all Mafia in New Jersey anyway." Here's this wet-behind-the-ears 17-year-old kid thinks he wants to be a labor lawyer—"Better go and be a poet—you're too sensitive." (*laughs*) We had different lives.

Burroughs has a totally different view: "the earth is doomed, why help the stupid Homo-Sap?" or, you know, "why not acknowledge Homo-Sap has brought on his own demise?" compared to Kerouac's "save the lamb" or my "free the junkies" or something. (*laughs*) Everybody's got their own style, but there's a substrate of agreement on widening the area of consciousness, new vision, religious consciousness. Buddhism, existence as suffering—that's the base, everybody agrees. What to do about it—whether you go out marching in the street, which Kerouac didn't want to do—but one problem was that since he was so dependent on his mother and drinking, and his mother was so much against his friends and anti-Jewish, he wasn't allowed to bring me home. After a while Burroughs wasn't allowed, Lucien Carr, Neal Cassady, his girlfriends, anybody. So that immediately limited him socially, so he couldn't do anything out front in public that his mother would disapprove of, exactly. Or even write anything that his mother might read, like getting his cock sucked by me or something like that, so he wouldn't put that in. So that limited what he could say, though in private, he was much more open.

CF: It seems like it was at that juncture—it might have been around '57 or so—that he was talking about, when you as a group began to recognize this momentum had occurred and were thinking of great things to do with that.

AG: I would say a real breach came in '58 or '59 when he went on television with Ben Hecht, and Hecht asked him about politics, about John Foster Dulles and Eisenhower. Kerouac said, "Our country is in good hands." They were just preparing the Vietnam war, overthrowing Guatemala—preparing for the catastrophes of the sixties, seventies, eighties, and Dulles refused to shake hands with Chou En-lai, a good guy, at the Geneva conference. They refused to sign the Geneva Agreement—Eisenhower and Dulles did—and they were cooking up horrors. And Kerouac said we were in good hands, and I didn't feel that. And I thought he was kind of betraying me and really got mad.

On the one hand, Kerouac believed in tending to your own personal matters and not getting involved with affairs of the state—there was a certain discretion about that that I liked too. It kept him out of the public eye and writing, preventing him from becoming a loudmouth. But I was mad at him because I thought he was being too specifically approving of the government. On the other hand, probably, if I had my druthers, I would have said something nice about Mao Tse-tung, so I'd have been wrong too. And as it happens, years later, I was marching around the White House denouncing the Shah of Iran, not realizing it would bring on a catastrophe greater than the Shah.

But this division into public and private, I think he disapproved of (Gary) Snyder, even his politics. I had to swear to Kerouac, I'd never approve of violence in any of these public demonstrations. Actually his big question was his mother's—"If they have a revolution, will they take away my house?" I said no, but actually the Maoists would have and the Stalinists would have, and I wasn't quite sharp enough on that. Burroughs always was; he was antibureaucratic from the 1930s on. I was kind of wishy-washy, I didn't have any particular ideal. My mother was Marxist, but I was anti-Communist because I was anti-ideological, so when I hit the Communist countries, I immediately reacted subversively. Probably Kerouac would have disapproved of that too like Gregory (Corso) did—"Why are you getting into trouble abroad instead of just being a poet?" But Gregory changed his view on the May King incident after he realized what had really happened.

STEVE SILBERMAN
December 16, 1996,   San Francisco

www.HotWired.com

Steve Silberman was turned on to Allen Ginsberg when he read the *Playboy* (see Paul Carroll) interview at age eleven. Later he read the *Gay Sunshine* (see Allen Young) interview, discovering "a huge permission to be sexual and emotional in whatever ways arose naturally in my heart."

He saw Ginsberg read poetry for the first time at Queens College in 1976, and "made a vow to be wherever Allen was going to be that summer, helping him out in whatever ways he needed." Then eighteen, Silberman sold a camera that his grandfather had given him and used the money to travel to the Naropa Institute in Boulder, Colorado. There he became one of Allen's secretaries at the Kerouac School at Naropa Institute, typing transcripts of journals and taking Allen's classes in the history of the Beat Generation. He later became one of the poet's teaching assistants.

Silberman, who had interviewed Allen for the *Whole Earth Review* in 1987, wanted to do an interview with him on the Web because, as he explained in 1999, "I saw online publishing as an extension of the samizdat, self-published, uncensored media lineage that Allen always championed. I wanted to bring Allen's voice and words into this spunky, insurgent new medium."

The HotWired interview was not Allen's first appearance on the site. His collaboration with painter Francesco Clemente, *Pastel Sentences*, was also published there, allowing net surfers to view Clemente's images while they heard Allen read the poems that had been inspired by those images in streaming audio. The talk with Allen was part of an ambitious series of interviews produced by HotWired editor John Alderman, which featured cultural notables Brian Eno, Laurie Anderson, Yoko Ono, Grateful Dead drummer Mickey Hart, DJ Spooky, and members of the New Wave band Talking Heads.

"Escorting Allen around the HotWired office that afternoon," Silberman recalled, "I felt like I was hosting a visionary dignitary from the past in a visit to the laboratory of the future."

The interview took place live in streaming audio from the HotWired studio on Third Street in San Francisco, which was used for live broadcasts to the Net. Silberman transcribed the interview and posted it on HotWired several days after the live broadcast, with an introduction that tells what happened after the interview, as Silberman gave Allen his first tour of the World Wide Web: "I immediately took Allen to Levi Asher's Literary Kicks site, to the page on his work there, clicking through links on Jack Kerouac's and Neal Cassady's names to demonstrate hypertext to him. He didn't say much. Then I took him to a search engine, where a search on the phrase 'allen ginsberg' called out 2,000 hits—probably the maximum. He looked at the list of all the pages built in his name. 'Thank God I don't know how to work this,' Allen sighed."

Silberman later explained that "after our interview, I took Allen to the Booksmith in the Haight-Ashbury for what turned out to be his final poetry reading in San Francisco. At the book-signing table, he kissed me and looked me in the eyes and said, 'Have a nice, sweet life.'

"I felt absolutely certain that it was the last time I would ever see him and that he wanted our twenty-year conversation on Earth to end with that blessing."

One hundred and ten days later, Allen Ginsberg died from liver cancer.

—DC

Steve Silberman: Hello. I'm very, very happy to have Allen with us today. It's hard to imagine the last several decades of public life without Allen's work. The publication of "Howl" in the late fifties was a huge gesture towards honesty and openness and sincerity in public discourse, and his poetry has influenced many generations of artists and musicians. Welcome to *HotWired*, Allen.

AG: Hi, Steve. As you know, or as you don't know—listeners, lookers—Steve Silberman and I are old friends, going back a decade or longer.

SS: I was Allen's student when I was 19, and I'm now 39, so . . .

AG: It was out at Naropa Institute, in Boulder, Colorado—the Jack Kerouac School of Disembodied Poetics. It's still going on. I'll be there this summer.

SS: Yeah. A very magical, creative community there. Allen is in San Francisco and performed last night at the Live 105 benefit for the Wilderness Society. How was that, Allen?

AG: Oh, that was a lot of fun. I haven't been in a big pop rock 'n' roll concert here in the United States before, as just another band, so to speak, or another act. I was right in the middle, at a good time, at around 9:08 I went on, so I was right in the middle of the show when everybody was in there, settled, and still not tired, because everybody was waiting for Beck who didn't get on till midnight.

I had a very good band—a pickup band here—Ralph Carney, that I'd worked with before, and one of Beck's guitarists sat in with me, and we had a drummer, and performed a version of "The Ballad of the Skeletons," which is now out on a CD from Mercury. A political poem, with very definite political statements about the far right and the monotheist theocratic Stalinists. And there were a lot of young kids there, lined up—there was an autograph thing, where you sit down and give out autographs. Eleven-, twelve-, thirteen-year-olds—it was fun. Some of them knew who I was, some of them just lined up for an autograph of what's supposed to be a star or something.

SS: Yeah. It's a great band you have on that recording: Philip Glass, Paul McCartney, Marc Ribot, and Lenny Kaye. It's a band that spans a couple of generations of great music.

AG: Yeah. Well, I've worked with Marc Ribot, and in *Musician* magazine, I think, he's listed as one of the hundred best guitarists worldwide.

SS: He's played with Elvis Costello. I think Carney has too, actually.

AG: Yeah, he's played with Tom Waits. Carney's played with all sorts of people, including me, and Beck, and I've been seeing McCartney on and off over the last few years, looking at his poetry. We were working on haikus. He's interested in that. Linda, his wife, was working with that.

SS: Were you at the original recording session for "All You Need Is Love?"

AG: No. That's the hotel room thing? I forget.

SS: No, I saw footage of the recording session for "All You Need Is Love," and I thought maybe you were there.

AG: No. I was there for a very, very interesting one with Lennon and the guy from the Stones—Jagger. In the late '60s, "Butterfly Fly Away," at the Abbey Roads studios—sitting in with Miles, who's a friend of mine and theirs.

SS: And your biographer, right?

AG: Yeah, he's editing my correspondence now. But I hadn't seen too much of McCartney, until he came to New York a couple years ago for *Saturday Night Live* on his world tour, and he remembered very clearly, because we had spent a few evenings together, and greeted me like a long-lost brother or friend. Invited me down to his place in England, and we got involved. I had written "The Ballad of the Skeletons," and I read it to him, and his daughter filmed me doing it, on a little 8-millimeter camera. And I had a concert with Anne Waldman, the British poet Tom Pickard, and about 13 other poets at the Royal Albert Hall a year ago. I asked McCartney for advice for a young guitarist who's a quick pick-up—a quick study—and he gave me some names. They sounded like older guys, like Jeff Beck. And he said, "But as you're not fixed up with a guitarist, why don't you try me, I love the poem," and I said, "Sure, it's a date."

So he showed up for the sound check. Actually, we rehearsed one night at his place. He showed up at 5 P.M. for the sound check, and he bought a box for his family. Got all his kids together, four of them, and his wife, and he sat through the whole evening of poetry, and we didn't say who my accompanist was going to be. We introduced him at the end of the evening, and then the roar went up on the floor of the Albert Hall, and we knocked out the song. He said if I ever got around to recording it, let him know. So he volunteered, and we made a basic track and sent it to him on 24 tracks, and he added maracas and drums, which it needed. It gave it a skeleton, gave it a shape. And also organ, he was trying to get that effect of Al Kooper on the early Dylan. And guitar, so he put a lot of work in on that. We got it back just in time for Philip Glass to fill in his arpeggios on piano.

SS: The last arpeggio is amazing.

AG: So it's a very interesting record. Mercury put it out with some new verses for "Amazing Grace" that Ed Sanders had ordered up, about the homeless. We did a clean version of "The Ballad of the Skeletons," seven minutes long, an original version with a few blue words, a four-minute version for radio play, and then the three minute "Amazing Grace," and it's out on the CD.

SS: And Gus Van Sant directed a video that's getting a lot of play on MTV.

AG: Yeah, that was amazing. Van Sant and I had been down to Princeton in a limousine together, and when we got to our hotel he opened up the back of the car, and there was his guitar case, and I said, "Oh, do you play?" and he said, "Yeah, I have a band in Portland." And I said, "Well, I need an accompanist." So we ran through it in his room, and he was a little nervous about it but pulled it off. He had his lecture on film, and I had a poetry reading, and I introduced him because he was staying over anyway, and he did a good performance. So he knew the thing inside out. Then when the MTV people requested a video, which was rare, Danny Goldberg and Mercury put out a little bit of money. I think 10 grand, which is nothing for videos. I don't know what Michael Jackson pays, or any normal band—$70,000, 60, 50. We pulled it in, I think, for 14. And they liked it so much on MTV, they started playing it on *Buzz Clips*, and now it's going to be playing at that film festival in Utah . . .

SS: Sundance?

AG: The Sundance Festival. Yeah, I was invited. Because it's really good. Have you seen it at all?

SS: No, I haven't.

AG: It's a great collage. He went back to old Pathé Satan skeletons, and mixed them up with Rush Limbaugh, and (Bob) Dole, and the local politicians, Newt Gingrich, and the president. And mixed those up with the atom bomb, when I talk about the electric chair—"Hey, what's cookin?"—you got Satan setting off an atom bomb, and I'm trembling with the Uncle Sam hat on. So it's quite a production, it's fun.

SS: He's a great director. I remember when I saw his first commercial release, *Mala Noche*, it was the first film I had ever seen where people smoking joints looked really like just people smoking joints, not like actors smoking fake joints. And that first film, *Mala Noche*, was a very honest portrayal of a gay relationship without being a sort of gay ghetto stereotype.

AG: Yeah, I like what he did with Burroughs in *Drugstore Cowboy*, and he used him later. I went up there once for a reading, and I ran into him, and he showed me the town. He showed me his old sites, where boys hang out and whatnot, and where he filmed things—the old hotel he used for *Drugstore Cowboy*. But what knocked me out was River Phoenix in *My Own Private Idaho*. I thought that campfire scene between River Phoenix and Keanu Reaves, where the hustler shows his heart, was really amazing.

SS: The dialogue in that scene was improvised by River Phoenix. It was not scripted.

AG: Apparently yes. So everything went very nicely for the record that we were doing, and then Mercury asked me to prepare a whole album next year, so now I got some work ahead.

SS: Great. Well, the great thing about that poem—you mentioned that it was a political poem, which of course it is—but it also reminded me of a traditional Buddhist meditation of visualizing yourself as a skeleton . . .

AG: Yeah.

SS: So it seemed to address the essential nature—shared nature—of humanity, at the same time that it highlighted the vanity of the Christian Coalition.[200]

AG: Also the vanity of human wishes to begin with. It's an old trick, to dress up archetypal characters as skeletons: the bishop, the pope, the president, the police chief. There's a Mexican painter—Posada—who does exactly that . . .

SS: *Día de los muertos*.

---

[200]Founded in 1989 by Televangelist Pat Robertson, the Christian Coalition is one of the most influential of all the far-right-wing Christian groups. They boast a large and dedicated staff of volunteers and claim responsibility for many victories of reactionary candidates.

AG: Yeah. Very very funny, when you get the bishop all dressed up—or the cardinal with his hat and staff with a skeleton head—or a skeleton president addressing mobs of skeleton heads.

SS: Yeah, there's a whole genre. Like little dioramas of whorehouses where both the whore and the john are skeletons.

AG: Yeah, it's an old, old—but it's also from the medieval days, too. I think probably the Spanish—when they came over to Mexico—brought that tradition with them, so it was an easy . . . maybe I should read that poem, I don't know.

SS: Go for it.

AG: Maybe it's just as well read as it is sung, and it has an interesting ending, too. It's called "The Ballad of the Skeletons."

> Said the Presidential Skeleton
> I won't sign the bill
> Said the Speaker skeleton
> Yes you will
>
> Said the Representative Skeleton
> I object
> Said the Supreme Court skeleton
> Whaddya expect
>
> Said the Military skeleton
> Buy Star Bombs
> Said the Upperclass Skeleton
> Starve unmarried moms
>
> Said the Yahoo Skeleton[201]
> Stop dirty art
> Said the Right Wing skeleton
> Forget about yr heart

---

[201]Yahoo—from Swift's *Gulliver's Travels*: a member of a race of brutes who have all the human vices, hence a boorish, crass, or stupid person.

Said the Gnostic Skeleton
The Human Form's divine
Said the Moral Majority skeleton
No it's not it's mine

Said the Buddha Skeleton
Compassion is wealth
Said the Corporate skeleton
It's bad for your health

Said the Old Christ skeleton
Care for the Poor
Said the Son of God skeleton
AIDS needs cure

Said the Homophobe skeleton
Gay folk suck
Said the Heritage Policy skeleton
Blacks're outa luck

Said the Macho skeleton
Women in their place
Said the Fundamentalist skeleton
Increase human race

Said the Right-to-Life skeleton
Foetus has a soul
Said Pro Choice skeleton
Shove it up your hole

Said the Downsized skeleton
Robots got my job
Said the Tough-on-Crime skeleton
Tear gas the mob

Said the Governor skeleton
Cut school lunch
Said the Mayor skeleton
Eat the budget crunch

Said the Neo Conservative skeleton
Homeless off the street!
Said the Free Market skeleton
Use 'em up for meat

Said the Think Tank skeleton
Free Market's the way
Said the Saving & Loan skeleton
Make the State pay

Said the Chrysler skeleton
Pay for you & me
Said the Nuke Power skeleton
& me & me & me

Said the Ecologic skeleton
Keep Skies blue
Said the Multinational skeleton
What's it worth to you?

Said the NAFTA skeleton[202]
Get rich, Free Trade,
Said the Maquiladora skeleton[203]
Sweat shops, low paid

Said the rich GATT skeleton[204]
One world, high tech
Said the Underclass skeleton
Get it in the neck

Said the World Bank skeleton
Cut down your trees
Said the I.M.F. skeleton[205]
Buy American cheese

[202]NAFTA—North American Free Trade Agreement
[203]Maquiladora—Foreign-owned factories operating on the Mexican side of the
U.S.–Mexican border producing goods mainly for the U.S. market.
[204] GATT—General Agreement on Tariffs and Trade
[205]I.M.F.—International Monetary Fund

Said the Underdeveloped skeleton
We want rice
Said Developed Nations' skeleton
Sell your bones for dice

Said the Ayatollah skeleton
Die writer die
Said Joe Stalin's skeleton
That's no lie

Said the Middle Kingdom skeleton
We swallowed Tibet
Said the Dalai Lama skeleton
Indigestion's whatcha get

Said the World Chorus skeleton
That's their fate
Said the U.S.A. skeleton
Gotta save Kuwait

Said the Petrochemical skeleton
Roar Bombers roar!
Said the Psychedelic skeleton
Smoke a dinosaur

Said Nancy's skeleton
Just say No
Said the Rasta skeleton
Blow Nancy Blow

Said Demagogue skeleton
Don't smoke Pot
Said Alcoholic skeleton
Let your liver rot

Said the Junkie skeleton
Can't we get a fix?
Said the Big Brother skeleton
Jail the dirty pricks

Said the Mirror skeleton
Hey good looking
Said the Electric Chair skeleton
Hey what's cooking?

Said the Talkshow skeleton
Fuck you in the face
Said the Family Values skeleton
My family values mace

Said the NY Times skeleton
That's not fit to print
Said the CIA skeleton
Cantcha take a hint?

Said the Network skeleton
Believe my lies
Said the Advertising skeleton
Don't get wise!

Said the Media skeleton
Believe you me
Said the Couch-potato skeleton
What me worry?

Said the TV skeleton
Eat sound bites
Said the Newscast skeleton
That's all Goodnight

SS: Thank you, Allen.

AG: The interesting line there, just in my mind right at the moment, is "Said the CIA skeleton / Cantcha take a hint?" referring back to the *San Jose Mercury News* revelations about CIA involvement with cocaine traffic, with the Contras selling coke in LA,[206] and the sort of general

---

[206]Journalist Gary Webb wrote a series of articles titled *Dark Alliance* for the *San Jose Mercury News* in 1996. The allegations, considered suspect by the nation's leading papers and forthrightly denied by government officials, later cost him his job and

denial you get in the *Washington Post* and the *New York Times*, trying to shift the analysis from what went on with the Contras and the CIA to what goes on with the *San Jose Mercury News* and the reporters! Sort of like "Cantcha take a hint?" "You don't have to prove it, you don't have to make such a big deal about it," you know. "Why are you making such a big deal about this when you can't prove it was a CIA decision at the top?" Although you *can*, really.

SS: You were chronicling CIA involvement in hard-drug trafficking in the Vietnam era.

AG: Since the '70s. Actually I wrote "NSA Dope Calypso" back in 1990, which covered this story which is now current in the newspapers but adds some stuff that you didn't find in the *Times* or in Walter Pincus's very restrained, "Cantcha take a hint?"-type of reporting in the *Washington Post*. So this poem was written from January to February 1990. The information is from a very famous investigation by Senator (John) Kerry and the Subcommittee on Terrorism, Narcotics, and International Operations, who nailed the CIA.[207] Nowadays they say, "Oh, but it was just stringers, you can't prove that the CIA had a deliberate policy." Oh, but you can prove. So this is the story:

> Now Richard Secord and Oliver North
> Hated Sandinistas whatever they were worth
> They peddled for the Contras to ease their pain
> They couldn't sell Congress so the Contras sold cocaine
>
> *They discovered Noriega only yesterday*
> *Nancy Reagan & the CIA*
>
> Now coke and grass were exchanged for guns
> On a border airfield that John Hull runs
> Or used to run till his Costa Rican bust
> As a CIA spy trading Contra coke dust

---

brought his research under great scrutiny. His articles and defense are now published as *Dark Alliance* (New York: Seven Stories Press, 1998).
[207]This subcommittee of the Senate Committee on Foreign Relations was later known as the Subcommittee on International Operations. The 1989 work Allen used in writing "NSA Dope Calypso" is generally referred to as *The Kerry Report*.

They discovered Noriega only yesterday
Nancy Reagan & the CIA

Ramón Milian Rodriguez of Medellín Cartel
Laundered their dollars & he did it very well
Hundreds of millions through U.S. banks
Till he got busted and sang in the tank

It was buried in the papers only yesterday
When Bush was Drug Czar U.S.A.

Milian told Congress $3,000,000 coke bucks
Went to Felix Rodríguez, CIA muck-a-muck
To give to the Contras only Hush Hush Hush
Except for Donald Gregg & his boss George Bush

Buried in the papers only yesterday
With Bush Vice President U.S.A.

Rodríguez met Bush in his office many times
They didn't talk business, they drank lemon & limes
Or maybe they drank coffee or they smoked a cigarette
But cocaine traffic they remembered to forget

Buried in the papers only yesterday
And Bush got in the White House of the U.S.A.

Now when Bush was director of the CIA
Panama traffic in coke was gay
You never used to hear George Bush holler
When Noriega laundered lots of cocaine dollar

Bush paid Noriega, used to work together
They sat on a couch & talked about the weather

Then Noriega doublecrossed his Company pal
With a treaty taking back our Panama Canal
So when he got into the big White House
Bush said Noriega was a cocaine louse

The Cold War ended, East Europe found hope,
The U.S. got hooked in a war on dope

Glasnost came, East Europe got free
So Bush sent his army to Panama City
Bush's guns in Panama did their worst
Like coke fiends fighting on St. Marks & First

Does Noriega know Bush's Company crimes?
In 2000 A.D. read the New York Times.

And that's actually the prediction of what the *New York Times*' reaction is going to be when the news comes out in 2000 A.D.—three or four years.

SS: And read it on the Web earlier than that.

AG: Back in '71 I worked on a book called *The Politics of Heroin in Southeast Asia*—Harper & Row—by Al McCoy, which we conceived together actually, on May Day, 1970, at Yale, when Jean Genet was there. I went to Washington and did a lot of research at the Institute for Policy Studies[208] and talked to a lot of ex-CIA agents and actually got to challenge Richard Helms, the director of the CIA. I made a bet with him, that if I was right—that the C.I.A. was dealing with dope—he'd have to sit and meditate an hour a day for the rest of his life, and if I was wrong, I'd give him my *vajra*, my diamond seal, a Tibetan ritual object. And he got to be in the newspapers, Flora Lewis's columns in *Newsday* and Jack Anderson's columns—the political columnist that was syndicated. Helms had to go in front of the American Society of Newspaper Editors and deny it all, and say that they had "a clean bill of health" from the Treasury Department Narcotics Bureau, which is what the CIA is claiming again!

But I had lunch with a guy named Walter Pincus, and an old college roommate, Joseph Kraft, who was one of those pundits that wrote iffy columns. Pincus is the one who wrote the story—the denying story—the elder sort of CIA specialist for the *Washington Post*. I gave them all the information I had, thinking that this was quite scandalous, that the CIA was involved with dope trafficking from the Tan

---

[208]Besides the McCoy book, see also the "Political Opium" section of *Allen Verbatim*.

Son Nhut Airport and the Plain of Jars, and that Madame Nhu, and later Marshall Ky, were involved. And he said, "Well, why are you worried about that? There isn't a matter of killing there? You're just worried about the drugs?" A very cynical attitude. And I said, "Well, I think if people knew what was going on, they would suspect the war even more." And he never did anything about it except make cynical remarks to me. He was the one that was assigned to negate the *San Jose Mercury News*. I wrote him about it the other day and asked him if he included the Kerry Subcommittee information that I got to make this little "NSA Dope Calypso" and have not gotten an answer yet.

The *Times*—I brought the same story to them, '71, about heroin, and they were very lackadaisical. I had lunch with a guy named C. L. Sulzburger, who was their foreign correspondent, of the Sulzburger family who owned them, and he said that he thought I was full of beans. Then he retired in '78 or so, a few years later, and he sent me a strange letter saying, "In going over my dispatches, I find that the information you gave me was accurate and real. I thought you were full of beans but I now apologize, and it was really true the story you had about CIA connections and opium trafficking."[209] But the *Times* never did run a big story about it until, in an editorial about a year and a half ago, they mentioned that the CIA had been nailed for dope trafficking in Indochina, but they've never had a story. It was a casual reference, maybe 25 years later. So I said "In the year 2000 A.D. read *The New York Times*," and get the story updated.

SS: Now even commentators like William Buckley talk about the legalization of drugs, which you proposed early on. California just passed a medical marijuana initiative so that people with AIDS and glaucoma can smoke marijuana. What do you think is driving the intense response against the use of marijuana, which even Clinton all but admitted to? I remember when I was a kid, I would hear people say, "Well, you know, in 20 years, all these lawyers who are in law school now, turning on, will be judges and so marijuana will be legal."

---

[209]The full text of the letter, on C. L. Sulzburger's letterhead, reads:
"Dear Allen,
April 11, 1978
I fear I owe you an apology. I have been reading a succession of pieces about CIA involvement in the dope trade in South East Asia and I remember when you first suggested I look into this I thought you were full of beans. Indeed you were right and I acknowledge the fact plus sending my best personal wishes."

AG: Well, that's slowly happening, apparently. So, who's against it? Well, there is a vested interest in there being a drug problem. First in the drug bureaucracy. From the street level narc to the highest reaches of government, the C.I.A. and even Donald Gregg, who was Vice President Bush's foreign security advisor, and later our ambassador to South Korea under Bush. There was corruption, and there has been continually all along from the top level up to this new CIA-Contra business. But back, all the way back, way back before that, it goes back to the '40s, during the war, when the OSS asked Thomas Dewey to let Lucky Luciano out of jail in New York to take over the Mafia in Sicily in order to fight the partisan Communists in Italy who beat out Hitler and the Fascists. They didn't want the Commies to have an infrastructure in Sicily, so they'd rather have the Mafia. From then on, Luciano was the lord of the drug trade, from Corsicans—the *Union Corse*—in Indochina through Marseilles, at a time when 80 percent of the world's illegal opium was coming from Indochina. Although the official story in America was that it was all coming from Turkey, but they only had one Treasury Department narc in Indochina, and about 20 or 30 in Turkey. The World Health Organization reported in 1971 that 80% of illegal heroin was all coming from Indochina.

So, a kind of strange thing going on within the government, down to, as you know, the street narcs who are on the take for their own reasons, corruption. We get that in New York, scandals like that every 20 years. Last one was in '71 when they had the Knapp Commission Report, saying that most of the corruption was endemic in the Narcotics Bureau in New York City, the largest in the world, especially in the Special Intelligence Unit, three successive heads of which were appointed by the Mafia. So you have at this point a $15 billion budget bureaucracy addicted to having an addiction problem. Simple as that. If the addiction problem was wiped out one way or another, then they all lose their jobs and like everybody else have to go to work. No side money.

There are the tobacco companies that don't want any kind of competition, and the indication of that rather general theory was that when there was a unified, single treaty around the world not to legalize marijuana, the head of the UN's Narcotics Committee was the head of the international tobacco trading board, a guy named Goldsmith or something like that. Of course they would have their motives also. My suggestion, rather than have all this critique, is that marijuana be legalized for a family farm, unadvertiseable cash crop to rehabitate the countryside. I remember when I was living up in Nevada City, there were lots of very intelligent, Harvard trained people who wanted to rusticate, get back to the country, who were able to support their local activities, school boards, with small cash crops of marijuana. Then the state helicopters came in, because the local sheriffs didn't care, knew it was all right—so it's a Big Brother thing too.

SS: Totally. There was a raid of many gardening supply shops called "Operation Green Merchant," where they took everyone's credit card numbers and names, and raided their households.

AG: Burroughs says the whole drug thing is an excuse for surveillance, international surveillance.

So I think marijuana's a simple matter. Then LSD I would give back to psychiatrists, and take it away from the army, which has power over LSD at the moment. You can only do experiments in hospitals under army auspices. Junk I would send back to the doctors, as an illness rather than a crime. Like I have to shoot insulin every day. If somebody took away the insulin, I'd be in convulsions. Junkies can be cured, and if they can't be cured, you can't punish them, you can't torture them—yet that's what's going on.

I think that once you took the cash nexus out of the whole junk problem syndrome, the black market would collapse, the Mafia involvement would collapse, and you'd get it back to a "minor medical problem," which, as Burroughs says, was what it was before World War I. So that leaves what? Cocaine. Certainly get the government out of the cocaine business to begin with. Get the drug companies out of the amphetamine business, 'cause they've been dumping amphetamines in Mexico for reimport into the United States. So there would be ways of ameliorating the problem that are sensible, that the neoconservative "Big Brother off our back people" would agree would be better— that Chicago economist, big Nobel prize winner—the one who advised Chile—I forgot his name.[210]

SS: A lot of the conservatives, actually.

AG: Yeah, the *libertarian* conservatives. That's the libertarian understanding. Then there's the right wing Stalinoid conservatives who

---

[210]Ginsberg is referring to Milton Friedman, Nobel Prize winner in economic science, 1976.

really want to impose a police state, basically. A monotheistic police state, with just one thought. That edges over to the monotheist Bible-thumping, right-wing-fundamentalist politician fundraisers and their various politics, which are quite ratty when you're lookin' at 'em. You know the history of the massacres in Guatemala? Our CIA- and army-subsidized colonels and military—trained in the US—were responsible for the murder of maybe 200,000 Guatemalan Indians, especially under the reign of a guy named Rios Mont, back in the '80s.[211] And Rios Mont's guru was none other than Pat Robertson. So that Bible-thumper's got a lot of mass murder on his conscience, actually. Ralph Reed now of Christian Coalition was his assistant, *those days*.

And Jesse Helms was always trying to justify (Roberto) D'Aubisson, the head of the hit squads in Salvador, and bring him to a kind of polite position in Washington, and whoever it was in the CIA or the state department was trying to rehabilitate Colonel François[212] and get the new presidents of Haiti to employ him, when, in a story in the *New York Times* in the same day, you read that he had shipped tons of cocaine to America. A Colonel François who we're sheltering at this point. And there was a new scandal of the CIA anti-drug military head of Venezuela, who was actually shipping tons of cocaine to America. And you have all this Contra cocaine mess, so it's an old story. If the government would actually get out of it—get out of it in every way—the pushing, stealing, robbing, and sucking off the budget—we might have a chance of calming the city streets, emptying out the prisons, and ameliorating all the racist application of the drug laws. So it's a big, big problem. And you might think, "It's just a minor thing the hippies are interested in," but remember—"crime in the streets," safety in the streets, "quality of life," drug problems—every time there's an election, that's the second biggest consideration, demagogic talking-point, particularly with the right wing. So it's not just a Ginsberg preoccupation—which it *is*, definitely, but it's a national issue that people keep jawboning whenever it gets to be time to become demagogues to get votes.

[211]See *New York Times*, March 11, 1999, "Clinton Apologizes for U.S. Support of Guatemalan Rightists."
[212]Joseph Michel François was involved in the coup that overthrew Jean Bertrand Aristide. See *New York Times*, March 8, 1997, late edition: "A Leader of Former Haitian Junta Is Charged With Smuggling Tons of Drugs to U.S."

SS: One thing that I appreciate in some poems in your last two books is—there's a passage in Kerouac, where he talks about how something that's changed in the behavior of people on the street is that they don't look in each other's eyes anymore, because they're afraid they'll be thought queer . . .

AG: Or dope fiends, or muggers.

SS: Right. In poems like "The Charnel Ground," and another poem called "May Day," you talk about the particulars of behavior in your neighborhood, being a good citizen of your block. How did "The Charnel Ground" get written?

AG: I read a little thing about my late guru, Chögyam Trungpa Rinpoche, a Tibetan lama who founded Naropa Institute that we mentioned before, and he said that the world is a charnel ground. Things die, and out of it flowers grow, animals feed, worms feed, new things arise, old things fall. It's an impermanent condition, like in a charnel ground—and that we should look on it, but not be afraid of it, appreciate the place where we are with all the facts. And I thought, well that's kinda interesting, I never have really written about my *neighborhood*. You know, what I do when I go out, take a right hand turn out of my front door and walk to First Avenue in New York, what do I see? What's going on with the bus system—tearing up the roadway to put in new pipes—there's a guy that I saw over and over again, with a reddened face, in and out of mental hospitals Peter Orlovsky told me, lived with his mother, he's out there shaking coins in a tin cup to get more money for wine, by a church door, every other day I'd see him there. So, all the icons. There was a dry cleaner's, with the door open, and there'd be a couple of old Puerto Rican winos lying there with the fumes of the dry cleaning place coming out. They were there pretty regularly, looking sick. They'd probably go home sometimes at night. So what were the specifics of my own neighborhood—seeing it as a charnel ground, both good and bad. It's interesting.

I realized it was the basis of an epic poem. I could go throughout New York City, all over, with memories that go back to 1944. But I just kept it to my block—lost the energy after that. Might go back to it.

SS: Allen, I want to ask you, I have some questions from the Net.

AG: Oh, great. There's somebody out there—good for you.

SS: "Mr. Ginsberg, what projects are you working on these days?"

AG: Mercury asked me to do a big album for next year, which might be a double album. Two projects musically. One is a complete recording of all of *Blake's Songs of Innocence and of Experience*. I have about 30 of them in the can now, ready, and 15 to go to complete it, or 14 more. And, also, several evenings of music at St. Mark's Church with 20–30 musicians, including Lee Renaldo, a drummer from Sonic Youth, and Lenny Kaye, and Mo Baron who plays digeridoo and horn, and Lenny Pickett, and a lot of great musicians, Steven Taylor, Ed Sanders.

On two occasions, Hal Wilner has organized these big musical fiestas with "Wichita Vortex Sutra" as the center of one, a 45-minute musical thing. And then when the "Ballad of the Skeletons" came out, same day as *Selected Poems*, we had another big evening, similar personnel. So those only need a couple of days in the studio to fix up and put out. So those two projects, and the possibility of doing an album of all my songs to come out fragmentarily. I have, on my desk, completed *Selected Essays*[213]—a gigantic manuscript covering 40 years, at least from '59 to '96, huge document. Cause I've written a lot of prefaces, critiques, reviews, expostulations, blurbs, anything.

SS: There are great essays in *The Poetics of the New American Poetry*, edited by Donald Allen.

AG: Yeah, some of the essays are there. There are also essays on politics, essays on drugs, essays on literature, essays on Buddhism, a lot of essays on Kerouac, lil' prefaces to Burroughs. So that's waiting for me to finish reviewing. It's all edited. And simultaneously, I have also edited and on my desk a quite large book of *Selected Interviews* from over a long period of time. It's all edited, but I have to go over it, read it. Miles, who I mentioned before, in London, is working on my *Selected Letters*. So for written projects, that's that. And also I have a huge mass of poetry I've accumulated since *Selected Poems*, but I think I'll wait on that and have a big, thick book of poems when I'm finished with the other three projects and the two recording projects.

Also, I've been working on a lot of photography. So I have a book of photographs that I think Bulfinch or Aperture has requested for several years. This time I'll take my own time and arrange it myself,

---

[213]The title of this volume, edited by Bill Morgan, is *Deliberate Prose*.

consulting Robert Frank, who's my mentor there. And a series of lith-
ographs I did at the Gemini GEL—a great, very elegant printing estab-
lishment in Los Angeles. I was there in residence for about a month
this year and produced six images which they'll make into a portfolio.
One of them was an illustrated "Ballad of the Skeletons," which they
made a special edition of 100. They cost $1,500 each, on this really
good paper, with a signed edition and what not. So those are out, and
there are five other images. Some collaboration with George Condo,
the painter. He did the cover of the *Selected Poems*, and he's a pal, and
also Hiro Yamagata helped, a Japanese guy whose skeleton skull,
Japanese style, is on the center of the "Ballad of the Skeletons" CD. So
I've had a lot of good luck and a lot of work—exhausting, actually. Then
recently this work with Eric Drooker came out this year, I think it was.

SS: *Illuminated Poems*.

AG: *Illuminated Poems*. A comic—not a comic book exactly—but some of
them are arranged as comic strips. Many of them were covers in the
New Yorker or illustrations of poems that were in the *Nation*, or things
that he cooked up as posters for the St. Mark's Poetry Project in New
York, my neighborhood club.

So there's a great deal of material around, *Collected Poems*, *Selected
Poems*, *Cosmopolitan Greetings*, "Ballad of the Skeletons," a four-CD
box set—*Holy Soul Jelly Roll 1949–1993*—poems and songs from Rhino.
There's a new "Howl" this year, with the Kronos Quartet, a really great
reading, a new reading, done this last year. All the experience that I've
had reading that particular poem was just put into one sort of perfect,
triumphant chant, with Lee Hyla's classical music performed by the
Kronos Quartet. There's also an opera, *Hydrogen Jukebox*, that came
out last year or the year before, with Philip Glass. We just cooked up
some new work. I was thinking of doing "White Shroud," a poem, with
David Mansfield, who wanted to do some work together.

SS: Who's David Mansfield?

AG: He is on this "Skeletons" record. He was one of the touring steady
musicians 20 years ago for the Rolling Thunder. He and I have got
together since and have played on stages and he's recorded with me. I
like him a lot. He's a really good all around vibraphone, guitar, fiddle,
dobro, pedal steel. He knows everything. He can play almost any-

thing—the mandolin—exquisite. He was part of these two big nights at St. Marks that we recorded for Hal Wilner.

Then also, I'd like to do Shelley's "Ode to the West Wind" and his "Hymn to Intellectual Beauty," and Hart Crane's "Atlantis"—which I'd read to Lester Young, I mentioned—with Philip Glass. Be the vocalist. Because they are great poems, and they're great vocalizations. And it's the kind of an ecstatic thing that gets Philip going. And me too. So those are what I've got in mind right now.

SS: Speaking of getting you going, I wanted to ask you—I know you have a lot of health problems and congestive heart failure—

AG: Yeah, right now. That's why I'm coughing.

SS: Right. What brings you joy right now?

AG: Making love to younger fellows, and I seem to be able to still—I can't get it up so easily—but certainly heart to heart naked is great. And I seem to have some sort of good karma that way.

SS: Well, you're famous. But you're also sweet.

AG: I'm told that I'm good in bed—a good lover. Also, writing poems, finishing poems and seeing it come to conclusion. Working on new songs. I have a new song, "Gone, Gone, Gone," and I got up in the middle of the night and recorded it. You know, just vocal. And I've got to go back and transcribe now. "Grey hair's all gone, everywhere's all gone." It's like old blues. That gives pleasure.

Finishing an artwork, seeing a new photograph that I've done, well printed. I still have about eight years of contact prints to scour through and refine. I've just skimmed the surface, but I have lots and lots of photos for this next book. I need a couple months of just looking at photos. Those are always a pleasure. That's a lot of fun—it's easy. Just look at something pretty, and decide what's clear, and ask my print maker if that really is clear and sharp enough to print up, 11 by 14 or even 16 by 20. I have a great printmaker, Sid Kaplan. Though photography is a financially losing proposition. It's an expensive hobby, but I love it, as a distraction from other things.

Finishing a new drawing. I have a lot of pleasures. Some physical—I like to cook. Some artistic. Some spiritual. Seeing my present lama advisor who'll be here in San Francisco this Thursday night actually, lecturing—Gelek Rinpoche, who has a center in Ann Arbor.

He does some advising for me, meditation advice, and for Phil Glass. Twice a year, Phil and I go off on a retreat with him and another 150 people. Phil and I are roommates, so we cook up more mischief. Last time I think we cooked up music for "The Weight of the World is Love" and for "The Cremation of Chögyam Trungpa" as a pair of things, pretty good.

But I want to get on to that Shelley, because it would be kind of interesting. "Ode to the West Wind." If you folks out there haven't had it in grammar school or high school, try it out. The key thing is to read it aloud, paying attention to the punctuation for your breathing instructions. Every bit of punctuation means a breath, whether it's a parenthesis, a comma, or a period. And you'll find it's easy to do. It's not like you, "Oh take me away." It's *"Oh*, take me *away."* That's not the actual text, but you notice that you have an "Oh," which is a big *"Oh*, take me *away,"* and you have a breath in between, so you're not losing your breath. Shelley was really sharp on that, measuring the breath itself. Which is what poetry does. Read it aloud, you get a buzz. What do you call it? Hyperventilating? It's amazing. If you do it as a mass thing in a classroom, everybody winds up dazzled and high.

SS: That's great. In the *Cantos*, Pound said, "What thou lov'st well remains, the rest is dross." What has remained for you, now at age 70?

AG: Well, a big pile of books, a big pile of records, a big pile of photographs, a big pile of drawings, a big pile of memories, of friends, imprints of their spirit on my own, imprints of their breathing and of their minds, like Kerouac. You know, you get an imprint from your family. You know what I mean by imprint? You're conditioned by growing up with them, and looking through their eyes at yourself and at other things. So I had the advantage, from the age of 16, of looking at myself through Burroughs's eyes, and Kerouac's, and soon after, Gregory Corso, Peter Orlovsky, and Gary Snyder, and Philip Whalen—now a roshi here, a Zen master in San Francisco, at the Hartford Street Zen Center.

So I had the real intellectual and emotional pleasure of having an intimate life with a lot of great artists—and still do—like Philip Glass, or Francesco Clemente, the painter, or Robert Frank, the photographer. I even wound up on stage with Yehudi Menuhin the other day. Philip had assigned me to read the "Sunflower Sutra" to music, to be conducted by Menuhin, and a string orchestra at Alice Tully Hall in Lincoln Center. It blew my mind, because I had remembered him as an unapproachable titan when I was young. Turned out to be a nice old Jewish guy, very sensitive and very elegant-handed, you know—his gestures. Very sharp and exquisitely gentle, European elegance. And his manners were very beautiful. He's 80, and he's lost a lot of his hearing and doesn't play anymore, but he conducts quite a bit.

I had a lot of good encounters with people like Tristan Tzara the Dadaist, Man Ray, Marcel Duchamp, and Jean Genet. Here in San Francisco, Genet and I went to Wooey Gooey Louie's restaurant, and in Chicago I took him to the Chicago bus terminal to see where all the boys hung out, all the hustlers. And Louis-Ferdinand Céline, the great French novelist, I went to visit with Burroughs in 1961 or '60.[214] So, I've had a very good life, especially great luck with teachers particularly Chögyam Trungpa Rinpoche and now Gelek Rinpoche. Both have great hearts. So there's a basic security to all that.

SS: Allen, I'd like to give you a chance to rest in between this and your next obligation. Do you have a short poem you'd like to read to close?

AG: Yeah, my next obligation, in case anybody's alive and livin', is a booksigning up at the Booksmith.

SS: The Booksmith on Haight Street in San Francisco.

AG: Yes, I do have one short, short poem. A kind of interesting one, but I've got to find it. Here it is, called "Autumn Leaves." This is from four years ago.

At 66, just learning how to take care of my body
Wake cheerful 8 A.M. & write in a notebook
rising from bed side naked leaving a naked boy asleep by the wall
mix miso mushroom leeks & winter squash breakfast,
Check bloodsugar, clean teeth exactly, brush, toothpick, floss,
    mouthwash
oil my feet, put on white shirt white pants white sox
sit solitary by the sink

---

[214]It was in 1958 that Ginsberg and Burroughs met Louis-Ferdinand Céline.

a moment before brushing my hair, happy not yet
to be a corpse.

SS: Thank you very much Allen, and thank you all for listening. It's
been our pleasure to have poet Allen Ginsberg on the *HotWired* net-
work today. Be well. Thanks.

AG: Ahh!

AFTERWORD

My work on this book began with a desire to put the best of Allen Ginsberg's interviews, a form of communication at which he excelled, before the world. This simple motivation has remained my guide throughout the editing of this volume, but because Allen Ginsberg died before this book was finished—and therefore before he could review it—my role necessarily was enlarged. I feel therefore that I owe it to the reader to explain how this volume was put together.

I had begun talking to Bob Rosenthal, Allen's secretary, in 1992 about the possibility of editing a book of Allen's interviews. Bob told me that Allen had wanted such a volume for a long time and that in fact Allen regarded the interview as part of his art. Discussions about how such a book should be edited continued between Bob and myself for many months. Allen had given hundreds of interviews, and Bob had carefully collected 160 in Allen's Union Square office when we began discussing this book. (By the time I had finished the final draft of this book, between what was added to Allen's office files and material I had added, I had a collection of 352 Ginsberg interviews in print form.) That there was such a huge quantity of material that Allen considered part of his art meant that there were many editorial questions to consider. As we continued our talks, Bob let me know that he was keeping Allen abreast of our discussions.

At the end of June 1993 the three of us had lunch at the Book Friends Café on 18th Street to discuss editing Allen's interviews together into a book or books. I made some notes as we talked and wrote an account of our conversation immediately after the meal. I mention this because it is the only account I know of where Allen explains how he saw the interview as an art form. As we discussed approaches to editing the book, the question arose. I give Allen's answer as I recorded it at the time:

Allen said that there were really two things. First is to treat the interviewer as a future Buddha: to speak truthfully and fully and to really try to answer the question, not to try to avoid it. After all, the interviewer is doing him a service too by interviewing him, transcribing his words, and helping him to get his message out. Secondly, he tries to compose the sentences in oral speech so that they have literary ring and clarity like good prose: composed "on the tongue." This is hard to do. Bob asked him if this is something he is only able to do after much practice and experience? Allen said yes, and that sometimes he could do it and sometimes not. Allen said it is best when the interviewer is up to the task and (therefore) there is a real rapport.

As a final point, Allen said that he then tried to "stitch together the syntax" so that the overall literary effect was pleasing.

It is interesting that Yves Le Pellec noted all of the above characteristics when he interviewed Allen on August 1, 1972:

He was very courteous and, though I was meeting him for the first time, willing to offer as much help and information as he could. I was struck and charmed to see how thoroughly and scrupulously he answered any question. He is a long-winded talker, picking his words carefully, I would say almost with delectation, as if they were ripe cherries, and associating them in long sentences without losing the thread of his exposition.

Not long after the June lunch I began formally working on the book. I was given a set of keys to Allen's office, and I would often go there to photocopy interviews as well as to do bibliographic and other research. Sometimes I would run into Allen there and we would have a short talk about my progress, but such encounters and exchanges were rare. (Allen did a lot of his work at home, and when he went to the Union Square office it was often late at night to work alone, probably to avoid the office phone's incessant ringing.)

How was the selection for this volume made? The primary criterion I used in selecting interviews was Allen's concept of the interview as art. In other words, I tried first and foremost to select the interviews that are the most brilliant, that glow with Allen's unique genius as evinced in the originality and depth of his point of view, as well as with his generosity, candor, human warmth, and literary expressiveness. Second, I strove for balance in subject matter, so that while some interviews in this selection may not be the equal of Allen at his very best according to the first criterion, they are still engaging and more than competent discussions that contain unique and important material. Third, I kept a chronological balance in mind, and a couple of interviews were included on the basis of being the best available from their era, so that the reader knows what Allen said at all stages of his public career.

I have, in general, not edited heavily because of the overall excellence of these interviews. Indeed, most of my editing consisted of cutting out material that either was repetitious or that did not move the interview forward. Naturally there were some exceptions to this general rule: while I wanted to respect the integrity of the interviews, I did cut material whenever I felt it best served Allen and the reader. Deletion of material from published interviews has not been indicated by ellipses both so as not to distract the reader and because the intent of this volume is not to be a critical edition of Allen Ginsberg's interviews; publication data is given for each interview, however, so that the interested reader can compare previous versions of interviews with those in this volume. I hope the reader will bear the omission of this editorial device in mind and when a question may not seem to flow from the preceding discussion, not to assume that this is the fault of the interviewer.

Naturally I made the changes that Allen had indicated on his own interview copies. While a number of the interviews in this volume have never been published at all, Allen's files contained versions of some interviews that were fuller than the published versions. I have restored some of these unpublished sections when I judged the material to be of sufficient interest. In other instances, transcriptions were poor or inaccurate, and I acquired copies of tape recordings and in one instance a videotape to correct and add to the versions Allen had. All of these steps seemed necessary to me to make this edition as accurate as possible and to do justice to Allen's work.

At the June 30, 1993 lunch Allen had said that he would like each interview in the book to have an introduction giving essential information about the interview, such as when the interview was done, the circumstances of the interview, who transcribed it, and where it was published. Questionnaires were therefore sent to the interviewers, and this information has been included when available.

As I gave the manuscript a final reading in preparation for submitting it to the publisher, however, I realized that there would be many references in the interviews that readers might not understand without some context, so it seemed desirable to add a bit more explanation to a number of the introductions to the individual interviews, usually information on Allen's biography or about social or political events the interviews comment upon. The information contained in the introductions to each of the individual interviews is only intended to provide such necessary context and not to give full summaries of Allen's activities up to the time of any particular interview an introduction precedes. Readers who wish to know more about Allen's life in general can resort to several biographies of him.

Allen had also told me that he wanted the interviews to be annotated. In an attempt to both minimize the number of notes and to make using the book's critical apparatus convenient, a list of names is found at the back of the book giving the dates of the person listed and some essential information on that person. Generally, names have not been included in the biographical list either when they are so well known that their identities are common knowledge (e.g., Nixon, Shakespeare) or when they occur only once in the book and are sufficiently identified in the interview itself.

When a person is mentioned in the interviews only by his or her first or last name, the person's other name has been inserted in parentheses on the first mention or if their name has not been used for some interval. This was also done on occasions where a lack of clarification could result in confusion, such as when there are two people with the same first name. Again, very well-known figures are generally given without first names in parentheses if they were mentioned only by last name in the interviews.

To avoid confusion between the titles of poems and those of books taken from the titles of poems, quotation marks and a roman font are used to refer to poems while book titles are given in italics; thus, the poem "Howl," but the book *Howl* (or *Howl and Other Poems*).

While I knew Allen had health problems, I never really envisaged his not living long enough to be able to go over this book. If he had, I have little doubt that he would have wanted some changes, but it is of course impossible to say how many or of what kind. Obviously the selection of interviews may have been different, and certainly numerous corrections and additions to the text, now lost for all time, would have been made, just as I have seen on the many copies of interviews Allen did work on. Allen's work on the copies of the interviews in his office is evidence of just how seriously he took interviews: no mistake was too small to correct, as Allen not only expanded on answers and corrected words mistranscribed, but corrected spelling and punctuation, indicated where he wanted lines of poetry to break, and supplied and even corrected bibliographic data. Although Allen did not have a chance to review this book, that he did edit so many of his interviews so carefully means that most of the interviews here have benefited from his having gone over them.

While my ultimate satisfaction from working on this book would be that if Allen could read this volume he would be pleased with it, it is also my fond hope that readers of this book will get as much pleasure and gain as much insight as I did from encountering Allen Ginsberg's spontaneous mind.

David Carter
Greenwich Village
1993–2000

ACKNOWLEDGMENTS

While editing this book has been more demanding than I had imagined it would be, the rewards of doing so were also greater than I had expected and sometimes came in surprising ways. So many people were so very generous to me that it is a pleasure to thank them here. I am afraid that over the course of seven years I have forgotten or lost track of some favors I should acknowledge here, so to any persons I have failed to include, I offer my sincere apologies.

First among those I want to thank for making this book (and my role in it) possible is Allen Ginsberg. Although my personal contact with him was limited, he was always generous and kind to me, but never so much so as when he agreed to entrust me with editing this book. Thank you for giving me this opportunity, Allen; and thank you for all you taught me through your interviews; and thanks yet again for giving so generously of yourself for all of your public life, including by giving all the hundreds of interviews you did, Whitman father, generous soul.

While it is true that my debt to Allen is primary, in many important ways my debt to Bob Rosenthal is even greater. From several points of view Bob is the true father of this book, for if the Ginsberg office is a cottage publishing industry, he is its editorial director. Bob was such a great help so many times and in so many ways, to catalog all the things he did for this book would make a very long list indeed if I could remember them all. Therefore, rather than try to create such a list I will simply say that because of his kindness, patience, warmth, discernment, tact, good humor, and grounded good judgment, Bob did so much to make working on this book the pleasurable experience that I will always fondly remember it as. I find it hard to believe that in the eight years we have worked together on this book, that with all the other responsibilities that he has had, that he never once heard my voice on the phone and sounded unhappy to be hearing from me with yet another request. The proverb says that the apple does not fall far from the tree, and I feel that Bob not only shares the core values Allen held dear but expresses those values in the way he works on Allen's behalf.

The next person in Allen's office I must acknowledge is Peter Hale, whose own cheerful enthusiasm for this book has been a wonderful source of support. So often over the last years my phone would ring and Peter would be on the other end asking, "Do you know about this interview?" It is a pleasure to try to remember some of the many services Peter has provided, from making copies of interviews to tracking down addresses and phone numbers, even to coming over to my apartment to teach me Buddhist *vipasyana* meditation when I was going through a personal rough patch. Having known Allen and his circle for many years, Peter also saved me from numerous possible misinterpretations. But as helpful as Peter has been from the start, he performed yeoman's labor on this book in the last half year before it went to press. During this time he executed many demanding tasks ably and without ever complaining (even when he probably had very good right to). While his greatest responsibilities were compiling the book's back material (researching and writing almost all of the notes created for this book as well as the entirety of the biographical list), he also helped greatly with administrative chores, all the while remaining his pleasant sunny self.

Bill Morgan has also been a model of collegiality as it is for many years now that he too has patiently answered my every query and selflessly shared his unique knowledge of the Ginsberg universe that he acquired while serving for many years as Allen's archivist and bibliographer. There is no doubt that without Bill Morgan being the Virgil to the Ginsberg manuscript collection that he is, because of the years he spent organizing and cataloging it, even attempting to negotiate it would truly be Hell for researchers. Especially helpful for this book was Bill's collection of citations of Allen's interviews that he gave me.

This seems a fitting place to acknowledge that this book would not have been possible without the efforts of many previous Ginsberg scholars, and so I want to acknowledge my special debt to Bill Morgan for his two superb published bibliographies of works by and about Allen, indispensable for this book: *The Response to Allen Ginsberg*

1926–1994: A Bibliography of Secondary Sources and *The Works of Allen Ginsberg 1941–1994: A Descriptive Bibliography*; to Bob Rosenthal for his database of Allen's interviews; and to Peter Hale for his unpublished Ginsberg audiotape bibliography. Special kudos to Allen's biographers (in English), Michael Schumacher, Barry Miles, and Jane Kramer. Michelle Kraus's and George Dowden's early Ginsberg bibliographies also proved to be useful resources in my research.

My profound thanks go to President Václav Havel for writing the preface to this work. I know that if Allen were alive, he would be very pleased that a man both he and the world hold in high esteem penned the words that are a prelude to Allen's own.

I would also like to thank Allen's friend, Josef Jařab, for approaching President Havel about writing the preface.

It is also a pleasure to thank Edmund White for agreeing to write the introduction to this book. White has also interviewed Allen: He did so for his acclaimed biography of Jean Genet, and I have no doubt that it would have pleased Allen to know that one of America's leading writers and the keeper of the flame for Jean Genet penned the introduction to this volume.

My very dear friend Arlo McKinnon performed a more valuable service than his place in this list suggests by carefully reading the entire manuscript when it was in an early form. I needed not only an objective eye but someone with good literary judgment to tell me if I was on the right track. Arlo's intelligent, sensitive, and close reading of the book was invaluable to me in editing this volume.

It is also my distinct pleasure to warmly thank Allen's colleague and neighbor Simon Pettet for giving the most recent draft of this work the kind of detailed reading that only someone with the most profound knowledge of both Allen's work and of poetry in general could give. Simon painstakingly catalogued a long list of suggestions and queries, both saving me from many mistakes and substantially improving the manuscript.

Thanks here to my brother, William C. Carter, for reading a portion of the earlier manuscript and for making valuable suggestions, as well as for his most helpful support at a critical juncture early in my work on this volume. Thanks as well are due my sister-in-law Lynn for lending her own editorial expertise on the question of permissions as well as for her general support. John Selfridge also provided valuable advice and support as he has always generously done.

At HarperCollins, my thanks go to Terry Karten and her assistants Megan Barrett, Melissa Dease, and Andrew Proctor.

At the Wylie Agency, I would like to thank Andrew Wylie, Sarah Chalfant, and Jeff Posternak.

To my dear friends J.-P. Bochêne and Olivette Halton, Tracy Turner, John Freed, David Tsang, Chuck Peach, Pauline Park, J., and Mark Christianson, thanks for your loving support.

For generous and patient assistance with computer advice, I would like to thank computer consultant Lindsey Ottman. Thanks also to E. E. Krieckhaus, my ever kind neighbor, for providing critical computer backup as the final version of this manuscript was being turned in.

For assistance in locating interviewers I would like to thank the Authors Guild, the American Society of Journalists and Authors, the Newspaper Guild, James H. Kaye at the Writers Guild of America, the PEN American Center, the Science Fiction Writers of America, and the Romance Writers of America.

For research assistance and for advice in various matters I would like to thank Bob Dylan expert Paul Williams, the National Writers Union, the American Poetry Archives, Jeffrey Katz, Director of Libraries at Bard College, Jay Kempen at the Washington University Archives, Department of Special Collections, the Museum of Television and Radio, Scott Bramlett, and Ed Sanders.

For assistance in obtaining copies of published and unpublished interviews in various formats, I would like to thank many of the interviewers who provided me with copies of their interviews; Linda Long, Project Archivist Steven Mandeville-Gamble, and Peter H. Whidden at the Stanford University Library of Special Collections; William R. Massa, Jr. and Mary La Fogg at the Manuscripts and Archives collection of Yale University Library; the New York Public Library, especially for assistance with interlibrary loans; the Butler Library at Columbia University; the New York University Library; Gordon Ball for loaning me his copy of the "Death of Ezra Pound" broadcast along with accompanying correspondence and clippings; Pierre Joris; and Studs Terkel. Special thanks to Michael Schumacher for sending me copies of many important interviews as well as for saving me from several errors by reviewing the introductions to the individual interviews in this book. Michael also informed me of the existence of a number of interviews that I was unaware of.

Thanks to Bill Morgan, Simon Pettet, Jaqueline Gens, Gina Pelli-cano, Kathy Laritz, Debbie Burr, and Gelek Rinpoche at Jewel Heart for assistance given Peter Hale in researching the notes and bio-graphical list for this volume.

While it is a pleasure to thank the several individuals who have examined this manuscript in various drafts, any errors found in the text are ultimately my sole responsibility.

Finally, while I know the interviewers in this volume made their work available because of what Allen Ginsberg meant to them, on behalf of myself and the Allen Ginsberg Trust, I wish to thank them for sharing what Allen gave them with the rest of the world.

BIOGRAPHICAL LIST

Abbott, Berenice (1898–1991). American photographer known for photo documentation of New York City in the thirties; she also preserved the works of Eugène Atget.

Amram, David (b. 1930). American French horn player, improviser, and composer of both classical and jazz works that incorporate music from various ethnic sources.

Ananda (6th century B.C.). First cousin of the Buddha and one of his principal disciples.

Antler (b. 1946). American poet; author of *Factory* and *Ever Expanding Wilderness*.

Apollinaire, Guillaume (1880–1918). Poet who participated in the French avant-garde movements and circles at the beginning of the twentieth century. Author of *Alcools*.

Artaud, Antonin (1896–1948). French dramatist, poet, actor, and Surrealist.

Ashberry, John (b. 1927). New York School poet, author of *Rivers and Mountains* and *Self Portrait in a Convex Mirror*.

Atget, Eugène (1857–1927). Influential French photographer of the early twentieth century.

Ayler, Albert (1936–1970). American tenor saxophonist and a major influence in free jazz.

Baez, Joan (b. 1941). American folk singer and songwriter, a leading voice of the 1960s along with Bob Dylan.

Bateson, Gregory (1904–1980). English-born American cultural anthropologist, husband of Margaret Mead. His works include *Naven* and *Mind and Nature*.

Beck (b. 1970). Singer songwriter; emerged in the mid-1990s.

Beck, Jeff (b. 1944). English virtuoso electric guitarist.

Belson, Jordan (b. 1926). Filmmaker. Some of his works include *Mandala* and *Northern Lights*.

Bhaktivedanta Swami (1896–1977). Founder of the Hare Krishna movement in the United States.

Blackburn, Paul (1926–1971). American poet; author of *In On Or About the Premises* and *The Cities*.

Blaser, Robin (b. 1925). Canadian poet; colleague of Jack Spicer and Robert Duncan; author of *The Holy Forest*.

Blavatsky, Helena Petrovna (1831–1891). Russian spiritualist, author, and Theosophist.

Bremser, Raymond (Ray) (1934–1998). Beat-associated poet; author of *Poems of Madness* and *Angel*.

Breton, André (1896–1966). French poet, essayist, critic, promoter, and one of the founders of the Surrealist movement.

Brown, Jerry (b. 1938). Former governor of California; frequent presidential candidate.

Buchanan, Pat (b. 1938). Politician, journalist, advisor to three American presidents (Nixon, Ford, and Reagan) who made a bid for the Republican presidential nomination in 1992 and 1996 and ran for president on the American Reform party ticket in 1996 and 2000.

Bunting, Basil (1900–1985). British poet; author of epic poem *Briggflats*; peer of Ezra Pound; associated with the Vorticist movement.

Burroughs, William S. (1914–1997). Author of *Naked Lunch* and *Junky*; seminal figure of the Beat generation.

Carpenter, Edward (1844–1929). English writer and gay forefather, greatly influenced by Walt Whitman; author of *Toward Democracy*.

Carroll, Jim (b. 1950). New York writer and poet; author of *The Basketball Diaries*.

Cassady, Carolyn (b. 1923). American writer, artist, and author of *Heartbeat: My Life with Jack and Neal*.

Cassady, Neal (1926–1968). Author of *The First Third* and inspiration for Ginsberg and central characters in Kerouac's work.

Castro, Raul (b. 1931). Fidel Castro's brother. He became his brother's chief associate in Cuban affairs.

Cayce, Edgar (1877–1945). Considered a psychic; performed healings and life readings that often involved tracing past lives back to Atlantis.

Céline, Louis-Ferdinand (1894–1961). French writer and physician.

Cézanne, Paul (1839–1906). French post-Impressionist painter.

Chandler, Raymond (1888–1959). Irish-born American detective novelist, creator of the private detective Philip Marlowe. Some of his noted works include *The Big Sleep* and *Five Murderers*.

Cherry, Don (1936–1995). Trumpeter, band leader, jazz musician, protégé of Ornette Coleman and Free Jazz exponent, he went on to develop an eclectic style known as "world music" with collaborations ranging from Indian music to punk rock.

Chomsky, Noam (b. 1928). American linguist and political activist.

Christian, Charlie (1916–1942). American jazz guitarist; one of the early improvisers on electrically amplified equipment. His style led toward early bebop, developing away from swing jazz.

Clausen, Andy (b. 1943). Belgium-born American poet. His works include *40th Century Man* and *The Iron Curtain of Love*.

Cleaver, Eldridge (1935–1998). Black militant and American novelist. His autobiographical novel *Soul on Ice* is a classical account of a black person's alienation in the Unites States.

Clemente, Francesco (b. 1952). Italian painter.

Coleman, Ornette (b. 1930). American jazz saxophonist, composer, and bandleader who was a main initiator of "free jazz" in the late fifties. Known for his "harmolodic theory," which broke improvisation away from the fixed harmonic patterns of modern jazz.

Coltrane, John (1926–1967). American jazz saxophonist, bandleader, and composer influential in jazz in the sixties and seventies.

Cope, David (b. 1948). American poet in the objectivist tradition of Reznikoff and Williams. Author of *Silences for Love*.

Corso, Gregory (b. 1930). American poet; author of *Gasoline* and *The Happy Birthday of Death*; early figure in the Beat Generation.

Costello, Elvis (b. 1955). English singer songwriter; rose to popularity in the late seventies; coined the term New Wave for music.

Creeley, Robert (b. 1926). Poet associated with the Black Mountain School; author of *For Love* and *The Gold Diggers*.

Crosby, Caresse (1892–1970). American publisher and poet; as an American literary expatriate in Paris, established publishing imprint of Editions Narcisse and Black Sun Press.

Dahlberg, Edward (1900–1977). Boston-born American writer; author of *Bottom Dogs*, *From Flushing to Calvary*, and the autobiography *Because I Was Flesh*.

Dellinger, David (b. 1915). One of the Chicago Seven and an antiwar pacifist.

Diem, Ngo Dinh (1901–1963). Vietnamese political leader who became president and dictator of South Vietnam from 1955 until his assassination.

DiPrima, Diane (b. 1934). Poet associated with the Beat Generation and San Francisco author of *Revolutionary Letters*.

Donovan (Leitch) (b. 1946). Scottish-born English folksinger-song-writer popular for such songs as "Mellow Yellow" and "Season of the Witch."

Doolittle, Hilda (commonly known as H.D.) (1886–1961). Pennsylvania-born poet; writer friend of Ezra Pound. A major imagist poet, she also wrote plays, novels, and children's stories.

Dorn, Ed (b. 1929). American poet; author of the four-volume epic poem *Slinger*; other collections include *The Newly Fallen* and *Abhorrences*.

Drooker, Eric (b. 1958). New York poster artist and illustrator.

Dudjom Rinpoche (1904–1987). Head of the Nyingmapa order of Tibetan Buddhism, Ginsberg met him in Kalimpong, India, in 1962. In Ginsberg's words, "who sucked air through his teeth in sympathy calming my fears of LSD hallucination and advised 'If you see anything horrible don't cling to it if you see anything beautiful, don't cling to it.' "

Duncan, Robert (1919–1988). American poet; taught at Black Mountain College.

Dylan, Bob (b. 1941). Nineteen-sixties folk revival singer-songwriter who evolved into a rock icon of the sixties and seventies.

Eichmann, Adolf (1906–1962). German Nazi official; head of the Jewish Department; tried and convicted in Israel for Nazi war crimes.

Eldridge, Roy (1911–1989). American trumpeter; member of the Fletcher Henderson orchestra in its last year and later of Gene Krupa's band in the forties.

Elliot, Ramblin' Jack (b. 1931). Folksinger; protégé of Woody Guthrie; and later a figure in the early sixties folk music revival.

Evans-Wentz, Walter Yeeling (1878–1965). Buddhist scholar and translator.

Fearing, Kenneth (1902–1961). American novelist and poet; editor of *Partisan Review*; his books include *Stranger at Coney Island* and *New and Selected Poems*.

Ferlinghetti, Lawrence (b. 1919). American poet and painter; author of *A Coney Island of the Mind*; publisher of City Lights Books.

Fiedler, Leslie (b. 1917). American literary critic, educator; author of *Love and Death in the American Novel*; in 1964 he joined the faculty of the State University of New York at Buffalo.

Flaubert, Gustave (1821–1880). French novelist; author of *Madame Bovary*.

Forcade, Tom. Underground Press Syndicate founder.

Frank, Robert (b. 1924). Swiss-born American photographer; close friend of the Beats.

Gampopa (1079–1153). One of the central saints in the lineage of the Kagupa School of Tibetan Buddhism.

Gaudier-Brzeska, Henri (1891–1915). French artist; one of the earliest abstract sculptors; and an outstanding exponent of the Vorticist movement. Friend of Ezra Pound, who became his propagandist and patron.

Gelek, Rinpoche (b. 1939). Kyabje or Ngawang Gelek Rinpoche; Tibetan-born friend and teacher to Allen Ginsberg and the founder of Jewel Heart Tibetan Buddhist centers.

Gingrich, Newt (b. 1943). Georgia representative; U.S. House Speaker, 1995–1998.

Giorno, John (b. 1936). American poet and innovator in performance poetry. His works include *The American Book of the Dead* and *Grasping at Emptiness*.

Goldberg, Danny (b. 1950). Music Business executive and ACLU activist. At the time of the Silberman interview, he was president of Mercury Records.

Goodman, Paul (1911–1972). Novelist, poet, playwright, psychologist, philosopher.

Gregory, Dick (b. 1932). Macrobiotic nutritionist and author of many dietbooks, including *Natural Diet for Folks Who Eat; Cookin' with Mother Nature!*

Guevara, Che (1928–1967). Prominent figure and guerrilla warfare tactician in the Cuban Revolution and later guerrilla leader in South America.

Guthrie, Arlo (b. 1947). Son of Woody and musician in his own right, best known for *Alice's Restaurant*.

Hakuin (1686–1769). Zen priest, writer, and artist. He helped revive Rinzai Zen Buddhism in Japan.

Hammond, John, Sr. (1910–1987). Music producer of such acts as Fletcher Henderson's Orchestra, Billie Holiday, and Benny Good-

man; produced Bob Dylan's first two releases and guided many
more careers.

Hartley, Marsden (1877–1943). U.S. painter who, after extensive trav-
els had brought him into contact with a variety of modern art move-
ments, arrived at a distinctive, personal type of Expressionism.

Hawkins, Coleman (1904–1969). American tenor saxophone jazz
musician who helped establish his instrument as one of the most
popular instruments in jazz.

Hecht, Ben (1894–1964). American novelist, playwright, and film
writer.

Hollo, Anselm (b. 1934). Finland-born poet, translator, journalist,
editor, teacher. Author of *Near Miss Haiku*.

Holmes, John Clellon (1926–1988). Author of *Go*, the first Beat novel.

Huncke, Herbert (1915–1996). Arguably the writer who gave the Beat
Generation its name; longtime friend of Ginsberg; a Times Square
hipster in the early 1940s who introduced Kerouac, Ginsberg, and
Burroughs to the subterranean culture of New York City as well as a
character appearing in their work. His stories are available in *The
Herbert Huncke Reader*.

Hyla, Lee (b. 1952). American modern classical composer.

James, Henry (1843–1916). American novelist and naturalized Eng-
lish citizen from 1915; his works include *Daisy Miller*, *The Portrait of
a Lady*.

James, Nehemiah "Skip" (1902–1969). Mississippi country blues
legend.

Johnson, James Weldon (1871–1938). American poet, diplomat, and
anthologist of black culture. Author of *Autobiography of an Ex-Col-
ored Man*. He collaborated with his brother, Rosamond Johnson, as
a lyricist, including "Lift Every Voice and Sing," considered the
"black national anthem."

Johnson, Robert (c. 1911–1938). Widely influential American blues
composer, guitarist.

Jones, Elvin (b. 1927). American jazz drummer and bandleader who
developed a polyrhythmic style for the drum set combining differ-
ent meters with the hands and feet.

Jones, Leroi, also called Imamu Amiri Baraka, (b. 1934). Playwright,
poet, novelist, and essayist; his works include *Blues People* and *Pref-
ace to a Twenty Volume Suicide Note*.

Jordan, June (b. 1936). African-American poet, essayist, novelist. Her works include *Who Look at Me* and *Things That I Do in the Dark*.

Kahn, Herman (1922–1983). American strategist, futurologist, and physicist known for controversial studies of nuclear warfare. Author of *Thinking About the Unthinkable* and *The Emerging Japanese Superstate*.

Kaufman, Bob (1925–1986). American poet; author of *The Ancient Rain* and *The Golden Sardine*.

Kaye, Lenny (b. 1946). Musician, producer, music critic, guitarist of Patti Smith Group.

Kerouac, Jack (1922–1969). Novelist, poet, and primal member of the Beat circle of writers and artists. His works include *On the Road*, *The Dharma Bums*, and *Mexico City Blues*.

Kesey, Ken (b. 1935). Author of *One Flew over the Cuckoo's Nest*; organizer of Merry Pranksters' LSD trips.

Khan, Ali Akbar (b. 1922). Sarod player and composer who actively presents Indian music to westerners. In 1967 he founded a music school in Marin County, California.

Kinsey, Alfred (1894–1956). American zoologist and student of human sexual behavior; his books include *Sexual Behavior in the Human Male* and *Sexual Behavior in the Human Female*, based on 18,500 personal interviews, two of whom were Ginsberg and Huncke.

Kline, Franz (1910–1962). American Abstract Expressionist painter.

Koch, Kenneth (b. 1925). Poet of the New York School.

Konitz, Lee (b. 1927). American jazz alto saxophonist; a leading figure in cool jazz.

Kooper, Al (b. 1944). Keyboard player, cofounder of The Blues Project, and often a session man for Bob Dylan; he also enjoyed a long solo career.

Koppel, Ted (b. 1940). Television anchorman and talk show host on the ABC network.

Korzybski, Alfred (1879–1950). Polish-born American scientist and philosopher. Author of *Science and Sanity*.

Ky, Nguyen Cao (b. 1930). South Vietnamese military and political leader.

Kyger, Joanne (b. 1927). Poet, traveled with Ginsberg, Orlovsky, and Snyder in Japan and India; author of *The Japan and India Journals* and *Places to Go*.

Laforgue, Jules (1860–1887). French Symbolist poet; contemporary of Arthur Rimbaud and one of the inventors of *vers libre* (free verse). Author of *L'Imitation de Notre-Dame la Lune* as well as many articles on art criticism.

Lamantia, Philip (b. 1927). American Surrealist poet; author of *The Blood of the Air.*

Leary, Timothy (1920–1996). American psychologist and drug activist, known for the statement "tune in, turn on, drop out."

Lennon, John (1940–1980). Musician; member of the Beatles.

Lenya, Lotte (1900–1981). Austrian actress-singer popular for performing the music of her husband, Kurt Weill, and his longtime collaborator Bertolt Brecht.

Levertov, Denise (1923–1998). Poet associated with the Black Mountain School. Author of *Here and Now, Relearning the Alphabet,* and many other collections, she credited William Carlos Williams as a major influence.

Limbaugh, Rush (b. 1951). American right-wing talk show host; author of *The Way Things Ought to Be* and *See, I Told You So.*

Longfellow, Henry Wadsworth (1807–1882). American poet, said to be the most famous of the nineteenth century.

Lord, Sterling (b. 1920). New York literary agent; represents Kerouac.

Lorde, Audre (1934–1992). African-American poet, essayist, and autobiographer.

Luciano, Lucky (1896–1962). Italian-born American organized crime chief in the 1930s.

Marinetti, Filippo Tommaso (1876–1944). Italian-French prose writer, novelist, poet, and dramatist, he wrote the first Futurist manifesto.

McClure, Michael (b. 1932). American poet associated with the San Francisco Renaissance and Beat poets, author of *Hymns to St. Geryon and Other Poems.*

McKay, Claude (1890–1948). Jamaican-born poet and novelist; author of *Home to Harlem.*

McReynolds, David (b. 1929). Director of the War Resistors League.

Meher Baba (1894–1969). Spiritual master born in India, he observed silence for forty-four years, saying, "I have come not to teach, but to awaken." His followers believe him to be the incarnation of God in our time.

Milarepa (1035–1135). Revered as the greatest poet-saint in Tibetan history; founder of the Kagüpa School, Chögyam Trungpa's school. See page 399.

Millett, Kate (b. 1934). Writer, political activist, artist; first known for her Columbia University Ph.D. dissertation, published as *Sexual Politics*, which brought her to national attention in the feminist movement.

Mingus, Charles (1922–1979). American jazz composer, bassist, bandleader, and pianist.

Monk, Thelonious (1917–1982). American pianist and composer who was among the first creators of modern jazz.

Muste, Abraham John (1885–1967). Dutch-born American antiwar Christian pacifist activist and minister.

Niedecker, Loraine (1903–1970). Poet loosely associated with the Objectivists.

Ochs, Phil (1940–1976). American folk singer; contemporary of Bob Dylan in the early 1960s folk revival. Unlike Dylan, Ochs continued to write protest songs throughout his career.

O'Hara, Frank (1926–1966). New York poet and central figure of the New York School poets; author of *Second Avenue* and *Lunch Poems*.

Olson, Charles (1910–1970). American poet associated with "projective verse" and central figure of the Black Mountain School; author of *The Maximus Poems*.

Oppenheimer, Joel (1930–1988). Poet associated with the Black Mountain and New York Schools; the first director of the Poetry Project at St. Mark's Church, New York City.

Orlovsky, Peter (b. 1933). Poet and longtime companion of Allen Ginsberg. Author of *Clean Asshole Poems and Smiling Vegetable Songs*.

Padgett, Ron (b. 1944). American poet associated with the New York School; his collections include *Great Balls of Fire*.

Parker, Charlie (1920–1955). American alto saxophonist, composer, and bandleader. The father of the modern jazz style known as bebop and considered by many the greatest improviser in jazz history.

Parra, Nicanor (b. 1914). Chilean poet and originator of "antipoetry." Author of *Antipoems*.

Peret, Benjamin (1899–1959). Dada-Surrealist poet and writer; author of *Death to the Pigs and Other Writings*.

Phoenix, River (1970–1993). Teen movie star who emerged in the mid-1980s with lead roles in *Stand By Me, Mosquito Coast,* and *My Own Private Idaho*.

Piaf, Edith (1915–1963). French singer and actress internationally famous for her interpretation of the chanson, or French ballad.

Podhoretz, Norman (b. 1930). A Columbia University classmate of Ginsberg, he was the editor-in-chief of *Commentary* for thirty-five years; his most recent book, *Ex-Friends*, chronicles his adversarial relationship with Ginsberg.

Powell, Bud (1924–1966). American jazz pianist who emerged in the mid-1940s.

Prévert, Jacques (1900–1977). French poet and screenwriter.

Ransom, John Crowe (1888–1974). Literary critic, poet; founder of the *Kenyon Review* and author of *The New Criticism*.

Reeves, Keanu (b. 1964). Movie star know for his roles in *Bill & Ted's Excellent Adventure, My Own Private Idaho,* and *Speed*.

Reich, Wilhelm (1897–1957). Austrian-born American psychiatrist, author; founded the Orgone Institute and invented the orgone box, for which he was jailed for violating Food and Drug Administration laws and alleged fraud.

Rexroth, Kenneth (1905–1982). American poet and essayist; influential leader of the San Francisco Renaissance.

Rimbaud, Arthur (1851–1891). French Symbolist poet; famous for *A Season in Hell* and *Illuminations*.

Rios-Montt, Efrain (b. 1926). Guatemalan dictator who rose to power in a 1982 coup lasting seventeen months. Claiming himself a "born again" Christian reformer, and backed by President Reagan, his campaigns were responsible for the destruction of native villages and the killing of tens of thousands.

Rivers, Larry (b. 1923). American painter, though originally a professional musician; kin to the New York School of poets.

Robertson, Pat (b. 1930). A Protestant evangelist who created the Christian Broadcasting Network; ran an unsuccessful bid for the Republican presidential nomination in 1988.

Rubin, Jerry (1938–1994). Radical anti–Vietnam War activist; one of the Chicago Seven and author of *Do It!*

Rudd, Mark (b. 1947). Radical activist who sparked the 1968 Columbia University student strike and later prominent in the Weathermen.

Rukeyser, Muriel (1913–1980). American poet concerned with social and political problems.

Rusk, Dean (1909–1994). U.S. secretary of state during the John F. Kennedy and Lyndon Johnson administrations who consistently defended the United States's participation in the Vietnam war.

Safire, William (b. 1929). Journalist, speechwriter, and special assistant to President Richard Nixon, currently a columnist for the *New York Times* who won the Pulitzer Prize for commentary in 1978.

Sakaki, Nanao (b. 1923). Japanese Zen Buddhist poet.

Sanders, Ed (b. 1939). Poet, writer, and musician. Founder of the musical group the Fugs.

Sangharakshita, Bhikshu (b. 1925). Buddhist poet; author of *Crossing The Stream*.

Santamaria, Haydee. Cuban revolutionary; author of *Moncada* and *Memories of the Attack That Launched the Cuban Revolution*.

Schapiro, Meyer (1904–1996). Lithuanian-born American art historian, teacher, and critic.

Schwartz, Delmore (1913–1966). American poet, short-story writer, and literary critic.

Selby, Hubert, Jr. (b. 1928). Author of *Last Exit to Brooklyn*.

Shankar, Ravi (b. 1920). Indian musician, sitar player, and composer.

Shariputra (6th century B.C.). One of the ten principal disciples of the Buddha.

Shearing, George (b. 1919). English-born American jazz pianist.

Shivananda Swami (185?–1934). Teacher to Satchitananda Swami. Visited by Allen Ginsberg, Peter Orlovsky, Gary Snyder, and Joanne Kyger, Rishikesh, 1962. Ginsberg quotes him as telling him, "Your own heart is your Guru."

Smart, Christopher (1722–1771). British poet.

Smith, Harry (1923–1991). Hermetic filmmaker, archivist, bibliophile, painter, and musicologist. Noted for compiling the influential *Anthology of American Folk Music*.

Smith, Jack (1932–1989). Filmmaker and performer whose *Flaming Creatures* was censored in New York City.

Smith, Patti (b. 1946). Rock singer, songwriter. Her first record, *Horses*, remains one of the most popular rock albums.

Snyder, Gary (b. 1930). Poet associated with the San Francisco Renaissance; author of *Riprap* and *Mountains and Rivers Without End*.

Solomon, Carl (1928–1992). Friend of Allen Ginsberg; dedicatee of *Howl*, author of *Mishaps Perhaps*.

Spengler, Oswald (1880–1936). German philosopher and historian who wrote *The Decline of the West*.

Spicer, Jack (1925–1965). Poet (*One Night Stand and Other Poems*) who with Robin Blaser and Robert Duncan in the late forties envisioned a "Berkeley Renaissance."

Steiglitz, Alfred (1864–1946). American photographer. A passionate advocate of modern art, he coordinated exhibitions in three New York City galleries. One of these, the American Place, was a favorite hangout for William Carlos Williams.

Suzuki Daisetsu Teitaro (1870–1966). Japanese Buddhist scholar and an important modern interpreter of Zen in the West.

Suzuki Shunryu (1905–1971). Japanese Zen master of the Soto school who came to the United States in 1958 and founded several Zen centers, including the Zen Center in San Francisco.

Taylor, Cecil (b. 1933). American jazz pianist and composer. A major figure in free jazz.

Taylor, Stephen (b. 1955). A guitarist with a Ph.D. in ethnomusicology, Ginberg's arranger and accompanist since 1976.

Thieu, Nguyen Van (b. 1923). President of the Republic of Vietnam (South Vietnam) from 1967 until the republic fell to North Vietnam in 1975.

Toomer, Jean (1894–1967). American poet and novelist associated with the Harlem Renaissance, author of *Cane*.

Trilling, Diana (1905–1996). Friend of Ginsberg; U.S. writer; and wife of Lionel Trilling; a member of the circle of writers and critics known in the thirties, forties, and fifties as the New York intellectuals.

Trilling, Lionel (1905–1976). Critic and writer; Ginsberg's professor at Columbia University.

Tristano, Lennie (1919–1978). American jazz pianist and a major figure of cool jazz.

Trungpa, Chögyam (1939–1987). Tibet-born founder of the Naropa Institute and Shambhala centers; author of *Cutting Through Spiritual Materialism* and *Meditation in Action*.

Tzara, Tristan (1896–1963). Romanian-born French poet and essayist best known for founding Dada.

Ungaretti, Giuseppi (1888–1970). Italian poet and founder of the Hermetic movement in Italian poetry.

Van Doren, Mark (1894–1972). Critic, writer; professor to Ginsberg at Columbia University.

Waits, Tom (b. 1949). Innovative American singer, pianist, songwriter, and composer.

Waldman, Anne (b. 1945). New York poet and cofounder of the Jack Kerouac School of Disembodied Poetics with Allen Ginsberg. Author of *Fast Speaking Woman*.

Welch, Lew (1926–1971). American poet; author of *I Remain*.

Weston, Edward (1886–1958). American photographer whose style dominated the early part of the twentieth century.

Whalen, Philip (b. 1923). Poet associated with the San Francisco Renaissance; author of *Scenes of Life at the Capital*; longtime friend of Ginsberg; currently a Zen *sensei* at Hartford Street Zen Center in San Francisco.

Wieners, John (b. 1934). Poet associated with the San Francisco Renaissance; author of *The Hotel Wentley Poems*.

Young, Lester (1909–1959). American tenor saxophonist who came on the scene in the mid-thirties in the Kansas City jazz world with the Count Basie band. He significantly added to the basis of the modern jazz solo.

Zappa, Frank (1940–1993). Prolific American rock, jazz, and avant-garde classical musician. The Mothers of Invention was his band for five years, which put out the albums *Freak Out!*, *Lumpy Gravy*, and *We're Only in It for the Money*. They disbanded in 1969.

Zukofsky, Louis (1904–1978). Editor and Objectivist poet.

The Allen Ginsberg Trust thanks the following for generously granting permission for their interviews to be used in this volume: Simon Albury, Michael Aldrich, Guy Amirthanayagam, William F. Buckley, Jr., Nancy Bunge, Peter Barry Chowka, Tom Clark, Ekbert Faas, Stephen Foehr, Mary Jane Fortunato, Clint Frakes, Thomas Gladysz, Michael Goodwin, Josef Jařab, Yves Le Pellec, John Lofton, Fernanda Pivano, Paul Portugés, Bill Prescott, Michael Schumacher, Steve Silberman, and Allen Young.

INDEX

Abbott, Berenice, 525, 529–30
"Aether," 19
Alcott, Bronson, 264
Allen, Don, 253, 359
Allen Ginsberg: Photographs, 523
Amram, David, 435, 436, 437
Anderson, Sherwood, 281, 294, 321
Antler, 501
Apollinaire, Guillaume, 105, 146, 504
Aquinas, Saint Thomas, 34, 40, 60
Artaud, Antonin, xii, xiii, 25, 32, 146,
    286, 460
Arthur, Gavin, xiii, 317, 318
Ashbery, John, 6, 52, 269
Atget, Eugène, 525
"Autumn Leaves," 569–70
Ayler, Albert, 277

Baez, Joan, 392
Ball, Gordon, 343–44
"Ballad of the Skeletons, The,"
    552–57, 566
Baraka, Amiri (LeRoi Jones), 269, 277,
    348, 501, 507–8, 518
Barbusse, Henri, xii, 274, 350
Barry, Ernie, 9–16
Barzun, Jacques, 64
Bateson, Gregory, 156, 157, 194, 463
Baudelaire, 283
Beatles, 57, 69, 74–75, 130–31, 180,
    329, 349–50, 430, 452, 458–59,
    465, 505, 533, 535–36

"Beginning of a Long Poem on The
    States," 137
Belson, Jordan, 271
Berrigan, Ted, 501
Berryman, John, 501
Bhaktivedanta, Swami, 75, 203, 386,
    457
Blackburn, Paul, 6, 508
Blake, William, xv, xvi, xviii, 18, 20,
    26–27, 31–44, 48–49, 57–58, 65,
    74, 90, 124, 142–44, 155, 162, 176,
    182, 188, 222–24, 236, 247, 257,
    264–66, 270–71, 279, 291, 307, 325,
    339, 341, 376, 385, 396, 407, 427,
    431–33, 436, 459–60, 468, 471–73,,
    511, 515; *Songs of Innocence and
    Experience* (recording), 259–63,
    434, 439, 565
Blaser, Robin, 295
Boehme, Jacob, 263, 266, 279
Bowen, Michael, 453, 454
Brancusi, 199, 267
Bremser, Raymond, 6, 508
Breton, André, 509
Brooks, Gwendolyn, 499
Brown, Richard "Rabbit," 429
Brown, Sir Thomas, 370
Bruce, Lenny, xviii, 64, 76–77,
    84–86
Buber, Martin, 9, 47
Buckley, William F., Jr., 76–102, 256,
    289, 560

Bunting, Basil, xviii, 20, 119, 120, 122,
    138, 199, 302, 351–52
Burroughs, William S., xiv, xviii, 4, 5,
    9, 21–22, 24, 35, 50, 57, 62, 65, 72,
    76–77, 83, 86, 91, 103, 136, 153, 162,
    180, 190–91, 198, 207, 210, 227,
    229–31, 265–66, 270–71, 281,
    283–84, 286, 291–93, 296–97, 301,
    307, 309, 339–41, 350, 377, 379,
    410, 421–22, 439–41, 446, 459–60,
    500–1, 503, 507, 527–28, 540, 543,
    551, 568

Carpenter, Edward, xiii, 310, 317–18
Carroll, Jim, 501
Cassady, Carolyn, 513, 528
Cassady, Neal, xiii, xv, 3, 24, 36, 62,
    115, 261, 270, 272, 278, 287,
    289–90, 293–94, 304, 312, 317–18,
    322–25, 440, 455–56, 500–1
Caster, Cyril, 260, 261
Castro, Fidel, xvii, 17, 329, 331–32, 535
Cayce, Edgar, 270
Céline, Louis-Ferdinand, xii, xiii,
    xviii, 4, 20, 83, 147, 274, 284, 288,
    291, 297, 350, 370, 504, 569
Cézanne, Paul, 26–31, 448
Chandler, Raymond, 293
"Change: Kyoto-Tokyo Express, The,"
    10, 48, 53
Chaplin, Charlie, 433
"Charnel Ground, The," 564
Cherry, Don, 260, 261, 277
Chomsky, Noam, 534
Christian, Barbara, 514
Christian, Charlie, 276, 508
Chuang-Tzu, 353
Clash (band), 441–42
Clausen, Andy, 501, 532
Cleaver, Eldridge, 508
Clemente, Francesco, 546, 568
Cohn, Al, 51, 276, 277, 437
Coleman, Ornette, 269, 277, 457
Coleridge, 44, 264, 266

Coltrane, John, 269
Commoner, Barry, 194
Cope, David, 501, 510
Corso, Gregory, xviii, 3, 4, 5, 9, 24, 51,
    103, 115, 127, 173, 199, 286, 332,
    375, 401, 422, 435, 440, 453, 487,
    500–2, 510, 527, 532, 535, 540, 543,
    545, 568
Crane, Hart, 125, 133, 294, 296, 313,
    396, 428, 445, 525
Creeley, Robert, 6, 16, 49–50, 52, 114,
    132, 146, 255, 302, 359, 452, 501,
    504, 510
Crosby, Caresse, 504

Dahlberg, Edward, 115
Daley, Richard, 160, 174, 181, 184–85,
    187–88, 212, 328
D'Aubisson, Roberto, 563
Death to Van Gogh's Ear!, 173
De Kooning, Willem, 269, 322
Dellinger, Dave, 153, 200–1, 221,
    230–33
De Quincy, 44
Dickinson, Emily, 245, 254, 444,
    446–47, 511
Diem, Ngo Dinh, 93, 94
Diggers, 92, 195
DiPrima, Diane, 454, 507
Donne, John, 154
Donovan, 69, 349
Dorfman, Else, 527
Dorn, Ed, 52
Dorough, Bob, 261
Dostoevsky, xiii, 35, 288, 370, 433,
    543
Dryden, John, 511
Duchamp, Marcel, 504, 569
Dulles, John Foster, 94, 99, 153, 497,
    498, 544
Duncan, Robert, 6, 16, 115, 147, 199,
    271, 295–96, 302, 306, 386
Dylan, Bob, 54, 69, 76, 153, 157–58,
    174, 180, 208, 302, 349–50, 376,

Dylan, Bob (*con't.*)
    378, 389–96, 430, 435–37, 441,
    458, 501, 504–5, 520, 533

"Ego Confession," 388–89, 404–6
Eichmann, Adolf, 295
Einstein, 74, 461, 463, 485–86, 506
Eliot, T.S., 112, 117, 199, 247, 256, 353,
    511–12, 519
Elliott, Ramblin' Jack, 392
Emerson, Ralph Waldo, 59, 254, 266,
    447, 504
Emery, Clark, 351
Empty Mirror, 71, 118, 237, 357, 406, 423
Epstein, Jason, 22
Evans-Wentz, Walter Y., 384–85

"Face the Nation," 132, 137
Fall of America, 54, 133–35, 137–38,
    396
Fass, Bob, 211
Faulkner, William, 5, 103
Fearing, Kenneth, 20
Federn, Dr., 292
Fenollosa, Ernest, 248–49
Ferlinghetti, Lawrence, xviii, 3–4,
    9–10, 201, 274, 296, 501, 509
Fields, W. C., 409, 433
First Blues (book), 437–39
First Blues (record), 434–39
Flaubert, Gustav, 116–17, 365
Foran, Thomas, xi, 203–6, 209, 211,
    213, 214–18, 222–24, 229, 235–42
Frank, Robert, 9, 261, 440, 524–31,
    566, 568
Freud, 292
Frobstein, Congressman, 99
Fuller, Buckminster, 180, 190, 207

Gaudier-Brzeska, Henri, 199
Gelek Rinpoche, 569
Genet, Jean, xii–xiii, xiv, 20, 57, 81,
    83–84, 227, 230, 231, 321, 329, 504,
    559, 569

Gillespie, Dizzy, 146, 268, 276, 508
Ginsberg, (father), 55–56, 400
Ginsberg, Naomi (mother), xv
Giorno, John, 501
Glass, Philip, 548, 549, 566–69
Goodman, Paul, 180, 190, 194, 299
"Gospel of Noble Truths," 389
Graham, Bill, 455
"Green Automobile, The," 335
Gregory, Dick, 195
Guevara, Che, 328
Guthrie, Woody, 157, 208, 211

Hammett, Dashiell, 21
Hammond, John, Sr., 437–38
Hartley, Marsden, 525
Havel, Václav, 532–33
Hayden, Tom, 200, 506, 517
H.D. (Hilda Doolittle), 115
Hecht, Ben, 544
Heisenberg, 491
Helms, Jesse, 563
Helms, Richard, 379, 559
Hemingway, 103, 116, 316
Heraclitus, 265, 491
Hicks, Dr., 65–66
Hitler, 50, 467
Ho Chi Minh, 96, 97
Hoffman, Abbie, 159, 180–82, 200,
    206–8, 211–12, 217, 237, 348, 506,
    517, 542
Hoffman, Julius, xi, 200
Holmes, John Clellon, 290, 506,
    513–15
Homer, 156
Hotel Wentley poems, xviii
Housman, A. E., 31
Howard, Mel, 389
"Howl," 3, 6, 19, 20, 22–23, 30, 46, 52,
    55, 65, 84, 118, 122, 131, 141, 146,
    149, 173–74, 248–49, 258, 290–91,
    296, 313, 370, 447, 469, 489–90,
    493, 505, 520, 566
Howl and Other Poems, 3, 4

Hughes, Langston, 518
Humphrey, Hubert, 184–85, 188, 382–83
Huncke, Herbert, xviii, 36, 55, 60–63, 65, 99, 282, 507, 514
Huxley, Aldous, xiii, 195, 286

Indian Journals, 10, 263, 386
"In Society," 238–40
Irwin, Professor, 165–66

Jacquet, Illinois, 20, 508
Jagger, Mick, 158, 549
James, Henry, 82, 279
James, William, 257, 431, 473
Jarrell, Randall, 501
"Jimmy Berman Rag," 437
Johnson, James Weldon, 518–19
Johnson, Lyndon B., 45n, 50, 99, 152, 153, 160, 184, 200, 227, 382
Johnson, Robert, 520
Jones, Elvin, 262
Jones, Leroi. *See* Baraka, Amiri
Jordan, June, 509
"Journal of Night Thoughts," 358
Joyce, James, 5

"Kaddish," 9, 19, 46, 50, 53, 56, 118, 122, 151, 258, 311, 370, 385
Kafka, Franz, 40n, 291
Kahn, Herman, 98
Kaplan, Sid, 526, 527, 567
Kaufman, Bob, 507, 514
Kaye, Lenny, 548
Keats, John, 51, 428, 485
Kennedy, John F., 50, 466
Kerouac, Jack, xiii, xv, xviii, 3, 5, 20–21, 24–25, 35, 50–51, 55, 62, 83, 91, 103, 114–15, 122–24, 132, 134, 145–47, 149, 160, 176, 185, 197–98, 247–48, 251–56, 268, 270, 274–98, 300, 304–8, 317–18, 323, 330, 366–70, 372–73, 380, 385, 392–94, 408, 421–22, 424, 436–37, 440,

446–49, 459, 463, 500–10, 513–17, 520, 526, 528, 533, 540–45, 568
Kerry, John, 557, 560
Kesey, Ken, 272, 317, 454–56
Khan, Ali Akbar, 458
Kissinger, Henry, 280, 449, 534
Kline, Franz, 322
Koch, Kenneth, 6, 52, 269
Konitz, Lee, 277, 278
Koppel, Ted, 534
Korzybski, Alfred, 64
Kostelanetz, Richard, 535
"Kral Majales," 17, 151
Kunstler, William, 200, 213, 214
Kyger, Joanne, 10, 454, 457

Laforgue, Jules, 105
Laing, R. D., 292
Lamantia, Philip, 51–52, 295–96, 302, 440, 509–10
"Laughing Gas," 6n, 19
LaVigne, Robert, 322–24
Lawrence, D. H., 81
Leadbelly, 433, 443
Leary, Timothy, xviii, 9, 130, 165–66, 180, 190, 195, 205, 271, 295, 337, 453, 458, 514, 528
Lennon, John, 432, 505, 549
Levertov, Denise, 6, 16
Lewis, Sinclair, 464
Lindsy, Vachel, 31–32, 157, 296
Lion for Real, The, 40
Loka II, 410
Longfellow, Henry W., 104, 254
Loran, Earl, 27, 28
Lorca, García, 318–19
Lorde, Aurdre, 509
"Love Poem on Theme by Whitman," 240–42
Lowell, Robert, 6, 301, Robert, 501
Lowenfels, Walter, 517–18
LSD 25, 163
Luciano, Lucky, 561
Lumumba, Patrice, 50

McCarthy, Eugene, 182n, 185
McCartney, Paul, 548–49
McClure, Mike, 6, 25, 52, 274, 280,
    368, 440, 454, 501–2, 510
McKnight, Dean, 284, 286, 330
McLuhan, Marshall, 71, 73
McReynolds, David, 153
"Magic Psalm," 46
Mahesh Yogi, Maharishi, 459, 465
Mailer, Norman, 5, 92, 170–71, 513
Mansfield, David, 437, 566–67
Mao Tse-Tung, 13, 96, 97
"Marijuana Notations," 357
Marinetti, F. T., xii, 105
Martinelli, Sheri, 353–54
Marvell, Andrew, 246
Marx, Karl, 33, 285
Marx Brothers, 433
Matakrishnaji, Shri, 34–35
Maupassant, de, 365
Mayakovsky, Vladimir, 104
Melville, Herman, 279, 294, 446, 511
Menninger, Karl, 335
Menuhin, Yehudi, 568–69
Mercedes, Denise, 390, 391, 401, 437
"Mescaline," 312
Milarepa, 380, 399, 417
Miles, Barry, 535, 538, 540–41, 549,
    565
Mill, John Stuart, 41
Miller, Arthur, 539, 540
Miller, Henry, xiii, 57, 81, 296, 506,
    527–28
Millett, Kate, 265
Milton, 245, 427
Mind Breaths, 403, 412
Mingus, Charlie, 457
Monk, Thelonious, 70, 146, 268, 276,
    373, 440
Moore, Marianne, xii, 104, 110–11,
    117, 145, 247, 504
Muktananda, Swami (Paramahamsa),
    386
Muste, A.J., 153

Nashe, Thomas, 342
Neibuhr, Reinhold, 76
Newton, Isaac, 155
"New York to San Fran," 125–26
Nhu, Madame, 10, 13, 560
"Night-Apple," 237–38
Nixon, Richard, 93–94, 184, 188, 200,
    288, 382
"NSA Dope Calypso," 557–59, 560

Ochs, Phil, 153, 208, 211, 260
O'Hara, Frank, 6, 52, 115, 269
O'Hara, John, 21
Olson, Charles, 6, 16, 49, 52, 114–15,
    145–47, 153, 195, 207, 255, 269,
    302, 386, 461, 504
Oppenheimer, Joel, 6, 508
Orlovsky, Peter, xvi–xviii, 3, 9, 14, 47,
    54, 65, 124, 171–73, 199, 299, 308,
    312, 322–27, 337, 375, 377, 381,
    435–36, 440, 457, 499–500, 508,
    510, 540, 568

Padgett, Ron, 501
Paracelsus, 263, 266
Parker, Charlie, 70, 114, 146–48, 268,
    276, 373, 440, 508
Parra, Nicanor, 329, 535
Peret, Benjamin, 504
Perkoff, Stuart, 6
Pincus, Walter, 559–60
Planet News, 54, 103, 130, 133, 358
Plastic People, 533
Plato, 40, 433, 457, 473
Plotinus, 60, 263, 353
Plutonian Ode, 439
Podhoretz, Norman, 5, 313, 508,
    515–16
Poe, Edgar A., 31, 245, 254, 296, 446,
    511
Pope, Alexander, 511
Pound, Ezra, xii, xviii, 50, 81, 103–9,
    110, 112–13, 115–17, 119–23, 133–34,
    138, 142, 148, 157–58, 194, 199,

Pound, Ezra (*continued*)
 247–49, 254, 267, 281, 302, 336,
 346–54, 374–75, 407, 414, 429,
 504–5, 509, 568
Presley, Elvis, 146
Prévert, Jacques, 509
Proust, 5, 147, 252
Pythagoras, 263

Rabelais, 76, 370
Ransom, John Crowe, 117, 430
Ray, Man, 504, 569
Reagan, Ronald, 449, 466
Reality Sandwiches, 240
Reed, Lou, 532
Reich, Wilhelm, 292
"Returning North of the Vortex," 496
"Returning to the Country For a Brief
 Visit," 412–13
Rexroth, Kenneth, xviii, 274, 276,
 295–96, 440
Reznikoff, Charles, 425, 510
Rimbaud, xii, 35, 52, 105, 141, 198,
 283, 286, 291, 307, 460, 517
Rivers, Larry, 322, 440
Robertson, Pat, 563
Robinson, Edwin Arlington, 247
Roszak, Theodore, 396
Rubicheck, Dr. Jiri, 163, 180, 466
Rubin, Jerry, 159, 200, 204–6,
 209–12, 217, 382, 456, 517, 542
Rudge, Olga, 350, 351
Rusk, Dean, 99, 153
Russell, Arthur, 437, 439

Sad Dust Glories, 404
St. John of the Cross, 60, 473
St. John Perse, 52
Sakaki, Nanao, 377, 511, 532
Sandburg, Carl, 20, 180, 533
Sanders, Ed, 153, 208, 217–19, 501,
 550
Santamaria, Haydee, 329
Saroyan, William, 147

Satchitananda, Swami, 203
Schapiro, Meyer, 26
Schlesinger, Arthur, 280
Schwartz, Delmore, 501
Seale, Bobby, 200
Seeger, Pete, 376
Selby, Hubert, Jr., 506
Shakespeare, 81, 148, 154, 176, 365,
 369, 370, 384, 422, 427, 485
Shankar, Ravi, 458
Shearing, George, 276–78
Shelley, Percy, 152, 198, 246, 264,
 266, 296, 500, 568
Shepard, Sam, 378
Shivananda, Swami, 47–48, 341
Sholle, Jon, 261–62, 437, 439
Shumacher, E. F., 467
Sidney, Sir Philip, 55
Sims, Zoot, 51, 276–77, 437
Sinclair, John, 208–9, 222, 260
Smart, Christopher, 20, 148, 414, 511
Smith, Bessie, 443
Smith, Harry, 156n, 271, 295, 434–35
Smith, Jack, 440
Smith, Patti, 501
Snyder, Gary, xviii, 3, 6, 9–10, 50–52,
 114, 124, 137, 148, 170, 195, 199,
 202, 269, 271, 280, 300–1, 304–5,
 355, 368–69, 386, 422, 440,
 453–56, 458–59, 500–2, 508, 510,
 532, 544, 568
Socrates, 425–26, 432
Solomon, Carl, xviii, 22, 64–65, 489
Southern, Terry, 227, 229, 231
Spellman, A. B., 74
Spellman, Cardinal, 67, 94–95, 100,
 153
Spengler, Oswald, 64, 275, 291, 307,
 513, 517
Spicer, Jack, 271, 295
Stein, Gertrude, 103, 116, 147, 251,
 257, 297, 359
Stieglitz, Alfred, 267, 530
Stokowski, Leopold, 75

Strummer, Joe, 441–42
"Studying the Signs," 119–20
Sullivan, Harry Stack, 66
"Sunflower Sutra," 31, 36–37, 151, 251, 252, 568–69
Suzuki Roshi, 136, 346, 454–55, 459
Swift, Jonathan, 76, 297

Taylor, Cecil, 277, 388
Taylor, Stephen, 437, 439
"Television Was a Baby Crawling Toward That Deathchamber," 118–19, 127–30, 136, 151, 251
Thatcher, Margaret, 466
Thelin, Ron, 453–54, 461
Theobald, Robert, 190, 195
Thomas, Dylan, 157
Thompson, E. P., 467
Thoreau, Henry D., 59, 74, 254, 294, 299
Tolstoy, 288
Toynbee, 64
Trilling, Diana, 18
Trilling, Lionel, 56, 63, 136
Tristano, Lennie, 276–78
Truman, Harry S., 281
Trungpa, Chögyam, xviii, 338, 361, 366, 368, 379, 381–85, 388, 390–91, 399, 403–7, 410, 416, 436, 454, 541, 564, 569
Twain, Mark, 76
Tzara, Tristan, 504, 569

Ungaretti, Giuseppi, xii, 104–5
Updike, John, 76

Van Doren, Mark, 63, 278, 368–69, 423
Van Gogh, 448
Van Sant, Gus, 550–51
Vidal, Gore, 76
Voznesensky, Andrei, 71

Waldman, Anne, 388, 401, 452, 501, 532, 549

"Wales Visitation," 87–89, 163, 256
Waley, Arthur, 374
"Walking into King Sooper," 413
Wallace, George, 288
Warhol, Andy, 533
Watts, Alan, 453
Weaver, Raymond, 279, 385, 446
Weinglass, Leonard, 200–1, 204–11, 215–35
Welch, Lew, 269, 500–1, 510
Weston, Edward, 527–28
Whalen, Philip, 6, 16, 52, 114, 148, 199, 269, 271, 301, 359, 367, 440, 500–1, 507–8, 510, 568
Whitman, Walt, xiii, xviii, 13, 20, 24, 58–59, 74, 104, 110, 148, 166–67, 198, 242, 246–47, 266, 270, 279–80, 282, 294, 306–7, 310, 312–21, 348, 370, 396, 407–8, 414, 427, 446–47, 449, 483–85, 504–6, 511–12, 539
"Wichita Vortex Sutra," 54, 132, 135, 152, 251, 287, 364, 370–71, 565
Wieners, John, xviii, 6, 52, 510
Williams, William Carlos, xii, xviii, 18, 24, 71–72, 104, 106–8, 112–18, 127, 129, 146, 148, 158, 194, 247, 249, 255, 266, 269, 271, 280–81, 301–2, 365, 371, 375, 385, 402, 406–9, 414, 422–23, 425, 429, 448, 504, 510–11, 525, 530
Wittgenstein, 74
Wolfe, Thomas, 5, 21, 35, 147, 278, 370
Wordsworth, 41, 90, 109, 162, 246, 396

Yeats, W. B., 64, 119, 291, 309, 352, 512
Young, Lester, 20, 148–49, 268

Zábrana, Jan, ix, 499
Zappa, Frank, 533
Zukofsky, Louis, 115, 123, 302, 523